STRATHCLYDE UNIVERSITY LIBRARY

30125 000732122

KU-796-584

ANDERSONIAN LIBRARY

3 0 AUG 2000

B O U N D

STRATHCLYDE UNIVERSITY LIBRARY

30125 00073212 2

Proceedings of SPIE—The International Society for Optical Engineering

Volume 398

Industrial Applications of Laser Technology

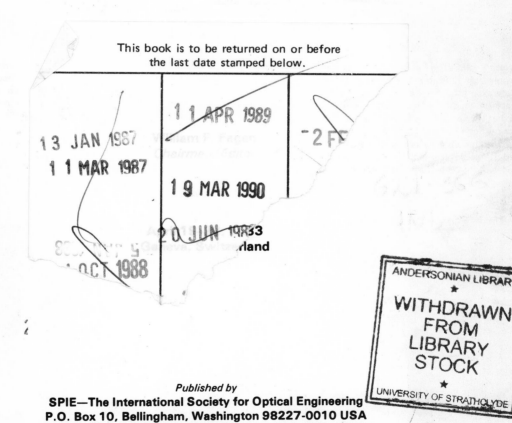

This book is to be returned on or before
the last date stamped below.

13 JAN 1987

1 1 MAR 1987

1 1 APR 1989

1 9 MAR 1990

2 0 JUN 19833

1 OCT 1988

ANDERSONIAN LIBRARY
★
WITHDRAWN
FROM
LIBRARY
STOCK
★
UNIVERSITY OF STRATHCLYDE

Published by
SPIE—The International Society for Optical Engineering
P.O. Box 10, Bellingham, Washington 98227-0010 USA
Telephone 206/676-3290 (Pacific Time) • Telex 46-7053

SPIE (The Society of Photo-Optical Instrumentation Engineers) is a nonprofit society dedicated to advancing engineering
and scientific applications of optical, electro-optical, and photo-electronic instrumentation, systems, and technology.

06050196

The papers appearing in this book comprise the proceedings of the meeting mentioned on the cover and title page. They reflect the authors' opinions and are published as presented and without change, in the interests of timely dissemination. Their inclusion in this publication does not necessarily constitute endorsement by the editors or by SPIE.

Please use the following format to cite material from this book:
Author(s), "Title of Paper," *Industrial Applications of Laser Technology*, William F. Fagan, Editor, Proc. SPIE 398, page numbers (1983).

Library of Congress Catalog Card No. 83-50303
ISBN 0-89252-433-2

Copyright © 1983, The Society of Photo-Optical Instrumentation Engineers. Individual readers of this book and nonprofit libraries acting for them are freely permitted to make fair use of the material in it, such as to copy an article for use in teaching or research. Permission is granted to quote excerpts from articles in this book in scientific or technical works with acknowledgment of the source, including the author's name, the book name, SPIE volume number, page, and year. Reproduction of figures and tables is likewise permitted in other articles and books, provided that the same acknowledgment-of-the-source information is printed with them and notification given to SPIE. **Republication or systematic or multiple reproduction** of any material in this book (including abstracts) is prohibited except with the permission of SPIE and one of the authors. In the case of authors who are employees of the United States government, its contractors or grantees, SPIE recognizes the right of the United States government to retain a nonexclusive, royalty-free license to use the author's copyrighted article for United States government purposes. Address inquiries and notices to Director of Publications, SPIE, P.O. Box 10, Bellingham, WA 98227–0010 USA.

Printed in the United States of America.

D
621.366
IND

INDUSTRIAL APPLICATIONS OF LASER TECHNOLOGY

Volume 398

Contents

INDUSTRIAL APPLICATIONS OF LASER TECHNOLOGY

Volume 398

Conference Committee

Chairman
William F. Fagan
AOL-Dr. Schuster GmbH, Austria

Co-Chairmen
Gordon M. Brown
Ford Motor Company, USA

László Erdélyi
AOL-Dr. Schuster GmbH, Austria

Armin Felske
Volkswagen AG, West Germany

Z. Füzessy
University of Budapest, Hungary

M. Murata
Mitsubishi Heavy Industries Ltd., Japan

P. Paulet
Centre Technique CITROEN, France

A. F. Rashed
King Abdul Aziz University, Saudi Arabia

P. Schuster
AOL-Dr. Schuster GmbH, Austria

Peter Waddell
University of Strathclyde, Scotland

Program Committee
A. E. Ennos
National Physical Laboratory, England

H. Kreitlow
Bremen Institute for Applied Radiation Techniques (B.I.A.S.), West Germany

Norbert Kroó
Central Research Institute for Physics, Hungary

J. McKelvie
Centre for Industrial Innovation, Scotland

Dieter Schuöcker
University of Technology, Vienna, Austria

Lorand J. Wargo
WEC Corporation

Session Chairmen
Session 1—Holographic Interferometry I
M. Murata, Mitsubishi Heavy Industries Ltd., Japan
A. E. Ennos, National Physical Laboratory, England

Session 2—Holographic Interferometry II
Gordon M. Brown, Ford Motor Company, USA
Peter Waddell, University of Strathclyde, Scotland

Session 3—Holography, Speckle, and Moiré Techniques I
H. Kreitlow, Bremen Institute for Applied Radiation Techniques, West Germany
P. Paulet, Centre Technique CITROEN, France

Session 4—Holography, Speckle, and Moiré Techniques II
Z. Füzessy, University of Budapest, Hungary
J. McKelvie, Centre for Industrial Innovation, Scotland

Session 5—Laser Measurement Techniques I
A. F. Rashed, King Abdul Aziz University, Saudi Arabia
William F. Fagan, AOL-Dr. Schuster GmbH, Austria

Session 6—Laser Measurement Techniques II
Armin Felske, Volkswagen AG, West Germany
P. Schuster, AOL-Dr. Schuster GmbH, Austria

Session 7—Laser Material Processing I
Lorand J. Wargo, WEC Corporation, USA
László Erdélyi, AOL-Dr. Schuster GmbH, Austria

Session 8—Laser Material Processing II
Dieter Schuöcker, University of Technology, Vienna, Austria
Norbert Kroó, Central Research Institute for Physics, Hungary

INTRODUCTION

This seminar, Industrial Applications of Laser Technology, formed part of the 1983 SPIE International Technical Conference/Europe. This was the first occasion when SPIE-The International Society for Optical Engineering sponsored a technical conference in Europe. As its title implies, the seminar deals with the remarkable diversity of applications of the laser in industry. Although just over 20 years old, the laser has shown itself to be a powerful tool not limited to the laboratory. The reader will find in these proceedings practical examples of how engineers and scientists have used the technology to improve the design, safety and efficiency of industrial components. Session topics include: the automatic quantitative analysis of holograms and specklegrams, the application of holographic interferometry in the automotive industry, surface flatness evaluation by laser measurement techniques, and laser material processing. The international representation of the contributed papers indicates the worldwide interest in this important area of modern optical technology. As chairman of this seminar, I would like to offer my congratulations and thanks to my fellow co-chairmen and all the members of the SPIE staff who did so much to make this conference a success.

William F. Fagan
AOL-Dr. Schuster GmbH, Austria

INDUSTRIAL APPLICATIONS OF LASER TECHNOLOGY

Volume 398

Session 1

Holographic Interferometry I

Chairmen
M. Murata
Mitsubishi Heavy Industries Ltd., Japan
A. E. Ennos
National Physical Laboratory, England

HOLOGRAPHIC INTERFEROMETRY AND SPECKLE METROLOGY:

A REVIEW OF THE PRESENT STATE

H.J.Tiziani

Institute of Applied Optics, University of Stuttgart

Pfaffenwaldring 9, D-7000 Stuttgart 80, W-Germany

Abstract

The review paper will describe the present state of holographic and speckle techniques as applied to industrial measurement. The basic principles of both techniques will be outlined with special attention given to their advantages and limitations. Current developments in the field will then be described including the testing of rotating automobile tyres by holographic interferometry and a heterodyne technique to gain an insight into noise generation mechanism. Methods for real-time holographic and speckle recording to facilitate the use of the techniques in an industrial situation will be discussed.

Introduction and historical background

Holographic interferometry and speckle techniques are widely used; the major applications are not in optics but in measuring mechanical displacement, vibration, stress and deformation. The discovery of holographic interferometry in 1964 by R.L.Powell and K.A.Stetson[1] was a mile stone in the development of the application of the laser in metrology. Five years later J.Leendertz[2], at the ICO meeting in Reading, presented a paper on interferometry with diffusely reflecting objects by means of laser speckles. In 1968 Burch and Tokarski[3] published a paper in Optica Acta on multiple exposures of speckle patterns with equal displacements between, which led to the development of speckle photography techniques.

In the field of holographic interferometry, double exposure, multiple and time average exposure techniques were introduced, as well as beam modulation and stroboscopic exposures[4,5]. In addition, fringe localizations together with fringe pattern analyses in three dimensional space have been investigated. By 1970, published material was available predicting the object motion accurately from the fringe pattern. Methods were established to extract vectorial object displacement from the fringes, their parallax and their localization. The difficulty with these fringe analyses, however, was the amount of calculation required in their applications. The search for simpler techniques of describing fringes in holographic interferometry initiated various studies on fringe analysis. The theories developed so far simplify the analysis and make it easier for the engineer to understand and apply it. Very often, however, they are too difficult to be of practical value for many engineering problems, although they can be very useful for special applications.

In recent years, matrix methods and tensor calculus have been introduced for fringe analysis, leading to a number of strain analysis techniques. Phase detection has been significantly improved to one part in 1000 by heterodyne interferometry[6]. Now at last, holographic interferometry is beginning to show its true potential in structural metrology.

The field of speckle interferometry and photography has passed through an interesting development phase. The first of the techniques were described as speckle interferometry and are analogous to classical interferometry in that they use interference between two or more randomly speckled fields[2]. Detection of this random interference is based upon the cyclic repetition of the combined speckle pattern with every 2π phase change between the fields, leading to interferometric sensitivity. By contrast, speckle photography has extended the range of displacement measurements. It is based on the displacement of speckles in the image plane. Speckle photography fills the gap between holographic interferometry and Moiré techniques.

Speckle techniques are supplementary to holographic interferometry. Fringe localization is easier, some of the movements can be separated by taking the image-plane, Fourier-plane[10] or defocused[11] speckle pattern. Speckle correlation, however, is lost by greater surface tilt. By contrast, the analysis of general three dimensional speckle displacements is not unimportant. Some of the mathematical procedures of holographic strain analysis are applied in speckle metrology. Both speckle metrology and holographic interferometry depend on electronic data acquisition and processing for application to genuine problems that occur in practice.

For strain analysis a "grid" technique is used, applying high-frequency reflective crossed grids to the surface to be studied. The grids are illuminated and produce fringes corresponding to displacements relative to the original grid[16].
Photographic emulsions are generally used for holographic interferometry and speckle photography. For real-time recording, thermoplastic material or photorefractive crystals can also be used.

Holographic and speckle systems will be used more frequently in the future, when they can be integrated with data read-out and processing systems.

Holographic interferometry of rotating objects

Holographic interferometry has now progressed to a point where the main problem is that of improving the means of extracting the required information from the fringe pattern. However, the analysis of deformation, stress and vibration of rotating objects requires unwanted rigid body rotation to be eliminated while preserving the information about the elastic object deformation.

Three methods have been used to carry out holographic interferometry and speckle techniques on rotating objects. These are stroboscopic, rotating plate, and image derotated holographic interferometry.

The stroboscopic method consist in making a hologram of the object while stationary. For the second exposure with strobed light, the rotating object is illuminated, the illumination being at the same angular orientation to the object as for the first exposure.

Rotating-plate holographic interferometry uses the holographic plate fixed to the rotating axis of the object, but this is not always possible. In addition, the rotating hologram itself will be subject to vibration or rigid body motion, hence complicating fringe analysis.

Image derotation is the most promising approach for the study of rotating objects with holographic or speckle techniques. In this method, the image of the rotating object is passed through, or reflected by, a prism rotating at half the rotational speed of the object, thus cancelling out the rotational motion. A Q-switch double-pulsed ruby laser is then used to produce a double-exposure hologram of the rotating object[8,9].

An experimental set-up used for image-derotated holographic interferometry is shown in fig. 1. Light from the double-pulse ruby laser is divided by a beam splitter and illuminates the object via a second beam splitter. The reflection of the object passes through the derotator prism to interfere with the reference beam on the holographic plate and form an image-plane hologram. For the alignment it is important that the axis of the derotator is collinear with the rotation axis of the object, otherwise optical-path length differences will produce bias fringes between the two laser pulses. The exact 2:1 ratio between the object and prism speed is achieved by mounting an encoder disk on the drive shaft of the object and relaying its signals to an electronic unit controlling the speed of the servo motor.

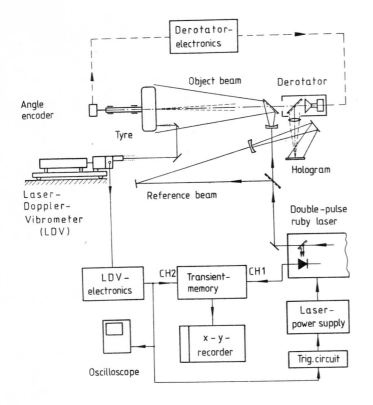

In addition, a laser Doppler vibrometer is used for the vibration analysis at a given point.

In our research we have studied the noise of rotating car tyres[10]. Fig. 1 shows the experimental arrangement with derotator and double pulse ruby laser. Fig. 2 shows the fringe pattern obtained by
a) pulse separations of 100 µs at n = 320 min^{-1} and a pulse width of 40 ns
and
b) by pulse separations of 50 µs.
The road contact was simulated by a toothed wheel. Fig. 3 shows a typical picture of the fringe analysis of the side wall (fig. 2b) of a rotating car tyre.

For noise analysis, a frequency analysis of vibration is required. A heterodyne technique can be used for the analysis of the amplitude and for the frequency of vibration at one or several points. In addition this facilitates fringe analysis of holographic interferometry . The two methods are therefore frequently used in parallel in our laboratory (fig. 1), the heterodyne technique for the analysis of the vibrations at a given point and the holographic or speckle techniques

Fig. 1 Arrangement for double pulse holography with derotator for noise analysis of a rotating car tyre.

Fig. 2a Fringe pattern obtained by double pulse holography of a rotating car tyre using an image derotator. The rotation is counter clock-wise at a speed of n = 320 min^{-1}.
Pulse separation 100 µs;
pulse width 40 ns

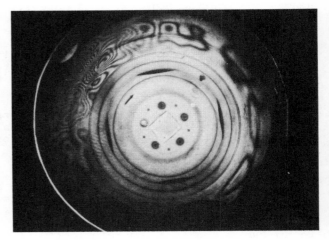

Fig. 2b Fringe pattern similar to fig. 2a, but with pulse separation of 50 µs at n = 320 min^{-1} in order to resolve the fringes near the contact zone.
The rotation is clockwise.

for the analysis of the spatial distribution of the vibration with reference to the movements of a single point, measured by using the Doppler frequency shift technique.

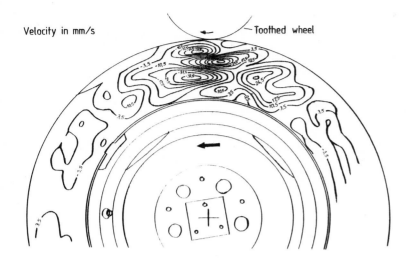

Velocity in mm/s — Toothed wheel

Fig. 3 Fringe pattern analysis
of fig. 2b of a typical
rotating car tyre; contour
separations Δv = 7 mm/s.

Heterodyne techniques applied to vibration analysis

The laser Doppler velocimeter is used to measure flow velocities of gases and liquids, using the light scattered from small particles suspended in the flowing medium. The speed of optically rough surfaces can be determined by similar methods.

The heterodyne interference or speckle techniques measure in-plane and out-of-plane displacements and vibrations of objects with diffusely scattering surfaces at high local and temporal resolution. We have been concerned primarily with out-of-plane vibration analysis.

A) Classical interferometry

In classical interferometry, phase differences of optical fields are transformed into detectable intensity variations. For two-beam interference, the two light fields are assumed to be

$$A_1 = a_1\cos(\omega_1 t + \phi_1) \qquad\qquad A_2 = a_2\cos(\omega_2 t + \phi_2)$$

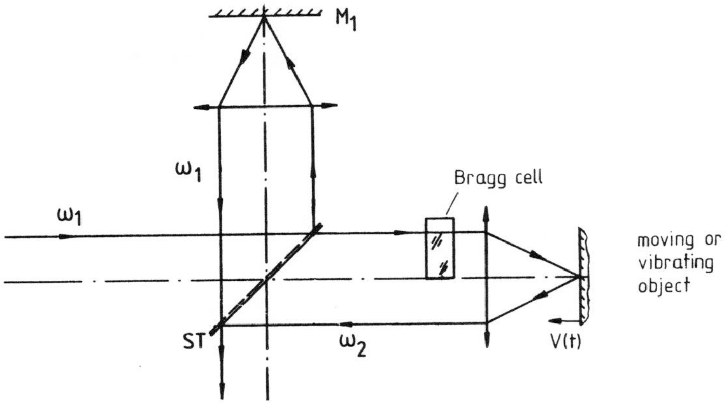

M₁

ω_1

ω_1

Bragg cell

moving or vibrating object

ST ω_2 V(t)

Fig. 4 Principle of a heterodyne
interference technique.

5

In classical interferometry the two optical frequencies ω_1 and ω_2 are identical ($\omega_1 = \omega_2$). Therefore, the intensity after the superposition is:

$$I_1 = |A_1 + A_2|^2 = |a_1|^2 + |a_2|^2 + 2a_1 a_2 \cos(\phi_2 - \phi_1)$$

The analysis of the phase difference $\phi_1 - \phi_2$ is not very accurate.

B) Heterodyne interferometry

In heterodyne interferometry, the two optical frequencies arechosen to differ by a small amount $\Delta\omega = \omega_2 - \omega_1$, say 40 MHz, for ease of electronic analysis. Superposition then yields a time dependent intensity. If the object point is moving at a speed of $v(t)$, the interference phase change is proportional to the displacement, namely $4\pi/\lambda\ v(t)\cdot t$. Hence, for $\omega_1 \neq \omega_2$ we find the intensity of the two interfering beams (fig. 4) to be

$$I = |a_1|^2 + |a_2|^2 + 2a_1 a_2 \cos(\Delta\omega t \pm 4\pi/\lambda\ v(t) t + \phi_2 - \phi_1)$$

and the frequency shift

$$\delta f(t) = 2/\lambda\ v(t)$$

which is proportional to the velocity v (parallel to optical axis) and can be regarded as a Doppler shift. For a vibrating object such as a car tyre the instantaneous velocity is measured.

For a harmonically oscillating object the phase change parallel to the optical axis is given by $4\pi/\lambda\ \rho \cos \Omega t$, which produces a frequency modulated output signal at the detector, with a carrier frequency of $\Delta\omega$ and frequency modulation of Ω; ρ is the amplitude of oscillation. The signal can be evaluated by the well-known frequency analysis techniques.

Fig. 5a shows the velocity for a single point of the side wall of the rotating tyre, obtained by using the derotator. Fig. 5b shows the corresponding frequency spectrum of fig. 5a. Such an arrangement is used for vibration analysis. Fig. 6 shows a comparison of results obtained with the Doppler-vibrometer (LDV) and independent measurements obtained by a microphone. Very close agreement is obtained between the noise detected with the microphone and the optical vibration analysis. No derotator was used for fig. 6 to avoid the Doppler effect due to the relative movement between the measured point and the source of vibration.

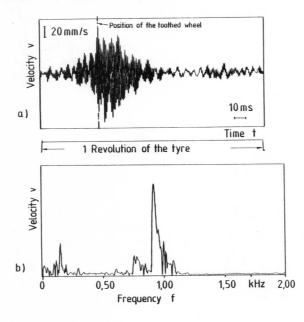

Fig. 5a Velocity of a single point of the side wall for one tyre revolution measured with the laser-Doppler-vibrometer through the image derotator

Fig. 5b Frequency spectrum of Fig. 5a

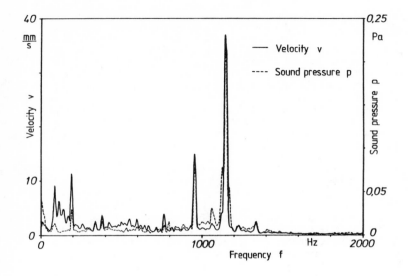

Fig. 6 Comparison of the analysis of a rotating car tyre and a microphone placed appropriately. It shows that the airborne sound is due to the structureborne sound.

Applications of speckle-techniques

A) Speckle photography

Speckle techniques are a useful tool for determining displacements, vibrations, deformations and contours of a wide range of optically rough surfaces. For speckle photography, an optically rough surface is illuminated with coherent light and photographed either in the image plane, in Fourier plane[11] or a defocused plane[12], depending on the application. The recorded image will have a speckled appearance. Exposing the image on photographic film before and after a small object movement (double exposure), pairs of practically identical speckles are recorded. Illuminating the developed double-exposed speckle pattern with a

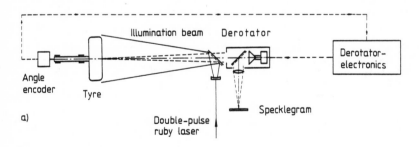

Fig. 7a Arrangement for image-plane speckle photography with double-pulse ruby laser through the derotator.

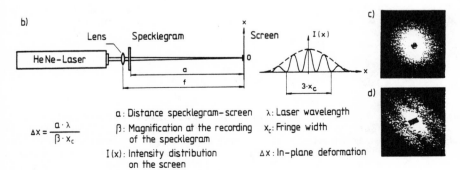

$$\Delta x = \frac{a \cdot \lambda}{\beta \cdot x_c}$$

a: Distance specklegram-screen

β: Magnification at the recording of the specklegram

I(x): Intensity distribution on the screen

λ: Laser wavelength

x_c: Fringe width

Δx: In-plane deformation

Fig. 7b Indicates schematically the display of the Young's fringes. The diffraction pattern in the Fraunhofer plane is shown without fringes in fig. 7c and with fringes in fig. 7d.

7

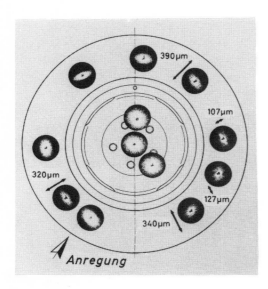

Fig. 8 Young's fringes obtained
from the speckle pattern
of the rotating car tyre,
recording through the
image derotator,
n = 850 min^{-1}

laser beam, Young's fringes are obtained in the Fraunhofer diffraction plane, with a separation inversely
proportional to the object displacement. Young's interference fringes occur only if the displaced speckles
remain correlated. A point-by-point analysis of the in-
plane motion of the displacement vector field can be
carried out even in the presence of small out-of-plane
movements (fig. 7). In fig. 8 Young's fringes of speckle patterns recorded through the derotator are shown
for a simulated disturbance of the car tyre. For tilt
analysis speckle recording in the Fourier plane or out-
of-plane speckle recording is suitable[5,11,12]. For applications where out-of-plane deformation of a surface
is to be measured, two methods can be used, i.e. double-
exposure holography or double exposure speckle photography with the camera deliberately defocused. In the
latter case the local surface tilt is measured and the
profile change can subsequently be obtained by integration. Errors can occur in speckle photography due to
focusing the camera incorrectly. The sensitivity of out-
of-plane measurements with defocused speckles can approach that of holographic interferometry, but without
requiring interferometric stability of the recording
apparatus. In addition it covers a wider measurement
range than the holographic technique[13], but additional
data processing is needed. The accuracy of the measurements is dependent upon instrumental setting and lens
performance. Additional complications occur for data
processing if the surface under examination is not flat.
Finally, the speckle shearing interferometry of Hung
(chapter 4 in[5]) permits the direct measurement of the
surface deformations.

Speckle photography is by now well understood and can be a useful tool in optical metrology. The limitations are those due to strong deformations (strain), rotation and tilt. For
example, deformation and tilt in the presence of translations lead to a limitation of speckle
photography due to loss of speckle correlation.

B) Analysis of double exposure speckle photographs

By measurements of the spacing and the direction of the Young's fringe patterns (fig. 7b)
for points on a square-mesh lattice, the two-dimensional strain field can be evaluated. Visual methods for the fringe analysis are time consuming and limited to small sample regions
and are heavily dependent on the skill of the operator. For these reasons electro-optical
read-out systems for automatic fringe analysis have been studied recently. Kaufmann[15] et al.
used the one-dimensional Fourier transformation of a small area of the speckle photograph
and formed Young's fringes on a self-scanning linear photodiode array. The output of the array is transfered to a computer where the displacement and strain fields are calculated.
For good contrast fringes speckle displacements were obtained to 0,1 μm standard deviation.

Bruhn and Felske[16] developed a fast two dimensional Fourier transform analysis of Young's
fringes using TV techniques together with image analysis methods in order to construct an
automatic fringe analysis system.

C) Electronic speckle pattern interferometric system

For recording interference pattern of diffuse reflecting objects, different names are
found in the literature, namely electronic speckle pattern interferometry (ESPI), TV speckle
interferometry, TV holography. ESPI is a holographic method usually employing a specular
reference beam as in conventional holography superimposed on the speckled object beam onto
a TV camera (see Butters & al. in reference 5, ch. 6). For a moving object, the intensity
on the photosurface varies cyclically, corresponding to a path-length change between refer-

ence and object beam by λ(λ beeing the wavelength of the laser). The video signal is processed, high pass filtered, rectified and displayed on the TV monitor. This electronic processing can be considered to correspond to the reconstruction in conventional holography.

On the monitor, fringe pattern occur depending on the subtraction of double-exposed speckle patterns or time-average fringes are produced of a harmonically oscillating object. Movements parallel to the line of sight are measured. The lateral shifts must be kept smaller than the mean speckle size. For in-plane strain measurements the object can be illuminated obliquely with two plane waves (reference 5, ch. 6). Fig. 9 shows a typical time-averaged fringe pattern of an oscillating membrane photographed from the TV monitor.

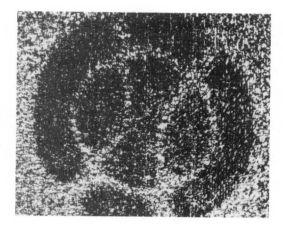

Fig. 9 Vibration analysis of the oscillating membrane, using an ESPI. Oscillation frequency 5,1 kHz.

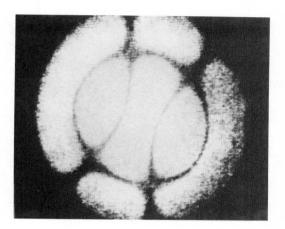

Fig. 10 Vibration analasis of the oscillating membrane analysed in fig. 9 but using a BSO crystal for recording. The fringes were photographed from the monitor.

Real-time holography and speckle recording

The storage media used for holographic interferometry and speckle photography are mainly photographic materials based on silver halide. Alternatves are photoresist, dichromated gelatin,photochromic- or thermoplastic materials, or photorefractive electro-optical crystals. Thermoplastic material is frequently used in holographic interferometry.

The most promising electro-optical materials are bismuth silicon oxide $Bi_{12}SiO_{20}$ (BSO) and bismuth germanium oxide $Bi_{12}GeO_{20}$ (BGO). In our laboratory we mainly used BSO for real-time holography and holographic interferometry and also for speckle techniques[17,18]. For BSO, the writing energy for a diffraction efficiency of 1 per cent is 0,3 mJ/cm^2 as compared with 30 mJ/cm^2 for $LiNbO_3$ (lithium niobate). The sensitivity is therefore comparable to that of the fine-grain Kodak 649 F spectroscopic emulsion, but less than for thermoplastic materials.

The physical mechanism for holography and speckle pattern recording and erasure in electro-optical crystals are drift and trapping of photoelectrons under illumination. The photo-induced space charge field changes the refractive index of the crystal via the linear electro-optic effect, leading to a refractive index variation in the crystal volume. Flooding with uniform illumination leads to erasure of the stored information by space charge relaxation. Consequently, reading out with the recording wavelength is destructive. For the analysis of the fringe pattern from holographic interferometry or speckle applications, TV techniques are useful. Fig. 10 shows the reconstructed time averaged fringe pattern of the oscillating membrane used for fig. 9 recorded with the BSO crystal and photographed from the TV monitor. The contrast and spatial resolution obtained with the BSO storage device are superior to that of the ESPI although the experimental arrangement was found to be simpler. Real-time recording and fringe displacement, together with a real-time fringe analysis technique for holography and holographic interferometry and speckle techniques can become a useful tool for the engineer.

R. Litschel's help in preparation of the paper and the photographs is gratefully acknowledged. In addition, I thank F. Höller for fig. 10.

References

1. Powell, R. L., Stetson, K. A., "Interferometric vibration analysis by wavefront reconstruction", J. Opt. Soc. Amer., Vol. 55, pp. 1593-1598. 1965.

2. Leendertz, J. A., "Measurement of surface displacements by interference of speckle pattern" (1969 meeting of ICO in Reading), Optical instruments and techniques 1969, Oriel Press Newcastle-on-Tyne. 1970.

3. Burch, J. M., Tokarski, J. M. J., "Production of multiple beam fringes from photographic scatterers", Optica Acta, Vol. 15, pp. 101-111. 1968.

4. Dainty, J. C. ed., "Laser speckle and related phenomens", Topics in Applied Physics, Springer Verlag, Vol. 9. 1975.

5. Erf, R. K. ed., "Holographic Nondestructive testing", Academic Press, New York. 1974

6. Dändliker, R., "Heterodyne holographic interferometry, Progress in Optics", ed. E.Wolf, North Holland publ. company, Amsterdam, pp. 1-84. 1980.

7. Waddel. P., "Stopping rotary motion with a prism", Machine Design, Vol. 43, Nr. 12 (May). 1973.

8. Stetson, K. A., "The use of image derotator in hologram interferometry and speckle photography of rotating objects", Ex. Mech., Vol. 18, pp. 67-73. 1978.

9. Fagan, W. F., Beeck, M. A., Kreitlow, H., "Practical application of image derotated holographic interferometry to vibration analysis of rotating components", SPIE-proceedings, Vol. 36, pp. 260-266. 1980.

10. Essers, U., Eberspächer, R., Liedl, W., Litschel, R., Pfister, B., Tiziani, H. J., Zeller, A., "Entwicklungslinien in Kraftfahrzeugtechnik und Straßenverkehr", Verlag TÜV, Rheinland, Köln. 1981 u. 1982.

11. Tiziani, H. J., "A study of the use of laser speckles to measure small tilts of optically rough surfaces accurately", Opt. Comm., Vol. 5, pp. 271-276. 1972.

12. Gregory, D. A., "Basis physical principles of defocused speckle photography: a tilt topology inspection technique", Opt. Laser Technol., Vol. 8, pp. 201-213. 1976.

13. Ennos, A. E., "Comparative accuracy of holographic interferometry and speckle photography for out-of-plane deformation measurement", Optics comm., Vol. 33, pp. 9-12. 1980.

14. Mc Donach, A., Mc Kelvie, S., Walker, C. A., "Stress analysis of fibrous composites using moiré interferometry", Optics and lasers in Engineering, Vol. 1, pp. 85-105. 1980.

15. Kaufmann, G. H., Ennos, A. E., Gale, B., Pugh, D. J., "An electro-optical read-out system for analysis of speckle photographs", J. Phys. E. Sci. Instrum., Vol. 13, pp. 579-584. 1981.

16. Bruhn, H., Felske, A., "Schnelle automatische Bildanalyse von Specklegrammen mit Hilfe der Fouriertransformation (FFT) für Spannungsmessungen", VDI Berichte, Nr. 309, pp. 13-17. 1981.

17. Marrakchi, A., Huignard, J. P., Herriau, J. P., "Application of phase conjugation in $Bi_{12}SiO_{20}$ crystals to mode pattern visualisation of diffuse vibrating structures", Opt. Comm., Vol. 34, pp. 15-18. 1980.

18. Tiziani, H. J., "Real-time metrology with BSO crystals", Optica Acta, Vol. 29, pp. 463-470. 1982.

Determination of 3-D displacement and strain by holographic interferometry for non-plane objects

R. Dändliker, R. Thalmann

Institut de Microtechnique de l'Université, CH-2000 Neuchâtel, Switzerland

Abstract

For the evaluation of 3-D displacement and strain of non-plane objects by means of double exposure holographic interferometry, errors induced by defocussing of the object surface and distortion due to perspective have to be taken into account. The importance of these errors is discussed. Numerical methods for correction and experimental results are presented.

Introduction

Double exposure holographic interferometry allows to measure displacement vector fields and strain tensor fields on diffusely scattering surfaces of solid objects. In the fringe pattern of the interferogram, the interference phase ϕ (fringe order) corresponds to the component of the displacement vector $\vec{u}$ along the sensitivity vector $\vec{g}$:

$$\phi = k\ (\vec{g}\cdot\vec{u}) \quad \text{with } \vec{g} = \vec{k}_S - \vec{k}_A \quad \text{and} \quad k = 2\pi/\lambda\ , \tag{1}$$

where $\vec{k}_S$ denotes the unity vector from the illumination source to the object point and $\vec{k}_A$ the unity vector along the observation direction (Fig.1a). To determine the 3-D displacement vector field, at least three fringe patterns with different sensitivity vectors, corresponding to three independent displacement vector components, have to be evaluated (see e.g. Ref.1). For composing the displacement vector correctly, it is important, that these components are known at exactly the same point of the object surface. Therefore, the three phase measurements in the interferogram should belong to the same displacement of the same object point, which is not easy to realize: Different directions of object illumination have to be stored on independent holograms, different directions of observation will usually change the perspective, and the interference fringes in the different images will rarely have their maxima and minima at the same position. Otherwise, if the observed interference phases in the three images correspond to different object points, they must be transferred by interpolation to a common set of object points.

In an ideal setup for the determination of 3-D displacement, the optical imaging system is fixed and the three displacement vector components are measured at one and the same image point. This requires interference phase measurements by fringe interpolation to become independent of the position of fringe maxima and minima. To get different sensitivity vectors $\vec{g}$, three double exposure holograms are recorded with different object illumination directions $\vec{k}_S$. For reasons of reproducibility and practical handling, the three holograms may be stored on the same holographic plate, spatially multiplexed in such a manner that the three sub-holograms have a common center of gravity [2]. This method yields high sensitivity for the transverse components of the displacement, but presents some experimental difficulties for the hologram recording.

An other possibility to get different sensitivity vectors and keeping the imaging system fixed is to change the observation direction $\vec{k}_A$ by scanning the aperture of the imaging lens during reconstruction [3]. The sensitivity for the transverse displacement components is limited by the aperture of the imaging lens or the size of the holographic plate, respectively, but only one double exposure hologram has to be recorded and this makes the experimental work easier. However, some difficulties arise for non-plane objects. In the areas out of exact focus a fixed point in the image plane corresponds to slightly different points on the object surface for different viewing directions (Fig.1a). In the following, we shall discuss this problem and present a solution to correct this defect.

Anyway, in order to get sufficiently high accuracy and spatial resolution, the interference phase has to be evaluated independently of fringe position and mean intensity variations. Such fringe interpolation is only possible with the help of two-reference-beam holography [4] and heterodyne [5] or quasi-heterodyne [6] methods. Changing the relative phase between the two reference beams linearly in time (heterodyne) or stepwise (quasi-heterodyne) shifts the fringes in the reconstructed interferogram, which allows to determine the interference phase accurately and at any desired point by opto-electronic measurements.

Determination of the displacement vector

As mentioned above, the determination of the 3-D displacement vector field by the method of different observation directions suffers under the defect that a fixed image point corresponds to different points on the object, if the surface is out of focus. This effect can cause important errors if not corrected. The measured phases must be interpolated on a common object point. Figures 1a and 1b illustrate this interpolation.

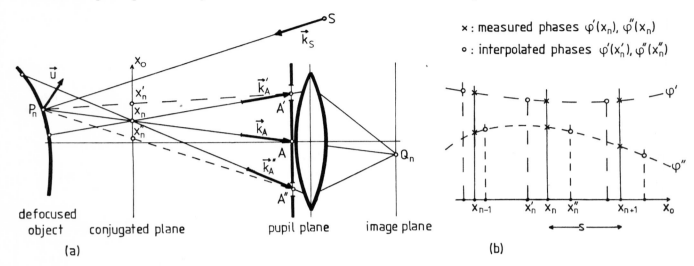

×: measured phases $\varphi'(x_n)$, $\varphi''(x_n)$

○: interpolated phases $\varphi'(x_n')$, $\varphi''(x_n'')$

defocused object   conjugated plane   pupil plane   image plane

(a)   (b)

Figure 1. (a) Optical setup showing the effects of defocussing for different observation directions $\vec{k}_A'$ and $\vec{k}_A''$. Q_n is a fixed point of observation in the image plane. (b) The interference phases ϕ' and ϕ'' for the different observation directions are measured on a regular array of sample points $(..,x_{n-1},x_n,x_{n+1},..)$. The phases at x_n' and x_n'', associated with the object point P_n, are obtained by quadratic interpolation.

The interference phases are measured in the image plane at the point Q_n to which corresponds the point x_n in the conjugated plane. The associated object point P_n for the mean observation direction $\vec{k}_A$ is given and can be calculated by the known geometry of the object and the imaging system. The points x_n', x_n'', x_n''' in the conjugated plane correspond geometrically to this common point P_n for the different observation directions $\vec{k}_A'$, $\vec{k}_A''$, $\vec{k}_A'''$. Physically, however, for each viewing direction the interference phases are measured for the fixed point x_n in the conjugated plane; i.e. one gets the phases $\phi'(x_n)$, $\phi''(x_n)$, $\phi'''(x_n)$ at the point x_n. To transfer the phases to the common object point P_n, one looks for the values of the phase function $\phi^{(i)}(x)$ at $\phi^{(i)}(P_n) = \phi^{(i)}(x_n^{(i)})$ by quadratic interpolation (Fig.1b). For an equidistant array of points $(x_{n-1},x_n,x_{n+1}$, separated by s) this yields , in one dimension, the interpolation formula

$$\phi^{(i)}(x_n^{(i)}) = \phi^{(i)}(x_n) + (x_n^{(i)} - x_n)\frac{\phi^{(i)}(x_{n+1}) - \phi^{(i)}(x_{n-1})}{2s}$$
$$+ \frac{1}{2}(x_n^{(i)} - x_n)^2 \frac{\phi^{(i)}(x_{n+1}) - 2\phi^{(i)}(x_n) + \phi^{(i)}(x_{n-1})}{s^2} \quad . \tag{2}$$

From a two-dimensional cartesian and equidistant array of points (x_n,y_n) in the conjugated or image plane, respectively, one gets the interpolated values $\phi(x_n^{(i)},y_n^{(i)})$ of the functions $\phi^{(i)}(x,y)$ by quadratic least square fitting in two dimensions over the nine nearest neighbours (Fig.2). The corresponding relations are

$$\phi(x_n^{(i)},y_n^{(i)}) = \phi_0 + \delta x_n \phi_{,x} + \delta y_n \phi_{,y} + \frac{1}{2}\delta^2 x_n \phi_{,xx} + \delta x_n \delta y_n \phi_{,xy} + \frac{1}{2}\delta^2 y_n \phi_{,yy} \tag{3}$$

with
$$\phi_0 = (5\phi_5 + 2\phi_2 + 2\phi_4 + 2\phi_6 + 2\phi_8 - \phi_1 - \phi_3 - \phi_7 - \phi_9)/9$$
$$\phi_{,x} = [(\phi_9 - \phi_7) + (\phi_6 - \phi_4) + (\phi_3 - \phi_1)]/(6s)$$
$$\phi_{,y} = [(\phi_9 - \phi_3) + (\phi_8 - \phi_2) + (\phi_7 - \phi_1)]/(6s)$$

$$\phi_{,xx} = (\phi_1 + \phi_3 + \phi_4 + \phi_6 + \phi_7 + \phi_9 - 2\phi_2 - 2\phi_5 - 2\phi_8)/(3s^2)$$
$$\phi_{,xy} = (\phi_9 - \phi_7 - \phi_3 + \phi_1)/(4s^2)$$
$$\phi_{,yy} = (\phi_1 + \phi_2 + \phi_3 + \phi_7 + \phi_8 + \phi_9 - 2\phi_4 - 2\phi_5 - 2\phi_6)/(3s^2)$$

Remember, that the points $(x_n^{(i)}, y_n^{(i)})$ are given by the object and the imaging geometry, i.e. by the effects of defocussing (Fig.1a).

Finally the phases $\phi^{(i)}(P_n)$ are converted into the components of the displacement vector $\vec{u}(P_n)$ by inversion of the relations (see e.g. Ref.1)

$$\phi^{(i)}(P_n) = k \, (\vec{g}^{(i)} \cdot \vec{u}(P_n)). \tag{4}$$

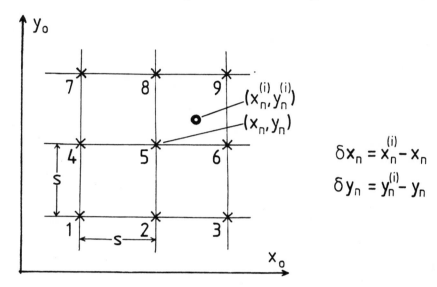

Figure 2. Two-dimensional regular array of points in the conjugated plane. The value of the interference phase $\phi(x,y)$ at the point $(x_n^{(i)}, y_n^{(i)})$, associated with the object point P_n (Fig.1a), is calculated by a two-dimensional quadratic least square fit.

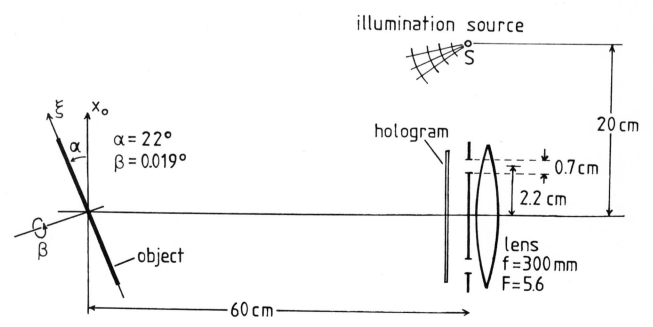

Figure 3. Optical arrangement of the experiment.

The importance and the success of this correction scheme have been verified for a special test object, shown in Fig.3. It is a metal plate, tilted with respect to the optical axis by the angle α, which undergoes a pure in-plane rotation by the angle β. The geometry of the optical arrangement is sketched in Fig.3. The hologram has been stored with two reference beams. The interference phases have been measured in the image plane of the lens with the heterodyne method [5]. The accuracy of fringe interpolation in this experiment was about $\lambda/100$. To get the different sensitivity vectors, a circular aperture with a diameter of 7 mm has been moved on a circle with a radius of 22 mm, corresponding to f/5.6, to three different positions in the pupil plane. Figure 4 shows the experimental result for the in-plane displacement components $u_x(y)$ and $u_y(x)$. Without correction, the y-component of the displacement vector along the defocused x-axis is systematically about 17% too small. The x-component of the displacement vector along the non-defocused y-axis serves as reference for the true rotation of the plate.

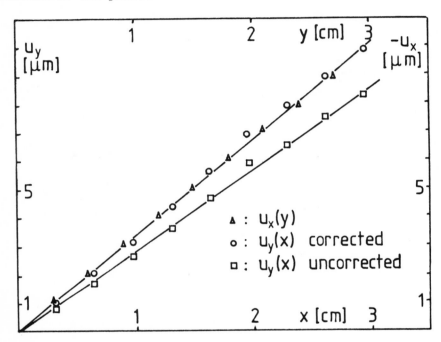

Figure 4. Measured in-plane displacement componants $u_x(y)$ and $u_y(x)$ of the rotated plate, shown in Fig.3.

Determination of the strain tensor

To calculate the in-plane derivatives of the displacement, i.e. strain and rotation, the effect of surface distortion due to perspective for non-plane objects has to be taken into account additionally. In general, the surface points P_n form an irregular array. Therefore, it is appropriate to calculate the derivatives in the tangential plane at each point by two-dimensional least square fitting of the displacement vector components through the point and his four nearest neighbours. The object point P_n is taken as the origin of a new coordinate system (ξ, η), tangential to the object surface at this point. Here, the function $u(\xi, \eta)$ of the displacement vector components are fitted by

$$u(\xi, \eta) = u_0 + \xi \frac{\partial u}{\partial \xi} + \eta \frac{\partial u}{\partial \eta} \ . \tag{5}$$

This function is known on an irregular array of five sample points in and around P_n: $u_i = u(\xi_i, \eta_i)$ (Fig.5). Least square fitting yields the first derivatives by solving the following system of linear equations:

$$\overline{A}_{lm} \, a_m - \overline{b}_m = 0 \ , \tag{6}$$

where $\overline{A}_{lm}$ is the symmetrical 3×3 matrix $\overline{A}_{lm} = A_{il} \, A_{im} = A_{li}^T \, A_{im}$ (i=1,..,5)

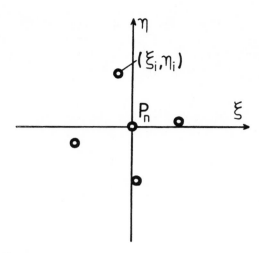

Figure 5. Irregular array of sample points, where the displacement vector is known. The in-plane derivatives of the displacement componants at P_n are obtained by a two-dimensional least square fit over five points.

and A_{im} is the 3×5 matrix

$$A_{im} = \begin{pmatrix} 1 & \xi_1 & \eta_1 \\ \vdots & & \vdots \\ 1 & \xi_5 & \eta_5 \end{pmatrix}.$$

The vector a_m of the three unknowns is defined as $a_m = (u_0, \frac{\partial u}{\partial \xi}, \frac{\partial u}{\partial \eta})^T$.

The vector of the constants is $\overline{b}_m = b_i A_{im}$, with $b_i = u(\xi_i, \eta_i)$.

Now, in-plane strain ε and rotation Ω are given by:

$$\varepsilon_{\xi\xi} = \frac{\partial u_\xi}{\partial \xi} , \quad \varepsilon_{\xi\eta} = \varepsilon_{\eta\xi} = \frac{1}{2}(\frac{\partial u_\xi}{\partial \eta} + \frac{\partial u_\eta}{\partial \xi}) , \quad \varepsilon_{\eta\eta} = \frac{\partial u_\eta}{\partial \eta} ,$$

$$\text{and} \quad \Omega = \frac{1}{2}(\frac{\partial u_\eta}{\partial \xi} - \frac{\partial u_\xi}{\partial \eta}).$$

(7)

For the test object of Fig.3, the calculus for the derivatives given above is not necessary along the ξ- and the η-axes, since the object is plane and the distortion due to perspective is avoided by appropriate choice of the sample point array. Anyway, without applying the corrections for the evaluation of the displacement vector, as described in the previous chapter, one gets besides an in-plane rotation of $\Omega = 3.1 \cdot 10^{-4}$, which is about 8% too small, also a non-zero componant of in-plane strain $\varepsilon_{\xi\eta} = \varepsilon_{\eta\xi} = -0.57 \cdot 10^{-4}$, which is in contradiction with the pure rotational motion of the object.

Conclusions

It has been shown that geometrical effects due to defocussing and distortion by perspective can cause important errors in the evaluation of the displacement vector field on objects of arbitrary surface shape. These errors can be corrected by appropriate interpolation of the measured interference phases corresponding to the different displacement vector components, so that they belong always to a common point of the object surface. The numerical methods described in this paper have been successfully applied to and verified for a test object showing typical errors of about 17% due to defocussing. The 3-D displacement vector field has been evaluated from a single double exposure hologram using different observation directions through the same imaging lens and by heterodyne detection of the interference phase.

References

1. Vest, C.M., <u>Holographic Interferometry</u>, chapter 2, John Wiley & Sons, 1979.
2. Ek, L., Majlöf, L., "Evaluation of 3-D Displacement by Holography and Computer", The Topical Meeting on Hologram Interferometry and Speckle Metrology, June 2-4, 1980, North Falmouth, Mass., USA.
3. Dändliker, R., "Holographic Interferometry and Speckle Photography for Strain Measurement: a Comparison", <u>Optics and Lasers in Eng.</u>, Vol. 1, pp. 3-19, 1980.
4. Dändliker, R., Marom, E., Mottier, F.M., "Two-Reference-Beam Holographic Interferometry", <u>J. Opt. Soc. Am.</u>, Vol. 66, pp. 23-30, 1976.
5. Dändliker, R., "Heterodyne Holographic Interferometry", in <u>Progress in Optics</u>, Vol.XVII, (ed. E. Wolf), North Holland, Amsterdam, 1980.
6. Dändliker, R., Thalmann, R., Willemin, J.-F., "Fringe Interpolation by Two-Reference-Beam Holographic Interferometry: Reducing Sensitivity to Hologram Misalignement", <u>Opt. Commun.</u>, Vol. 42, pp. 301-306, 1982.

Measurement of 3D-displacement by regulated path length interferometry

Z. Füzessy

Institute of Physics, Technical University Budapest
H-1521 Hungary

Abstract

Using several hologram plates for measurement of 3D-displacement a problem arises that the object looks different from different directions, so that the fringes become difficult to identify. It will be shown that recording three or more interferograms on a single plate the components of displacement vector can be determined to a higher accuracy. A 3-D displacement measurement has been reported using technique of regulated path length interferometry resulting in three independent interferograms on a single plate.

Introduction

Hologram interferometry is an accurate measuring technique and provides displacement measurement with high sensitivity. In simple cases where the direction of displacement is known a priori, measurement can be carried out with an interferometer of simple construction and interferograms in such cases can easily be evaluated. When the changes are 3-D and heterogeneous, the usefulness of hologram interferometry is determined by its ability or inability to measure all components of the displacement vector with sufficient accuracy.[1]

The three unknown components of a displacement vector can in principle be determined with equal accuracy using a multiinterferogram technique[2] where the component of displacement lying along the bisector of the angle between illumination and observation directions can be determined by measurement of fringe order number.

This paper describes a measurement of a real three-dimensional motion. The interferograms with three different sensitivity vectors have simultaneously been recorded on a single plate by regulated path length interferometry. A brief survey of the technique will be given with respect to measuring 3-D displacement.

Regulated path length interferometry

In the approach mentioned above for determination of the three components of displacement vector at least three interferograms are needed. When three separate plates are used for simultaneous recording of the interferograms observation of object and fringes from different directions result in complicated operations for identification of object surface points on different holographic images. Such difficulties can be avoided by projecting the real image from the interferograms back onto the object.[3]

Using the same view (one observation direction) for all images a more accurate identification is possible. To obtain interferograms with different sensitivity vectors the illumination directions should be different for each interferogram.[4,5]

Simultaneous recording and one single observation direction are typical for regulated path length interferometry.[5,6] The interferograms are recorded on a single plate by three object and three reference beams. Independent interferograms (no cross-talk among them) result from noncorrelation between beams not belonging together. The noncorrelation can be realized by optical path length differences that are larger than the coherence length of the laser used in the experiment.

Experimental conditions

The regulated path length interferometry has been applied to the measurement of 3-D displacement of an object shown on Figure 1. The dimensions of the object are 135x55x215 mm. The x-y plane of the coordinate system assigned to the surface is the large side of the object facing to us. The x axis is parallel to the lower edge with positive direction from left to right. The y axis is parallel to the left edge of the side mentioned with positive direction upward. The distance between the axes and the edges is 10 mm.
The upper part ($y \geq 40$mm) of the object can rotate around an axis parallel to z one at x=-10mm. Similarly it can rotate around an axis parallel to y; the axis is in x-y plane at x=35mm. The object has been mounted on two rotating tables.
The object surface points for which the requirements $-10 \leq x \leq 125$ (here and further on data are in mm), $-10 \leq y < 40$ and $-55 \leq z \leq 0$ were fulfilled did not move during the measurement.

Between the two exposures the remaining surface points have been rotated around the axis with coordinates x = -10, y = 40 due to a momentum parallel to -z axis. The points for which $35 \leq x \leq 90$, $40 < y \leq 165$, $-55 \leq z \leq 0$ have been rotated around an axis (x=35, z=0) too, due to a monumentum parallel to -y axis. So these points have been moved in three dimensions.

The projection of the experimental setup upon y-z plane (the plane of the holographic table) is shown on Figure 2. The light source was He-Ne laser of power 50 mW, type HNA-188 VEB Carl Zeiss Jena. The interferometer consists of 5 beamsplitters, 6 microscope objectives with pinholes, 12 mirrors and a hologram plate holder. The three object beams are denoted by O_1, O_2 and O_3, the reference beams by R_1, R_2 and R_3. The interferograms have been recorded on Agfa-Gevaert 10E75 plates. The path length difference between beams not belonging together was 40-80 cm.

The important geometrical data are as follows: coordinates of the illumination point sources are m_1(-255,2; 517,2; 413,4), m_2(-639,2; 121,6; 151,4), m_3(-181,7; 120,3; 769,9) and that of the observation point M(-250,8; 115,6; 560,5). It means that the beams illuminating the origin include the following angles with x, y, z axes: $\alpha_1=121,1°$, $\beta_1=43,71°$, $\delta'_1=31,08°$, $\alpha_2=166,7°$, $\beta_2=79,51°$, $\gamma_2=76,67°$ and $\alpha_3=76,72°$, $\beta_3=81,35°$, $\gamma_3=13,28°$. Furthermore, the origin is seen from the observation point from a direction determined by angles $\alpha=114,1°$, $\beta=75,78°$ and $\delta=24,10°$

Figure 1. The object.

During the first exposure the plate was illuminated by three object and three reference beams. Then the object was rotated in a way described above. During the second exposure the plate was also illuminated by six beams. The L_x and L_z components of displacement vector at point

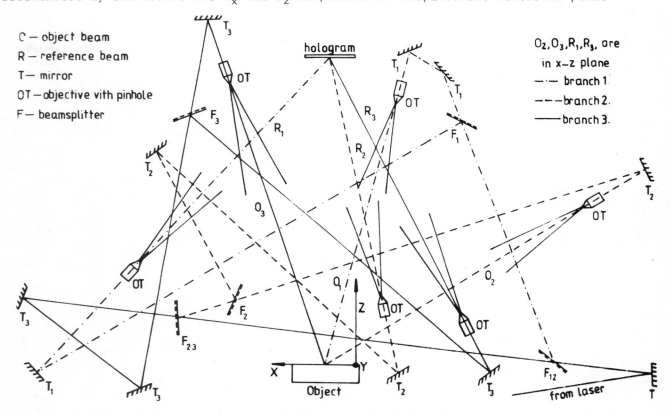

C — object beam
R — reference beam
T — mirror
OT — objective with pinhole
F — beamsplitter

hologram

O_2, O_3, R_1, R_3 are
in x-z plane
— · — branch 1.
— — — branch 2.
——— branch 3.

Figure 2. The experimental setup.

18

P(121; 200; 0) were measured by dial indicators. Their accuracy estimated was about 8%.

Results and evaluation

Interferograms recorded in such a way are presented on Figure 3. The axes of rotation

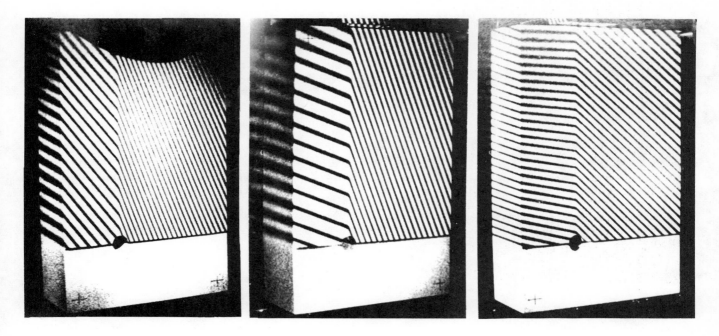

Figure 3. Interferograms with different mean sensitivity vectors.

are clearly visible on the interferograms. The density of fringes on interferograms is different according to different sensitivity vectors.

The components of displacement vectors are determined using the fundamental equation of double exposure hologram interferometry:

$$\underline{S}\underline{L} = \lambda \underline{N} \tag{1}$$

where $\underline{S}$ is the sensitivity matrix determined entirely by the geometry, $\underline{N}$ is a column vector of fringe order numbers, $\underline{L}$ is the displacement vector to be determined and $\lambda = 632,8$ nm is the wavelength of the light.

The components of displacement vectors of object surface points along the straight line y=1,05z+70, x=-10 within the interval -27,5 z 0 are shown on Figure 4. Values of the components shown belong to four independent measurements of fringe order number. Points on the line have not moved in direction of the z axis. The components L_x and L_y have been calculated from starting point z = -27,5 along the line through each 5 mm.

The components of displacement vectors of object point y=x+80; z=0 are reported on Figures 5 and 6. The coordinates of starting point are (-10; 69; 0) those of the end point (125,5; 205; 0). Figure 5 demonstrates the components L_x , while the components L_y and L_z are shown on Figure 6. Points belonging to section of straight line $-10 \leq x \leq 35$ have not moved towards the z axis. The values of components L_x and L_z measured by dial indicators at point P were 22,0 /um and -11,5 /um respectively.

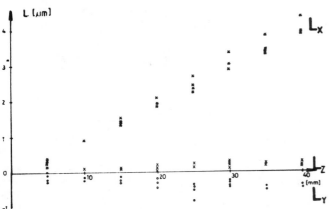

Figure 4. Components of displacement vectors (y=1,05z+70; x=-10).

The calculated $L_x=22{,}59\ \mu m$ and $L_z=-10{,}43\ \mu m$.

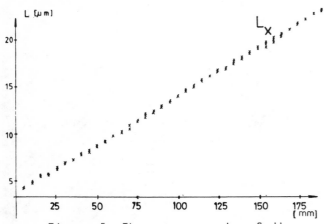

Figure 5. The x components of displacement vectors (y=x+80; z=0)

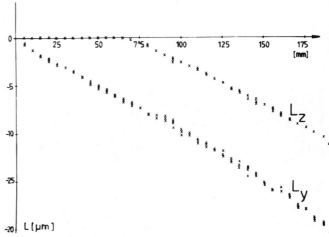

Figure 6. y and z components of displacement vectors (y=x+80; z=0)

There is a lack of symmetry in the interferometer. The components of displacement vectors are measured to different accuracy. The upper limit of absolute error of the measurement can be calculated by

$$\Delta u \leq \sum_{i=1}^{n}\ \left|\frac{\partial u}{\partial x_i}\right|\ \ \Delta x_i \tag{2}$$

where u is a function of quantities x; Δx_i is the absolute error of x_i. For our case (2) leads to an expression as follows:

$$\frac{\Delta L_k}{\lambda\ \Delta N}=\sum_{i=1}^{3}\ \ \left|\left[(\underline{\underline{S}}^T\ \underline{\underline{S}})^{-1}\ \underline{\underline{S}}^T\right]_{k,i}\right|\ \ \ ;\ \ k=x,y,z, \tag{3}$$

where ΔL_k is the absolute error of k-th displacement component, ΔN is the error of fringe order estimation (the same for each interferogram). The right hand side of equation (3) is the sum if absolute values of elements standing in k-row of inverse matrix composed from the sensitivity matrix in equation (1). Deriving the equation (3) an assumption was made stating that the error arising from the inaccuracy of geometry is much smaller than that of fringe order estimation.

The quantity $\frac{\Delta L_k}{\lambda\ \Delta N}$ is shown on Figure 7. for the straight line y=x+80; z=0 as a function of the distance from the starting point. Errors belonging to the same component of displacement vector differ slightly from each other: variation of the sensitivity vector is negligible along the straight line. This fact is represented by equidistancy of interference patterns on Figure 3.

Calculating the absolute error of components it was supposed that the estimation of fringe order number is better than 0,25 fringe. So at the point P(121; 200; 0) $\Delta L_x=0{,}28\ \mu m$; $\Delta L_y=0{,}54\ \mu m$ and $\Delta L_z=0{,}23\ \mu m$.

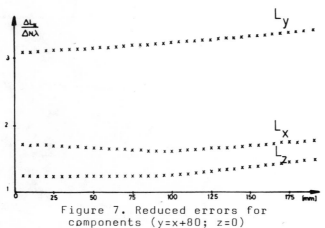

Figure 7. Reduced errors for components (y=x+80; z=0)

Conclusion

A 3-D displacement measurement has been reported using technique of regulated path length interferometry. One observation direction i.e. the same view for each interferogram and simultaneous recording are the inherent properties of the regulated path length interferometry.

For identification of object surface points on different holographic images there is no need of marking and evaluation of interferograms can easily be automized.

References

1. Matsumoto, T., Iwata, K., Nagata, R., Measuring accuracy of three-dimensional displacements in holographic interferometry, Appl. Optics, Vol. 12 pp. 961-967, 1973.
2. Ennos, A.E., Measurement of in-plane surface strain by hologram interferometry, J. Phys. E., Vol. 1, pp. 731-734. 1968.
3. Matsumoto, T., Iwata, K., Nagata, R., Distortionless recording in double-exposure holographic interferograms, Appl. Optics, Vol. 12, pp. 1660-1662. 1973
4. Hung, Y.Y., Hu, C.P., Henley, D.R., Taylor, E., Two improved methods of surface displacement measurements by holographic interferometry, Optics Communications, Vol. 8, pp. 48-51. 1973.
5. Füzessy, Z., Methods of holographic interferometry for industrial measurements, Periodica Politechnica, ME, Vol. 21, pp. 257-263. 1977.
6. Füzessy, Z., Abramson, N., Measurement of 3D-displacement: sandwich holography and regulated path length interferometry. Appl. Optics, Vol. 21, pp. 260-264. 1982.

Automatic Evaluation of Holographic Interferograms by Reference Beam Phase Shifting

Werner Jüptner, Thomas M. Kreis, Horst Kreitlow

Bremer Institut für angewandte Strahltechnik, BIAS
Ermlandstr. 59, 2820 Bremen 71, Germany

Abstract

In most cases experimental investigations in fracture mechanics using holographic interferometry demand a spatial resolution better than a fraction of the wavelength of light. High accuracy can be achieved by phase measurements rather than mere fringe counting. A variation of the phase is performed by heterodyne techniques using a frequency shift of the reference waves or by phase shifting of the reference waves during reconstruction. A method was proposed to determine the three dimensional deformation field of object surface points by solving three systems of equations, each one defined by the data of four phase shifted fringe patterns. By means of this method fringe pattern disturbances can be eliminated, an accurate determination of interference phases, even between fringe maxima, is performed, and the sign of the deformation vector can be achieved. Furthermore, this method is well suitable to be automated in the data generation, recording, and evaluation processes, respectively. These aspects will be discussed and for the first time results of an experimental application of this method in practice, containing all the necessary intermediate steps, are presented.

Introduction

Holographic interferometry is a method to determine the three dimensional deformation fields or strain and stress fields of surfaces of technical objects. This is established by a quantitative evaluation of the fringe patterns. A determination of interference maxima and minima and a numerical interpolation of the intermediate values is only sufficient if the fringe order varies continuously and monotonously. A valuation of local inhomogeneities, strain and stress measurements, and some other applications require a high density of evaluation points besides a high accuracy in determining the deformations /1/. The density of evaluation points results in a high amount of data to be processed in acceptable time. Thus the generation, recording, and processing of the holographically produced data have to be automated /2/. Another problem arising in the quantitative evaluation of holographic interference patterns is the determination of the direction of interference order variation, i. e. increasing or decreasing. All these problems can be solved simultaneously by measuring phase shifts in the interference pattern.

The use of phase variations of the interference patterns in the quantitative determination of deformation fields has been described in a number of contexts. In heterodyne techniques /3/, the interference phase is measured electronically at the beat frequency of the two frequency shifted reference waves. This gives highest accuracy, but requires high electronical and optical effort and proceeds pointwise, thus time consuming. Reconstructions with phase shifted reference waves have been used to compensate the varying background intensity /4,5/ by methods of image processing. The method dealt with here, first described in /6,7/, uses four reconstructions with three arbitrary mutual phase shifts, which have to be constant, but need not to be known quantitatively. Recently the same method was described, using three reconstructions, however, with the phase shifts to be known exactly in advance /8/.

In this paper we will discuss the optical arrangement, define the theory for the evaluation of the interferograms and present results of applications of the method using four reconstructions with arbitrary mutual phase shifts.

Optics and Electronics

Reference beam phase shifting can be performed during reconstruction in double exposure as well as in real time holographic interferometry. When double exposure holographic interferograms are dealt with, reference beam phase shifting requires a holographic setup with two reference waves, fig. 1: The first exposure of an ordinary double-exposure hologram is made with one reference wave, the second with the other. The interference pattern is produced by reconstruction with both reference waves, the phase shift is introduced by a shift of the reflecting mirror in one of the reference waves. This mirror is fixed to a piezo-electric mount, whereas the other mirror is stationary. Normally the reference sources have to be well separated in order to yield an interference pattern not overlapped by the undesireable cross-reconstructions, produced by reconstructing the object state recorded with the first reference wave by the second reference wave and vice versa. On the other hand, if the two reference sources are close together, all reconstructions overlap, thus decreasing the fringe contrast, but for the benefit of reducing the sensitivity to hologram misalignment /8/, fig. 2. The achieved phase shift is controlled by a photodiode, recording the interference between the two reference waves in an ordinary Michelson interferometer /4/.

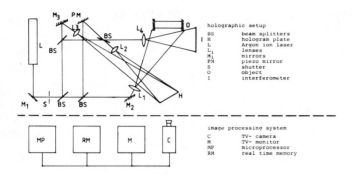

Figure 1. Arrangement with separated
reference sources

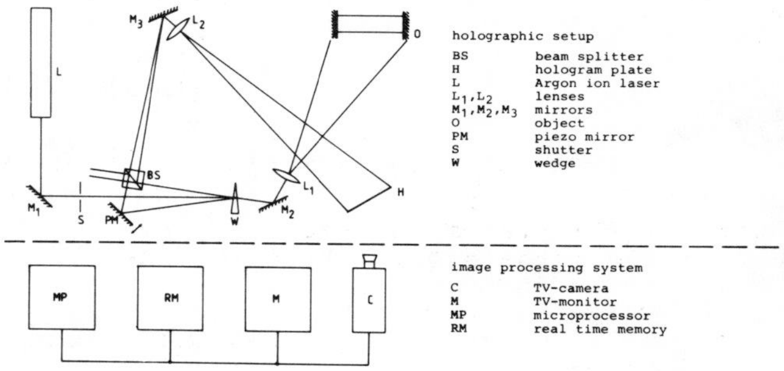

Figure 2. Arrangement with reference
sources close together

In real time holographic interferometry the deformed object itself is used to deliver directly the first object wave field needed to interfere with the holographically reconstructed wave field /9/. Only one shiftable reference wave is needed, fig. 3. Besides the simplicity of the holographic setup, in this case no problems with overlapping cross-reconstructions can occur.

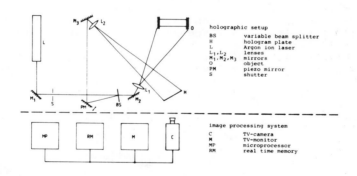

Figure 3. Arrangement for real-time
holographic interferometry

In a calibration process the dependence of the displacement of the piezo-electric mount as a function of voltage is recorded. Then in the linear range of this function a well reproducible and constant phase shift can be established.

For fringe evaluation the reconstructed images are recorded by a TV-camera, signal of which is digitized and quantized by an analog to digital converter, or recorded by a CCD-array or a line of photodiodes, if even higher resolution is required. The digitized image data are further processed by a microcomputer to obtain the desired deformation field. Along one line in the interferogram, the intensities of the four consecutively produced reconstructions are recorded and evaluated. In the same manner the next lines are processed. This way the total object surface is evaluated fully automatically. One evaluation line is sufficient, if the object itself has a one-dimensional structure, like cantilevers or tensile test specimens.

or three dimensional evaluation the same procedure is performed three times using three holographic interferograms produced with different illumination or observation directions. Advantageously, different illuminations of the object and one fixed observation direction is used, taking the three exposures on photo-thermoplastic film by one stationary positioned holo-camera, thus yielding fast and reliable results.

Theory

In double-exposure, stroboscopic, or real-time holographic interferometry the light intensity observed at point P of the interference pattern during reconstruction is of the form

$$I(P) = I_m(P) \cdot \cos^2(\vartheta(P)/2)$$
$$= I_m(P) \cdot (1 + \cos(\vartheta(P)))/2 \tag{1}$$

with $I_m(P)$ the mean intensity around P caused by the varying background intensity, and $\vartheta(P)$ the interference phase, defined by

$$\vartheta(P) = \frac{2\pi}{\lambda} \vec{d}(P) \cdot (\vec{b}(P) - \vec{s}(P)) \tag{2}$$

where $\vec{b}$ and $\vec{s}$ are unit vectors taken along the directions of observation and illumination, respectively. $\vec{d}(P)$ is the displacement vector at P.

The interference pattern reconstructed with the additional phase shift ϕ, can be written

$$I(x,y) = I_0(x,y) \cdot \cos(\phi + \vartheta(x,y)) + I_g(x,y) \tag{3}$$

where $I(x,y)$, the observed fringe intensity, is given by I_0, the maximum intensity possible at $P=(x,y)$, $I_g(x,y)$ the local mean intensity at (x,y), ϕ the performed phase shift, constant for all x,y and $\vartheta(x,y)$ the interference phase, due to the deformation. Because of evaluating line by line, the argument y can be omitted in the following.

The four reconstructions with arbitrary but constant mutual phase shifts used for quantitative evaluation yield the system of equations

$$I_1(x) = I_0(x) \cdot \cos(\vartheta(x)) + I_g(x)$$
$$I_2(x) = I_0(x) \cdot \cos(\phi + \vartheta(x)) + I_g(x)$$
$$I_3(x) = I_0(x) \cdot \cos(2\phi + \vartheta(x)) + I_g(x) \tag{4}$$
$$I_4(x) = I_0(x) \cdot \cos(3\phi + \vartheta(x)) + I_g(x)$$

This system is solved for the four unknowns I_0, I_g, ϕ, ϑ by:

$$\cos\phi = \frac{I_1(x) - I_2(x) + I_3(x) - I_4(x)}{2 \cdot (I_2(x) - I_3(x))} \tag{5}$$

$$\vartheta(x) = \arctan \frac{I_1 - 2I_2 + I_3 + (I_1 - I_3) \cdot \cos\phi + 2 \cdot (I_2 - I_1) \cdot \cos^2\phi}{\sqrt{1 - \cos^2\phi} \cdot (I_1 - I_3 + 2(I_2 - I_1) \cdot \cos\phi)} \tag{6}$$

$$I_0(x) = \frac{I_2 - I_1}{\cos(\phi + \vartheta) - \cos\vartheta} \tag{7}$$

$$I_g(x) = I_1 - I_0 \cdot \cos\vartheta \tag{8}$$

Normally, the determination of I_0 and I_g is not necessary. One way to proceed is to solve eqs. (5) and (6) for each point successively. The better way is to calculate first $\cos\phi$ according to eq. (5) for all points along the lines of evaluation and check whether this value remains constant. If it is nearly constant, eq. (6) can be solved, otherwise one can take an average over all calculated $\cos\phi$-values to determine the interference phase by eq. (6). The same solution is obtained by taking intensities 2,3,4 in eq. (6) instead of intensities 1,2,3.

The interference phase achieved by eq. (6) is indeterminate to a factor of 2π. To convert the discontinuous phase modulo 2π into the continuous phase distribution, at each point one has to determine first, whether a discontinuity occurs. There the absolute difference between the phase at this point and the phase at the preceding points must be greater than, e. g. $0.5*2\pi$. If a discontinuity is detected, the slope of the phase distribution over the last few preceding points has to be calculated, if it is positive, 2π must be added, if it is negative, 2π must be subtracted from the phase distribution at all following points. By cumulation of these 2π-terms, at each point an even integer multiple of 2π has to be added or subtracted. Another way to overcome the discontinuities is to look at the second solution obtained by the intensities 2,3,4.

Thus, the interference phase distribution over the whole line of evaluation is reached, which now together with the wavelength of the laserlight and the sensitivity vectors $\vec{b}-\vec{s}$, given by the geometry of the holographic arrangement, can be used to calculate the deformation field along this line. By performing this method line by line the deformation field of the whole object surface may be determined automatically.

Experimental Results

To confirm the validity of the proposed method, experimental tests were performed. The test object was a metal plate clamped at two opposite sides. A bending load was applied outside the center of the plate, to obtain an unsymmetrical deformation field. A 300 mW single frequency Ar-ion-laser with a wavelength of .5145 μm was used for object illumination. The holographic interferogram was produced by real-time technique according to the arrangement of fig. 3, the resulting fringe patterns recorded by a vidicon TV-camera, whose signal was fed to the MBS-III image processing system /10/, fig. 4.

image processing system   laser

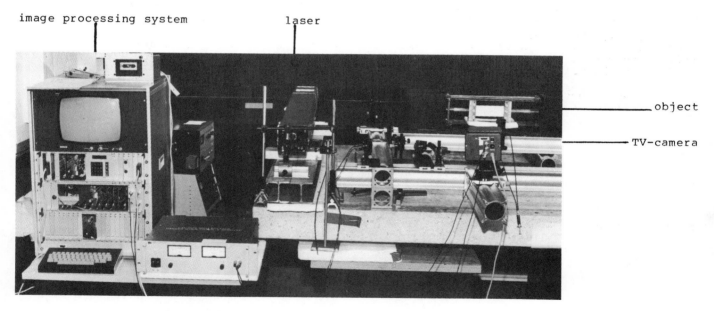

object

TV-camera

Figure 4. Electronical and optical setup

The analog image signal was digitized into 128 lines with 512 pixels per line, quantized into 256 gray-levels, and stored in RAM-memory. Figs. 5 to 8 show the interferograms reconstructed with relative phase shifts displayed at the monitor. The phase shifts between the records were produced by the reference mirror mounted on a piezo-electric driver, the voltage of which was varied constantly in the linear range, thus achieving constant but unknown phase shifts of the reference wave. Fringe evaluation was performed line by line, the actual evaluation line being marked bright in the display. The intensity distributions along this line are displayed below the interferograms. Quantitative evaluation was performed along the 256 central values of these distributions corresponding to the test object within the 512 pixels along the whole line. The plots of these intensities are shown in figs. 9 to 12.

Figure 5. Interferogram, first record

Figure 6. Interferogram, phase shift Φ relative to fig. 4

Figure 7. Interferogram, phase shift 2Φ relative to fig. 4

Figure 8. Interferogram, phase shift 3Φ relative to fig. 4

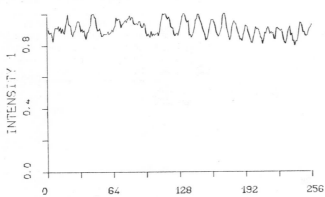

Figure 9. Intensity distribution along evaluation line

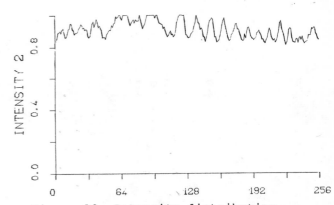

Figure 10. Intensity distribution with phase shift Φ

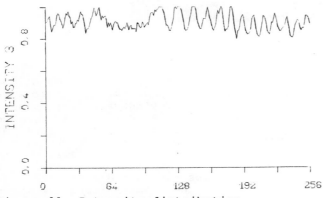

Figure 11. Intensity distribution with phase shift 2Φ

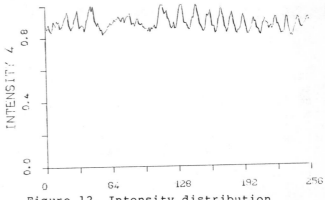

Figure 12. Intensity distribution with phase shift 3Φ

As one can see, these intensity distributions have not been recorded under optimized conditions, thus exhibiting low contrast, severe disturbances by speckle noise, diffraction patterns of dust particles and variing background intensity, as well as clipping the intensity at some points where the saturation of the analog-digital-converter is reached. These poor quality patterns have been chosen to prove the power of the proposed method.

From the four intensities the relative phase shift is determined automatically by the processor system, according to eq. (5), see fig. 13. The interference phase modulo 2π, calculated pointwise using eq. (6) is shown in fig. 14.

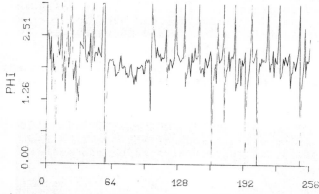

Figure 13. Calculated phase shift $\phi(x)$

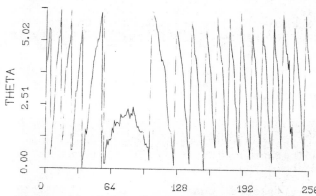

Figure 14. Determined interference phase modulo 2π

The continuous interference phase distribution along the evaluation line under consideration was attained by means of the above mentioned process, fig. 15. This interference phase distribution shows only minimal fluctuations even for the poor quality patterns that were used. Furthermore the experimentally established nonmonotonous phase variation has fully been recognized by this evaluation method. The normal component of the sensitivity vector was used to calculate the actual normal-displacement, fig. 16. Since it was assumed that the plate only is displaced in its normal direction, one illumination direction suffices to determine holographically the displacement. The results prove the capability of the proposed method to measure deformations automatically with high spatial resolution and accuracy even for image qualities far from optimum. As described above the determination of three dimensional deformation fields requires the same experimental procedure three times, only using three different illumination directions.

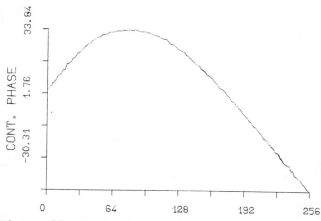

Figure 15. Evaluated continuous interference phase ϑ

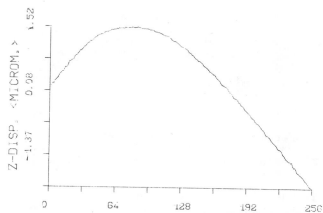

Figure 16. Determined displacement in normal direction

Discussion of errors

In performing the proposed method a number of systematic and statistical errors may be introduced. To investigate the effects of the various error sources, the method has been simulated additionally by computer programs.

One systematic error is introduced by not optimized and not constant phase shifts: Optimum phase shifts are taken in the range from .6 to 2 radians, as shown by computer simulations. Phase shifts not constant as well as variations of I_0 and I_g between recording the intensities cause a fluctuation in the phase shift calculated from these intensities. Particularly at points, where the denominator of eq. (5) is near 0, sharp erroneous peaks may occur, see fig. 13. On the other hand, the determination of the interference phase ϑ according to eq. (6) is very insensitive to falsely calculated phase shifts, thus leading to a high accuracy even for suboptimum measuring conditions.

The disturbances of the holographic interference pattern itself, that remain constant during phase shifts, are determined and compensated implicitly in the solution of eqs. (5) and (6).

Conclusions

The proposed method of reference beam phase shifting used for quantitative evaluation of holographic interference patterns has a number of advantages compared to common evaluation techniques:

- It measures phase distributions instead of counting fringes, thus yielding a higher spatial resolution and accuracy, as needed particularly in strain and stress analysis.

- It determines the direction of interference order variation and thus, the sign of the deformation vector.

- It compensates disturbances of the interference pattern, resulting from speckles, background intensity variations, or diffraction patterns.

- It can be performed with double exposure as well as with real time holographic interferometry. As shown, the real-time technique avoids the disturbing cross reconstructions.

- It does not require the previous knowledge of the phase shifts.

- It is well suitable to be automated in recording, reconstruction, and quantitative evaluation of the holographic interferograms using video- or CCD-techniques and digital image processing, thus delivering the deformation field for all points of the object surface in a fast and comfortable way.

- Fastest measurements of three-dimensional deformation fields are achieved with this method when employing e. g. a holo-camera using photo-thermoplastic film. The three components of the deformation vector can be determined even with one camera position if three illumination directions are installed.

References

/1/ Kreis, Th., Fischer, B., Jüptner, W., Sepold, G., Automatisierte Auswertung holografischer Interferenzmuster bei der Untersuchung von Zugproben, Proceedings of Laser 81 Conference Munich, Springer-Verlag, 1981

/2/ Kreis, Th., Kreitlow, H., Jüptner, W., Automatisierte digitale Verarbeitung holografischer Interferenzmuster, Informatik Fachberichte Vol. 29, Springer-Verlag, 38-47, 1980

/3/ Dändliker, R., Auswerteverfahren für die holografische Dehnungsmessung, Tagungsband Frühjahrsschule 78 "Holografische Interferometrie in Technik und Medizin", Hannover 1978

/4/ Kreis, Th., Kreitlow, H., Quantitative Evaluation of Holographic Interference Patterns Under Image Procesing Aspects, SPIE-Proceedings Vol. 210, 196-202, 1979

/5/ Lanzl, F., Schlüter, M., Video-Electronic Analysis of Holographic Interferograms, Tagungsband Frühjahrsschule 78 "Holografische Interferometrie in Technik und Medizin", Hannover 1978

/6/ Jüptner, W., Automatisierte Auswertung holografischer Interferogramme mit dem Zeilen-Scanverfahren, Tagungsband Frühjahrsschule 78 "Holografische Interferometrie in Technik und Medizin", Hannover 1978

/7/ Fischer, B., Geldmacher, J., Jüptner, W., Untersuchungen zur automatisierten Erkennung und Verarbeitung holografischer Interferenzmuster mit dem Zeilen-Scan-Verfahren, Proceedings Laser 79 Conf., IPC Science and Technology Press Ltd., 412-425, 1979

/8/ Dändliker, R., Thalmann, R., Willemin, J.-F., Fringe Interpolation by Two-Reference-Beam Holographic Interferometry: Reducing Sensitivity to Hologram Misalignment, Optics Communications Vol. 42, 301-306, 1982

/9/ Hariharan, P., Oreb, B.F., Brown, N., Real-time holographic interferometry: a microcomputer system for the measurement of vector displacements, Applied Optics Vol. 22, 876-880, 1983

/10/ Kreitlow, H., Kreis, Th., Entwicklung eines Gerätesystems zur automatisierten statischen und dynamischen Auswertung holografischer Interferenzmuster, Proceedings Laser 79 Conf., IPC Science and Technology Press Ltd., 426-436, 1979

HOLOGRAPHIC MEASURING OF THE DEFORMATIONS OF VARIOUS INTERNAL COMBUSTION ENGINE PARTS

P. Paulet

Service Métrologie, Centre Technique CITROEN, 2 route de Gisy, 78740 VELIZY (France)

Abstract

To increase the torque and power of an engine one needs to carry out form modifications and matérial distribution. This has been studied through different methods and especially holography. A study of the relative mouvements in the areas of the helved throws of the multi-parts crankshaft, bearing on two supports in V and submitted to a deflecting force, was obtained by measuring point by point in the real image field of a double exposure hologram. The crankshaft being positioned in a cylinder-block equipped with one cylinder and one cylinder head, the gas pressure is simulated by an oil pressure slowly increasing. Real time holograms have been performed on three different cylinder-blocks and a 16 mm film (duration 10') shows how interference fringes appear in the virtual image when the oil pressure varies. This film has permitted to the motor mechanical engineer to remedy to the cylinder-block breakings during preseries.

Introduction

The hereafter measuring has been performed upon inquiry of our motor mechanical chief engineer of our Research Department and deals with a 650 c.c. air-cooled all-aluminium boxer engine fitting our Visa model. In order to insure a safety margin, the engine capacity was increased up to 780 c.c. and corresponding engines were submitted to fatigue testing on bench and road. Within a short period, cracks appeared on the crankcase ; an exhaustive study of the crankcase and of the crankshaft was undertaken by calculating and computing and with the help of every measuring technique, holography included. This work was performed in 1979.

Measuring crankshaft

This part is made of 3 components, assembled by hooping (see Fig. 1). Breaks having occured on assembled crankshaft during fatigue testing, two questions arose :

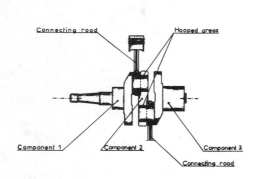

Figure 1

Figure 2

1) Were there any relative sliding of components in the hooped areas ?
2) Were flexions not too great?
   By using holography we could answer both questions.

1) Relative sliding

The mechanical set-up (see Fig. 2) is a rigid assembly made of 2 V's. On each of them, two inserted small bars reduce the contact between set-up and crankshaft to a 1 mm generating line. A piezoelectric cell measures the applied bending effort. The holographic set-up (see Fig. 3) includes an Argon laser. It has been carried out so that the sensibility vector[1] varies little. The reference wave lighting the holographic plate is even, which makes the real image restitution easier by turning the hologram 180°. The crankshaft axle is parallel with the hologram plane. We realized double exposure (see Photo 1).

One can see that the fringes are parallel but in different directions and with different fringe spacing according to components 1, 2 and 3. They are very close to the object surface and indicate rotations.

For preciseness's sake we realized sandwich holograms of the object in position 2 (see Fig. 3 and photo 2) after rotation through an angle of 19° to see the helved areas better. The photo 3 shows a double exposure holographic interferometry with fringes spaced and directed differently according to either parts 1, 2 and 3. The same sandwich hologram tilted by an angle +36° in relation to an horizontal axis allows to obtain the zero order on the left side of the crankshaft and tilted by an angle of 25 to obtain the zero order on the right side (photo 5). Qualitatively it can be said that the extreme parts of the crankshaft have opposed rotations in relation to the central part.

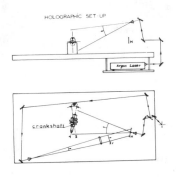

Figure 3

Photo 1

One can reckon[2] the rotations of the object surface :

$$\alpha = \frac{1}{2} \, arctg \, \frac{d}{D} \sqrt{\left(\frac{n}{\sin \gamma}\right)^2 - 1} \qquad d = 1.3 \, mm \, ; \, n = 1.5 \text{ (see Fig. 4)}$$
$$D = 960 \, mm$$

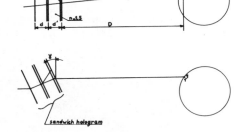

Figure 4

Photo 2

For component 1 one obtain : $\alpha = + 1'$
For component 3 one obtain : $\alpha = - 42''$

Photo 3

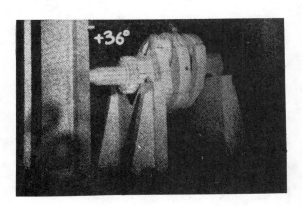

Photo 4

The rotations are small and several measurements on preseries crankshaft have shown there was no relative sliding in the two helved areas, the measured rotations being very small and due to bending torsion.

Photo 5

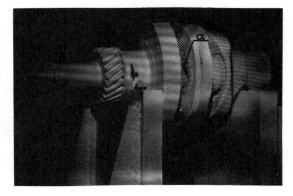

Photo 6

However, we constructed a crankshaft with less tightened helved areas and we made a real time hologram of it. This permitted to follow the phenomenon when the bending load increases progressively. Photo 6 was taken when the sliding occured. One can see fringes on ring marked (a) that are identical to those on the right side of the crankshaft. The central component had little rotated (very large fringe spacing) and the left component had rotated in a medium way (middle spaced fringes by comparison with central and right components).

These measures showed together with other testing that the technic of this assembly is not mastered and that investigations should be oriented towards the roughness of the helved areas.

2) Bending

The holographic arrangement is the same as above (see Fig. 3), the crankshaft being in position 1. Several double exposures are carried out because, in the case of bending under real load, the displacements are too great.

Real time holograms have shown that the deformation increased continuously as a fonction of load. Measuring has been carried out in the conjugate real image reconstructed out of double exposure holograms of the object under 500 daN and 1 000 daN. The displacement vector component in the object plane, perpendicular to the viewing direction, should be measured.

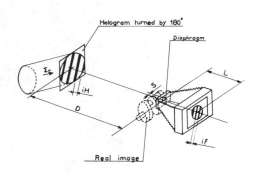

Figure 7

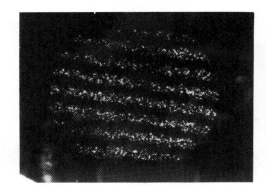

Photo 7

A 0.6 mm diaphragm is placed in the real image on the point of the object surface were one wishes to perform the measuring and one takes photographs of the "YOUNG fringes" appearing in the hologram plane with a 700 mw laser and a 3 000 ASA film the exposure time is 10-15 minutes. We measured 34 points of the deformation curve of 4 crankshafts respectively. This means that the duration of measuring was very long. It is at this time that we decided to automatize the measuring by using holographic interferometry with electronic phase detection[2], the corresponding investment being justified by the here reported results.

The displacement vector component in a plane perpendicular to the viewing direction is :

$$\jmath = \lambda L / if$$

Photo 8

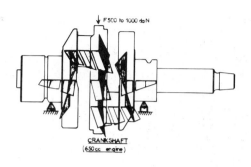

Figure 8

We admitted as bounds of experimental errors in measurement :
ΔD = 1 mm $\Delta \gamma$ = 0,5°
ΔL = 1 mm Δif = 0,2 mm   and as maximum final error ±5 %

The still blocks supporting the object become deformed under load and a measuring in these supporting points allows to substract the displacement from the deformation curve. (see F8)

Cylinder-block deformation

The all-aluminium cylinder-block, die cast in two parts is assembled by bolts. The first fatigue tests revealed the weakness of the left half-cylinder block. The testing with strain gages sticked on breaking areas gave low stress values and did not permit to clarify the appearences of cracks during the fatigue testing. Through the holographic global method one investigated the 3 cylinder-blocks which differ on their respective ribbing.

Photo 9

Die cast standard
cylinder-block
Crack areas marked (a)

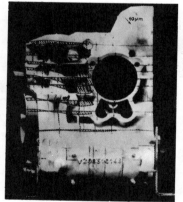

Photo 10

Die cast cylinder-block
n° 105 differing from
the former only by 2
strengthening ribs marked (a)

Photo 11

Precise sand cast
cylinder-block.
Its structure was
reviewed in order to
increase its rigidity

Mechanical set-up

A cylindrical bar whose rigidity is equivalent to the crankshaft one is loaded through a rod simulating the connecting rod and bearing on the piston pin (see Fig. 9). Oil is introduced into the combustion chamber with the help of a pump. By varying the oil pressure one simulates the internal efforts of the engine. The mock rod is equipped with strain-gages allowing to measure the efforts. This arrangement has two advantages :
    - The efforts remain inside the system, what eliminates the global object dis-
      placements,
    - The cylinder-block is loaded in conditions nearby real functioning.

A squaring, pencil-plotted on the object surface allows to locate the sections on which one wishes to do measuring (see photo 9, 10 and 11).

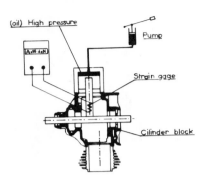

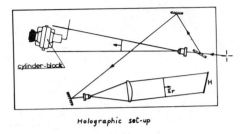

Holographic set-up

Fig. 9   Fig. 10

Holographic arrangement and measuring

This arrangement was similar to the one used for crankshaft (see Fig. 10 and Photo 12). Seven doubled exposed holograms were carried out according efforts increasing from 400 daN to 400 daN up to 3000 daN for each cylinder-block respectively.

Photo 12

By manual counting performed on photographs of the virtual object image, one plotted the deformation curve of each section (see photo 9, 10 and 11). These deformation curves represent the locus of the displacement vectors component in the viewing direction, that is almost perpendicular to the object surface. These values are accurate within some percentage points but accuracy is not peculiarly searched after. However the deformation profile and direction were particularly useful for the determination of the object areas to strengthen.

A 16 mm film of the virtual image of 3 holograms in real time was taken when oil pressure is increasing in cylinder. This film shows the deformations of 3 cylinder-blocks and its qualitative utilization has enabled the mechanical engineers to better understand the engine behaviour and to infer the weak engine areas.

This study has revealed an area of big deformation amplitude corresponding to the localization of the observed cracks (see Photo 9). In the heavely lodded area of the central crankshaft bearing it was also necessary to add an horizontal rib (see Photo 10) and to increase the canvas.

The 650 c.c. preseries cylinder-block has been modified taking into account the above results and the fatigue testing has confirmed the good behaviour of its parts.

Acknowlegments

The author thanks Mr Dupin for discussing the subject with him and for his precious help as also Messrs Szuplewski and Duprix for the skillful experimental word.

References

1) Collier, R. Burckhardt, C. Lin, L. optical Holography, Academic Press 1971
2) Dändleker, R. Quantitative strain measurement through holographic interferometry. Int. Conf. on appl. of holog. Jerusalem, Israel, Aug. 23 1976
3) Abramson, N. Sandwich hologram interferometry : Some practical calculations. Appl. Opt. Vol 14, Nr 4, p 981 1975

Fringe modification with amplification in holographic interferometry and application of this to determine strain and rotation

D. Cuche and W. Schumann

Laboratory of Photoelasticity, Swiss Federal Institute of Technology
Rämistrasse 101, CH 8092 Zürich
Switzerland

Abstract

Fringe modification is carried out so that the geometrical change is much greater than its effect on the fringe function. The equations denoting the optical path difference, the fringe vector and the fringe interspace are deduced, and then applied to determine strains and rotations on the surface of an opaque body.

Introduction

Holographic interferometry has long been used to determine strain and rotation of an object (see e.g. ref. 1-5). For more flexibility in the evaluation of the fringes, a geometrical modification can be applied to the holographic arrangement during the reconstruction. The change of the wave-length will not be considered in this paper although it also gives more flexibility. Thus the recording of the two states of the object in double exposure necessitates two reference sources [6,7] or two different holograms[8].A displacement of one of the two reference sources or holograms in the reconstruction will directly modify the fringes [9]. However, this displacement, which can be interpreted, in the second case, as a repositioning error, is of the same order of magnitude as the displacement of the object and therefore difficult to be performed with sufficient precision. Consequently, holographic interferometry with modification can become even more interesting if an amplification occurs which will reduce the effects on the fringes of the applied modification [10] or of the repositioning error [11]. With a particular construction, the two effects can be reduced simultaneously. This is achieved by having one hologram and two close reference sources which are moved as a rigid body through the medium of a mirror. The displacement of the hologram, or of the two joined sources, is of an order of magnitude greater than its effect on the fringes. The equations denoting the optical path difference, the fringe vector and the fringe interspace are deduced, paying particular attention to the modification terms. An experimental application of these equations is carried out for the determination of all six components of strain and rotation. The principal value to be measured is the displacement of the mirror, while the resulting modification appears mainly in the change of the fringe direction.

Optical path difference with determination of the displacement

The object, illuminated by the source S, is recorded on the hologram {H} with the first reference source Q (Fig. 1). After being deformed, the object, still illuminated by the source S, is recorded on the same hologram {H} with the second reference source Q'. In the reconstruction, the corresponding light waves, formed by diffraction, represent the object in its initial and deformed state simultaneously. If the two reference sources Q and Q' are displaced from their original places, it results in the two object positions being modified. The point P on the object and its corresponding P' on the deformed object act as point-sources for spherical waves. On the hologram, the complete amplitude of these two waves can be denoted by U and U'. The two corresponding amplitudes in the reconstruction are

$$\tilde{U} = \tilde{V}V^*U \quad , \quad \tilde{U}' = \tilde{V}'V'^*U' \tag{1}$$

where V and V' are the amplitudes of the waves on the hologram emitted by the reference sources Q and Q'. The sign ~ expresses the corresponding amplitude in the reconstruction, and the sign * expresses the complex conjugate. The interference between the two reconstructed waves of the object occurs in the direction of observation **k** at the point $\tilde{K}$ (Fig. 1). Extending the line $\tilde{K}P$ and $\tilde{K}\tilde{P}'$, we obtain the points H and H' on the hologram where the arguments of Eq. (1) describe the condition of interference identity:

$$\text{At H}: \frac{2\pi}{\tilde{\lambda}}\tilde{p} + \tilde{\phi} = \frac{2\pi}{\tilde{\lambda}}\tilde{q} + \tilde{\psi} - \left[\frac{2\pi}{\lambda}q + \psi\right] + \frac{2\pi}{\lambda}p + \phi,$$

$$\text{At H'}: \frac{2\pi}{\tilde{\lambda}}\tilde{p}' + \tilde{\phi}' = \frac{2\pi}{\tilde{\lambda}}\tilde{q}' + \tilde{\psi}' - \left[\frac{2\pi}{\lambda}q' + \psi'\right] + \frac{2\pi}{\lambda}q' + \phi'. \tag{2}$$

p is the length between P and H, $\tilde{q}'$ is the length between $\tilde{Q}'$ and H', ϕ' is the phase at P' and ψ is the phase at Q. The distance HH' is small when compared,for example,to the distance

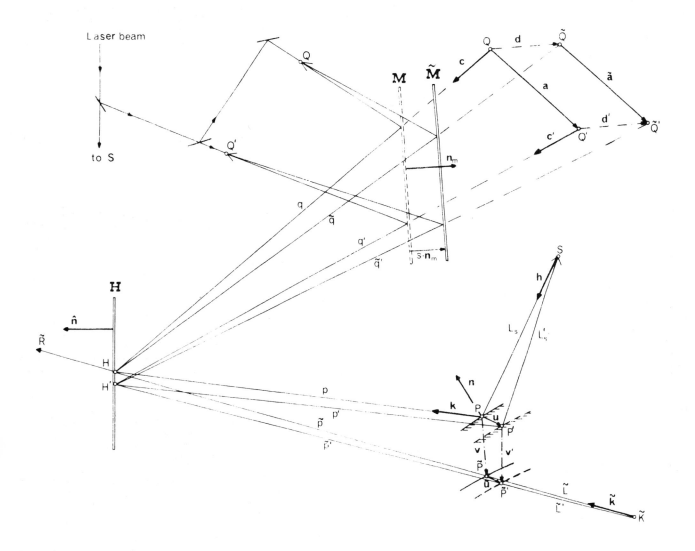

Figure 1: Displacement of the sources Q,Q' produced by the motion of the mirror **M**. Here, only a translation of the mirror (and not a rotation) is represented (**d** = 2 s· **n**$_m$). This translation produces a shift in the holographic image and a modification of the fringe pattern.

between the sources, as the rays $\widetilde{K}\widetilde{P}H$ and $\widetilde{K}\widetilde{P}'H'$ must enter the pupil of the observer, without which the interference fringes could not be seen. We only consider in this paper a geometrical modification, so that the wave length λ will be the same in the reconstruction, ($\widetilde{\lambda} = \lambda$). The phases' differences, $(\widetilde{\psi} - \widetilde{\psi}') - (\psi - \psi')$, remain at zero, because the modification of the sources will be controlled by a mirror. The optical path difference is for the point $\widetilde{K}$:

$$D = (\frac{\lambda}{2\pi} \, \widetilde{\phi} - \widetilde{L}) - (\frac{\lambda}{2\pi} \, \widetilde{\phi}' - \widetilde{L}') \, , \qquad (3)$$

where $\widetilde{L}$ is the distance $\widetilde{P}\widetilde{K}$ and is defined as positive when $\widetilde{K}$ lies behind the surface. The unknown phases $\widetilde{\phi}$ and $\widetilde{\phi}'$ can be eliminated by introducing Eqs. (2) in Eq. (3). The phase ϕ at P and ϕ' at P' can be replaced by the phase at the source S and the lengths L_s and L_s', so that Eq. (3) becomes:

$$D = (p - \widetilde{p}) - (p' - \widetilde{p}') + (L_s - L_s') + (q' - \widetilde{q}') - (q - \widetilde{q}) - (\widetilde{L} - \widetilde{L}'). \qquad (4)$$

By introducing the unit direction vectors **k** , **h** , **c** , the displacement **u** = PP' , the small source modification **d** = QQ and the corresponding small object modification **v** = P$\widetilde{P}$, then Eq. (4) takes the form:

$$D = \mathbf{u} \cdot (\mathbf{k} - \mathbf{h}) + \mathbf{d}' \cdot \mathbf{c}' - \mathbf{d} \cdot \mathbf{c} + O(\frac{v}{p} \cdot u) + O(\frac{|\mathbf{d} - \mathbf{d}'|}{q} d) + O(\frac{d^2}{q^2} \cdot a) \, . \qquad (5)$$

We have already in this equation a separation of the term appearing in standard holography, **u·g** (with sensitivity vector **g = k − h**), from the term of the modification, **d'·c' − d·c** (which depends only on the position and displacement of the two sources). Nothing has yet been stated about an amplification process, which we can now make appear by relating **d'** with **d** and **c'** with **c**. If one leaves the distance $|QQ'| = a$ moderately small ($a \ll q$), then **c'** can be expressed by **c, a**, the projector $\mathbf{C} = \mathbf{I} - \mathbf{c} \otimes \mathbf{c}$, and the super-projector $\mathbf{\mathcal{C}} = \mathbf{C} \otimes \mathbf{c} + \mathbf{c} \otimes \mathbf{C} + (\mathbf{C} \otimes \mathbf{c})^T$:

$$\mathbf{c'} = \mathbf{c} - \frac{1}{q}\mathbf{Ca} - \frac{1}{2q^2}\mathbf{a\mathcal{C}a} . \tag{6}$$

The super-projection is necessary here for precise measurements because $|\mathbf{a}|$ is considered only as moderately small. By leaving this distance $|\mathbf{a}|$ constant during the modification ($\tilde{Q}\tilde{Q}' = QQ' = a$), then, we have

$$\mathbf{d'} \simeq \mathbf{d} + \boldsymbol{\omega} \times \mathbf{a} = \mathbf{d} - \boldsymbol{\Psi}\mathbf{a} , \tag{7}$$

where $\boldsymbol{\omega}$ is the small rotation vector and $\boldsymbol{\Psi}$ the corresponding skew-symmetric tensor. Such a "rigid body" motion of the two sources can be easily achieved if a mirror is placed between them and the hologram (Fig. 1). Then the displacement and the rotation of this mirror induces a shift of the two sources while $|\tilde{Q}\tilde{Q}'| = |\mathbf{a}|$ remains constant. With (6) and (7), Eq. (5) becomes:

$$D = \mathbf{u} \cdot \mathbf{g} + \mathbf{a} \cdot \left[\boldsymbol{\Psi}\mathbf{c} - \frac{1}{q}\mathbf{Cd} \right] - \frac{1}{2q^2}(\mathbf{a\mathcal{C}a}) \cdot \mathbf{d} . \tag{8}$$

The equation (8) (which can be compared to Eq. 22, Ref. 12) does not contain the length $\tilde{L}$ and therefore is valid for any point on direction **k**. As **g** is the sensitivity vector for **u**, one can designate $-(\mathbf{Ca}/q + \mathbf{a\mathcal{C}a}/(2q^2))$ as a sensitivity vector for **d**, and $-(\mathbf{a} \otimes \mathbf{c})$ as a sensitivity tensor for the rotation tensor $\boldsymbol{\Psi}$. For a better view of the amplification process, let us divide Eq. (8) by a typical distance of the holographic construction; this distance, ℓ, can be q or p:

$$\frac{D}{\ell} = \frac{\mathbf{u}}{\ell} \cdot \mathbf{g} + \frac{\mathbf{a}}{\ell}\left[\boldsymbol{\Psi}\mathbf{c} - \mathbf{C}\frac{\mathbf{d}}{\ell}\frac{\ell}{q} \right] + \cdots . \tag{9}$$

In order to modify or compensate for the standard term, $(\mathbf{u}\cdot\mathbf{g})/\ell$, the modification term must be of the same order of magnitude. q/ℓ is of a finite order and a/ℓ is assumed to be of a small order $\delta (\delta \ll 1)$. If u/ℓ is also assumed to be of another small order $\varepsilon (\varepsilon \ll 1)$, then $|\boldsymbol{\Psi}\mathbf{c}|$ and d/ℓ must be of an order ε/δ, otherwise their influences on D would be either negligible or dominant. In addition, ε must be much smaller than $\delta(\varepsilon \ll \delta)$, otherwise Eq. (7) and the development in Eq. (5) would not be correct. In the following table, the terms involved are grouped with respect to their order:

$O(1)$	$O(\delta)$	$O(\varepsilon/\delta)$	$O(\varepsilon)$
$\frac{p}{\ell}, \frac{q}{\ell}, \frac{L_R}{\ell}, \ldots$	$\frac{a}{\ell}$	$\frac{d}{\ell}, \|\boldsymbol{\Psi}\mathbf{c}\|, \frac{d_H}{\ell}, \frac{v}{\ell}, \ldots$	$\frac{u}{\ell}, \frac{\|\mathbf{v'}-\mathbf{v}\|}{\ell}, \frac{\|\mathbf{d'}-\mathbf{d}\|}{\ell}$ $\mathbf{e} \cdot (\boldsymbol{\nabla} \otimes \mathbf{u})\mathbf{e}$
Lengths of the holographic set-up	Relative distance of the two sources	Motion of the sources as a "rigid body". Motion of the hologram. Absolute displacements of the holographic images.	Object deformation. Relative displacement of the holographic images and of the sources.

One should notice that the order δ depends on the geometry of the holographic set-up, whereas ε depends on the amount of the object deformation. δ has a minimum value, $\delta_{min} = a_{min}/\ell$: The reconstructed object, $\tilde{U} = \tilde{V}V*U$, and its cross-reconstruction, $\tilde{\tilde{U}} = \tilde{V}'V*U$, must not correlate [11]. Thus, corresponding areas on these two images must not overlap when they are observed with an instrument (or the eye) of aperture radius $\mathring{r}$, placed at a distance $\tilde{L}$ from the images (see Eq. (4.49) ref. 9). Then we have the equation

$$\left| \hat{M}\mathbf{Ca}_{min} \right| = \frac{2\mathring{r}q}{\mathring{L} - p} \mathbf{k}\cdot\mathring{\mathbf{n}} \tag{10}$$

where $\hat{M} = \mathbf{I} - (\hat{\mathbf{n}} \otimes \mathbf{k})/(\hat{\mathbf{n}}\cdot\mathbf{k})$ is an oblique projector, $\hat{\mathbf{n}}$ is the unit vector normal to the hologram, and $\mathring{\mathbf{n}}$ is the unit vector normal to the plane of the pupil. If these two images were to overlap each other, the resulting fringes would be localized on the hologram [6], for which reason the distance p is included in (10). In brief, the developped equations are valid as long as

$$1 >> \delta > \delta_{min} \quad \text{and} \quad \delta >> \varepsilon . \tag{11}$$

The hologram can be moved (Fig. 2) instead of the sources and the optical path difference becomes:

$$D = \mathbf{u} \cdot \mathbf{g} + \frac{1}{q} \mathbf{a} \cdot \mathbf{C}\mathbf{d}_H + \frac{1}{2q^2}(\mathbf{a}\mathbf{C}\mathbf{a}) \cdot \mathbf{d}_H \tag{12}$$

which is similar to Eq. (8), and where $\mathbf{d}_H$ is the displacement of the hologram point H. As already mentioned, this point is where the sight-line (direction $\mathbf{k}$) intersects with the hologram plane. It is worth pointing out that $\mathbf{d}_H$ could also be considered as a repositioning error of the hologram, as will be employed here, and the equation (12) can be used to estimate this repositioning error. The term $a/q = \delta$ is also present in Eq. (12) and thus d_H/ℓ has an effect only if it is of order ε/δ. Therefore, the influences on the fringes from the shift of the reference sources and of the hologram are both reduced by the order δ. However, we shall apply Eq. (8) instead of Eq. (12) for different practical reasons: Firstly, the displacement $\mathbf{d}$ is easier to measure than $\mathbf{d}_H$ (because one needs to know the hologram point H precisely in order to determine $\mathbf{d}_H$, and also that $\mathbf{d}_H$ is a function of $\mathbf{k}$) and secondly, the repositioning of the hologram is more accurate if the plate-holder remains fixed to the optical table. We shall also consider in the experiment only a translation of the sources, i. e. a translation of the mirror. We notice that no modification is achieved when carrying out an in-plane rotation or an in-plane translation of the mirror. The similarity between Eq. (8) and Eq. (12) is also present in sandwich holography when, in order to modify the fringe pattern, the reference source is displaced instead of the holograms. The two holograms are interchanged between the two exposures [10], so that the repositioning error of one hologram relative to the other will not be reduced by the amplification term.

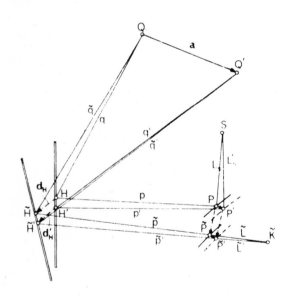

Figure 2: Modification produced by the displacement of the hologram

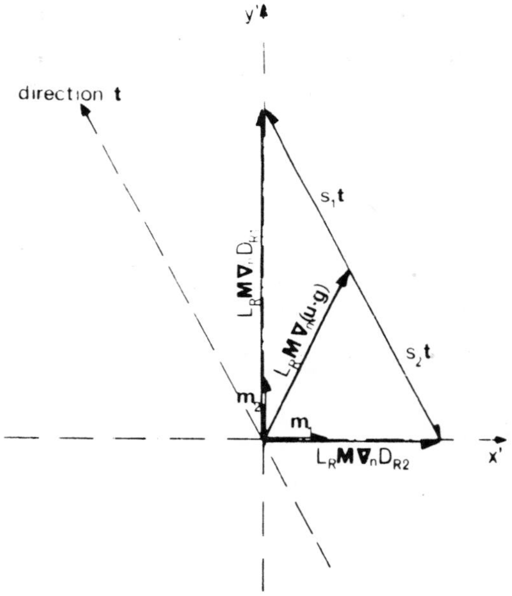

Figure 3: "Rotation" of the fringe vector produced by the modification.
$L_R\mathbf{M}\bar{\mathbf{\nabla}}_n(\mathbf{u}\cdot\mathbf{g})$: initial fringe vector.
$L_R\mathbf{M}\bar{\mathbf{\nabla}}_n D_{R1}$ and $L_R\mathbf{M}\bar{\mathbf{\nabla}}_n D_{R2}$: modified fringe vectors for two different modifications.

The displacement of the point P on the surface of the object can be calculated by using Eq. (8). The modification term in (8), $m = \mathbf{a} \cdot \left[\mathbf{\Psi}\mathbf{c} - \mathbf{C}\mathbf{d}/q - \mathbf{a}\mathbf{C}\mathbf{d}/(2q^2) \right]$, depends on the direction $\mathbf{k}$, however it is still necessary to measure D in three or more directions in order to obtain $\mathbf{u}$. The method of Aleksandrov and Bonch-Bruevich [13] (sometimes called the "dynamic method"), combined with the least-square method [3], are applied here:

$$\lambda\Delta n_i = \Delta D_i - \Delta m_i = \mathbf{u} \cdot (\mathbf{k}_o - \mathbf{k}_i) = \mathbf{u} \cdot \mathbf{f}_i \quad i = 1,2,3,4,5... \tag{13}$$

with $\Delta m_i = m_o - m_i$ and $m_i = m(\mathbf{k}_i)$. $\Delta D_i/\lambda$ is the counted number of fringes between the directions $\mathbf{k}_o$ and $\mathbf{k}_i$, and is equal to Δn_i when no modification is done. The observation direction $\mathbf{k}_o$, $\mathbf{k}_i$ must be carefully chosen and must not lie on a cone, in order that the $\mathbf{f}_i$'s are not on the same plane. The vectors $\mathbf{u}$ and $\mathbf{f}_i$ can be represented by their components according to a given coordinate system. For example, a cartesian system can originate from the object point P, with its direction z along the normal to the surface, and the direction x, y in the tangent plane to the surface. Thus u_x becomes

$$u_x = \lambda \frac{\begin{vmatrix} \Sigma \Delta n_i f_{xi} & \Sigma f_{xi}f_{yi} & \Sigma f_{xi}f_{zi} \\ \Sigma \Delta n_i f_{yi} & \Sigma f_{yi}f_{yi} & \Sigma f_{yi}f_{zi} \\ \Sigma \Delta n_i f_{zi} & \Sigma f_{yi}f_{zi} & \Sigma f_{zi}f_{zi} \end{vmatrix}}{\Delta} , \qquad \Delta = \begin{vmatrix} \Sigma f_{xi}f_{xi} & \Sigma f_{xi}f_{yi} & \Sigma f_{xi}f_{yi} \\ \Sigma f_{xi}f_{yi} & \Sigma f_{yi}f_{yi} & \Sigma f_{yi}f_{zi} \\ \Sigma f_{xi}f_{zi} & \Sigma f_{yi}f_{zi} & \Sigma f_{zi}f_{zi} \end{vmatrix} \qquad (14)$$

and u_y and u_z are expressed in similar equations. In practice, the modification term Δm_i is particularly useful to determine accurately the non-integer value of Δn_i[14] (while $\Delta D_i/\lambda$, equal to the observed number of fringes passing over the point P, remains an integer) and to define the sign (+ or −)[15] of the displacement $\mathbf{u}$ (when $\Delta D_i = 0$, the sign of Δm_i establishes the sign of $\mathbf{u}$). m_i can easily be measured experimentally by modifying with the mirror while always looking in the same direction ($\mathbf{k}_i = \mathbf{k}_o$). However, the flexibility created by the modification term becomes much more apparent in calculating the strain and rotation when the derivative of D is considered.

The derivative of D and determination of strain and rotation

The derivative of D, with the observer $\tilde{R}$ as a fixed point, shows the angular variation of D and contains the fringe vector, Thus,

$$dD_{\tilde{R}} = d\tilde{\mathbf{r}} \cdot \boldsymbol{\nabla}_{\tilde{n}} D_{\tilde{R}} = d\tilde{\phi}\, \tilde{\mathbf{m}} \cdot \tilde{L}_{\tilde{R}} \tilde{\mathbf{M}} \boldsymbol{\nabla}_{\tilde{n}} D_{\tilde{R}} , \qquad (15)$$

where $\tilde{\mathbf{m}}$ denotes an unit vector perpendicular to the viewing direction $\tilde{\mathbf{k}}$, $\tilde{L}_{\tilde{R}}$ the distance $\tilde{R}\tilde{P}$, $\tilde{\mathbf{M}}$ an oblique projector and $\boldsymbol{\nabla}_n$ the derivative operator on the surface of the object. We especially consider the modification as being very small compared to the optical paths, so that $\tilde{R} \simeq R$, $\tilde{\mathbf{k}} \cong \mathbf{k}$, $\tilde{\mathbf{M}} \cong \mathbf{M}$, $\tilde{L}_{\tilde{R}} \cong L_R$ and $\boldsymbol{\nabla}_{\tilde{n}} = \boldsymbol{\nabla}_n$, otherwise, the duality concept[16] would be useful. The fringe vector can then be developed with Eq. (8) and $\mathbf{M}\boldsymbol{\nabla}_n = \{(L_R - p)/L_R\}\hat{\mathbf{M}}\boldsymbol{\nabla}_{\hat{n}}$ to give

$$L_R\mathbf{M}\boldsymbol{\nabla}_n D_R = L_R\mathbf{M}\boldsymbol{\nabla}_n(\mathbf{u}\cdot\mathbf{g}) + \left(\frac{L_R - p}{q}\right)\hat{\mathbf{M}}\left(-\mathbf{C}\boldsymbol{\psi}\mathbf{a} + \frac{1}{q}\mathbf{aC}\mathbf{d} + \frac{1}{2q^2}(\mathbf{aCa})\mathbf{d}\right). \qquad (16)$$

In the derivation, $\mathbf{a}, \boldsymbol{\Psi}$ and $\mathbf{d}$ are constant since they do not change when the direction of observation varies. The superprojector $\mathbf{C}$ can be written alternatively, as $\mathbf{C} = \overset{2}{\mathbf{C}} \otimes \overset{2}{\mathbf{c}} + \overset{3}{\mathbf{C}} \otimes \overset{3}{\mathbf{c}} + \overset{4}{\mathbf{C}} \otimes \overset{4}{\mathbf{c}}$ and similarly the hyper-projector $\mathbf{C} = 2[\overset{2}{\mathbf{C}} \otimes \overset{2}{\mathbf{c}} \otimes \overset{2}{\mathbf{c}} + \overset{3}{\mathbf{C}} \otimes \overset{3}{\mathbf{c}} \otimes \overset{3}{\mathbf{c}} + \overset{4}{\mathbf{C}} \otimes \overset{4}{\mathbf{c}} \otimes \overset{4}{\mathbf{c}} + \overset{2}{\mathbf{C}} \otimes \overset{3}{\mathbf{c}} \otimes \overset{4}{\mathbf{c}} + \overset{3}{\mathbf{C}} \otimes \overset{4}{\mathbf{c}} \otimes \overset{2}{\mathbf{c}} + \overset{4}{\mathbf{C}} \otimes \overset{2}{\mathbf{c}} \otimes \overset{3}{\mathbf{c}}] - [\overset{2}{\mathbf{C}} \otimes \overset{3,4}{\mathbf{C}} + \overset{3}{\mathbf{C}} \otimes \overset{4,2}{\mathbf{C}} + \overset{1,4}{\mathbf{C}} \otimes \overset{2,3}{\mathbf{C}}]$. $L_R\mathbf{M}\boldsymbol{\nabla}_n D_R$ is the fringe vector with modification which we call the "modified fringe vector" and contains less terms than the corresponding one in the sandwich holography method, Eq. 25, ref. 12. $L_R\mathbf{M}\boldsymbol{\nabla}_n(\mathbf{u}\cdot\mathbf{g}) = L_R\mathbf{M}(\boldsymbol{\nabla}_n \otimes \mathbf{u})\mathbf{g} - (L_R/L_S)\mathbf{M}\mathbf{H}\mathbf{u} - \mathbf{K}\mathbf{u}$[9], is the fringe vector without modification which we name "initial fringe vector" and contains the deformation tensor $(\boldsymbol{\nabla}_n \otimes \mathbf{u})$. For the determination of the six components of this tensor, we must firstly calculate the vector $L_R\mathbf{M}\boldsymbol{\nabla}_n(\mathbf{u}\cdot\mathbf{g})$ for three or more directions $\mathbf{k}$. The three terms in Eq. (16) are all vectors in the same plane due to the presence of the oblique projectors $\mathbf{M}$ and $\hat{\mathbf{M}}$, this plane being perpendicular to the direction of observation $\mathbf{k}$. So that, for the control of the modified fringe vector, two independent parameters are enough, e.g. one rotation and one displacement. By applying a particular modification, $\boldsymbol{\Psi} = \boldsymbol{\Psi}_o$ and $\mathbf{d} = \mathbf{d}_o$, we can make the modified fringe vector disappear completely and then write the vector equation,

$$L_R\mathbf{M}\boldsymbol{\nabla}_n(\mathbf{u}\cdot\mathbf{g}) = \frac{L_R - p}{q}\hat{\mathbf{M}}\left[\mathbf{C}\boldsymbol{\psi}_o\mathbf{a} - \frac{1}{q}\left(\mathbf{aC} + \frac{1}{2q}\mathbf{aCa}\right)\mathbf{d}_o\right] . \qquad (17)$$

Thus, the fringe vector is equivalent to the modification term, and the fringes mark a "saddle" or a "summit" in P. We can expect, during this type of modification, an improvement of the fringe contrast and motion of the point of localization towards the observer.

However, as already mentioned, we will only consider in this paper the change of one parameter, i.e. the displacement of the mirror. The modified fringe vector can be rotated but not reduced to zero, so that it can take a direction $\mathbf{m}_1$ with one displacement of the mirror s_1 and another direction $\mathbf{m}_2$ with a displacement s_2 (Fig. 3 and 4). $\mathbf{m}_1$ and $\mathbf{m}_2$ are, as pointed out before, unit vectors perpendicular to $\mathbf{k}$. Therefore,

$$0 = \mathbf{m}_1 \cdot L_R\mathbf{M}\boldsymbol{\nabla}_n D_{R1} = \mathbf{m}_1 \cdot L_R\mathbf{M}\boldsymbol{\nabla}_n(\mathbf{u}\cdot\mathbf{g}) + s_1\mathbf{m}_1 \cdot \mathbf{t}$$
$$0 = \mathbf{m}_2 \cdot L_R\mathbf{M}\boldsymbol{\nabla}_n D_{R2} = \mathbf{m}_2 \cdot L_R\mathbf{M}\boldsymbol{\nabla}_n(\mathbf{u}\cdot\mathbf{g}) + s_2\mathbf{m}_2 \cdot \mathbf{t} \qquad (18)$$

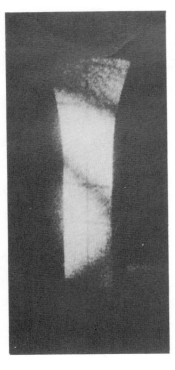
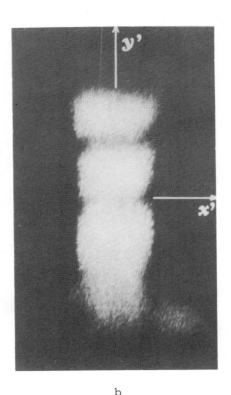
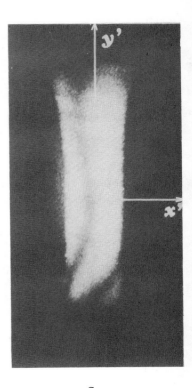

a   b   c

Figure 4: Fringes. The camera was focused on the fringes and not on the object.
a) no modification   b) the fringes are modified so   c) the fringes are modified
                     that they become parallel   so that they become pa-
                     to the x' axis   rallel to the y' axis

The vector $\mathbf{t}$ in (18) is defined by

$$\mathbf{t} = \frac{2(L_R - p)}{q^2} \hat{\mathbf{M}} (a\mathbf{C} + \frac{1}{2q} a\mathbf{C}a)\mathbf{n}_m \qquad (19)$$

and called the "modification vector", where $\mathbf{n}_m$ is the normal to the mirror and s_1, s_2 are the displacements of the mirror along $\mathbf{n}_m$. We have replaced the displacement of the sources, $\mathbf{d}$, by the displacement of the mirror, $\mathbf{s}$, which are related to each other by $\mathbf{d} = 2s\mathbf{n}_m$ (Fig. 1). The unit vectors $\mathbf{m}_1$ and $\mathbf{m}_2$ can be placed, so that they are perpendicular and corresponding to the axis x' and y' (Fig. 3). In order that x', y', z' form a cartesian system, the axis z' must be parallel to $\mathbf{k}$. In this system, Eq. (18) expresses simply the components of the vector $L_R \mathbf{M} \nabla_n (\mathbf{u} \cdot \mathbf{g})$:

$$L_R \mathbf{M} \nabla_n (\mathbf{u} \cdot \mathbf{g}) \triangleq \left\{ \begin{array}{c} - s_1 t'_x \\ - s_2 t'_y \\ 0 \end{array} \right\} . \qquad (20)$$

If a cartesian system x, y, z originates from P with the axis z parallel to $\mathbf{n}$ and x, y on the level surface of the object, then the deformation tensor $\nabla_n \otimes \mathbf{u}$ can be expressed by the matrix

$$\nabla_n \otimes \mathbf{u} \triangleq \begin{bmatrix} \varepsilon_x & (\gamma_{xy}/2 + \Omega) & - \omega_y^r \\ (\gamma_{xy}/2 - \Omega) & \varepsilon_y & \omega_x^r \\ 0 & 0 & 0 \end{bmatrix}$$

which contains the values we are looking for: Strains ε_x, ε_y; shear $\gamma_{xy}/2$; out-of-plane rotation ω_x^r, ω_y^r and in-plane rotation Ω. Therefore, we need to know the matrix to pass from the system x',y',z' into the system x,y,z, so that the components of the vector $L_R \mathbf{M} \nabla_n (\mathbf{u} \cdot \mathbf{g})$ of Eq. (20) can be transformed into components of the second system. The directions z' and z have already been defined (z'//$\mathbf{k}$, z//$\mathbf{n}$), and the directions x and y are drawn on the object surface. One can position y' parallel to $\mathbf{K}\mathbf{e}_y$ where $\mathbf{e}_y$ is the unit vector in the direction of y (Fig. 5); x' is hence determined and the transformation matrix is written:

$$R = \begin{bmatrix} \dfrac{k_z}{\sqrt{1-k_y^2}} & -\dfrac{k_x k_y}{\sqrt{1-k_y^2}} & k_x \\[2ex] 0 & \sqrt{1-k_y^2} & k_y \\[2ex] -\dfrac{k_x}{\sqrt{1-k_y^2}} & -\dfrac{k_y k_z}{\sqrt{1-k_y^2}} & k_z \end{bmatrix}$$

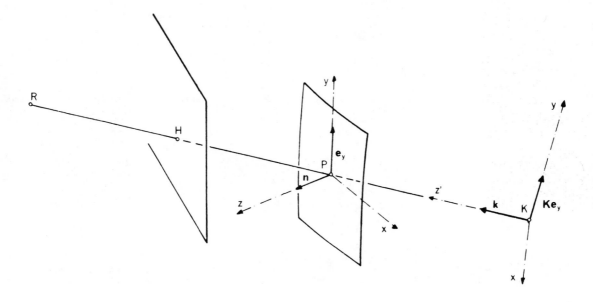

Figure 5: Relation between the two cartesian systems x',y',z' and x,y,z.

In the system x,y,z, Eq. (20) becomes

$$L_R \mathbf{M}\, \boldsymbol{\nabla}_n (\mathbf{u}\cdot\mathbf{g}) \triangleq \left\{ \begin{array}{l} \dfrac{k_x k_y}{1-k_y^2} t_y\, (s_2 - s_1) - s_1 t_x \\[2ex] \qquad\qquad\qquad - s_2 t_y \\[2ex] \dfrac{k_y k_z}{1-k_y^2} t_y\, (s_2 - s_1) - s_1 t_z \end{array} \right\} \tag{21}$$

whereas the oblique projector **M** is expressed by the matrix:

$$\mathbf{M} \triangleq \begin{bmatrix} 1 & 0 & 0 \\[1ex] 0 & 1 & 0 \\[1ex] -\dfrac{k_x}{k_z} & -\dfrac{k_y}{k_z} & 0 \end{bmatrix} \ .$$

In the development of the initial fringe vector, the deformation tensor is included, so that we have the vector equation:

$$\mathbf{M}(\boldsymbol{\nabla}_n \otimes \mathbf{u}\,)\mathbf{g} = \frac{1}{L_R}(L_R \mathbf{M}\boldsymbol{\nabla}_n(\mathbf{u}\cdot\mathbf{g}) + \mathbf{K}\mathbf{u}) + \frac{1}{L_s}\mathbf{M}\mathbf{H}\mathbf{u} \ . \tag{22}$$

After the matrix notation in the x,y,z system this equation becomes:

41

$$
\begin{bmatrix} 1 & 0 & 0 \\ 0 & 1 & 0 \\ -\dfrac{k_x}{k_z} & -\dfrac{k_y}{k_z} & 0 \end{bmatrix}
\begin{bmatrix} \varepsilon_x & (\gamma_{xy}/2+\Omega) & -\omega_y^r \\ (\gamma_{xy}/2-\Omega) & \varepsilon_y & \omega_x^r \\ 0 & 0 & 0 \end{bmatrix}
\begin{bmatrix} g_x \\ g_y \\ g_z \end{bmatrix}
= \frac{1}{L_R}
\begin{bmatrix} \dfrac{k_x k_y}{1-k_y^2} t_y (s_2-s_1) -s_1 t_x \\ -s_2 t_y \\ \dfrac{k_y k_z}{1-k_y^2} t_y (s_2-s_1) -s_1 t_z \end{bmatrix} +
$$

$$
+ \frac{1}{L_R}
\begin{bmatrix} 1-k_x^2 & -k_x k_y & -k_x k_z \\ -k_x k_y & 1-k_y^2 & -k_y k_z \\ -k_x k_z & -k_y k_z & 1-k_z^2 \end{bmatrix}
\begin{bmatrix} u_x \\ u_y \\ u_z \end{bmatrix}
+ \frac{1}{L_s}
\begin{bmatrix} 1 & 0 & 0 \\ 0 & 1 & 0 \\ -\dfrac{k_x}{k_z} & \dfrac{k_y}{k_z} & 0 \end{bmatrix}
\begin{bmatrix} 1-h_x^2 & -h_x h_y & -h_x h_z \\ -h_x h_y & 1-h_y^2 & -h_y h_z \\ -h_x h_z & -h_y h_z & 1-h_z^2 \end{bmatrix}
\begin{bmatrix} u_x \\ u_y \\ u_z \end{bmatrix}
\quad (23)
$$

which produces two different equations instead of the three expected, as the third line is simply a linear combination of the first two lines:

$$
\begin{aligned}
\varepsilon_x g_x + (\gamma_{xy}/2+\Omega) g_y - \omega_y^r g_z &= b_x \\
(\gamma_{xy}/2-\Omega) g_x + \varepsilon_y g_y + \omega_x^r g_z &= b_y
\end{aligned}
\quad (24)
$$

with $\quad b_x = \dfrac{1}{L_R} \left[\dfrac{k_x k_y}{1-k_y^2} t_y (s_2-s_1) - s_1 t_x + u_x - k_x (\mathbf{u}\cdot\mathbf{k}) \right] + \dfrac{1}{L_s} \left[u_x - h_x (\mathbf{u}\cdot\mathbf{h}) \right]$

and $\quad b_y = \dfrac{1}{L_R} \left[-s_2 t_y + u_y - k_y (\mathbf{u}\cdot\mathbf{k}) \right] + \dfrac{1}{L_s} \left[u_y - h_y (\mathbf{u}\cdot\mathbf{h}) \right] \quad .$

Both equations in (24) contain three unknowns and can be solved separately. When the initial fringe vector is determined, according to (20), for more than three different sensitivity vectors $\mathbf{g}_i$, the equations in (24) are overdetermined and the least-square method [3] can be applied. From the first equation of (24) we then obtain:

$$
\varepsilon_x = \frac{\begin{bmatrix} \Sigma b_{xi} g_{xi} & \Sigma g_{xi} g_{yi} & \Sigma g_{xi} g_{zi} \\ \Sigma b_{xi} g_{yi} & \Sigma g_{yi}^2 & \Sigma g_{yi} g_{zi} \\ \Sigma b_{xi} g_{zi} & \Sigma g_{yi} g_{zi} & \Sigma g_{zi}^2 \end{bmatrix}}{\Delta}, \quad \text{with } \Delta = \begin{bmatrix} \Sigma g_{xi}^2 & \Sigma g_{xi} g_{yi} & \Sigma g_{xi} g_{zi} \\ \Sigma g_{xi} g_{yi} & \Sigma g_{yi}^2 & \Sigma g_{yi} g_{zi} \\ \Sigma g_{xi} g_{zi} & \Sigma g_{yi} g_{zi} & \Sigma g_{zi}^2 \end{bmatrix}
\quad (25)
$$

and similar expressions for $\gamma_{xy}/2+\Omega$ and $-\omega_y^r$, as well as for $\gamma_{xy}/2-\Omega$, ε_y and ω_x^r from the second equation of (24).

At this point, one should look closer at the modification vector $\mathbf{t}$ defined by Eq. (19). The components of $\mathbf{t}$ can be obtained indirectly by measuring separately the vectors $(\hat{\mathbf{n}}, \mathbf{n}_m, \mathbf{a}, \mathbf{c}, \mathbf{k})$ and the lengths (L_R, p, q) present in $\mathbf{t}$. They can otherwise be obtained directly by recording the object again, but this time without deforming it. According to (16), the fringe vector is then parallel to the modification vector $\mathbf{t}$, and (15) simply becomes

$$
\frac{dD_R}{d\phi} = s\,\mathbf{m}\cdot\mathbf{t} \quad .
\quad (26)
$$

The fringes are only produced by the modification and are almost straight and equally spaced (Fig. 6), since $\mathbf{t}$ changes slightly when $\mathbf{k}$ varies. The derivation, $dD_R/d\phi$, can then be replaced by a difference $\Delta D_R/\Delta\phi$ and (26) expresses the length of the vector $\mathbf{t}$:

$$
|\mathbf{t}| = \frac{\lambda \cdot |\Delta n|}{|s| \cdot |\Delta\phi|} \quad ,
\quad (27)
$$

when Δn and $\Delta\phi$ are measured perpendicularly to the fringes [2]. The angle between the fringe at P and the axis x' is measured, and the direction of $\mathbf{t}$ is determined by a rapid estimation of Eq. (19), so that the components $t_x', t_y', (t_z'=0)$ are obtained, and the components t_x, t_y, t_z are calculated by applying the transformation matrix. The direct and indirect method can be combined together to achieve a greater precision in the determination of $\mathbf{t}$.

6.a   6.b

Figure 6: Fringes only produced by the modification. They are localized on the hologram (camera focused on it) and always of visibility one. The vector **t** is perpendicular to the fringes. The modification is produced by the translation s of the mirror: in a) s = 1000 μm, in b) s = 5000 μm. The fringe direction remains the same while **t** is a constant for the same observing direction **k**.

Measurements

A tensile test was carried out with an aluminium bar of cross-section A = 2x6 mm (Fig. 7). However, the three components of the strain ($\varepsilon_x, \varepsilon_y, \gamma_{xy}/2$) and the three components of the rotation ($\Omega, \omega_x^r, \omega_y^r$) are considered unknown for the measurements and the application of the above equations as they would be generally. The mechanical relations between the components of strain in this simple test permit a rapid verification of the results, as we must get $\varepsilon_x = -\nu\varepsilon_y$, $\varepsilon_y = F/(A \cdot E_{A1})$ and $\gamma_{xy} = 0$. The probe is thin and long, so that a perfect positioning of the force on the center line of the probe is impossible and the fixations of the two ends of the probe produce an unavoidable torsion. These two effects are shown up mainly by the out-of-plane rotations ω_x^r and ω_y^r. The force F is produced by a weight of 50 g* placed on a balance of an equal lever, so that the force is 80 times greater than the weight. At first, the displacement **u** of the point P must be determined according to Eqs. (13) and (14), since the projection **Ku** is present in the calculation of the stresses and rotations.

Two tests (Fig. 8) have been carried out. In the second, in order to decrease the out-of-plane rotation ω_x^r, a metal plate was placed behind the bar to prevent it from moving backwards. In the first test, the displacement of the point P was u_x = -2.7 μm, u_y = -3.5 μm and u_z =-6.4 μm; in the second test it was u_x = -0.8 μm, u_y = -3.8 μm, u_z = -2.1 μm. The components of **u** in the x and z direction confirm that the rotation differs from one test to the other. Afterwards the fringes are modified so that they become parallel to the x' axis for a displacement s_1 of the mirror and parallel to the y' axis for a displacement s_2 of the mirror. These measurements are repeated for different observing directions **k**. In the first test, the point P is observed from five different directions **k** through the four corners and the center of the hologram plate (5 x 4 inches Agfa Gevaert 8E75 holotest plate). In the second test, four more directions **k** were employed, two of which correspond to the viewing direction of P with a mirror placed horizontally in front of the object. The two others

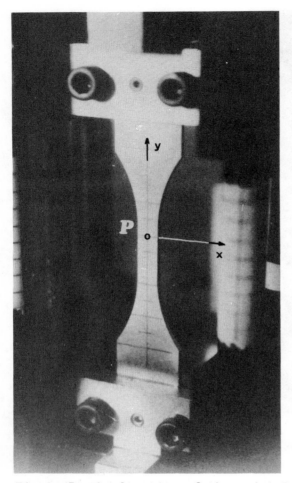

Figure 7: The location of the point P on the aluminium bar where displacement, strain and rotation are determined.

Figure 8: Holographic set-up used to perform the experiments.

Table 1: Measured Values

		κ_x	k_y	k_z	$s_1(\mu m)$	$s_2(\mu m)$	$L_R(mm)$	t_x	t_y	$b_x \cdot 10^4$	$b_y \cdot 10^4$
1st test	1	0.491	0.469	0.734	-3010	2520	748.5	-0.0571	0.0599	-0.976	-2.015
	2	0.610	0.271	0.745	-3970	2910	710	-0.0342	0.0504	-1.016	-2.087
	3	0.437	0.292	0.851	-1750	2970	697	-0.0605	0.0534	-1.005	-2.294
	4	0.144	0.394	0.908	-680	2760	748.5	-0.1295	0.0617	-1.008	-2.285
	5	0.349	0.184	0.919	-1190	3190	682	-0.0725	0.0512	-1.048	-2.426
2nd test	1	0.491	0.469	0.734	-2640	-360	748.5	-0.0571	0.0599	-1.461	0.261
	2	0.610	0.271	0.745	-3980	-270	710	-0.0342	0.0504	-1.433	0.150
	3	0.437	0.292	0.851	-1760	-250	697	-0.0605	0.0534	-1.357	0.151
	4	0.144	0.394	0.908	-870	-320	748.5	-0.1295	0.0617	-1.479	0.232
	5	0.349	0.184	0.919	-1340	-270	682	-0.0725	0.0512	-1.368	0.155
	6	0.205	-0.543	0.814	-860	-470	777	-0.1144	0.0512	-1.240	-0.357
	7	0.357	-0.432	0.828	-1560	-340	721	-0.0734	0.0437	-1.472	-0.261
	8	0.674	0.460	0.578	2210	-440	752	-0.0537	0.0501	-1.287	0.264
	9	0.723	0.312	0.616	2230	-360	711	-0.0443	0.0432	-1.218	0.179

$h_x = 0.465$ $\qquad$ $h_y = 0.104$ $\qquad$ $h_z = -0.879$ $\qquad$ $L_s = \infty$

correspond to the viewing directions with a mirror placed perpendicularly on the side of the object. Finally, we determine the unit vector **h** of the collimated beam ($L_s = \infty$) that illuminates the object while the components of **t** and the value L_R are measured for each observing direction **k**. (Table 1). At this point, all the necessary information has been gathered, so that Eqs. 24 and 25 can be solved. We then obtain the sought after components of strain and rotation compiled in table 2. The strain components $\varepsilon_x, \varepsilon_y, \gamma_{xy}/2$ (intrinsic values) and the in-plane rotation Ω are practically the same, whereas the out-of-plane rotations differ as was expected. A further comparison can be made between the generally accepted elastic constants of aluminium and those obtained from the two tests. In tensile tests the value of Poisson's ration is simply expressed by $\nu = -\varepsilon_x/\varepsilon_y$ and the value of Young's modulus by $E = \sigma_y/\varepsilon_y = F/(A \cdot \varepsilon_y)$. Some attention must be paid to the evaluation of the force F due to friction in the balance system which could make F a slightly larger force than the 4 kg* expected. The comparison is shown in table 3. Another important result of those two tests is that the intrinsic strains can be found with sufficient accuracy with whatever rotation the object and the whole mechanical system has undergone during the test. Moreover, the precision can be improved by more accurate measurements of the geometrical characteristic of the holographic set-up, and by an extensive use of mirrors to obtain very different viewing directions.

Table 2: Results

	1st test	2nd test
ε_x	$-0.20 \cdot 10^{-4}$	$-0.18 \cdot 10^{-4}$
ε_y	$0.56 \cdot 10^{-4}$	$0.61 \cdot 10^{-4}$
$\gamma_{xy}/2$	$0.02 \cdot 10^{-4}$	$-0.04 \cdot 10^{-4}$
Ω	$-0.04 \cdot 10^{-4}$	$-0.03 \cdot 10^{-4}$
ω_x^r	$-1.37 \cdot 10^{-4}$	$0.03 \cdot 10^{-4}$
ω_y^r	$0.59 \cdot 10^{-4}$	$0.83 \cdot 10^{-4}$

Table 3: Comparison on ν and E

	1st test	2nd test	values of Al
ν	0.357	0.295	0.33
E (kg*/mm^2)	5950	5460	6000-7000

Conclusion

The results in this paper show that a fringe modification, combined with an amplification effect, give the necessary flexibility for an exact evaluation of the state of deformation from holographic fringe patterns. For any viewing direction, the initial fringe vector, which contains the sought after deformation gradient $\nabla_n \otimes u$, can be determined by compensating its two components in turn, by using two modification vectors $s_1 t$ and $s_2 t$. s_1 and s_2 are measured mirror displacements and **t** is a vector which can be obtained by simple matrix calculation from some geometrical parameters of the holographic set-up. The amplification produced by the two close reference sources makes the set-up less sensitive to repositioning and measurement errors. The determination of the initial fringe vector is repeated with several viewing directions, so that the strains and rotations can be extracted from a system of linear equations. It must be noted that fringe visibility could vanish in a large modification as it is not controlled in the method described here. However, the out-of-plane rotation of the mirror (two additional modifying components) could be used to control the fringe visibility and also the fringe interspace. The results of the experiments show the validity of the method in determining strain and rotation without restriction on the state of deformation.

References

1. Walles S., Ark. f. Fysik 40 no 26 299-403 (1970)
2. Bijl D. and Jones R., Opt. Acta 21 105-118 (1974)
3. Stetson K.A., Appl. Opt. 14 2256-2259 (1975)
4. Stetson K.A., Optik 31 576-591 (1970)
5. Charmet J.C. and Montel F., Rev. Phys. Appl. 12 603-610 (1977)
6. Tsuruta T., Shiotake N. and Itoh Y., Jpn. J. Appl. Phys. 7 1092-1100 (1968)
7. Ballard G.S., J. Appl. Phys. 39 4846-4848 (1968)
8. Gates J.W.C., Nature 220 473-474 (1968)
9. Schumann W. and Dubas M., Holographic interferometry, Springer Verlag Berlin (1979)
10. Abramson N., Appl. Opt. 13 2019-2025 (1974)
11. Dändliker R., Thalmann R. and Willemin J.F., Optics communications 42 No. 301-305 (1982)
12. Dubas M. and Schumann W., Opt. Acta 24 1193-1209 (1977)
13. Aleksandrov E.B. and Bonch-Bruevich A.M., Sov. Phys.-Tech. Phys. 12 258-265 (1967)
14. Dändliker R., in: Progress in optics, Vol. 17 1-84 (1980)
15. Hovanesian J.D., Hung Y.Y. and Plotkowski P.D., Proceedings of the 1981 Spring meeting SESA
16. Schumann W., J. Opt. Soc. Am. 71 525-528 (1981)

Computer-based evaluation of holographic interferograms

E. Müller, V. Hrdliczka, D.E. Cuche

Institute for Machine Tool Construction and Manufacturing Techniques
Swiss Federal Institute of Technology
CH 8092 Zurich, Switzerland

Abstract

The computer-based evaluation of holographic interferograms described in this paper allows a complete quantitative evaluation of the displacement field for:
- a) a single holographic interferogram
- b) holographic interferograms simultaneously taken from different points in order to compute three-dimensional displacements
- c) holographic interferograms reconstructed with a laser light different from that used to take the interferogram (e.g. pulsed holography with a Ruby laser reconstructed with a He-Ne laser)

Introduction

Quantitative evaluation methods of holographic interferograms have been developed based on the theory of holography and on holographic interferometry. The static[1] and the dynamic[2] methods are particularly well known. Both methods allow the determination of displacement vectors between two states of an object with the holographic interferometry. Although both methods are already known for a long time, they have only been used in a few cases to determine three-dimensional displacement vectors for some selected object points.

The use of pulsed lasers with short exposure time drastically reduces stability requirements for the holographic set-up. Clumsy vibration-isolated working tables are no longer necessary and even moving objects can be holographed.

As holograms taken with a pulsed Ruby laser are normally not reconstructed with an expensive continuous wave Ruby laser but with a less expensive He-Ne laser, image modifications are produced during reconstruction. These modifications resulting from the change in wave-length complicated in the past an exact evaluation of pulsed holograms.

Modifications can be compensated by the use of modern equipment. The costs for quantitative evaluations of holographic interferograms are reduced efficiently, extending the range of application of holographic interferometry.

In the first part of this paper, image modifications caused by the reconstruction with other wave-lengths are described. The determination of a displacement field, based on the evaluation of three interferograms taken simultaneously follows in the second part.

Image modifications

When the holographic set-up for the reconstruction is changed, the reconstructed image is modified. Compared to the real object, the image appears displaced, distorted and of a different size. Furthermore, reconstructed object points are astigmatic. When the direction of incidence of the reconstruction beam is different, or when another wave-length is used, the resulting diffraction angle changes, causing the indicated modifications[3-6].

Displacement, distortion and size alteration of the reconstructed image can be corrected by compensating for the modified direction of observation. Calculation of the astigmatic image range allows to estimate the attainable image quality as well as to calculate the original view point.

Usually, the modification of the direction of observation in hologram reconstruction already can be seen for a small change of wave-length. The modified direction of observation can be determined by comparing a spherical wave $\tilde{U}$ to the wave actually produced and by applying the condition of interference identity[4].

According to Fig. 1, following phase difference results at the hologram point H:

$$\theta(H) = \frac{2\pi}{\tilde{\lambda}} (\tilde{p} - \tilde{q}) - \frac{2\pi}{\lambda} (p - q) . \tag{1}$$

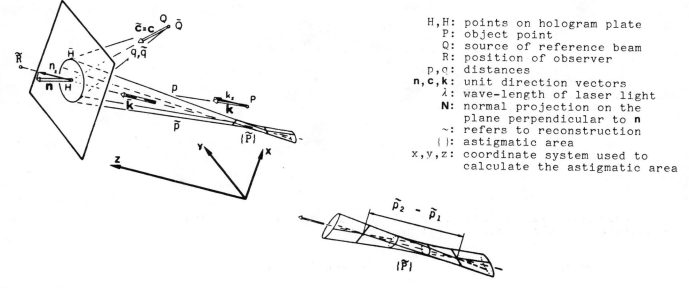

H,H: points on hologram plate
P: object point
Q: source of reference beam
R: position of observer
p,q: distances
n,c,k: unit direction vectors
λ: wave-length of laser light
N: normal projection on the plane perpendicular to **n**
~: refers to reconstruction
{ }: astigmatic area
x,y,z: coordinate system used to calculate the astigmatic area

Figure 1.   Holographic set-up showing the modification of the direction of observation $\tilde{k}$ and the astigmatic area of the point $\{\tilde{P}\}$ due to a change in wave-length during reconstruction.

The phase difference θ should be stationary in the neighborhood of the hologram point H. This condition leads to the development of θ in a Taylor series around H.

Setting the first differential of the phase difference to zero gives the basic equation of the modified direction of observation and the equations governing a pure wave-length alteration ($\mathbf{c} = \tilde{\mathbf{c}}$).

$$\mathbf{N}\left[\frac{1}{\tilde{\lambda}}(\tilde{\mathbf{k}} - \tilde{\mathbf{c}}) - \frac{1}{\lambda}(\mathbf{k} - \mathbf{c})\right] = 0, \qquad \mathbf{N}\tilde{\mathbf{k}} = \mathbf{N}\left[(1 - \frac{\tilde{\lambda}}{\lambda})\mathbf{c} + \frac{\tilde{\lambda}}{\lambda}\mathbf{k}\right], \quad \mathbf{N}\mathbf{k} = \mathbf{N}\left[(1 - \frac{\lambda}{\tilde{\lambda}})\mathbf{c} + \frac{\lambda}{\tilde{\lambda}}\tilde{\mathbf{k}}\right], \qquad |\tilde{\mathbf{k}}| = 1. \quad (2)$$

The change in the direction of observation depends on the reconstructed hologram area, on the direction of the reference beam and on the wave-length alteration. Even for an observer placed on the hologram itself, the direction of observation is changed at every single object point. Therefore, the modified direction of observation causes a distortion and a size alteration of the reconstructed image. This is the reason why distortion-free magnification of a large image cannot be obtained when changing the wave-length for the reconstruction.

Setting to zero the second differential of the phase difference gives an equation to determine the astigmatism of the reconstructed object point. Fig. 1 shows the astigmatic area of this point $\{\tilde{P}\}$ which is described by

$$\left|\tilde{\mathbf{T}} - \frac{1}{\tilde{p}}\tilde{\mathbf{K}}\right| = 0 \tag{3}$$

where

$$\tilde{\mathbf{T}} = \frac{\tilde{\lambda}}{\lambda}\frac{1}{p}\tilde{\mathbf{M}}\mathbf{K}\tilde{\mathbf{M}}^T + (1 - \frac{\tilde{\lambda}}{\lambda})\frac{1}{q}\tilde{\mathbf{M}}\mathbf{C}\tilde{\mathbf{M}}^T, \qquad \mathbf{C} = \mathbf{I} - \mathbf{c} \otimes \mathbf{c}, \qquad \mathbf{K} = \mathbf{I} - \mathbf{k} \otimes \mathbf{k}, \qquad \tilde{\mathbf{M}} = \mathbf{I} - \frac{\mathbf{n} \otimes \tilde{\mathbf{k}}}{\mathbf{n} \cdot \tilde{\mathbf{k}}},$$

$$\tilde{\mathbf{K}} = \mathbf{I} - \tilde{\mathbf{k}} \otimes \tilde{\mathbf{k}}, \qquad \tilde{\mathbf{M}}^T = \mathbf{I} - \frac{\tilde{\mathbf{k}} \otimes \mathbf{n}}{\mathbf{n} \cdot \tilde{\mathbf{k}}}.$$

This equation determines the distance $\tilde{p}$ and, for a pure wave-length change, can be expressed in the coordinate system of Fig. 1 ($\tilde{k}$ parallel to z-axis and x-axis parallel to the normal projection of **n** on the plane perpendicular to $\tilde{k}$)[5]. The two eigenvalues $\tilde{p}_1$ and $\tilde{p}_2$ of this equation limit the astigmatic area of the reconstructed point $\{\tilde{P}\}$. The length of this area and the distance $\tilde{p}$ from its center to the hologram are given by

$$\tilde{p}_2 - \tilde{p}_1 = \frac{\sqrt{\{A\}^2 - 4\{B\}}}{\{B\}}, \qquad \qquad \tilde{p} = -\frac{\{A\}}{2 \cdot \{B\}} \tag{4}$$

47

with

$$\{A\} = \frac{\tilde{\lambda}}{p\lambda}[\frac{1}{n_z^2}(1 - k_z^2) + 2k_z\frac{k\cdot n}{n_z}] + \frac{1}{q}(1 - \frac{\tilde{\lambda}}{\lambda})[\frac{1}{n_z^2}(1 - c_z^2) + 2c_z\frac{c\cdot n}{n_z}] \quad ,$$

$$\{B\} = \{[\frac{\tilde{\lambda}}{p\lambda}]^2[\frac{k\cdot n}{n_z}]^2 + [\frac{1}{q}(1 - \frac{\tilde{\lambda}}{\lambda})]^2[\frac{c\cdot n}{n_z}]^2 + \frac{\tilde{\lambda}}{p\lambda}[\frac{1}{q}(1 - \frac{\tilde{\lambda}}{\lambda})] \cdot$$

$$[(c_y^2 - 1)(k_x - k_z\frac{n_x}{n_z})^2 + (k_y^2 - 1)(c_x - c_z\frac{n_x}{n_z})^2 + \frac{1}{n_z^2}(2 - c_y^2 - k_y^2) - 2(-k_xk_y + \frac{n_x}{n_z}k_yk_z)(-c_xc_y + \frac{n_x}{n_z}c_yc_z)]\}.$$

Fig. 2 shows some examples of holographic set-ups and expected image modifications when reconstructing with a different wave-length.

set-up 1   set-up 2   set-up 3

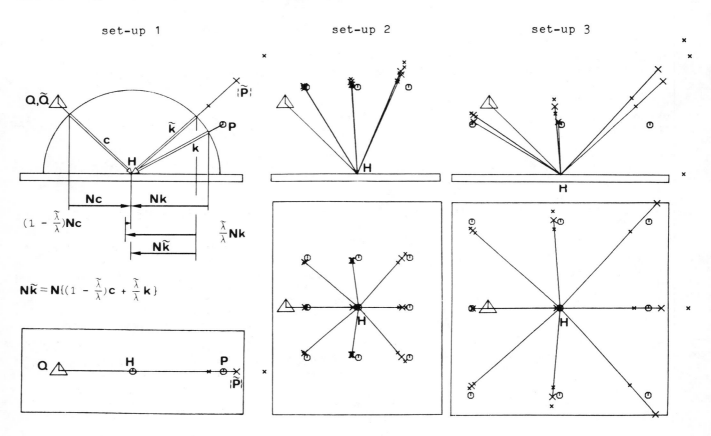

List of symbols:
P   ☉   :object point
{P} ·X· :reconstructed and modified astigmatic object point
Q   △   :source of reference and reconstruction beam
     ▭   :hologram plate
H   :hologram point used to reconstruct the object point

Figure 2.   Modification of object points holographed with a Ruby laser ($\lambda = .6943\mu$m) and reconstructed with a He-Ne laser ($\tilde{\lambda} = .6328\mu$m). The picture shows two normal projections of each holographic set-up. A geometric calculation of the modified direction of observation $\tilde{k}$ for a single object point is seen on the left. The picture in the center represents the position and astigmatism of the reconstructed points of a small object. The picture on the right shows the same set-up with a large object close to the hologram plate. Distortion and displacement of the reconstructed image are clearly seen.

As shown in Fig. 3, the view point of the reconstructed image R̃ does not coincide with the view point of the original object. Similarly to the conversion of the reconstructed and modified image points into original object points, the adequate astigmatic view point {R} of the object can be determined with the position of R̃[7]. If we define the length RH by f and R̃H by f̃, the values of f and of its astigmatic interval are given as in (3) by

$$\left| T - \frac{1}{f} K \right| = 0, \qquad T = -\frac{\lambda}{\tilde{\lambda}} \frac{1}{\tilde{f}} M \tilde{K} M^T + \left(1 - \frac{\lambda}{\tilde{\lambda}}\right) \frac{1}{q} MCM^T, \qquad M = I - \frac{n \circ k}{n \cdot k}, \qquad M^T = I - \frac{k \circ n}{n \cdot k}. \qquad (5)$$

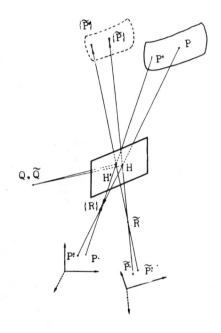

H,H' :points on hologram plate

P,P' :original object points

{P̃} , {P̃'} :reconstructed and modified points

P·,P·! :points on image plot

P̃·,P̃·! :reconstructed and modified picture points (photograph)

Q,Q̃ :source of reference and reconstruction beam

{R} :view point of original object

R̃ :view point of reconstructed image

Figure 3.   Image modifications and determination of view point when the wave-length used to reconstruct the hologram is changed.

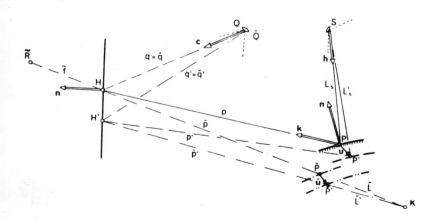

H,H': points on hologram plate
P,P': object points
Q,Q̃: source of reference and reconstruction beam
R̃: position of observer
S: point source illuminating the object
L,p,q: distances
c,h,n,k: unit direction vectors
g: sensitivity vector
u: displacement vector
K: localisation point of the interference fringes
~: refers to reconstruction

Figure 4.   Modification of interference fringes due to a change in wave-length.

A change in wave-length when reconstructing not only modifies the object points but also the interference fringes. Using Fig. 4, we can determine the optical path difference D at the point K (location of the interference fringes) and the phases of the light waves at P̃ and P̃'. Insertion and transformation gives the equations D= u·(k − h) = u·g for the optical path difference. This result shows that for the calculation of the deformation it is necessary to know the original directions of illumination and observation. Furthermore, the modification of the interference fringes by a change in wave-length for reconstruction is similar to that of the object points.

Determination of three-dimensional displacement field

With a double pulsed Ruby laser we simultaneously take three holographic interferograms during the exit of a cutting tool. We then reconstruct and photograph each single interferogram (Fig. 5), measure the holographic set-up geometry and locate the view points of the photographic exposure. These view points lead to the astigmatic view points {R} of the original object.

P is the central point of the object and $\tilde{F}1$ to $\tilde{F}4$ are fixed points. The picture also contains the order of the interference fringes. On top are the spring elements attached to a stable pillar and resting on the tool clamping device. Two accelerometers (Acc1 and Acc2) are fixed on both sides of the clamping device to check the three-dimensional displacement vectors calculated from the interferograms.

Figure 5. Picture of the reconstructed hologram H2.
The cutting tool is clamped on a dynamometer screwed on the table of a milling machine. The workpiece rotates with the spindle and the cut is activated by a vertical table feed.

Once the three-dimensional geometry of the object is defined in the calculator, a view of the object can be plotted. The central point of the astigmatic area {R} is used as view point of this calculated image. The orders of the interference fringes are determined with the two spring elements shown in Fig. 5. The use of large hologram plates allows an observation from different directions to check the fringe orders.

The photographs of the reconstructed image are digitized on a digitizing table to determine the position of the fixed points (characteristic object points) and of the measurement points (points located on interference fringes), which are then stored together with their fringe orders. Since the positions of the fixed points are known on the plotter image and on the photographs, we now can plot the measured points.

This plotting of the measurement points corresponds to a global correction of image modifications. The difference between object outline and digitized points illustrates the image distortions due to the reconstruction with a He-Ne laser (Fig. 6).

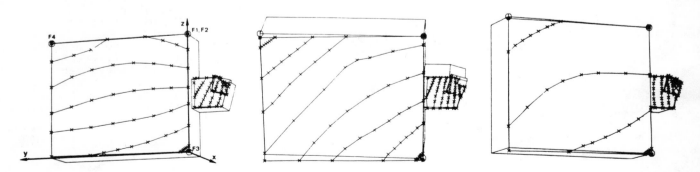

Figure 6. Plotter images of the cutting tool and clamping device. The digitized points are plotted with the help of the fixed points. Lines connecting the measured points correspond to the interference fringes on the photographs.

The compensation of image modifications finally gives an exact plot of the measured points on the object images (Fig. 7).

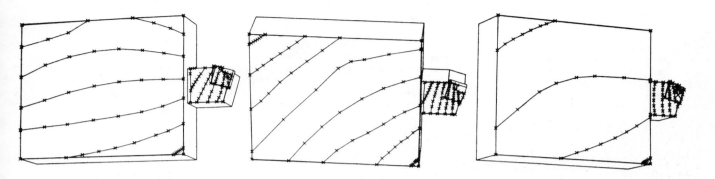

Figure 7.   Modification-free plotter images of the cutting tool, clamping device and measured points. A comparison with Fig. 6 clearly shows that without compensation large position errors occur which in reality do not exist.

From these pictures we can now determine the original points corresponding to the measured points. This is done by projecting the image on the object. When the deformations are calculated, a projection of the deformation vector on the corresponding sensitivity vector of the holographic set-up is found for each measurement point.

Three projections of the deformation vector at the same object point are needed to determine a three-dimensional deformation vector. This requires a conversion from interference fringe coordinates (measurement points) to grid point coordinates. The fringe orders of the grid points can be calculated from the fringe orders of adjacent measurement points [8].

The graphical representation of the three-dimensional displacement vectors in Fig. 8 clearly shows the rigid body motion and the deformation of the object between both exposures.

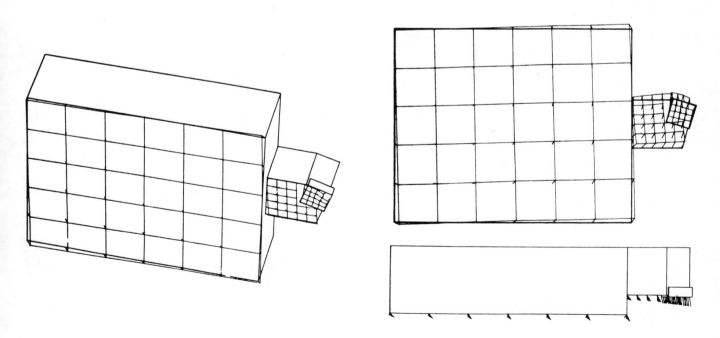

Figure 8.   View of the object with grid points and three-dimensional displacement field (magnification 2000): a perspective view on the left and two normal projections on the right.

The holographic interferometric determination of three-dimensional deformations was assessed with two triaxial accelerometers (Fig. 5) fixed on the clamping device. A difference on the order of 10% was observed between both techniques [5]. This alternative determination clearly shows the possibilities and limitations of both measuring methods [9].

Conclusions

The new method describt in the paper establishes a simple relation between reconstructed hologram image and original object. If image modifications are compensated, an exact evaluation of pulsed holograms is possible. Moreover, the image quality of holographed objects reconstructed with another light can be estimated and improved. It is then easy to calculate the corresponding displacement vectors and to display them graphically.

References

1. Ennos A.E., Measurement of In-Plane Surface Strain by Hologram-Interferometry. J.Sci. Instrum. 1, 731-734 (1968).
2. Alexandrov E.B., Bronch-Bruevich A.M., Investigation of Surface Strains by the Hologram Technique. Sov.Phys.-Tech.Phys. 12, 258-265 (1967).
3. Meier R.W., Magnifications and Third-Order Aberrations in Holography. J.Opt.Soc.Am. 55, 987-992 (1965).
4. Schumann W., Dubas M., Holographic Interferometry. Springer Verlag Berlin (1979).
5. Müller E., Evaluation of Holographic Interferograms. Diss. ETH 7246 (in preparation).
6. Dändliker R., Hess K., Siedler Th., Astigmatic Pencil of Rays Reconstructed from Holograms. Israel Jour. of Tech. 18, 240-246 (1980).
7. Schumann W., Duality Property in Holographic Imaging. J.Opt.Soc.Am. 71, 525-528 (1981).
8. Laporte M., Elaboration Rapide de Cartes Gravimetriques. Geophys. Prospecting 10, 238-257 (1962).
9. Müller E., Looser W., Gygax P.E., Comparison of Modal Analysis and Holography. Annals of the CIRP, 29, 1, 397-402 (1980).

Holographic Measurement of Three Dimensional Displacements by Means of a Holographic Plate

Hugo Rytz

Institute for Applied Physics
Swiss Federal Institute of Technology
8093 Zuerich, Switzerland

Abstract

With a well known holographic setup we can determine the three dimensional displacement vector with only one single holographic plate. The probe must be imaged by a lens onto the photoplate and holographically recorded with a plane reference wave. Using an unexpanded laser beam, we can reconstruct the interesting spots of the object whose displacement we want to determine. Due to the displacement we can see interference fringes on the real image. The shape of the fringes are tightly bound to the displacement of the spot that we are reconstructing. The direction and the curvature give us the three absolute coordinates of the displacement vector.

Introduction

The holographic measurement of displacement has become a very strong tool to analyse the properties and behaviour of an object under stress. To record the displacement is a simple task, one takes with a holographic setup an exposure before and one after applying the deformation to the object of interest. In the reconstruction stage of the hologram one compares both virtual images, and due to the displacement of the object between the exposures one can observe interference fringes on and in front of the image. In a general case these kinds of fringes give a qualitative information of the displacement of the whole object. To get the quantitative information, especially the three components of the displacement vector of a particular point of the object, one has to apply a more careful examination of the fringes.

There are many methods proposed how to obtain the three dimensional displacement vector from one single hologram. One method is to observe the virtual image of the hologram from different points and to measure the fringe shift [1,2,3]. From an overdetermined set of equations one can obtain the three displacement components.

The method proposed in this paper is based on the ideas of Gates [4] and Boone [5]. They examined the real image of the hologram. If one filters out one point of the object (whose displacement is of interest) with a pinhole, one obtains a new set of interference fringes. These fringes can be observed on a screen parallel to the holographic plate. From the shape of these fringes one can calculate the three displacement components. A more theoretical treatment is given by Liu [6,7].

Principles of Method

The basic setup is sketched in Fig.1; to record the hologram we use a well known imaging holographic setup. The illumination and reference waves are plane. We make a double exposure hologram with one exposure before and one after the displacement of the object. To reconstruct the hologram we use the conjugate reference beam. The real image of the imaged object is located close to the hologram. The interference fringes which can be seen there, are interpreted by Sikora [8]. Now let us look on the real image of the imaging lens. If we reduce the diameter of the reconstruction wave down to the unexpanded laser beam, we are able to observe a kind of inteference fringes in the pupile of the lens. Since we are using the unexpanded beam for reconstruction, we reconstruct only a small area of the real image of the object. Effectivly this is the same as filtering out the point with a pinhole. This means, that the fringes, which are visible in the pupile of the lens, follow the same law as described by Boone [5]. In this paper we will outline only the main steps of the mathematical treatment.

Let us assume that the area of the object, which is reconstructed, has undergone a uniform displacement. In this case we consider the amplitude distribution the in reconstruction area as a convolution between an original wavefront $U(\vec{x})$, scattered by the object, and two deltafunctions $\delta_1(\vec{x})$ and $\delta_2(\vec{x}+\overline{\Delta x})$. The far field pattern of this amplitude distribution can be expressed as the farfield pattern of the scattered wavefront multiplied by the farfield pattern of both deltafunctions:

$$A(\vec{x}) = \widetilde{U}(\vec{x}) \cdot (\overline{\delta_1(\vec{x}) + \delta_2(\vec{x} + \Delta\vec{x})})$$
$$= U(\vec{x}) \cdot const \cdot \cos(\vec{k} \cdot \Delta x) \qquad (1)$$

where $\vec{k}$ = the wave vector, $|\vec{k}| = 2\pi/\lambda$
        $\Delta\vec{x}$ = the displacement vector

The argument of the cos-function can easely be expressed as parametric equation of a set of cones with the parametric value m :

$$\vec{k} \cdot \Delta\vec{x} = 2\pi m \qquad (2)$$

where m = parametric value between $\emptyset .. \Delta x/\lambda$
          (physically known as fringe number)

If the set of cones intersects a plane, the pattern field can be seen as a set of conic sections. The intensity distribution in the plane can be calculated from (1):

$$I(x,y) = I_o \cdot (1 + \cos(k \cdot \Delta x \cdot \cos(\theta(x,y)))) \qquad (3)$$

where I_o = uniform intensity derived from $U(\vec{x})$
        $\theta(x,y)$= angle between observation direction and axis of displacement (see Fig.2)

From the shape of the fringes we can determine the displacement vector:

- the axis of symmetry gives the displacement direction parallel to the sectional plane.

- the bending of the fringes gives the information about the angle θ_o between the perpendicular of the plane and the axis of the displacement vector.

- the period of the fringes combined with the above mentioned angle gives us the amount of the displacement.

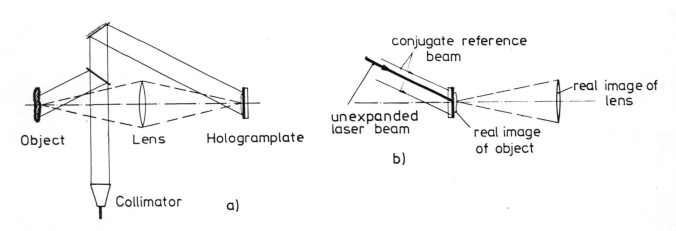

Figure 1. Holographic setup to record and reconstruct a
          double exposure hologram
          a) Recording setup
          b) Reconstruction setup

The quantitative interpretation of the fringes patterns ,given by Boone [5] can also be adapted to this method. The beam $\vec{k1}$ gives fringe number m1, beam $\vec{k2}$ fringe m2 (see Fig.2). The number of fringes between the two beams are m2-m1 and are countable. We rewrite (2) in a more practicable form:

$$\lambda \cdot m = \Delta x \cdot \cos(\theta) \qquad (4)$$

and insert $\vec{k1}$ and $\vec{k2}$ into (4)

$$\lambda m_1 = \Delta x \cos(\theta_o)$$
$$\lambda m_2 = \Delta x \cos(\theta_o + \nu) \qquad (4a)$$

The subtraction of both expressions gives us

$$\Delta x = \frac{\lambda \cdot (m_2 - m_1)}{\cos(\theta_0 + \gamma) - \cos(\theta_0)} \qquad (5)$$

For the angle γ we can write

$$\gamma = \arctan(D/2d) \qquad (6)$$

where D is the diameter of the pupile
      d is the distance hologram-lens

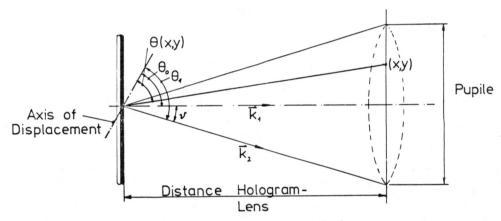

Figure 2. Geometric setup for calculation of the displacement

To calculate the displacement we have to perform the following steps:

1. to establish the axis of symmetry

2. to determine the out-of-plane displacement angle θ_0. This is not a trivial
   task, because one has to analyze the bending of the fringes. Most applicable
   it is to compare the fringes with a synthetically calculated pattern for
   different angles drawed on a transparent paper.

3. to count the interference fringes between two distinguished points on the
   symmetry axis.

   With the above determined numbers we can calculate the detected displacement. This
method gives only the amount of the displacement and its axis but by no means we can
detect the direction (for example if the inplane displacement was up or down with respect
to the hologram).

   In Fig.3 we can see the interference fringes resulting from a predetermined
displacement of a flat probe. As mentioned above, these are the real images of the lens.
In Fig.3a the probe has undergone only an inplane displacement. The bending of the
fringes, which should be parallel, are due to the curvature of the image field. In Fig.3b
the probe has undergone a pure out of plane displacement. We can see that the fringes are
concentric as the theory told us. Fig.3c shows the fringes resulting from a combined
inplane and out of plane displacement. In this case, if we want to calculate the amout of
the displacement, we have to determine the angle θ_0 from the bending of fringes. For
this task we compared the fringes with a prepared pattern drawn on a transparent paper.
The accuracy of the determined angle θ_0 was within 5 degrees.

 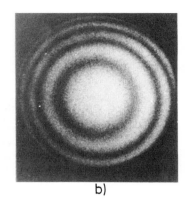 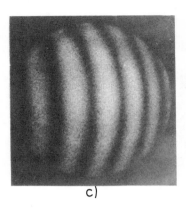

a)   b)   c)

Figure 3. Measured displacement of a probe.
   a) inplane displacement of 10 µm.
   b) out of plane displacement of 100 µm.
   c) combined inplane displacement of 10 µm and out of
      plane displacement of 20 µm.

Limitations

The upper and lower limit of the displacement which can be measured is a matter of the speckle correlation and geometric setup. The farfield of the amplitude-distribution $U(\vec{x})$ of the aperture is speckled and a detection of interference fringes will become impossible, if the speckle size is in the same magnitude as the fringe spacing. This limitation is also well known when using the speckle photography to measure inplane displacements.

But even with an unspeckled farfield we can calculate a limit of the detectable displacement. Since we can only see the fringes in the area of the pupile of the imaging lens, the size of this area involves the limits. To calculate the lower limit, let us look at Fig.2. The beam $\vec{k1}$ gives us the fringe number m1, $\vec{k2}$ gives fringe m2. We must see in the pupile at least one whole fringe, so we can say

$$m_2 - m_1 = 1 \tag{7}$$

We calculate Δx_{min} from Eq. (5)

$$\Delta x_{min} = \frac{\lambda}{\cos(\theta_o + v) - \cos(\theta_o)} \tag{8}$$

Due to the nonlinearity of the denominator of (8) we see that the lower limit depends of the angle θ_o or in other words to the inplane/outplane displacement ratio. Fig.4 shows this dependance for a particular application. The figure tells us that the method measures inplane displacements with much more accuracy than pure out-of-plane displacements.

To determine the upper limit of the detectable displacement we have to remember that (3) is only true, when the aperture is comparable to the amount of the displacement. Usually the aperture is larger. In this case we must integrate (3) over the area of the aperture:

$$I(x,y) = I_o \int_{Aperture} (1 + \cos(k \ x \cos(O(x,y,\eta,\xi)))) d\eta \, d\xi \tag{9}$$

where η, ξ are coordinates in the plane of the aperture
$\theta(x,y,\eta,\xi)$ the angle θ as function of x,y and η,ξ

The cos-argument is now no more a function of x,y only but it also depends from the aperture. It includes the angle under which the aperture is seen. If this angle exceeds φ (angle between two beams $\vec{k1}$ and $\vec{k2}$, giving two neighbouring fringes, see (10)), then the intensity will become uniform over the x,y plane; in other words, the fringe contrast decreases to zero. If a certain aperture is given (e.g. diameter of the laser beam) then the upper limit of the displacement, which can be detected, is given by:

$$\arctan(a/2d) \leq \tfrac{1}{4}\varphi$$
$$= \tfrac{1}{4}\arcsin\frac{2\cdot\Delta x\cdot\cos(\theta_o)}{\lambda} \qquad\qquad (10)$$

where a = diameter of the aperture.

We may reduce the aperture to increase this limit, but at the cost of increased speckle size preventing further detection of fringes.

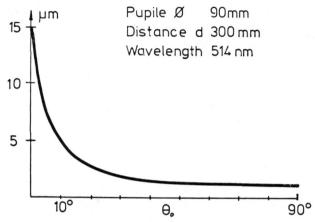

Figure 4. Lowest detectable displacement versus out of plane angle

Application

We have used this method to detect the behaviour of up to 10 hours old concrete. The probe of concrete was a cube with 4 in. edge length. After removing the casing, we took a first exposure of a small spot at the top of the cube and several minutes later a second one. Between both exposures the probe has been deformed by his own weight. Due to the very small deformation we applied an extra horizontal displacement of known amount between the two exposures. The double-exposure was repeated several times.

When we are interpreting the deformation, we subtracted the horizontal displacement. This gaves us the pure vertical setting. As expected this setting is about 0.5 μm at the first double exposure and decreased with advanced time.

We compared two different types of concretes, one pure type and one reinforced with synthetic fibres. Fig.5 shows the different deformation velocity of both types of concrete.

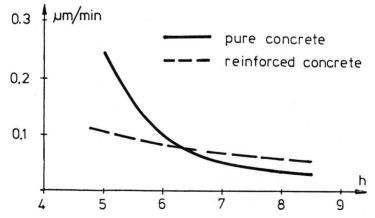

Figure 5. Deformation speed versus time of setting of concrete

Acknowledgments

The author wishes to thank
    Prof. H.Melchior for his managment support
    Commission for Promote the Scientific Research (KWF), Berne, for the financial support.

References

1. Aleksandrov,E.B. et al., "Investigation of Surface Strains by Hologram Technique", Soviet Physics, Technical Physics, 12(2) (August 1967)

2. S.K.Dhir and J.P.Sikora, "An improved Method Obtaining the General-Displacement Field from a Holographic Interferogram", Experimental Mechanics, Vol 12, No.7 (July 72)

3. V.Fossati Bellani and A.Sona, "Measurement of Three-Dimensional Displacement by Scanning a Double-Exposure Hologram", Appl. Opt., Vol 13, pp. 1337ff (1974)

4. J.W.Gates, Proceedings of the "symposium sur les applications de l'holographie", Besancon 1970

5. P,M.Boone, "Determination of Three Orthogonal Displacement Components from One Double-Exposure Hologram", Optik, Vol 37, No.1

6. H.K.Liu et al., Proceeding of SPIE, Vol 92, (1976)

7. H.K.Liu et al., "A Practical Method for Holographic Interference Fringe Assessment", Optical Engineering, Vol 16, No. 2, (March 1977)

8. J.P.Sikora, "A Three-Dimensional Displacement Analysis from an Image-Plane Hologram", Experimental Mechanics, Vol 18, No.3, (March 1978)

Automated digital analysis of holographic interferograms of pure translations

A. Choudry[*][♦], H.J. Frankena[*] and J.W. van Beek[¶]

[*]Department of Applied Physics, and [¶]Department of Mechanical Engineering,
Delft University of Technology, P.O. Box 5, 2600 AA Delft, The Netherlands.
[♦]Computer Science Department, University of Amsterdam,
P.O. Box 41882, 1009 DB Amsterdam, The Netherlands.

Abstract

Holographic interferometry is a versatile technique for non-tactile measurement of changes in a wide variety of physical variables such as temperature, strain, position etc. It has a great potential for becoming an important metrologic technique in industrial applications. For holographic interferometry to become more attractive for industrial practice the problem of quantitative analysis of the patterns and thereby eliciting reliable values of the relevant parameters has to be addressed. In an attempt to calibrate the technique of holographic interferometry and ascertain the reliability of the subsequent digital analysis, we have chosen precisely known translations as a basis. Holographic interferograms taken from these are analysed manually and by digital techniques specially developed for such patterns. The results are promising enough to indicate the feasibility of automated digital analysis for determining translations within an acceptable accuracy. Some details of the evaluation techniques, along with a brief discussion of the preliminary results are presented.

Introduction

For the determination of small displacements or deformations of objects with the aid of holographic interferometry, fringe counting is the primary operation. For background information and current progress in the field refs. 1 and 2 may be consulted. Quantitative analysis comprises of the insertion of the counted numbers in formulae. For simple (e.g. single-component) displacements, this process, albeit a laborious undertaking, can be carried out without any special tools. Reliable and accurate results from such simple examples are useful for calibrating the holographic measuring technique.

However, if the displacements and the shape of the object are more complicated, the deformation determination at many surface points will demand intricate measurements and evaluations. Without some degree of automation the processing time may become unacceptably long. Dedicated detection techniques (e.g. involving solid-state detector arrays) coupled to a digital computer then seem to provide an appropriate system. In the present paper we concentrate on the latter aspect: digital analysis. To arrive at a proper metrologic technique, the method has to be calibrated accurately. For this purpose mechanical systems have been developed to produce precise displacements, along with a holographic set-up to produce the corresponding interferograms, as also software to perform the digital analysis. Here, only a part of a larger program[3], for calibrating various pure translations and rotations, will be described. Further details of the calibration program will be published elsewhere. We restrict our account to in-plane translations of a plane object, generated by a specially designed spring mechanism[3] in which kinematical corrections compensate spurious motions. The interferograms thus obtained are analysed manually as well as with the aid of an image processing system and a tailored software package. This package[4] also includes noise management techniques for isolating the desired information from the background.

Calibration procedure

As is common practice in quantitative length measurements, the calibration is carried out by comparing the subject (which can be a standard, a method or an instrument) to a better known ("higher") one. In our problem the knowledge of the mechanical system, generating the displacement, acted as the higher standard. A theoretical model for the formation of an interferogram, leading to formulae for the fringe locations generated by a given deformation, was scrutinized and extended for our optical set-up[3]. Experimental verification confirmed its accuracy. Using this model, the interferogram geometry was derived as the next-lower standard.
The results obtained by actual fringe counting and insertion into the formulae, as well as those from an automated digital process, were now calibrated to that derived standard. Starting point for both were the interferograms, obtained in the virtual image of double exposure holograms taken from a plane, diffusely reflecting, screen which was translated laterally. The virtual image was preferred over the other one for reasons of superior fringe visibility.
Utilizing a specially developed translation mechanism, as shown in Fig.1, and a conventional

optical set-up for holographic interferometry, as shown in Fig.2, we obtained interference patterns consisting of straight, parallel, equidistant lines located at infinity. Without going into mathematical details we note that the fringe separation decreases with increasing displacement (see Fig.3).

Figure 1. Mechanism to generate pure translations.

Figure 2. The optical configuration.

Photographs of these patterns, provided that they are all obtained with the same enlargement, carry the proper information for the calculations.

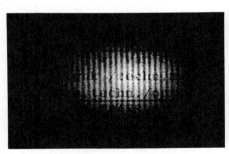 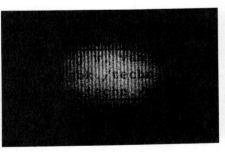 

a. Translation $\delta=(100\pm5)\,\mu m$.   b. Translation $\delta=(150\pm5)\,\mu m$.   c. Translation $\delta=(200\pm5)\,\mu m$.

Figure 3. Interferograms for pure lateral translations of a plane screen.

For convenience (and in view of the fact that photographic negatives were used for manual analysis and that the maxima can be detected more accurately) the light intensity maxima will be identified with the fringes.
Due to the limited eye-resolution, fringe interpolation reliable to 0.1 fringe separation could not be made for the interferograms of Fig.3. For more accurate manual analysis we have,

therefore, generated two interferograms of smaller translations, namely, $\delta=(30.0\pm0.2)\mu m$ and $\delta=(60.6\pm2)\mu m$, as shown in Fig.4.

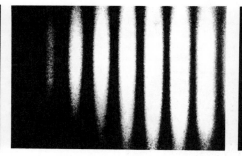

KALIBRATIE-HOLOGRAM
diss. j.w. van beek
werktuigbk./techn.nat.
fijnmech.techn./optica
ZUIVERE TRANSLATIE

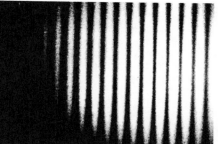

a. Object (rotation axis indicated for dynamic interpretation).
b. Translation $\delta=(30.0\pm0.2)\mu m$.
c. Translation $\delta=(60.6\pm0.2)\mu m$.

Figure 4. Enlarged presentation of the interferograms used for manual analysis.

Holographic interferograms of displacements can be evaluated quantitatively following either the 'static' or the 'dynamic' method. Fringe counting is basic to both the methods and big procedural differences notwithstanding, they lead to results of about the same accuracy. For calibration, when displacements are a priori known to a high degree of accuracy, both methods can be simplified to the mere measurement of fringe separations. This as well as counting fringes from the interferograms (recorded photographically) will often lead to the problem of interpolation between integers. Visually, a reliable estimation as low as a tenth of a fringe is possible provided that the fringe pattern matches certain resolution properties of the eye. While electronic systems allow the assessment to a hundredth of a fringe, they can also be used for a wide variety of fringe patterns without the limitation on resolution mentioned above.

For various values of the displacement, the theory gives a linear relation (within the accuracy of the present account) with the reciprocal of the fringe distance (spatial frequency). For actual interferograms deviations from this relationship occur, even for the average values taken over many pairs of adjacent fringes within each pattern. Such deviations are characteristic of the uncertainty in the calibration value of the pertinent configuration. The images of Fig.3 have been subjected to the automated analysis. To this effect the pattern was scanned by a vidicon and digitized to a raster of 512x512 pixels, each 1 byte deep. Hardware and software developed around a HP-1000 host computer was used to carry out the analysis of the digitalized image. In general, this analysis proceeds in three steps[4], viz.,

1. Noise management,
2. Fringe position (intensity maxima) determination,
3. Numerical calculation.

A digital image of a physical scene as made by a physical system, always contains noise. Some of the noise sources are fairly well known, e.g. turbulence, sensor noise, digitalisation noise etc. To cope with such noise some well-known techniques, such as temporal smoothing (frame averaging) and spatial smoothing (moving average) are utilised. However, interferograms made with coherent light, such as holographic interferograms, have additional noise due to the specific nature of the light (coherent) and the optical system used. This noise could broadly be divided into the following components:

Beam noise. In almost all patterns the fringe visibility is nonuniform over the entire image. This is generally due to 'shading' (variation in the average intensity) and 'modulation' (variation in the fringe contrast i.e. in the peak-to-valley ratio). Normally this is not considered as 'noise' but since it does effect the fringe maxima determination we considered it to be so. The 'visibility' variation over the pattern can in large part be ascribed to the beam intensity profile and partial coherence of the beam. In images produced by a given laser and optical set-up the fringe pattern has a characteristic shading and modulation. Due to these, there is a discrepancy between the fringe 'intensity-maxima' and the fringe 'true-position'. We have developed a technique to estimate this discrepancy and take appropriate measures to remove its effects. A detailed description of the technique will be published elsewhere.

Coherent light noise. Coherent light contributions scattered by the various surface irregularities interfere at the detector. This interference produces a 'speckle' pattern, which results in spurious peaks 'spikes' in the pattern. These spikes can seriously hamper the task of maxima determination and for automated analysis some sort of 'spike-management' becomes mandatory. For not too severe cases of speckle ordinary smoothing e.g. low-pass

filtering, is adequate. However, strong spikes, even a few isolated ones, call for special techniques. To isolate such spikes from a 'well-behaved' signal we have developed a technique (C-Transform[4]). In this method all the intensity maxima are found and grouped according to the relationship they bear to the neighbouring maxima. It turns out, in most of the cases of practical interest studied by us, that the spikes can be isolated and 'locally-filtered-out'. Details of this technique will be published elsewhere.
The diffraction rings produced by the dust particles, scratches etc., on the optical surfaces, are another source of noise. Mostly these are low contrast, high frequency rings and the smoothing applied for speckle removal can eliminate them. However, if high contrast rings with 'wide-spacing', i.e. comparable to the fring spacing, are present, again special measures have to be taken.

Missing fringes. In a given pattern the fringes occur with a certain regularity. Due to various reasons, e.g. dust on the photographic plate, shadows of optical element supports etc., a particular fringe may become locally obscured. In manual analysis this may not be a serious problem to detect and avoid, however, in automated analysis it is a serious factor. If unaccounted, a 'missing fringe' will lead to wrong fringe order assignments which usually will result in large deviations in the final calculations. If missing fringes are suspected then the automated analysis must incorporate some provisions for addressing this problem. The technique of C-Transform, as used above for spike removal, has been very helpful in detecting missing fringes.
To arrive at a physical interpretation of a pattern one needs to know the position at which the phase function of the interference intensity distribution assumes integer values. Without the 'noise', as discussed above, these positions are identical with the intensity 'absolute-extrema'. Assuming that noise has been adequately taken care of (as discussed above), in digital analysis there is another problem which obstructs the solution further. In practice it is quite common to find 'squared-off', instead of sharp, peaks and valleys in the pattern. This 'flattening' of the intensity extrema could be due to the photographic process itself (under- or over-exposure), or due to coarse digitalisation. Under such conditions the position of the intensity extrema is not unique and some measure has to be taken to arrive at a unique fringe position. There are several methods possible for this. One could, for example, take the mid-point of the 'flat-region', or the center of gravity under the 'flat-region', as the 'true-extrema'. More elaborate schemes, e.g. fitting polynomial curves etc. could be devised if so desired. Such a 'rule' for determining the extrema from flat-tops will have to be decided on a case-by-case manner since no general 'rule' seems to be universally applicable. Having determined the fringe positions (intensity extrema) one has to fit a physical model to the fringe positions. The image processing i.e. noise management and maxima determination in this case, and numerical fit usually form an iterative-feedback-chain. A set of image processing parameters e.g. filter type and size, extrema-rule etc., is chosen and the numerical fit is made. If systematic and/or large deviations result then a different set of image processing parameters is chosen. The analysis is considered successful if the deviations in the fit are uniformly random. In this sense image processing and numerical calculations are inseparable ingredients of the overall analysis. In our case this has to be the linear fit as mentioned earlier for manual processing. Again, the deviations from an exactly straight line indicate the method's inaccuracy. This error, as compared to the one caused by the mechanical system, determines the reliability of the automated method.

Results and discussion

The simplified situation, mentioned in the previous section, was first measured manually. From theory it is known that, if d is the distance between adjacent fringes and δ the translation, then the product $d\delta$ is constant[3]. The fringe distance d can be measured from Figs. 4b and c (actually performed on the photographic negatives). The fringe parallelism allows for an averaging process over a number of distances. This led to the values:

$$\delta = (30.0 \pm 0.2)\mu m,$$
$$d = (3.0 \pm 0.03)mm \text{ (average over 6 fringe separations)},$$

and

$$\delta = (60.6 \pm 0.2)\mu m,$$
$$d = (1.45 \pm 0.01)mm \text{ (average over 9 fringe separations)}.$$

Correspondingly,

$$(d\delta) = (90.01 \pm 1.50) \times 10^{-9} m^2,$$

and

$$(d\delta) = (87.87 \pm 0.89) \times 10^{-9} m^2.$$

The absolute uncertainties overlap partially, while the average values differ by 2.4%. This means that, if one of the translations δ is chosen as a standard, the value of the second as calculated from the interferogram shows a deviation of 2.4% with respect to the actually generated value.
A second way of calibration, suitable for manual operation, implies the rotation of the detection system around an axis in the object plane. Fig.4, known as the "dynamic interpreta-

tion method". During the rotation a number of fringes passes this system's viewing axis. From the equation[3],

$$\psi_2 - \psi_1 = \underline{\delta} \cdot (\underline{e}_2 - \underline{e}_1) 2\pi/\lambda,$$

in which $\psi_2 - \psi_1$ represents the counted number of fringes passed and $\underline{e}_2 - \underline{e}_1$ the angle of rotation, one component of the displacement $\underline{\delta}$ can be calculated. This has been elaborated in the case $\delta = (60.6 \pm 0.2)\mu m$, yielding the value $\delta = 61.6 \mu m$. This means a deviation of 1.65% as compared to the calibration value 60.6 μm.
In the case of the calibration value $\delta = (30.0 \pm 0.2)\mu m$, a similar procedure led to $\delta = 30.3 \mu m$, meaning a deviation of 1%.
These deviations (2.4% for measuring fringe separation and 1% for the dynamic fringe counting method) show that best obtainable result for manual method is in the order of 1% per component of the actual displacement. Careful investigation shows an absolute limit for such uncertainties of 0.01 μm (for translation) or 0.01" (for rotation).

Starting from Fig.3, we now consider the digital analysis. The object of pattern analysis is to measure the fringe separation by finding the intensity extrema. As usual, a scan line is chosen for processing the intensity distribution. A typical intensity distribution along such a line is shown by the thin line in Fig.5. A very prominent background, due to the laser beam profile is discernible. Before attempting to find the extrema, this background has to be removed. This is done by first finding the absolute maxima and then, through the C-Transform, the average frequency of the pattern is estimated. From this a filter window width is selected (equal to average fringe separation). With this filter the 'signal' is smoothed out and only the background remains. The thick line in Fig.5 shows the background. By subtracting the background from the full signal (and adding a constant level and multiplying by a factor of two) the 'corrected' signal, as shown in Fig.6 is obtained. Applying this procedure to the entire pattern in Fig.3a, the background substracted pattern of Fig.7 results. Since the C-Transform can operate automatically this 'correction' can be implemented in an automatic method for the analysis of such patterns.

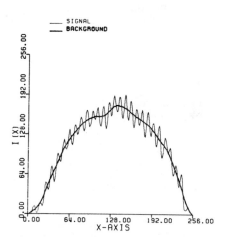

Figure 5.   Signal and background
from the translation
pattern.

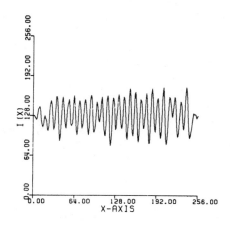

Figure 6.   Background subtracted
signal.

For quantitative analysis the positions of the intensity maxima can be determined. A straight line fit to these extrema positions and fringe order was made before and after the background correction, as shown in Fig.8. The thin line is for the raw data and the thick line is for the corrected data. The straight line fit parameters thus obtained can be used for quantitative comparison of different displacements as discussed below.

We now consider the three patterns of Fig.3, numbered by an index J (J = a,b,c), corresponding to displacements Q(J) of 100, 150 and 200 micrometers. Each Q(J) has an experimental error E(J) which depends on the optical configuration. This is expressed as a percentage of the displacement and is shown in the table below. For each pattern, choosing an arbitrary unit of length, one can define a periodicity P(J) as the number of fringes per unit length. Next, one defines a ratio R(J) = P(J)/Q(J). In general, R(J), for a given point on the measured surface is a constant as long as the optical configuration does not change. The magnitude of this constant is a function of, among others, the illumination and observation directions. If we assume that for all patterns:

- the illumination and observation directions are constant over the entire object, and that
- these patterns are produced by a pure translation,
then the value of R(J) is constant over the entire surface for all the patterns.

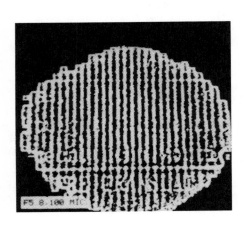

Figure 7. Background subtracted pattern.

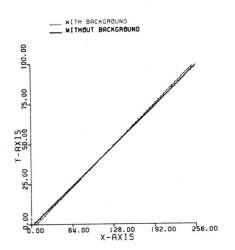

Figure 8. Straight line fit.

Each of these patterns was processed and a straight line was fitted, as discussed above. The slope of each line is directly proportional to the periodicity P(J). As a measure of goodness of fit the standard deviation S(J) was also calculated for each fit. This deviation is calculated as a fraction of the fringe separation. From Q(J) and P(J), R(J) was calculated. These results are shown below:

J	Q(J)	E(J)	P(J)	R(J)	S(J)
a	100	5.0%	.0278	.000278	.071
b	150	3.3%	.0429	.000286	.057
c	200	2.5%	.0558	.000279	.064

First of all we note that the spread of the data points, as measured by S(J), is less than a tenth of a fringe in all the patterns. Since the experimental error, E(J), is the smallest for J = c, one can take R(c) as a standard and calculate Q(a) and Q(b). With R(c) as standard, Q(a) = 99.82 micrometers, and Q(b) = 153.78 micrometers. These values are well within the experimental errors E(J), such that it can be concluded that the digital analysis has modified these inaccuracies only slightly. A maximal variation of about 2.5% in R(J) is also well within the experimental limits. Thus within these limits the above results have established that the displacements were pure translations of the specified Q(J).

Outlook

The automated analysis as described in this paper is not only applicable to translations, but it is also suitable for rotations. Interferograms of pure rotations, often showing similar patterns of straight fringes have already been analysed manually. The optical and mechanical set-up enabled us to reach a minimal uncertainty of 1% in the calibration value for the holographic interferometry described here. Automated analysis of this has still to be carried out.
Complex displacements (deformations) will result in more complicated interferograms. Application of automated analysis to such cases is under further investigation.
A more accurate calibration value may be expected if we are able to improve our initial standard, the pure translation (or rotation) mechanism. This necessitates sufficient control of the ambient and the photographic processing parameters. The latter can be avoided if the pattern is projected directly onto sensors.
The only way to improve the "higher" standard is by designing still more accurate displacement mechanisms, based on kinematic construction principles. Further improvements of the calibration value of the holographic system include very accurate determination of optical parameters (beam direction, parallelism), control over the influence of the photographic steps and of other detectors, and software development providing a better fringe interpolation possibility.

Acknowledgement

The digital processing part of this work was supported by a grant from Stichting voor Fundamenteel Onderzoek der Materie (FOM) the Netherlands, under project nos. 79.79.89 and 79.81.89. The authors would also like to thank the Pattern Recognition Group at the Delft University of Technology for the continued support.

References

1. Dändliker, R., Optical Methods in Mechanics of Solids, Ed. A. Lagarde, Sijthoff and Noordhoff, Alphen a/d Rijn, the Netherlands, 1981.

2. Vest, C.M., Opt. Eng., Vol. 19, p. 654. 1980.

3. Beek, J.W. van, Kwantitatieve verplaatsingsmeting door interferentie-holografie in het virtuele beeld, Ph.D. Thesis, Delft Univ. of Technology, 1982. (in Dutch).

4. Choudry, A., Proc. 2nd Scand. Conf. on Image Proc., Helsinki, June 18-21, 1981.

65

Automated interferometric evaluation of optical components
at the European Southern Observatory (ESO)

A. Choudry*$, H. Dekker*, and D. Enard*

*European Southern Observatory
Karl Schwarzschild Str. 2
D-8046 Garching bei Munchen, W. Germany

$Computer Science Department
Univ. of Amsterdam
P.O.Box 41882, NIKHEF-H-327
1009 DB Amsterdam, The Netherlands

Abstract

In an attempt to approach the theoretical limits of performance, modern telescope design places higher and higher demands on the quality of the components. Digital image processing is playing an increasingly important role in at least two distinct aspects of this endeavor. On the one hand analysis of patterns from optical quality control tests e.g. interferometry, Shack-Hartmann etc., is being made more quantitative through the digital analysis. Furthermore, active-optics, to correct for some of the more important defects in the optics and telescope maintenance, is envisaged to become an integral part of the new telescopes. Digital image processing is an essential ingredient of the active-optics approach of ESO's New Technology Telescope (NTT). If its full potential is realised it would considerably relax the otherwise most stringent requirements on optical quality and the telescope maintenance infra-structure. There is thus substantial impetus given to the development of digital techniques to meet these goals. Here we shall discuss the development of digital techniques for the analysis of interferograms originating from interferometric tests conducted to check the quality of optical components. Light passing through the component is made to interfere with a coherent reference beam. The resulting interferogram is digitally analysed to yield the fringe positions and order. To this data a fit of pseudo-Zernike polynomials is made to quantitatively evaluate the various optical aberrations.

Introduction

The use of interferometric methods for testing the quality of optical components is wide spread. The quantitative analysis of these interferograms, if attempted manually-visually, is tedious and prone to subjective biases. There is a gradual trend to analyse these patterns digitally. Over the years numerous attempts have been made in this direction. So far, a universally applicable digital technique has not emerged and the automated digital analysis of interferograms remains a major problem, Vest (1). We propose to divide the techniques developed so far into three broad classes, namely, 1. Hybrid, 2. Interactive and 3. Turn-Key.

The hybrid techniques, such as Bellani (2), Biederman (3), Fischer (4), Lanzl (5), Kreitlow (6), use an optical processing step prior to the digital analysis. In this manner a considerable improvement of the S/N ratio, as also ease of analysis is attained. However, these techniques demand non-trivial modifications of the usually employed interferometric set-up and are thus limited in their applications and not suitable for real-time applications.

Interactive techniques digitally analyse the pattern as it is and are thus more versatile. The interferogram is scanned by a suitable scanner and then with operator asistance a digital analysis is carried out, Augustyn (7), Bernal (8), Jones (9), Tichener (10), Yatagai (11). The operator involvement renders these techniques unsuitable for complete automation. To achieve a hands-off operation Turn-Key techniques are sought. In this mode of operation once the analysis process is started no further operator asistance is required. For certain applications this mode is a sine qua non. This includes such applications as, on-line process and/or quality control in industrial practice, active-optics control of telescopes etc. Such techniques are, as a rule, highly dedicated since much reliance is placed on the a-priori knowledge of the problem. Special provisions have to be implemented for developing the algorithms and streamlining their execution. Holographic interferometric studies of turbulence in convective heat transport and lateral displacements have been carried out in Turn-Key mode, Choudry (12), (13).

In this paper we shall discuss the work done for developing an interactive (and eventually a Turn-Key) technique for analysing interferograms of optical surfaces to evaluate the extent of various aberrations and the subsequent degradation of the image. Such considerations of image quality are essential to modern telescope design and may play a vital role in active-optics control, Wilson (14). Our study begins with producing an interferogram of the surface using the usual interferometric techniques. The fringe pattern is digitally analysed to find the fringe positions and their order. This step constitutes the major part of the image processing operation. The accuracy, in the presence of the ubiquitous noise, realised here is to a large extent responsibele for the ultimate worth of the entire operation. From the fringe data collected here the wave front is reconstructed by making a numerical fit, using quasi-Zernike polynomials as orthogonal eigenfunctions. The Zernike coefficients so determined bear a direct relationship to the various geometrical aberrations. The same results can also be used to calculate the surface morphology which can help in further processing the optical surface. Some details of the method employed along with the major problems encountered will be discussd next.

Method

To assemble a pool of a-priori knowledge the first step was to write a program to simulate interferograms, wave fronts and surfaces with known optical aberrations i.e. known Zernike coefficients. The necessity of this step in aberration analysis has been discussed by Geary (15). The interferograms so generated can be used for determinig the fringe positions and order and thus become the touchstones for the subsequent digital analysis and numerical fitting. To make it more realistic a simulation program should include noise due to, at least, speckle and parasitic rings which are almost always encountered in coherent light interferometry. In particular, speckle as a noise is the main culprit in image degradation and should be paid special attention. Apart from its optical manifestation, speckle is being actively pursued in percolation theory of disordered solids, Weinrib (16), (17), which provides easy to implement models of optical speckle. The present study at this stage does not include a speckle model in its simulation program.

Besides providing a priori knowledge about the patterns the simulation model can also serve as a qualitative check of the final analysis. After an experimental interferogram has been analysed and the corresponding Zernike coefficients determined, one can use these as input to the simulation program and create a theoretical interferogram. A visual comparison of the original interferogram and the reconstructed interferogram would reveal any anomalies encountered in the analysis. Fig.1 shows typical interferograms generated by the simulation program along with the corresponding fringe positions. Fig.2 shows a 3-D representation of an interferometric wavefront and the associated surface.

In Fig.3 the interferometric set-up used for testing the components is shown. The interferometer is a Tinsley, Model 102B. The collimator is a Newton type parabola of 30 cm diameter. An interferogram produced here can either be photographed for off-line analysis or it can be scanned by a TV-Camera into a raster of 1-byte deep 512x512 pixels. This is stored on disc and accessed by a VAX 11/780 operating under VMS. A DeAnza image analyser serves as the main image display and monitor. Once the image is in the computer memory an interactive program is started. This program is MENU-directed and offers the operator several choices e.g. filter, extrema determination, contrast variation, etc. To expedite the determination of the appropriate operations and parameters the first step is to make four 'reduced' copies of the image by 'left bottom veto', i.e. starting from the origin a neighborhood of 2x2 pixels is replaced by the single 'left-bottom' pixel. If desired, the 'left bottom' pixel could also be taken as the average of the entire 2x2 neighborhood. This would of course take more time than the straightforward veto. This reduced image is copied in all the quadrants to create four identical images of 256x256 pixels in the four quadrants. Different operations can now be applied to different quadrants and their effects can then be easily compared in the composite image. Fig.4 shows the effect of different filters and subsequent extrema determination. The left bottom quadrant (3rd quadrant) is the original interferogram. In quadrants 4.2.1 filters of increasing window width have been applied and the fringe maxima determined. The locus of extrema, called 'extrema-skeleton', is displayed as an overlay in Fig.4. The effect of filtering on smoothing the 'extrema-skeleton' of the fringes is readily seen. This is further illustrated in Fig.5. Here the grey value distribution along a typical scan line is shown. The left hand part of the graph shows the unfiltered signal whereas the right hand part is the filtered signal. By such diagnostic probing one can determine the proper filter for a given interferogram. This filter is then scaled up for the full (512x512) image and applied to it. For the purpose of this analysis of interferograms, one needs to find only two quantities, namely, fringe position and fringe order.

Fringe position is identified as the fringe intensity extrema. In general the fringe intesity I(x) along the x-axis in an interferogram is given by

$$I(x) = A(x)+B(x)Cos (kW(x)) \qquad (1)$$

Here A and B represent background i.e. 'shading' and 'modulation', respectively, and W(x) is the physically interesting wavefront phase (here measured in units of inverse wavenumber, k being the wavenumber of the laser light used in the experiment). In a companion paper in these proceedings we have briefly discussed the role of background terms in fringe position determination. Here we assume both A and B to be constant. Now the fringe maxima are given by those values of x at which W is an integer. Let these values be denoted by x(n) and thus

$$W(x(n)) = n, \quad n = 1,2,3, - - \qquad (2)$$

Since the absolute value of n, the fringe order, is not known we have arbitrarily chosen n to be a positive integer. In reality W is a function of both x and y i.e. W(x,y). The y-value is chosen by selecting a scan-line designated by a scan-line number m (in our case m = 1,2,3 - - 512). The object of digital image processing is to determine the triad-structured data string (x(n),m,n) for the pattern. We shall now briefly discuss the determination of x(n) and n.

In analogue signal processing the maxima are given as the roots of the equation

$$dI(x)/dx = 0 \qquad (3)$$

There is no digital operation that can replace the above exactly and hence one has to formulate ad hoc definitions of maxima seeking operations e.g. local absolute maxima, maxima in a given neighborhood, maxima of a fitted analytical curve etc. Furthermore, noise makes this process even more cumbersome by requiring some sort of 'smoothing' prior to maxima determination. In our interactive package the operator is given the choice of selecting the filtering and maxima determination algorithm's parameters. A few details of such a digital algorithm are disussed below. For reasons of simplicity we consider only a one dimensional case. Its generalisation to two dimensions is quite straightforward.

In an NxN pixel image let the real grey value g(x) at pixel x (x = 1,2,3, - - - N), be bounded by 0 and 1. An i-bit integer word I(x) is generated by an ADC at each pixel x, such that

$$I(x) = g(x)(2**i) \qquad (4)$$

If there are too many small amplitude spikes then relief could be obtained by scale reduction, or in the language of sampling theory, non-adaptive decimation, as follows.

$$ilog2:= (i-k)log2 \qquad (5)$$

or

$$ilog2:= ilog2-logk \qquad (6)$$

where := is the usual ALGOL data flow. Note that in this case eq (6) is the simple division of I(x) by k (a rather time consuming operation) whereas eq.(5) implies the truncation of I(x) at the kth LSB (Least Significant Bit), a much more efficient operation involving only k-shifts. Of course one can develop much more sophisticated and adaptive decimation rules. We restrict ourselelves to this single parameter (k) decimation. It can be built into the interactive package as a MENU-parameter. Now onwards we shall assume I(x) to be decimated if chosen. In the following all the variables and operations are integer unless otherwise stated.

We begin by defining a generalised forward finite difference D(x,r,s) of rth order and extent s at x (similar to the Newton's finite differences), by the following recursive relation

$$D(x,r,s) = D(x+s,r-1,s) - D(x,r-1,s) \qquad (7)$$

with the boundary condition

$$D(x,0,s) = I(x) \qquad (8)$$

In general one can use the above defined D's (with various values of r and s) to develop algorithms for noise management and extrema determination, however, we shall confine ourselves to the simple case of r = s = 1 and will not list them as explicit arguments of D. Of course in a more sophisticated algorithm they can also become MENU-parameters. Next we define two simple digital operations, namely, radixed entier ERD(a,b) and digital Kronecker-delta DEL(a,b) as follows

$$ERD(a,b) = ENT(a/b) \tag{9}$$

where ENT is the Entier, and

$$DEL(a,b) = 2/(ERD(a,b) + ERD(b,a)) \tag{10}$$

Note that the hardware realisation of DEL is a simple AND gate. Now we define a one bit variable SGN(x) through the following recursive relation

$$SGN(x) = DEL(0,D(x))SGN(x-1)+$$
$$(1-DEL(0,D(x)))(1-ERD(D(x),ABS(D(x))))/2 \tag{11}$$

with the condition that the recursion begins with x = b, where b is the smallest integer for which

$$(1-ERD(D(b),ABS(D(b))))/2 = +1 = SGN(b) \tag{12}$$

With these definitions it can be shown that the 'analogue' eq (3) has the following digital counterpart

$$SGN(x)+SGN(x-1) = 0 \qquad iff \ x \ is \ an \ extrema \tag{13}$$

The above can now be used to find the extrema in I(x). A further analysis will show that there is an alternate form of eq (11), which is a 'forward' recursive relation, i.e. it follows ascending values of x. One can also write a 'backward' recursive relation based on the descending values of x. Let these two be called 'Right' and 'Left' algorithms respectively and let x(R,n) and x(L,n) be the pixel values of the nth extrema as found by the R and L algorithms respectively, and let

$$j = x(R,n)-x(L,n) \tag{14}$$

then, without proving it here, it turns out that j is a measure of the 'flatness' of the extrema. In particular j = 0 implies a sharp single pixel peak. Value of j, as a MENU-parameter, can be used to discriminate against sharp spikes due to speckle. Needless to say that there are several such algorithms with the associated discriminants that can be used to deal with different types of noise and fringe characteristics.

Determination of n, the fringe order, is the easiest to formulate, however, its assignment to various fringes in a pattern proves to be most intractable. Part of the reason lies in the fact that whereas maxima determintion is largely a numerical problem, fringe order determination lies in the domain of true pattern recognition in that it demands that a whole fringe be first isolated and then ranked according to the 'pattern' formed in the neighborhood e.g. closed fringes, partly shadowed fringes etc. To solve this problem we have analysed various approaches based on segmentation, run-labelling, contour following, connectivity propagation, C-Transform (13) etc. Some details of this study will be published elsewhere. At this juncture suffices it to say that unless the pattern is very simple some operator intervention should be expected in assigning the proper fringe orders.

Having determined the fringe positions and the corresponding fringe orders, a numerical fit of quasi-Zernike polynomials is made to the data string (x(n),m,n). This is based on the standard least square method. However, it should be kept in mind that the condition number of the matrix involved here is generally very large and thus singularities may be encountered. In case such singularities are expected then techniques such as SVD (singular value decomposition) may be called for. In practice it may become necessary to go through an iteration between image processing i.e. filtering, maxima determination etc,. and numerical fit to arrive at an accurate, stable result.

Results and Conclusions

The interactive package developed so far has the potential to considerably ease the task of quantitative analysis of interferometric evaluation of optical components. For such techniques to become totally integrated into optical shop testing there are at least two distinct aspects that have to be further investigated. These are, quantitative accuracy and stability of the analysis and Turn-Key mode of operation. The former refers to the fact that interferograms are generally noisy and do require 'smoothing' through decimation, filtering etc. However, this noise management is not always without penalty. Every smoothing operation employed does produce the side-effect of distorting the signal. If this distortion is severe enough, or of such a critical nature (e.g. shifting the maxima position), then the worth of the ultimate numerical results might become dubious. Furthermore, buried in the high frequency noise, which is generally filtered out, there may even be some genuine interferometric information and while cleaning up the signal by filtering out the high frequency components one may be throwing out the baby with the water. This can only be tested with a systematic study of the noise characteristics e.g. by simulation, analysis of patterns from surfaces of known properties (as checked by independent means). This is similar to the calibration of the digital-holographic-interferometric technique described in the companion article, Choudry (13).

For routine shop use (and active-optics for the telescope maintenance) the Turn-Key mode is axiomatic. As mentioned earlier, Turn-Key mode requires special dedicated software. In this case one has to, for example, automatically evaluate the noise extent and signature in a given interferogram and then select the appropriate filter. For fringe order it calls for automatically isolating the fringes and ranking them according to the correct fringe order. To carry out all of these operations without operator intervention during the execution, the software has to be written and organised in a manner quite different from the usual approach e.g. for the interactive mode. Since no intermediate I/O is required bulk of the computations could be carried to the lower levels of the host computer. To further expedite the overall execution and reduce demands on the host computer's size and sophistication it may even be possible to formulate the algorithms in such a way as to reduce a large part of the coding to a microprogrammable reduced instruction set and thereby offer the possibility of running the Turn-Key version on a small stand-alone computer.

For off-line analysis an interactive package is now operational and a series of tests will soon begin to check its accuracy and its relation to S/N ratio. As a next step an on-line interactive version is planned. As a final step an on-line Turn-Key version is envisaged. To this effect, automated filtering and extrema determination seem quite feasible, however, fringe order assignment would involve a more fundamental study of the problem from the pattern recognition point of view.

Acknowledgments

We would like to thank the Computer Staff of ESO for their cooperation. One of the authors (AC) would like to acknowledge Stichting FOM, The Neteherlands, for the support given under Project nos. 79.79.89. and 79.81.89 for the development of the digital techniques in an earlier phase. AC would also like to thank the Pattern Recognition group at the Delft University of Technology for their continued support.

References

1. Vest, C.M.
Holographic NDT; Status and Future
NBS-GCR-81-318. Washington D.C. 1981

2. Bellani, V.F. and Sona, A.
Appl. Opt. 13(1974)1337

3. Biederman, K. and Ek, L.
Appl. Opt. 16(1977)2535

4. Fischer, B., Geldmacher, J. and Juptner, W.
Proc. Laser '79, Optoelectronic Conf.
Munich, July 2-6, 1979, pp 412-425.

5. Lanzl, F. and Schluter, M.
SPIE 136(1977)166

6. Kreitlow, H. and Kreis, T.M.
SPIE 210(1979)196

7. Augustyn, W.H., Patterson, J.S. and Rosenzweig, D.N.
SPIE 46(1974)144

8. Bernal, E. and Loomis, J.
SPIE 126(1977)143

9. Jones, H.D.
Computer reduction of holographic interferograms.
Sandia Labs. SAND77-8236, 1977.

10. Tichenor, D.A. and Madsen, V.P.
Opt. Engg. 18(1979)469

11. Yatagai, T., Nakadate, S. and Saito, H
Opt. Engg. 21(1982)432

12. Choudry, A.
Appl. Opt. 20(1981)1240

13. Choudry. A, Frankena, H.J. and van Beek, J.W.
SPIE 398(1983)0000

14. Wilson, R.N.
Optica Acta 29(1982)985

15. Geary, J.M. and Holmes, D.
Opt. Engg. 18(1979)39

16. Weinrib, A.
Phys. Rev. B. 000(1983)000 (preprint)

17. Weinrib, A. and Halperin, B.I.
Phys. Rev. B. 000(1983)000 (preprint)

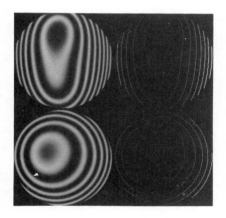

Fig. 1 Simulated
interferograms and
extrema skeletons.

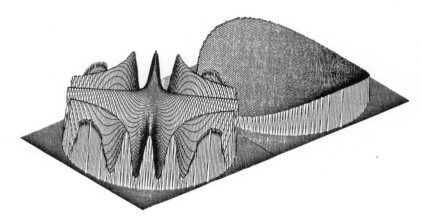

Fig. 2 Simulated wavefront and
surface profile.

Fig. 3 Interferometer.

Fig. 4 Extrema skeletons with different filters.
Left   bottom   original image
Right bottom   3x3 filter
Left top   4x4 filter
Right top   5x5 filter

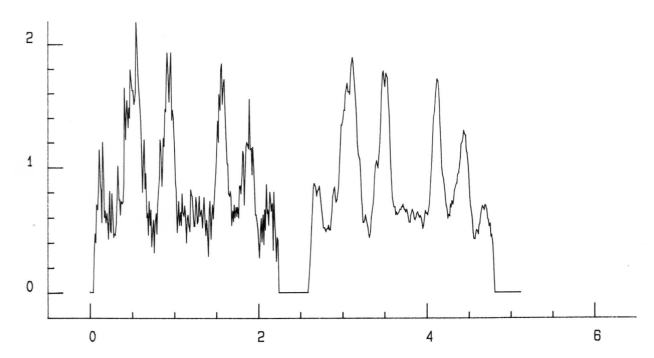

Fig. 5 Grey value distribution along a scan line.
Left half, unfiltered signal: right half, filtered signal.

INDUSTRIAL APPLICATIONS OF LASER TECHNOLOGY

Volume 398

Session 2

Holographic Interferometry II

Chairmen
Gordon M. Brown
Ford Motor Company, USA
Peter Waddell
University of Strathclyde, Scotland

APPLICATION OF HOLOGRAPHIC INTERFEROMETRY

TO PRACTICAL VIBRATION STUDY

M. Murata and M. Kuroda

Mitsubishi Heavy Industries, Ltd. Nagasaki Technical Institute
1-1 Akunoura-machi, Nagasaki 850-91 Japan

Abstract

This paper describes a brief summary of applications of holographic interferometry to practical vibration study in Nagasaki Technical Institute of MHI. The applications of vibration mode measurement are concerned with steam turbine blades, compressor impeller, internal combustion engine, car body and car brake disk. The techniques of holography contained herein are (1) the time average method giving contour fringes of vibration amplitude, (2) the phase modulation method providing information on the relative phases of vibration, and (3) the double pulse method that offers the possibility of visualizing the vibration pattern of internal combustion engine in operating condition, and the transient vibration pattern of an object excited by impact force.

Introduction

Recently, laser holography has been widely used as one of the powerful measurement techniques in the field of industrial measurement. Especially, the applications of holographic interferometry to vibration analysis have been prevalent. For example, continuous-wave laser holography (cw holography) has been applied to the resonant vibration modes analysis of turbomachinery components such as power plant steam turbines and aircraft engines [1] , etc.. And also, pulsed-wave laser holography (pw holography) has been applied to the noise control investigation of structures of internal combustion engines and vehicles, etc. [2 - 6]

The purpose of the present paper is to describe the applications of cw and pw holography to the practical vibration studies held in Nagasaki Technical Institute of MHI.

The cw holography has been mainly used to measure resonant vibration modes of blades and vanes of turbomachinery such as steam turbines, gas turbines and compressors. Although, there are various methods in the cw holography for vibration analysis of vibrating object, we have mainly used (1) the time average method giving the contour fringes of vibration amplitude, (2) the phase modulation method providing information on the relative phases of vibration, and (3) the time averaging double exposure method [7] which provides the information on relative phase of vibration as well as amplitude contours.

The pw holography has been mainly used to measure resonant vibration modes of turbomachinery components and various products in the factory environment, to investigate the compound vibration modes of internal combustion engines in operating condition, and to study the vibration modes of squealing disk brakes. Also, the pw holography has been used to obtain the transient vibration pattern of internal combustion engines excited by impact force.

Applications of cw holography

Experimental set up of the cw holography

A typical experimental set up used in our laboratory is illustrated in Figure 1. The laser beam emitted from an Ar-ion laser (Coherent Radiation Model CR5) is reflected by the mirror M_1 and then divided by the beam splitter BS_1 into two beams, one is used to illuminate the vibrating object, the other is used as the reference beam.

The beam reflected by the beam splitter BS_1 is introduced into the Michelson interferometer. The mirror M_3 mounted on the piezo-electric transducer (Tropel Model 254 PZ) and   used to modulate the reference beam.

The typical experimental set up can be used to make holograms by three kinds of methods, namely, the time average method, the phase modulation method and the time averaging double exposure method. [7]

The time average method

The hologram by this method is made by a reference beam reflected from the stationary mirror M_3. Then, we can obtain the reconstructed image of intensity $I_1(r)$ given by

$$I_1(r) = I_0(r) \ J_0^2(p). \qquad (1)$$

Here, $I_0(r)$ is the intensity at the point r on the stationary object, $J_0(p)$ is the zero-order Bessel function of the first kind, and p is given by

$$p = 2\pi \ a(r) \ (\cos \theta_1 + \cos \theta_2)/\lambda \qquad (2)$$

where $a(r)$ is amplitude of a point at r on the object, λ is the wavelength of the light used, θ_1 is the angle between the directions of the illuminating beam and the object vibration, and θ_2 is the angle between the viewing direction and the direction of the object vibration. As well known, the function $J_0^2(p)$ in

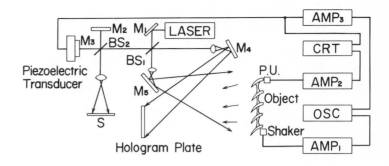

Figure 1. Experimental set up of cw holography for practical vibration studies.

equation (1) is the characteristic fringe function of the time average method which depends only on the vibration amplitude of the object.

The phase modulation method

The method requires the phase of the reference beam to be modulate sinusoidally in synchronization with the vibration at a fixed point on the object.

For the phase modulation of reference beam, the Michelson interferometer shown in Figure 1 is utilized. When the mirror M_2 is inclined slightly, parallel interference fringes are produced on the screen S. If the mirror M_3 is vibrated in synchronization with the object vibration, the intensity of the observed fringes becomes

$$I_f(\ell) = K\{1 + J_0(m) \cos (2\pi\ell/h)\}. \qquad (3)$$

Here, ℓ is the distance measured normal to the fringes, h the spacing of the fringes, K a proportional constant, and m the degree of phase modulation given by

$$m = 4\pi b/\lambda, \qquad (4)$$

with vibration amplitude b of the mirror M_3. The contrast of the fringes is given by the value of the zero-order Bessel function for m. The contrast vanishes when m is equal to the zeros of the Bessel function. By increasing the vibration amplitude b of the mirror M_3, the degree of phase modulation causing the first, the second and the third vanishing of the fringe contrast are measured. Thus, we can get the amplitude b corresponding to m = 2.40, 5.52, 8.63 or 11.79, etc. .

The relative phase between the modulation of the reference beam and the object vibration is fixed to zero by adjusting the phase difference between the two outputs of the oscillator OSC.

Now the hologram in this method is recorded by use of the phase modulation reference beam. The intensity $I_2(r)$ of the image from the hologram becomes

$$I_2(r) = I_0(r) \ J_0^2(p-m). \qquad (5)$$

The intensity is proportional to the function

$$F(p,m) = J_0^2(p-m), \qquad (6)$$

which depends both on the vibration amplitude of the object and that of the mirror M_3 for the reference beam modulation, as given by equations (2) and (4), respectively.

Figure 2 shows the function $F(p,m)$ for m=0 and m=10. The curve for m=0 is the characteristic fringe function $J_0^2(p)$ of the time average method and that for m=10 is shown as an example of the characteristic fringe function of the sinusoidal phase modulation method.

The time averaging double exposure method

In this method a double exposure
hologram is recorded firstly by the
time average method and secondly by
the sinusoidal phase modulation method
where the value of the degree of phase
modulation m is equal to 7.015.

For the phase modulation of reference beam, the Michelson interferometer
is utilized as in the sinusoidal phase
modulation method. The property of
the zero-order Bessel function is utilized
that the mean between the second and
the third zeros is nearly equal to 7.015.
By increasing the vibration amplitude b
of the mirror, the values of the amplitudes corresponding to the second and
the third zeros of the fringe contrast
are measured. When the vibration amplitude b is set equal to the mean between
two values, the degree of phase modulation becomes nearly 7.015. The relative phase between the phase modulation
of the reference beam and the object
vibration is fixed zero by adjusting the
phase difference between the two outputs
of the oscillator OSC.

When a hologram is made by double
exposures where the first and second
exposure are given with and without
reference beam modulation, respectively,
we can obtain the reconstructed image
of intensity $I_3(r)$ given by

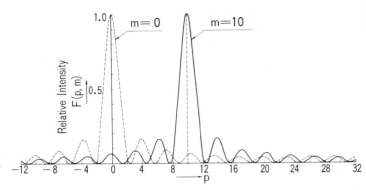

Figure 2. Characteristic function for the
sinusoidal phase modulation
method.

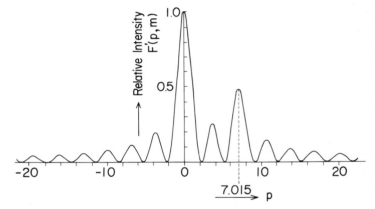

Figure 3. Characteristic function for the
time averaging double exposure
method.

$$I_3(r) = I_O(r) \ F'(p,m). \qquad (7)$$

where,

$$F'(p,m) = J_O{}^2(p) + J_O{}^2(p-m)/4 + J_O(p) \ J_O(p-m). \qquad m = 7.015. \qquad (8)$$

Figure 3 illustrates the function $F'(p,m)$. The curve represents the intensity variation
of the reconstructed image with The variable p   is proportional   to the amplitude
of the object vibration. The brightest fringe corresponds to the nodal line. The second
brightest fringe appears only on the positive p   that   corresponds to the object region
vibrating in the same phase with the mirror M_3 for the reference beam modulation.

The dark fringes correspond to the vibration amplitudes of

$$p = \begin{cases} 2.2, \ 5.1, \ 9.0, \ 12.1, \ 15.3, \ \ldots\ldots \\ -2.4, -5.4, -8, 5, -11.6, \ 14.7, \ \ldots\ldots \ . \end{cases} \qquad (9)$$

Typical examples of vibration mode of steam turbine blades

The vibration test by the cw holography has been carried out to determine the vibrational mode shapes of steam turbine blades for all resonance frequencies from about 100 Hz
to about 20 kHz, since this is the typical range over which excitation forces might be
present in steam turbines. The vibration modes of lower frequencies in the range are usually
measurable with conventional methods by use of an accelerometer, but those of higher frequencies of the range are very difficult to measure by other methods than holography.

The first step of experimental procedure for cw holography test is to tune the shaker
drive frequency until an accelerometer indicates a maximum in amplitude. Then, holograms
are made by the time average and the sinusoidal phase modulation method at each resonant
frequency. For the measurement of turbine blades with complicated mode shapes, the time
averaging double exposure method is used.

The reconstructed image obtained by the time average method is shown in Figure 4(a). It

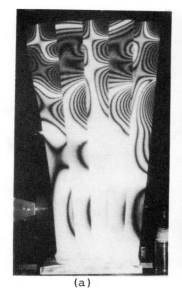  

| (a) | (b) | (c) |

Figure 4. Higher vibration mode pattern of blades group of steam turbine obtained by the time average method (a), by the sinusoidal phase modulation method (b), and by the time averaging double exposure method (c).

provides the contours of vibration amplitude . The brightest fringes correspond to nodes of vibration. However, the time average method gives no information on the relative phase of vibrating objects.

The reconstructed image obtained by the sinusoidal phase modulation method is shown in Fig.4(b). It provides the information on relative phase of vibrating objects. The brightest fringes show the regions which are vibrating in phase. Therefore we can determine the resonant vibration mode shape from the reconstructed images of both Figures 4(a) and 4(b). However, the comparison between the two images is very troublesome and requires much time in the case of complicated mode shapes.

The reconstructed image shown in Figure 4(c) is obtained by the time average double exposure method and provides information on the position of nodal lines, the contours of vibration amplitude and the relative phase of vibration. The brightest fringes show the nodal lines, and the dark fringes represent the contours of the vibration amplitude. The second brightest fringes which can be observed in certain regions formed between the nodal lines show that these regions are vibrating in the same phase as the mirror for the reference beam modulation. These regions show that the regions are on the positive p side of the characteristic fringe function of Figure 3. On the other hand, the regions without the second brightest fringes correspond negative p. They are vibrating 180°out of phase compared with the regions of the second brightest fringes.

The results of the phase determination are illustrated in Figure 5. The shadowed regions with dotted lines which represent the second brightest fringes are vibrating in phase ,while the other regions are vibrating 180° out of phase.

Applications of pw holography

Experimental apparatus and fringes interpretation in the pw holography

A schematic diagram of the apparatus of pw holography employed in our laboratory is illustrated in Figure 6. The pulsed laser beam emitted from a Ruby laser (Hardon Korad, Model special K1200QDH) is divided by a beam

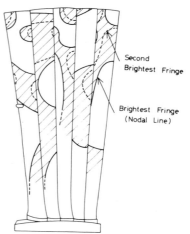

Second Brightest Fringe

Brightest Fringe (Nodal Line)

Figure 5. The schematic representation of Figure 4(c).

splitter BS into two beams, one of which is used to illuminate the vibrating object and the other is used as reference beam.

The essential of the optical system is to use the pulsed laser as light source instead of the continuous wave laser as shown in Figure 1.

Hologram is made by the double-pulse laser beam which is emitted at selected times during the vibration cycle of a vibrating object. The resulting fringes in the reconstructed image show the contours of the difference of displacements of the object at the two instants of time and are given by

$$I(r) = I_0(r) \cdot \cos^2[\frac{2\pi}{\lambda}d(r) \ (\cos \theta_1 + \cos \theta_2)], \ (8)$$

where $d(r)$ is difference of vibration amplitude of a point r at the two instants of time and the other notations are the same as those in equation (2). Consequently, the difference $d(r)$ of vibration amplitude at any region where dark fringes appear is expressed as

$$d(r) = (2n-1)\lambda/2(\cos \theta_1 + \cos \theta_2) \qquad (9)$$

where n is the order of fringes.

Vibration tests in factory environment

The pw holography has been employed to measure typical resonant vibration mode shapes of compressor vanes in a factory. The vanes were approximately 2.5m in diameter.

Figure 7 shows general view of the test equipment for the vibration studies of the compressor vanes. Typical test results are presented in Figure 8 (a),(b); the former representing the lower frequency vibration mode and the latter representing the higher frequency vibration mode.

As an example of the applications of pw holography to large scale products, the results of vibration mode analysis of the car body are presented in Figure 9. The size of the car body is approximately 4.5m in length and 1.4m in width. As shown in Figure 9, it is clear that the pw holography is applicable to the measurements of the amplitude distribution over the entire surface of about 4m long object at one time in the factory environment.

Measurement of propagating transverse wave in diesel engines

In the design and redesign of diesel engines, knowledge of the dynamic behavior of engine structures is extremely important from the viewpoint of the control of noises from the vibrating engine. As one of many measurement methods, the pw holography is utilized to measure propagating transverse waves in diesel engines excited by impact force. The optical arrangement used to make holograms is shown in Figure 6. As electronic trigger system to emit the laser pulse at the desired moment of time, Aprahamian's method [6] was used as shown in Figure 10. In Figure 10, the sequence of events is as follows:
    (1) The hammer is released from the magnet

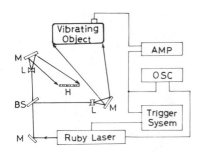

Figure 6. Experimental set up of pw holography for practical vibration studies.

Figure 7. Photograph of the experimental set up in the factory.

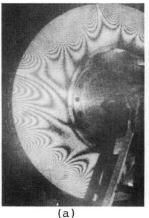 

(a)   (b)

Figure 8. Vibration mode of a compressor vane obtained by pw holography: typical examples of lower frequency mode (a) and higher frequency mode (b).

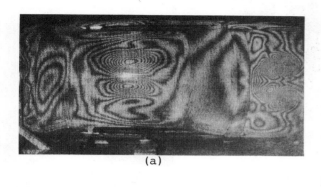

(a)

(b)

Figure 9. Typical examples of vibration
mode of a vehicle excited by
an electro-mechanical shaker.
Views of roof(a) and door(b).

Figure 10. An arrangement of trigger system
for making a hologram of propagating transverse waves in a
crank case.

C.H. : Cylinder Head
C.L. : Cylinder Liner
C.R. : Conn Rod
C.S. : Crank Shaft
M.B. : Main Bearing
F. G. : Force Gauge
ΔT :Time after Impact

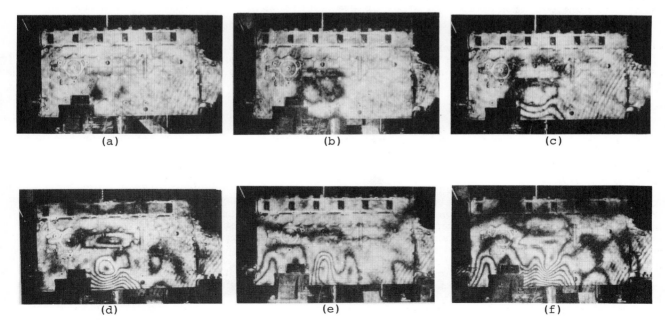

(a)   (b)   (c)

(d)   (e)   (f)

Figure 11. Reconstructed image showing the transverse waves in a crank case at
200(a), 224(b), 312(c), 424(d), 650(e), 754μ sec(f) after impact by
a hammer.

Figure 12. Photograph of the experimental set up in the test bench.

(a)   (b)

Figure 13. Typical vibration mode pattern of a diesel engine excited by firing at 3,000 rpm: lower (a) and higher (b) frequency vibration modes.

and gives an impact to the specimen.

(2) The hammer intercepts the light beam emitted from a He-Ne laser. The interception of the light beam causes the photocell to generate a signal. The signal enters the time-delay circuit and, after a suitable delay, the time-delay circuit sends out a signal to trigger operation of a Ruby laser.

(3) The hammer continues its motion until it makes contact with the steel pipe on the specimen. Timing of the contact of the hammer with the specimen is determined by the force gauge installed on the surface of the specimen.

(4) Emission timing of the double-pulse laser for exposing holograms is adjusted by the interval of the double pulses and time length of the time-delay circuit in Step 2.

Typical test results are shown in Figure 11. Figure 11(a) shows the transverse wave which appeared on the crank case at 200 μ sec after impact: the time after impact is monitored by use of the force gauge signals. The other pictures in Figure 11 show the propagation process of transverse wave excited by the impact.

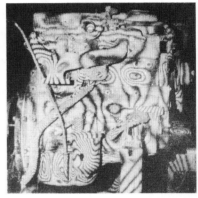

Figure 14. Example of vibration mode pattern of a diesel engine excited by motoring at 2,300 rpm.

These pictures provide us with informations on the dynamic behavior of transient vibration which can not be measured by means other than holography. The results have been utilized for the considering the methods for noise suppression of diesel engines.

Measurement of complex vibration modes of a running engines

To reduce noises of diesel engines, it is necessary to investigate actual vibrational behavior of their structures under operating conditions. The pw holography has been utilized to measure complex vibration modes of running engines which consist of various frequencies.

Figure 12 shows an experimental set up of the pw holography used. As a signal to trigger the double-pulse laser, the output signal of an accelerometer installed on the engine was used. The output signal was amplified and led to a spectrum analyzer. The analyzer detected the resonant frequencies in the output signal. A component of the output signal was passed through a RC filter and was utilized as a signal of the trigger system.

Typical test results are shown in Figure 13(a) and (b): the former representing the lower frequency vibration mode and the later representing the higher frequency vibration mode.

Figure 14 shows the higher frequency vibration mode of another type of engine under conditions of motoring operation.

The pictures in Figures 13 and 14 have been utilized for considering the noise reduction and for verifying the calculated resonant frequencies and mode shapes in engines design.

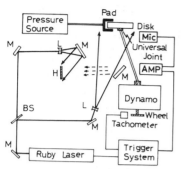

Figure 15. Experimental set up of pw holography to measure vibration modes of a squealing brake disk

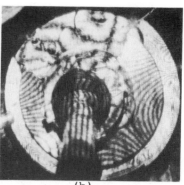

(a)   (b)

Figure 16. Photographs of a rotating brake disk(a) and a vibration mode of a squealing brake disk obtained by pw holography(b).

Vibration mode of a squealing disk brake

The stick-slip motion of brake pads of the brake disk gives rise to a squeal. The phenomenon which depends on friction force, temperature and sliding velocity,etc., is a self-excited instability of mechanical vibration of the components.

In order to investigate the characteristics of the phenomenon, the knowledge of the vibration mode shapes is required. Therefore, the pw holography was utilized to measure the vibration mode of a disk brake during operation on a test stand.

The experimental set up is shown in Figure 15. As an electric trigger system for emission of double pulses, the method adopted by Felske and Happe [5] was utilized. Holograms were made by using an optical arrangement shown in Figure 15, which gives the start signal only if two conditions are satisfied: the sound level must be sufficiently high and the speed of rotation must be less than 10 RPM. Figure 16(a) shows a photograph of the specimen. Figure 16(b) shows typical vibration mode of the brake disk during braking action. It indicate that the vibration mode shape is the 3 nodal diameter mode. The holographic data have been utilized for considerration of squealing noise reduction and evaluation of predicted mode shapes in the modification of brake structures.

Conclusion

The applications described in this paper have led us to the conclusions that holographic interferometry is an extremely useful technique for the practical vibration study of turbomachinery such as steam turbines, gas turbines and various compressors, and that it is also very powerful for noise diagnosis of internal combustion engines, car bodies and other vehicle components.

References

1. Hockley, B.S. and Butters, J.N., Holography as a Routine Methods of Vibration Analysis, j. Mech. Eng. Sci., Vol.12, PP.37-47, 1970.
2. Fagan, W.F., Beeck, M.A. and Kreitlow, H., The Holographic Vibration Analysis of Rotating Objects using a Reflective Image Derotator, Optices and Lasers in Engineering, Vol.2, PP.21-32, 1981.
3. Hockley, B.S., Ford, R.A.J. and Foord, C.A., Measurement of Fan Vibration Using Double Pulse Holography, ASME Paper No. 78-GT-111, 1978.
4. Felske, A. and Happe.A., Vibration Analysis by Double Pulsed Laser Holography, SAE Paper No. 770030, 1977.
5. Felske, A., Happe, G. and Matthai, H., Oscillations in Squealing Disk Brakes-Analysis of Vibration Modes by Holographic Interferometry, SAE Paper No. 780333, 1978.
6. Aprahamian, R., Evensen, D.A., Mixson, J.S. and Wright, J.E., Application of Pulsed Holographic Interferometry to the Measurement of Propagating Transverse Waves in Beams, Exp. Mech., Vol.11, PP.309-314, 1971.
7. Murata,M., A New Method of Holographic Interferometry for Practical Vibration Studies, Proc. ICO Conf. Opt. Methods in Sci. and Ind. Meas., Tokyo, PP.271-275, 1974.

Vibration analysis of automotive structures using holographic interferometry

G. M. Brown

Research Staff
Ford Motor Company
Dearborn, Michigan 48121 USA

R. R. Wales

Car Engineering
Advanced Vehicle Engineering and Technology
Ford Motor Company
Dearborn, Michigan 48121 USA

Abstract

Since 1979, Ford Motor Company has been developing holographic interferometry to supplement more conventional test methods to measure vehicle component vibrations. An Apollo PHK-1 Double Pulse Holographic Laser System was employed to visualize a variety of complex vibration modes, primarily on current production and prototype powertrain components. Design improvements to reduce powertrain response to problem excitations have been determined through pulsed laser holography, and have, in several cases, been put into production in Ford vehicles. Whole-field definition of vibration related deflections provide continuity of information missed by accelerometer/modal analysis techniques. Certain operational problems, common among pulsed ruby holographic lasers, have required ongoing hardware and electronics improvements to minimize system downtime.

Real-time, time-averaged and stroboscopic C. W. laser holographic techniques are being developed at Ford to complement the double pulse capabilities and provide rapid identification of modal frequencies and nodal lines for analysis of powertrain structures. Methods for mounting and exciting powertrains to minimize rigid body motions are discussed. Work at Ford will continue toward development of C. W. holographic techniques to provide refined test methodology dedicated to noise and vibration diagnostics with particular emphasis on semi-automated methods for quantifying displacement and relative phase using high resolution digitized video and computers. Continued use of refined pulsed and CW laser holographic interferometry for the analysis of complex structure vibrations seems assured.

Introduction

In 1978, state-of-the-art technologies for improving nondestructive testing (NDT) methodology were surveyed with particular emphasis on vibration and deflection analysis of vehicle components. To supplement established methodology largely based on measurements at individual points, such as accelerometer or directional microphone data, there is significant need to define whole field deformations for automotive structures. Also needed are improvements in experimental data for refinement of modal analysis testing and computer structural models. Optical interferometric techniques, of which laser holographic interferometry is the most sensitive, were investigated. Guided by early experimental efforts to employ holographic methods in vibration analysis of real world test subjects[1,2], we purchased a double pulse ruby laser holocamera (model PHK-1) developed by Rottenkolber GmbH and marketed by Apollo Lasers, Inc. The Rottenkolber/Apollo System (Figure 1) was among the first commercial double pulsed holographic laser systems to be offered with integral triggering controls to synchronize pulse timing to test object vibration. From the earliest applications studies--beginning with system delivery in 1979, through current use in survey studies of whole powertrain vibrations, higher order localized vibrations, and static deflections under load--pulsed laser holography has become a valuable addition to established NDT methods at Ford.

To realize the advantages of viewing complex modeshapes of vehicle structures in time-average and real-time interferograms, development of continuous wave holography test methodology began in 1981. A program was initiated to demonstrate C. W. capability for imaging vibration modeshapes of vehicle powertrains with high-resolution, large depth-of-field and large field-of-view. Efforts continue, with considerable success to date, in developing qualitative noise and vibration diagnostic capability and in initiating semi-automated methods for quantifying displacement and relative phase in interferograms.

Double pulse laser holographic interferometry (DPLHI)

Use of DPLHI to image vibration modeshapes on passenger car powertrain structures offers the opportunity to understand complex modal patterns that occur between accelerometer data points in a modal analysis. The whole field-of-view of normal displacement for a portion of the primary vibration half cycle is available for qualitative interpretation. For instance, the relative dynamic stiffness over all viewable points on a complex, nonlinear test structure may be assessed.

Laboratory considerations

A vital consideration to any vibration test methology is that the component or system be fixtured and excited in the laboratory in a manner as closely representing the in-vehicle loadings and boundary conditions as possible. For large, complex structures such as powertrains, the necessity of exciting vibration modes that are present in normal vehicle operation requires careful use of electromagnetic shaker input and control equipment. Since powertrains are mounted through elastomeric materials to vehicle body structures, duplication of this mounting in the holography laboratory is essential. However, elasto-meric mounting of massive test structures permits rigid body motions which degrade modeshape fringe patterns. Pulsed laser holography, with precise timing of laser firing to the monitored vibration response of test structures, can achieve good definition of major structure modes.

DPLHI is uniquely suited to the study of higher order modes that tend to be excited by internal components such as gearsets or bearing reactions. Those modes with the least damping will display re-occuring localized patterns of closed fringes in interferograms made at the instant of impacting the structure with a hammer. Again, the ability to make interferograms with variable pulse separations permits the hammer impact interferogram to capture the displacement character of relatively higher or lower orders.

In an idealized holography laboratory, powertrain structures should be self-powered through an absorbtion dynamometer and transmissions or transaxles driven by electric motors through decoupled shafts. Since holographic laboratories, including ours, usually do not have access to such equipment, the importance of verifying vibration measurements on laboratory test setups with prior measurements made in-vehicle is significant. High image quality double pulse interferograms are of little value as vibration data if the fringe patterns are summations of excess rigid body motion or other coupled and resonant modes not of interest.

Operational issues with DPLHI

Operational problems inherent in pulsed ruby holographic lasers have largely prevented turn-key operation. With the Apollo PHK-1 system, our experience at Ford has revealed the following, much of which applies to commercially available lasers of earlier vintage:
. Cavity alignment deteriorates with changes in room temperature
. Frequent triggering miscues cause laser firing at an undesired point in the vibration cycle
. Optimum sensitization and development of holofilm is difficult
. Limits to pulse separation prevent capture of full displacement fringe patterns for frequencies below 1000 Hz.

Considerable effort has been expended in the past two years to upgrade components of the PHK-1 system. The addition of a temperature controlled rear etalon greatly reduced multi-mode fringe occurrence. A thermoelectric camera air cooler and a unique cycle timer for power supply charging further reduced adverse effects of room temperature changes.

Powertrain studies with DPLHI

Three studies of deflections on powertrain components are presented to illustrate the most representative techniques employed.

Manual front wheel drive (FWD) transaxle. This study was performed to identify a source of gear noise occurring at 2700 RPM in 4th gear. The fixturing included an electric motor to drive the transaxle input shaft permitting the gears to generate the vibrations to cause localized case resonances. Figure 2 is a double pulse interferogram triggered by a filtered accelerometer signal. A characteristic bullseye was seen indicating a potential region of radiated noise on the case. A local stiffening rib was added to the housing as close to the bullseye as was feasible (Figure 3). From the original viewpoint, holograms now showed disruption of the local resonance to gear tooth contact excitation (Figure 4). The successful production implementation of the rib followed verification of its effectiveness in subsequent vehicle noise evaluations.

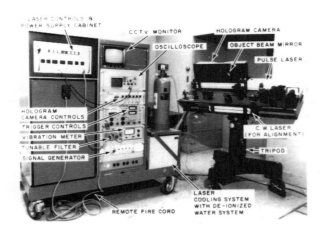

Figure 1. Ford double pulse holographic system (Apollo model PHK-1).

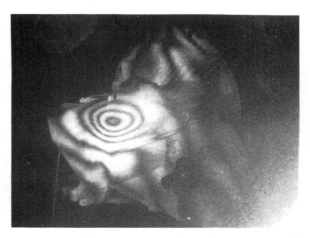

Figure 2. Interferogram of gear-excited resonance on transaxle housing.

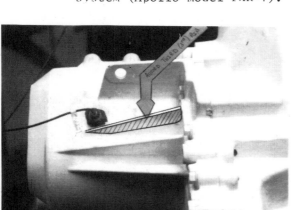

Figure 3. Stiffening rib added to reduce resonant response.

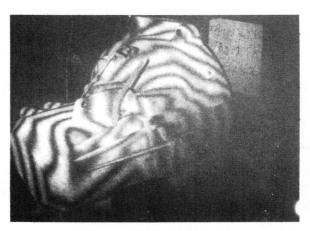

Figure 4. Resonant response eliminated under original excitation conditions.

Engine block noise prevention. To help circumvent potential noise radiating surfaces on a new engine block design, a holographic survey was conducted using both electromagnetic shaker and hammer blow excitations. A front wheel drive (FWD) powertrain was fixtured on vehicle engine mounts (Figure 5) and acceleration response spectrums generated to determine key modes. Holograms were made for these modes focussing on the side of the engine block and employing first surface mirrors to access block surfaces hidden by manifolds. To increase fringe count for low frequency modes, the holocamera was successfully operated in single pulse mode, retiming the lasing sequence to capture peak to peak modal displacement between charge cycles. The holograms below show modal deflection contours for the view of the exhaust manifold side of the engine in Figure 6. The single pulse mode interferogram of Figure 7 contains high contrast fringes with negligible rigid body motion. Figure 8 shows a complex transient response from hammer impact excitation. Repeated deflection patterns over a series of interferograms indicate regions for further design improvement. Noise evaluations are currently underway to determine if prototype engine blocks with rib structure added to attenuate deflections seen in the interferograms produce in-vehicle noise and thereby warrant production implementation.

Transaxle static deflections. This study was undertaken to establish regions of high deflection under torque loadings for a prototype manual FWD transaxle. A rigid fixture and torque loading apparatus was set up (Figure 9) and the holocamera positioned to view all surfaces including a top view by means of a first surface mirror (Figure 10). Again the laser was employed in single pulse mode with incremental changes in torque loadings made while recharging for the second pulse. Typical results are shown for two views of the final drive portion of the housing (Figures 11 and 12). These interferograms were employed in the later design phases of this transaxle to determine regions of the housing requiring increased stiffness. This was accomplished by small changes in casting wall thickness and selective internal ribbing.

Figure 5. FWD powertrain holographic test set-up.

Figure 6. Holocamera view of side of engine block under study.

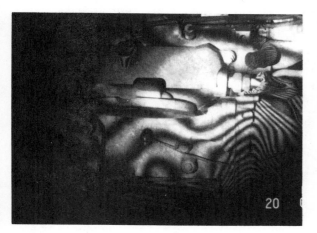

Figure 7. Single pulse mode interferogram of modal deformation at 145 Hz.

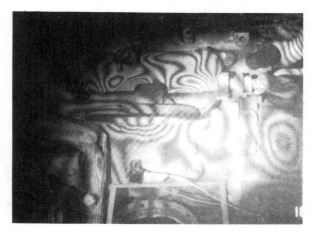

Figure 8. Double pulse (120 µs) interferogram of response to hammer impact.

Figure 9. Setup for study of torque load deflections of FWD transaxle.

Figure 10. Fixture and mirror positioned for imaging top view of transaxle.

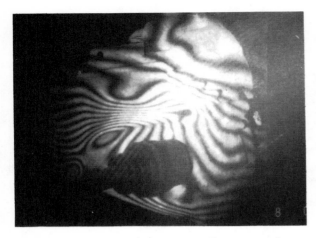

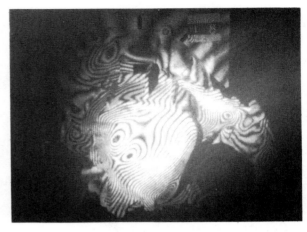

Figure 11. Interferogram of transaxle case (80 to 85 lb-ft torque).

Figure 12. Interferogram of tranxaxle case (85 to 112 lb-ft torque).

Continuous wave laser holographic interferometry (CWLHI)

Several limitations of DPLHI prompted the introduction of CWLHI for the study of powertrain vibrations. First, the maximum useful pulse separation of the double pulse laser is about 500 microseconds; thus, to observe maximum displacements of structures vibrating at less than 1000 Hz, two single pulses separated by a one minute cycle time are required. The two pulses must be synchronized with excitation waveform and separated by the cycle time plus half the vibration period; during this time, significant rigid body movement frequently occurs which limits the yield of useable holograms. Second, normal double pulse holography does not identify nodal regions as does time-average holography.[3,5,6] Third, the operational problems cited above severely limited the yield of useable interferograms. Fourth, the real-time[4,5,6] capability of CWLHI would permit the easy identification of structural resonances. Our objectives were to measure powertrain bending modes as well as attached component resonances.

Laboratory considerations

Holographic interferometer. A real-time/time-average holographic interferometer (Figure 13) was implemented which has: a large field-of-view (52°), sufficient holographic depth-of-field, simultaneous orthogonal views of a full size powertrain, efficient object light collection and the capability of working in a 50 to 3000 Hz frequency band. A Newport 4'x8'x18" pneumatic vibration isolation system is used as the stable platform. The optics and powertrain fixturing are mounted to the top surface; a Spectra Physics 125 He-Ne laser is placed on outrigger brackets along one edge of the surface plate. A variable ratio beam splitter is used to adjust beam ratios (2:1 to 6:1). The object beam is expanded thru a 60x spacial filter, reflected off a 5"x7" 50/50 beam splitter mounted directly in front of the hologram fluid gate and at 45° to the viewing axis. 3M retro-reflective paint coats the powertrain and sends the light back towards the point of origin with a ~10° spread. Normally, one to five second exposures are required using AGFA 10E75 in the fluid gate. Overexposure and underdevelopment is used for real-time holography to keep the hologram thin. The kinematic plate holder is removed for processing and can be tilt adjusted about two axes for a near zero order fringe condition. Initially, an etalon was not available for the SP-125 and an unequal path (6") Michelson interferometer was used in the reference beam to give a dual collinear spacially delayed reference beam which doubled the depth-of-field at the expense of lower hologram diffraction efficiency. Currently an airspaced intercavity etalon is used in a front extension of the SP-125; this provides adequate frequency stability and sufficient coherence length so that the reference path length need not be adjusted. Hologram reconstructions are usually recorded on 35mm Plus-X film or on video tape. Large first surface mirrors are mounted under the powertrain to facilitate simultaneous orthogonal views.

Powertrain mounting. Preferably, the powertrain should be mounted on normal elastomer engine mounts including significant vehicle support structures which are in turn rigidly mounted to the surface plate. Practically speaking, when this is done, low frequency rigid body motions prevent quality holograms in the 1-5 second exposure time. Thus, we have found it necessary to use ball or kinematic mounts (which permit bending but no rigid body movement), various thickness and durometer elastomer pads and normal engine mounts with appropriate vehicle structure in whatever combination that does not drastically compromise realistic mode shapes and frequencies. Mounting or boundary conditions significantly affect the modes and the need for realistic conditions cannot be over emphasized.

Accelerometers and spectrum analyzers are used to survey various mounting alternatives until some combination yields frequency spectra essentially the same as that on normal engine mounts and permits quality holography. It is also essential to decouple the powertrain from the holographic interferometer to prevent significant vibration of the optics and table.

Excitation. Convention B&K electrodynamic excitation equipment is used, including constant force feedback and frequency scanning. It is essential that the exciter be decoupled from the isolation table to minimize energy transfer to the optics. This has been accomplished by mounting the exciter on the floor or suspending it from an overhead crane. The exciter shaft is bolted to the powertrain; the attachment point and translation direction are chosen to simulate unbalanced engine forces. Location of the exciter can affect the mode shapes. Generally, in a complex structure, vibrations introduced in one direction can couple into vibrations in other directions.

Operational issues with CWLHI

At present, the few seconds exposure time often prohibits the making of quality holograms of powertrains supported by normal engine mounts. This situation dictates stiffening or damping the mounts until quality holograms are made; the degree to which this changes mode shapes and frequencies determines if the information is useable. This problem will be alleviated when a new 5 watt Argon-ion laser is incorporated. The powertrain can then be more uniformly illuminated over the whole field and the exposure time can be reduced to less than 1/30 second.

Powertrain studies with CWLHI

Studies of two different powertrain types are discussed and are representative of the present capabilities of CWLHI.

FWD transaxle powertrain. The objectives of this study were to (1) demonstrate the capabilities of CWLHI for measuring powertrain bending modes and attached component resonances and (2) investigate the effects of mounting and excitation. Figure 14 shows a FWD powertrain as seen from the hologram position. High quality holograms are obtained of a direct full side view and a partial bottom view thru a mirror. A dual spacially delayed colinear reference beam was used to double the normal 6" depth-of-field from a SP-125 laser. The powertrain is shown mounted on ball mounts instead of the normal engine mounts. The exciter was suspended from an overhead crane and attached at a rear valve cover bolt location. Real-time holograms, acceleration and force waveforms were used to identify modes as a function of exciter frequency. Figure 15 is a time-average interferogram of the first order bending mode at 154 Hz with 10 pounds of exciter force with a starter brace assembly attached. The corresponding mode without the brace attached occurs at 146 Hz (Figure 16). Both interferograms show a forced resonance of the attached air-conditioner unit. The bending mode is quite similar in both cases but considerably greater motion occurs in the bell housing and transmission without the brace. Figure 17 is a photograph of the same powertrain mounted on normal engine mounts attached to a subframe; a mirror again shows a partial bottom view. Considerable difficulty was encountered in obtaining readable interferograms due to significant rigid body motions during the one second exposure. Figure 18 shows a time average interferogram obtained after normal working hours using 4 pounds of force at 138 Hz with no starter brace. This shows the "free-free" first order bending mode and true nodal lines, whereas "apparent" nodes are seen in Figure 16 due to the ball mounts. Note that the motion of the starter and bell housing are similar except for "node" locations and amplitudes. We have concluded that the powertrains must be mounted on elastomeric mounts, preferably normal engine mounts, in order to get modes representative of in-vehicle conditions. Simple techniques for compensation of the unwanted low frequency rigid body motions have proved useless. Exposure time must be short enough to "stop" the low frequency motion while still long enough for time averaging the frequencies of interest.

Rear wheel drive (RWD) powertrain. CWLHI is being used to study the effects on first and second order bending of various structural designs on a four cylinder RWD manual powertrain. Of interest are: (1) starter braces, (2) magnesium vs aluminum transmission housing, (3) transmission to bottom of engine braces, (4) cast structural oil pan and, (5) drive shaft design. The overall set up is shown in Figure 13. Rubber pads were used for the front engine mounts while the normal rubber mount and structural cross member are used for the transmission extension mount. The drive shaft is attached and decoupled from the floor with foam rubber. Figure 19 is a photograph as seen from the hologram position. A large mirror assembly provides a nearly full bottom view. The exciter is attached vertically to the bottom of the transmission extension. Although these studies are not complete, we included some early results of a starter brace to show the capabilities of the technique. Figure 20 shows a time average interferogram of the first order bending mode at 93 Hz and one pound exciter force. Figure 21 is the corresponding second order mode at

109.5 Hz and one pound. Figure 22 shows the first order mode at 109.5 Hz and one pound as a result of attaching a starter motor brace. Note, that the significant motion of the starter motor in Figure 20 is reduced. Unfortunately, the direct view of the transmission extension appears dark in the interferogram; this results from an irrationally related low frequency low amplitude lateral movement.

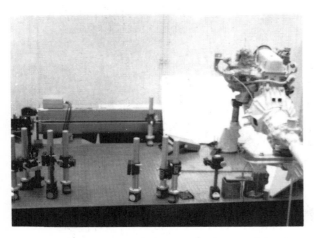

Figure 13. Ford CW laser holographic interferometer.

Figure 14. View of FWD powertrain on ball mounts as seen from the hologram.

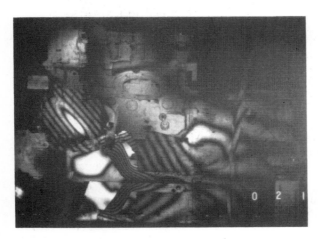

Figure 15. First order bending mode (with starter brace, 154 Hz, 10-lbs).

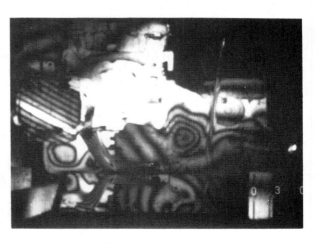

Figure 16. First order bending mode (no starter brace 146 Hz, 10-lbs).

Figure 17. View of FWD powertrain on rubber engine mounts and support frame.

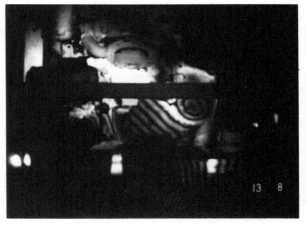

Figure 18. First order bending mode (no starter brace, 138 Hz, 4-lbs).

Figure 19. View of RWD powertrain on rubber mounts as seen from the hologram.

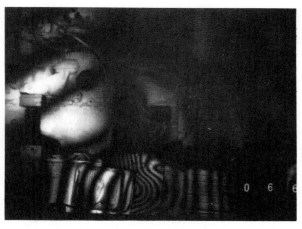

Figure 20. RWD powertrain first order bending mode (93 Hz, 1-1b).

Figure 21. Second order bending mode (no starter brace 109.5 Hz, 1-1b).

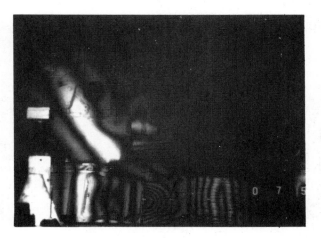

Figure 22. First order bending mode (with starter brace, 109.5 Hz, 1-1b).

Conclusions

The double pulse and continuous wave laser holographic methods provide significant qualitative information about mode shapes and structural deformation. This information has been used along with other data to indicate structural changes for improved performance. Future development efforts will emphasize semiautomated quantitative analysis using video/computer techniques. Continued use of these techniques as an integral part of product development seems assured.

Acknowledgements

The authors wish to thank their co-workers Mitchell Marchi and Carl Strauss of Advanced Vehicle Engineering and Technology for their participation in this work.

References

1. Hockley, B. S. and Butters, J. N., "Holography as a Routine Method of Vibration Analysis," Journal Mechanical Engineering Science, Vol. 12 No. 1, pp 37-47, 1970.

2. Felski, A. and Happe, A., "Vibration Analysis by Double Pulsed Laser Holography," SAE Publication No. 770030, 1977.

3. Powell, R. L. and Steson, K. A., "Interferometric Analysis by Wavefront Reconstruction," J. Opt. Soc. Am., Vol. 55, 1593-1598, 1965.

4. Stetson, K. A. and Powell, R. L., "Interferometric Hologram Evaluation and Real-time Vibration Analysis of Diffuse Objects," J. Opt. Soc. Am., Vol. 55, 1694-1695, 1965.

5. Brown, G. M., Grant, R. M. and Stroke, G. W., "Theory of Holographic Interferometry," J. Acoust. Soc. Am., Vol. 45, No. 5, 1166-1179, 1969.

6. Vest, C. M., Holographic Interferometry, John Wiley and Sons 1979.

Vibration analysis of an 8-cylinder V-engine by time-averaged holographic interferometry

H. G. Leis

Daimler-Benz AG, D-7000 Stuttgart 60 (Untertürkheim)

Abstract

Time-averaged holography is used to display and discuss the basic natural vibrations of the crankcase of a commercial vehicle.

Introduction

One challenge of today's development of motor vehicles lies in the reduction of their noise emission.

Under this aspect it is desirable to examine the natural vibrations of the engine, because these vibrations produce noise:

(1) directly by generating sound waves in the adjacent air,

(2) indirectly by exciting units connected with the engine.

This contribution describes the natural vibrations of the crankcase of an 8-cylinder V-engine - as a first step towards determining the vibration response of the complete engine. These engines are used in commercial vehicles.

Time-averaged holography [1] has proved to be an excellent test system for recording the vibrational behaviour of an object at constant frequency, especially for the natural vibrations of an object at its various natural frequencies. The local vibration amplitudes of the observed surface are made visible by an interference phenomenon in the form of contour lines of the amplitude superimposed on the object.

The dark and light interference fringes describe areas of constant component of the local vibration amplitude in the direction of the so-called sensitivity vector, which is defined as the angle bisector between the two directions of illumination and observation of the optical setup. This sensitivity vector is a function of the observed object point. For the setup used it can be regarded to be perpendicular to the image plane for the whole observed object surface.

The correlation between the interference order and the amount of the local vibration amplitude is presented in figure 2. The change of the local amplitude between two adjoining interferences fringes approximately [1],[2] is given by $\lambda/4$ with $\lambda(= 514$ nm), the wavelength of the used Argon-Laser.

Experimental setup

—Location of pick-up

—Point of application of force

—Mirror enabling the view of the one front end of the crankcase

Position 1   Position 2

Figure 1: Test setup

The crankcase is supported by 3 steel pins which are bolted to its front end and to 3 steel cylinders respectively. This assembly is put onto a 50 x 1000 x 1000 mm steel plate.

Two mounting positions are examined. Position 1 enables the following views: (1) one complete side of the crankcase, (2) the bottom part of the opposite side, (3) a portion of one front end via a mirror put between the supporting steel pins. Position 2 enables the views: (1) the two cylinder blocks forming the top of the crankcase and (2) another part of the front side via the mirror. The vibration generator is an electrodynamic shaker. The force is applied to the crankcase wall between two crankshaft bearings (see also Fig. 5). With constant amplitude and sweeping frequency of the exciting force, the various natural frequencies of the assembly correspond to a maximum input admittance, measured by means of a pickup which is positioned close to the point of application of force. To prevent the vibrating crankcase from exciting the optical setup for recording the holograms, the steel plate is isolated from the optical table by a 13 mm rubber mat.

The crankcase is painted in a white reflective colour: (1) to achieve a uniform reflection behaviour of the observed object surface, (2) to reduce the exposure time of the holographic plate.

Results

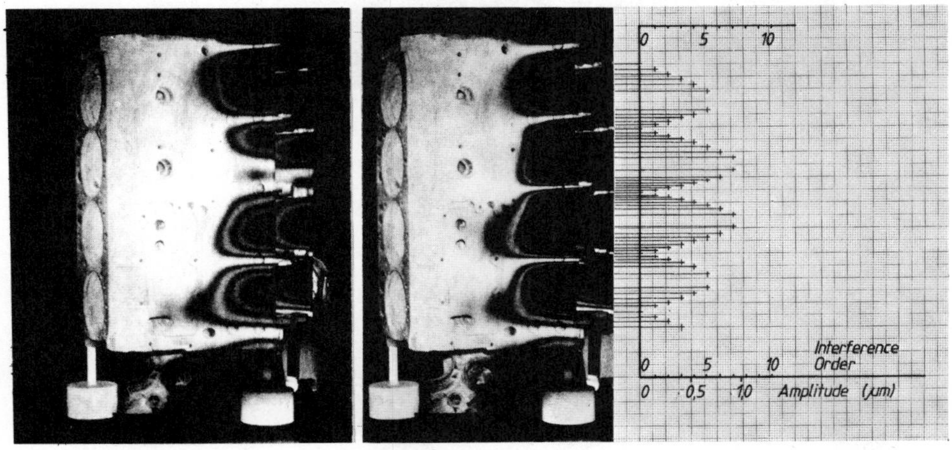

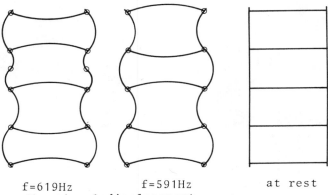

f=619Hz   f=591Hz

Figure 2: Diaphragm-type vibration of the crankcase wall between two crankshaft bearings

f=619Hz   f=591Hz   at rest
vibrational displacement
o indicating nodal points

Figure 3: Schematic vibrational behaviour of the lower edge of the crankcase and the crankshaft bearings

The contour lines indicate - with one exception at 619 Hz - that each wall between two crankshaft bearings vibrates in the basic mode of a rectangular membrane which is clamped on the side adjoining the cylinder block, hinged along the two crankschaft bearings and free at the bottom edge of the crankcase. The observable part of the front end crankshaft bearing also shows this fundamental vibrational mode.

To demonstrate this behaviour, the evaluation of the contour lines along the one bottom edge of the crankcase is plotted in the diagram: Nodal lines correspond to the areas of the object with maximum brightness.

On the cylinder block the vibration amplitudes are negligible relative to the amplitudes on the lower part of it. The exceptional vibrational behaviour at 619 Hz results from the coupled vibrations of the crankshaft bearings, schematically shown in figure 3.

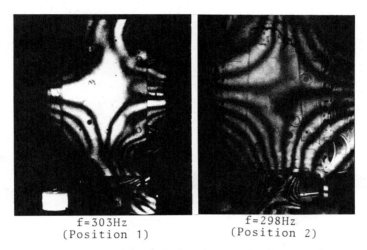

f=303Hz
(Position 1)

f=298Hz
(Position 2)

Figure 4: Torsional behaviour of the crankcase

The hyperbolic contour lines indicate a saddle-like vibration mode in position 1 and 2 - corresponding to a torsion of the crankcase. There are slight local disturbances of these contour lines at the lower edge of the crankcase. The pure torsional vibration is disturbed by the 3-point supporting mechanism of the crankcase. This support distinctly influences the vibration of the front end and of the supporting elements.

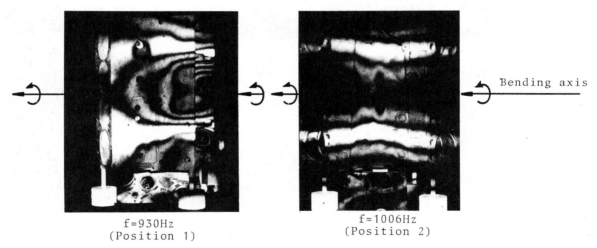

Bending axis

f=930Hz
(Position 1)

f=1006Hz
(Position 2)

Figure 5: Two bending modes of the crankcase

The contour lines at 930 Hz indicate a bending of the crankcase in position 1 about the indicated axis. This bending mode shows a decreasing amplitude from the lower to the upper part of the crankcase.

The point of application of force can be seen as a local antinode which is indicated by the concentric contour lines. At 1006 Hz, bending of the two cylinder blocks can be observed. The bending axis of the crankcase is turned by 90° relative to the bending axis at 930 Hz.

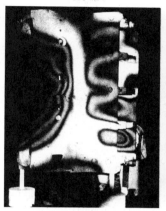

f=1090Hz
(Position 1)

f=1090Hz
(Position 2)

Figure 6: "Tuning fork" vibration of the 2 cylinder blocks

The two cylinder blocks vibrate in opposite phases like a tuning fork. The contour line pattern is not symmetrical on the two cylinder blocks which results from a certain asymmetry in the design and setup of the crankcase.

Applications

(1) This experimental approach for displaying natural vibrations can be used for the calibration of the corresponding finite element method calculations[3].

(2) The influence on the vibrational behaviour of design modifications or units mounted on the crankcase can be studied.

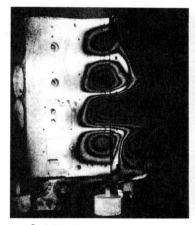
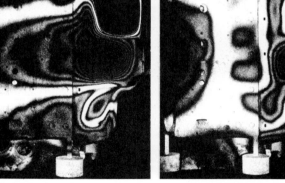

f=626;(591)Hz   f=924;(930)Hz   f=1095;(1090)Hz

Natural frequencies;frequencies without oil pan shown in brackets for comparison
Figure 7: Typical natural vibrations of the crankcase with mounted oil pan

As an example, figure 7 shows the influence of the mounted oil pan on typical vibrational modes. There is only a small influence of the oil pan on the natural frequencies and the natural vibrations of the crankcase. On the other hand, coupled vibrations of large amplitudes occur on the oil pan.

(3) Future research aims at a correlation between the holographically measured natural vibration modes with the corresponding natural frequencies on the one hand and the acoustically determined frequencies with a high noise level of a running machine on the other hand. From such a correlation design modifications result which are intended to reduce the high noise level at the corresponding frequency.

References

1. Collier, R. R., Burckhardt, C.B., Lin, L. H., Optical Holography, Academic Press New York 1971
2. Leis, H. G., Leipold, F., Stress-analysis by the combination of holographic interferometry and strain-gauge technology, Technisches Messen 49. Jahrgang 1982, Heft 7/8 p. 271-276
3. Dürr, W., Leis, H. G., Leipold, F., Eigenschwingungen eines Nutzfahrzeug-Motorblocks in V-Bauweise, Vergleich holografischer Messungen mit Finite-Element-Rechnungen, VDI-Berichte Nr. 456, 1982, p. 279-288

Possibility of extension of holographic interferometry for analysis of high amplitude vibration

Imre Péntek

Development Laboratory, Ganz Electric Works
Lövőház u. 39, H-1024 Budapest, Hungary

Abstract

The double pulse holographic technique, by means of so-called holographic sampling, makes the analysis of high amplitude vibration out of the holographic measuring range possible. In this paper the principle of holographic sampling as well as calculations simulating the real situation to verify the principle will be shown.

Introduction

Holographic interferometry is suitable for quantitative measurement of displacement in several times 10 μm that depends on the applied light-wavelength; no matter whether the displacement was caused by static or dynamic load. For this reason, sometimes it is necessary to bring about the load such a way that the displacement should be limited to this range. Thus, it is quite frequent that the obtained information does not reflect the real situation. By means of double pulse technique, of course it is also possible to measure unforced action. By this measuring method a part of the total displacement or in case of vibration, a part of the maximum deflection can be recorded in order to get information about a measured object. However, in general neither the maximum deflection nor the frequency and the harmonics can be determined. Very often it would be necessary to determine the total displacement field so as to ensure the success of the examination. As it can be seen in the following there is a possibility to reconstruct the total displacement field by means of holographic sampling. This way the holographic vibration analysis could be made complete resulting in much more extended industrial applications.

Principle of holographic sampling

The fact of recording a part of the high amplitude vibration by double pulse holographic technique - taking a sample - makes the reconstruction of the total displacement field possible. We focus our attention on such processes where the displacement of surface points is stationary in time and band limited in frequency.

When the maximum deflection of a surface point (y_{max}) is much higher than the maximum one (Δy_{max}) that can be obtained by holographic method, the defined ratio of displacement value (Δy) got by holography to interval (Δt) between double pulse

$$y'_n = \frac{\Delta y}{\Delta t} , \qquad (1)$$

so called holographic sample, is the tangent of the "e" straight (see Fig.1). It is easy to see that shifting away this straight in line with each other, it will be the tangent of the curve at t'_n abscissa point ($t_n - \Delta t/2 < t'_n < t_n + \Delta t/2$). It is not a big fault if this tangent is corresponded to the $((t_n - \Delta t/2)+(t_n + \Delta t/2))/2$ point. The error of principle arisen in case of every sample is

$$H_n = y'(t'_n) - y'(-\frac{(t_n - \Delta t/2)+(t_n + \Delta t/2)}{2}),$$

$$(2)$$

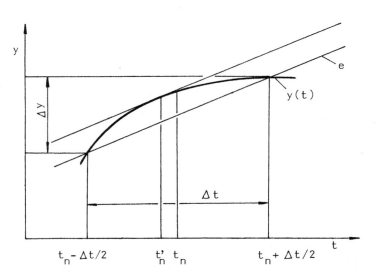

Figure 1. Principle of the holographic sampling.

where y'(   ) is the derivative value at the argument. In this manner the approximation made is that the y'_n difference ratio which equals to the derivative value ($y'=y'(t')$) belonging to the t'_n abscissa point, is corresponded to the t_n point. The shorter the Δt interval belonging to the double pulse, the more accurate the above supposition, namely if $\Delta t \rightarrow 0$ then $H_n \rightarrow 0$. However besides the technical difficulties the holographic sensitivity and the inclination of the displacement curve impose a limitation to this condition.

If the limiting frequency of band limited displacement is f_M then the smallest number of sampling points by which y' is specified for all t is achieved by taking $T_o = 1/2f_M$. So the holographic sample y' periodically, at intervals T_o sec apart, e.g., at times $t_n = nT_o$, where $-\infty < n < \infty$. By means of Shannon's interpolation formula the derivate function (velocity one) of the displacement can be restored

$$y'_H(t) = \sum_{n=-\infty}^{\infty} y'_n(nT_o) \frac{\sin (2f_M t-n)}{(2f_M t-n)}, \tag{3}$$

where sub H denotes that the restored function slightly deviates from the real situation in accordance with (2). We obtain the displacement function by integration of Eq. (3)

$$y_h(t) = \int y'_H(t) \, dt + c, \tag{4}$$

where H is described above and c is integration constant equal to the steady component ($<y(t)>$) of the displacement.

<u>Results</u>

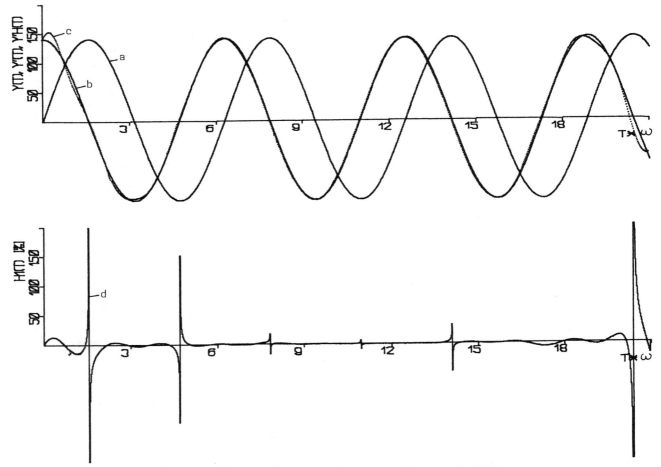

Figure 2. Simulation of the holographic sampling through computation
- curve "a" y(t)=140sin(t), vibration displacement
- curve "b" y'(t)=140cos(t), velocity function of the vibration displacement
- curve "c", velocity function reconstructed on the basis of the principle of holographic sampling. $T_o=0.72/\omega$. $T_p=0.09/\omega$. H=0%
- curve "d" error function $h_1(t)$.

To verify the correctness of the principle computations have been made for a few simple cases. In the course of the calculations a simple sinusoidal displacement has been assumed which can also be taken for one component of the band limited displacement. According to formula (3) the sampling of T_o period has to be carried out from $-\infty$ to $+\infty$ and this range was reduced to $(0-t=21/\omega)$ time range where ω is the angular frequency of the vibration. Thus the error of computations becomes considerably greater than that would be obtained as error of principle. Despite, the accuracy of the method is sufficient from practical point of view. The relative error function $h_1(t)$ is the basis for the error calculation

$$h_1(t) = \frac{y_H'(t) - y'(t)}{y'(t)} \ 100 \ \left[\%\right] \ ,\tag{5}$$

where $y'(t)$ is the derivate function of the displacement and $y_H'(t)$ is the derivate function obtained from the simulation of the holographic sampling.

Figure 2 shows the assumed displacement function $y(t)$, its derivate function $y'(t)$, function $y_H'(t)$ obtained on the basis of the assumed holographic sampling, and error function $h_1(t)$. The sampling period was $T_o=0.72/\omega$ and the double pulse time was $t=T_p=0.09/\omega$. The calculations have been based upon the assumption that the sampling can be carried out with absolute accuracy (the error of the holographic measurement is zero). Thus the error function

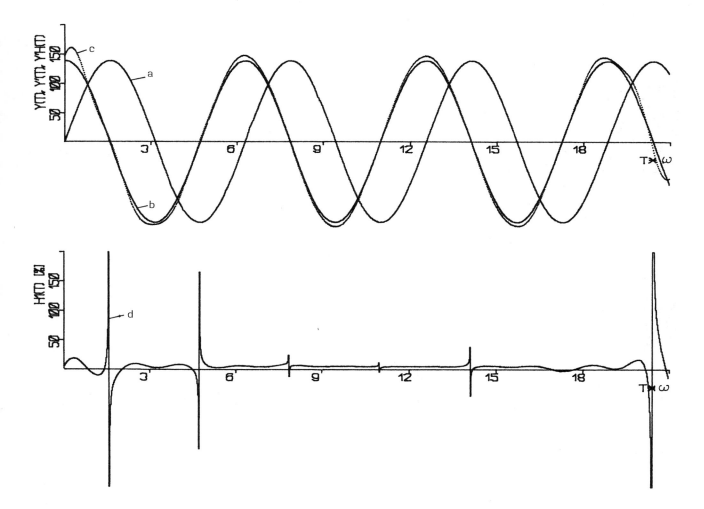

Figure 3. Simulation of the holographic sampling through computation
- curve "a" $y(t)=140\sin(t)$, vibration displacement
- curve "b" $y'(t)=140\cos(t)$, velocity function of the vibration displacement
- curve "c", velocity function reconstructed on the basis of the principle of holographic sampling. $T_o=0.72/\omega$. $T_p=0.09/\omega$. H= max. 10%.
- curve "d" error function $h_1(t)$.

is resultant of the errors coming from the principle of sampling and from the truncated interval. Due to the truncated interval the value of the error is considerably greater in the first cycle, and the last cycle of the interval. A few error values, belonging to Figure 2, can be found in column 2 of Table 1.

The displacement function y(t) and its derivative y'(t), shown in Figure 3, are the same as those shown in Figure 2. In the course of the reconstruction of $y'_H(t)$ it was taken into consideration that holographic sampling can be carried out within certain limit of error only. The deviation, coming from the error of the measurement, is H=max. 10% in the case of every sample in the direction of the increasing amplitude accidentally. Even with a measurement error, as great as can be seen from the error curve of Figure 3, practically sufficient results can be obtained.

An important practical question appears, namely, how many samples have to be taken - in addition to complying with the conditions of equation(3) - so as to obtain results of sufficient accuracy. Figures 4 and 5 show frequency and amplitude conditions being identical with those shown in Figures 2 and 3 as far as y(t) is concerned while the sampling period is twice as long (T_o=1.44/ω) i.e. the number of the samples is half as high as it was. For Figure 4 H=0% while for Figure 5 H=max.10%. The error curves and the data of the Table show that the accuracy is sufficient in this case, too.

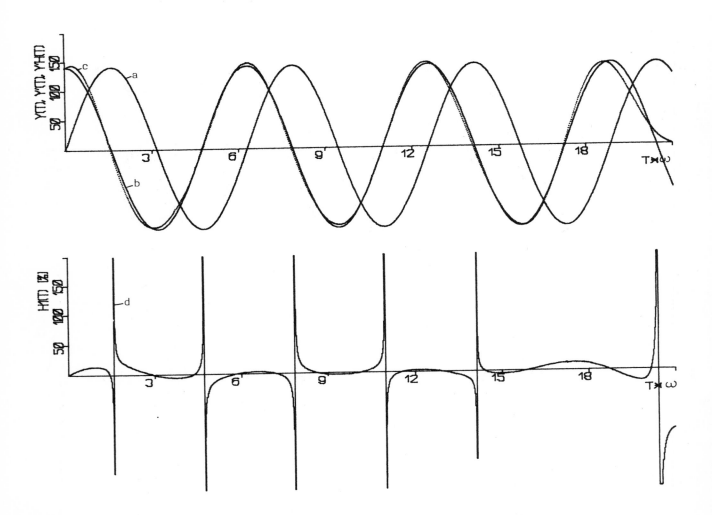

Figure 4. Simulation of the holographic sampling through computation
- curve "a" y(t)=140sin(t), vibration displacement
- curve "b" y'(t)=140cos(t), velocity function of the vibration displacement
- curve "c", velocity function reconstructed on the basis of the principle of holographic sampling. T_o=1.44/ω . T_p=0.09/ω . H=0%.
- curve "d" error function $h_1(t)$.

Table 1. Values of the error function vs. time for $T_p = 0.09/\omega$.

Time	$h_1 [T_o=0.72/\omega ;$ H=0%]	$h_1 [T_o=0.72/\omega ;$ H=max.10%]	$h_1 [T_o=1.44/\omega ;$ H=0%]	$h_1 [T_o=1.44/\omega ;$ H=max.10%]
6.28204	.809239	6.35017	3.56319	9.25549
6.39838	.346853	5.86238	3.85613	9.56452
6.51471	-.188797	5.29727	3.95482	9.66865
6.63105	-.682732	4.77621	3.8532	9.56143
6.74738	-1.03017	4.40968	3.54309	9.23427
6.86371	-1.15232	4.28079	3.01238	8.67441
6.98005	-1.00776	4.43331	2.24062	7.86023
7.09638	-.596179	4.86747	1.19194	6.75391
7.21271	.044118	5.54299	-.201408	5.28396
7.32905	.846217	6.38918	-2.07634	3.30598
7.44538	1.73368	7.32544	-4.74015	.495754
7.56172	2.65431	8.29667	-9.01276	-4.01173
7.67805	3.69327	9.39275	-17.9233	-13.412
7.79438	6.32326	12.1673	-58.2931	-56.0007
7.81765	8.55786	14.5247	-96.7053	-96.5241
7.84092	18.59	25.1082	-271.302	-280.717
7.86419	-16.986	-12.4231	349.485	374.191
7.88745	-3.08026	2.2469	106.947	118.322
7.91072	-.576311	4.88845	63.1806	72.1497
8.02705	1.71095	7.30143	20.1191	26.7214
8.14339	1.84005	7.43763	10.9579	17.0567
8.25972	1.54074	7.12188	6.57168	12.4293
8.37605	1.06894	6.62416	3.80957	9.51542
8.49239	.552989	6.07984	1.82954	7.42653
8.60872	.082678	5.58367	.317031	5.83091
8.72506	-.278988	5.20214	-.867767	4.58099
8.84139	-.497048	4.97206	-1.79395	3.60389
8.95772	-.564597	4.90081	-2.49831	2.8608
9.07406	-.49954	4.96946	-3.00158	2.3299
9.19039	-.33888	5.13895	-3.31596	1.99825
9.30673	-.130795	5.35848	-3.44954	1.85729
9.42306	.07504	5.57561	-3.40808	1.90104
9.53939	.235342	5.74471	-3.19607	2.1247
9.65573	.320329	5.83439	-2.81718	2.52441
9.77206	.317492	5.8314	-2.27388	3.09758
9.8884	.232364	5.74158	-1.56644	3.84389
10.0047	.085682	5.58683	-.691528	4.76692
10.1211	-.090825	5.40065	.361972	5.8783
10.2374	-.260932	5.22119	1.61877	7.2042
10.3537	-.390145	5.08485	3.13194	8.80055
10.4701	-.450585	5.02114	5.01402	10.7861
10.5864	-.421554	5.05172	7.52501	13.4351
10.7027	-.283756	5.19709	11.3637	17.4848
10.8191	.025557	5.5234	19.119	25.6664
10.9354	1.14366	6.70293	53.532	61.9708
10.9587	2.10411	7.71612	85.816	96.0293
10.9819	6.24833	12.0879	227.717	245.73
11.0052	-9.47037	-4.49414	-315.374	-327.212
11.0285	-2.91086	2.42559	-89.9923	-89.4423
11.075	-268.77	4.15799	-35.0835	-31.5154
11.1913	-.470742	4.99986	-11.4264	-6.55803
11.3077	-.206532	5.27858	-4.94573	.278908
11.424	-.080626	5.4114	-1.72414	3.67754
11.5403	-.031421	5.46333	.257947	5.76855
11.6567	-.030417	5.46438	1.59178	7.17573
11.773	-.05352	5.43998	2.51045	8.1449
11.8893	-.078656	5.41347	3.12332	8.79145
12.0057	-.088047	5.40355	3.48845	9.17663
12.122	-.071668	5.42085	3.64021	9.33672
12.2383	-.028391	5.46651	3.60135	9.29577
12.3547	.031749	5.52994	3.38871	9.07144
12.471	.093262	5.59482	3.01608	8.67831
12.5873	.135408	5.63931	2.49594	8.12959
12.7037	.139184	5.64324	1.84	7.43758

From the error functions it can be seen that the value of the errors suddenly increases in the vicinity of the zero passages of the velocity function (the environment of the maximums of the displacement function). This can be explained by the fact that on the basis of the principle of correspondence (see Fig.1) the error of the individual samples increases.

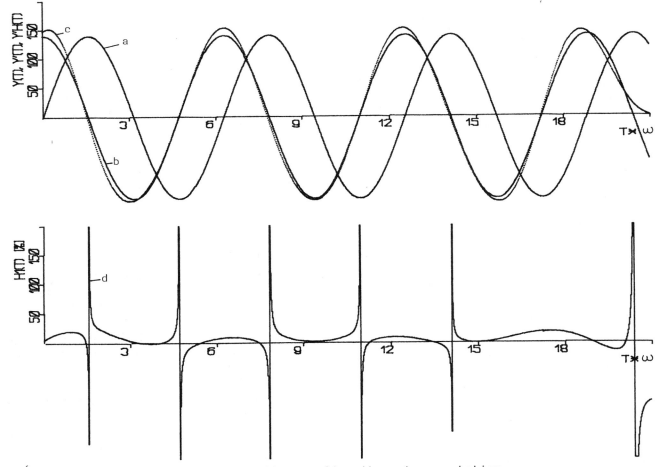

Figure 5. Simulation of the holographic sampling through computation
- curve "a" y(t)=140sin(t), vibration displacement
- curve "b" y'(t)=140cos(t), velocity function of the vibration displacement
- curve "c", velocity function reconstructed on the basis of the principle of holographic sampling. T_o=1.44/ω . T_p=0.09/ω . H=max.10%.
- curve "d" error function $h_1(t)$.

Conclusion

The principle of holographic sampling offers possibility for reconstruction of the band limited vibration of high amplitude outside of the range of holographic measurement so that all points of the surface are involved. With the help of the holographic sampling the velocity function of the displacement can be obtained and from this function, with the aid of the Fourier analysis, the harmonic components can be determined. The displacement function can be obtained from velocity function, with the help of integration, and the maximum displacements can be easily determined from the displacement function.

Reference

1. Shannon, C.E., Communication in the Presence of Noise, Proc. IRE, 37:10 (1949).

Comparison of vibration mode measurements on rotating objects
by different holographic methods

J.Geldmacher,H.Kreitlow,P.Steinlein,G.Sepold

Bremer Institut für Angewandte Strahltechnik (BIAS)
Ermlandstraße 59, 2820 Bremen 71, West Germany

0.Abstract

Vibration modes of rotating objects generally show different vibration behaviour compared to modes of the stationary object. Several methods can be used for holographic vibration analysis of rotating objects:

- Method with optimized holographic measuring arrangement
- Method with optical image derotator
- Method with laser triggering related to object location and phase of vibration
- Method with holographic interferometer rotating synchronously to the object

Optomechanical and optoelectronical devices developed to perform these methods as well as measurement results of vibration analysis on technical objects will be discussed.

1.Introduction

In mechanics of rotating objects it is very necessary to determine the actual vibration behaviour of the entire rotating component under real operating conditions in order to evaluate the dynamic strain or stress. With this knowledge the construction of rotating components that are heavily loaded and extremely stressed can be optimized.

A transfer of the vibration behaviour of non-rotating technical components on the vibration behaviour at self-excitation during rotation is not possible or only by taking in account considerable restrictions:

1. Both, the centrifugal force influence resulting from the object rotation, and the interaction of the excitation between the rotating component and the surrounding medium result in a displacement of the nodal lines of vibration.

2. A temporal change of the vibration modes takes place under certain boundary conditions. These non-stationary vibration modes appear for instance at radial compressor impellers:

   - as a result of changes of the mass flow rate at constant rotation speed
   - at resonances between the eigenfrequencies of the blades and the engine orders and
   - under non-stationary operating conditions of the compressor such as `rotating stall` or pumping.

Conventional measuring and testing methods to determine vibration behaviour of rotating objects have -besides a pointwise data acquisition- the disadvantage that it is difficult to transfer data, particularly at high rotation speeds. Therefore, optical measuring and transfering methods, mainly holographic interferometry, have lately been used to visualize modes of vibration.

Holographic interferometry has the advantage that on one hand the deformations in all surface points in the um-range are determined simultaneously and on the other hand this measuring is non-contacting and non-interacting. Determination of test results as well as the transfer of the results is carried out optically and thus free from disturbance.

In order to be able to evaluate vibration modes of rotating objects from the holographic interference pattern, it is necessary to eliminate the affect of object rotation on the formation of the interference pattern resulting from vibration. Suitable methods to visualize only the interference pattern are herewith presented.

These methods can principally be caused by object vibration distinguished either

   - by type of observation of the rotating object or
   - by controlling the exposure of the holographic plate:

1. The holographic measurement set-up is optimized in such a way that it is least sensitive as to measuring in-plane-rotations of the component, but has high sensitivity for out-of-plane vibrations.

2. The rotating component is stationary imaged in the holographic plate by an optical image derotator, whose prism rotates at half the speed of the object with rotation axes of prism and object, respectively, arranged collinear.

3. The holographic plate is only exposed in those instants of time when the object has the same spatial rotation location and two selected but different phase locations of the vibration.

4. The part of the holographic interferometer consisting of the holographic plate and the reference wave rotates at exactly the same speed as the rotating object with rotation axes of holographic interferometer and object arranged collinear.

2. Development and application of the holographic methods

The methods mentioned in 1.have been developed up to an extent where they can be used in the technical field under real operating conditions. At the present state of technique a pulsed laser was used for the methods 1.,2. and 3. Method 4. was tested using a CW-laser. Principally a CW-laser can also be used for method 2.

Employing the Ruby-Giant-Pulse-Laser the following conditions were complied with:

To prevent interference patterns in the shape of contour lines at the object, the ruby-laser oscillated each of the giant pulses in the same axial mode. For this purpose, the laser head including ruby rod and flash lamps was kept at constant temperature by a two-circuit cooling system.

In addition to that, an etalon and a resonance reflector were used, stabilized by oven heating at a temperature variation of less than $0,1^{o}C$. Objects with a diameter of up to 1000 mm could be illuminated so that the wave reflected by the object sufficiently modulates the reference wave needed to produce a hologram.

The two giant pulses needed for the holographic double exposure were switchable by a brewsterplate-pockelscell arrangement in the laser resonator. The giant pulse intervals were variable from 2 to 500 µs and above 3 sec in order to match the vibration amplitude and frequency as well as the vibration speed of the object. With optimal alignment of all resonator elements, outputenergies of >15 mJ per giant pulse with a pump energy of 1500 J were achieved.

2.1 Method with optimized holographic measuring arrangement

A low sensitivity of the holographic measuring arrangement relative to formation of interference fringes resulting from in-plane-rotations of the object is achieved by placing the rotating object in a tangent plane to the rotational symmetric ellipsoid, whose focal points are formed by the observation point and the object illumination source point, see fig.1 and /1/.

Computer programs have been developed to precalculate interference patterns resulting from in-plane-rotations of the object at given arrangement of object, illumination source point and observation point. By means of these programs, the holographic measuring arrangement can be optimized in such a way that interference fringes resulting from rotation are outside of the object surface, which is to be investigated, see object 1 in fig.2.

2.2 Method with optical image derotator

Usage of this holographic method to analyse vibration at rotating technical objects or components is possible if

- exact control of rotation speed of the derotator prism is guaranteed
- vibration modes should be determined in axial direction and
- only optical components of high quality as to parallelism, smoothness and optical density are used.

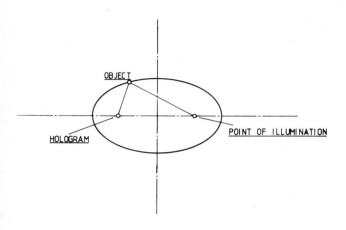

Fig.1.Holographic arrangement to reduce
measuring sensitivity as to in-plane-
rotation

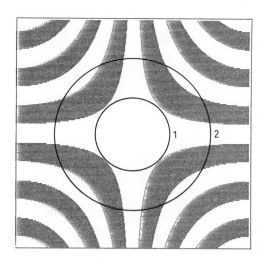

Fig.2.Precalculated interference pattern
for rotating plane objects 1 and 2
using a holographic arrangement
according to fig.1.

Essential advantage of this method is that short intervals (2 to 500 μs) between both
laser pulses can be used to illuminate the holographic plate, enabling a vibration
analysis with vibration modes that change as to time.

However, the use of this method is limited due to the size of the object. At large
rotating components, the reflected object wave can only be transferred through the
derotator prism by suitable optical device. Thus, the optical stationary object is imaged
smaller behind the derotator and near the photolayer, making it difficult or even
impossible to resolve very high densities of interference fringes.

Holographic measuring arrangement and performance of experiments

The holographic arrangement to measure vibration modes in axial observation direction
of the object consists of the following three functional units:

- ruby-giant-pulse-laser, see 2.
- optical image derotator
- holographic interferometer.

Optical image derotator: By means of optical image derotators using e.g. a
retroreflective prism, a spatially stationary object image is achieved, if the prism
rotates at half the object speed into the same direction of rotation as the object, and if
the rotation axis of object and prism are collinear, see fig.3 and /2/,/3/.
To produce the signal for controlling the rotation speed of the prism a He-Ne-laser was
directed onto black and white reference marks of same size fixed at the circumference of
the back side of the object. Periodically appearing intensity changes of the light
reflected at the reference marks when the object rotates were registered via a photo diode
and transferred as an electric pulse to the control unit of the image derotator, see
fig.4.

Holographic interferometers: To analyse vibration modes of the object in axial
observation direction the following holographic set-up was used, see fig.5.
The light emitted by the double pulse laser was divided into objectwave and
referencewave by a beamsplitter. The expanded objectwave was deflected via an additional
beamsplitter ST toward the rotation axis of the object in order to illuminate the object.
The light scattered by the object interferes with the expanded reference wave in the plane
of the holographic plate. As a result, the phase change of the object, which occured
between two exposures, was holographically stored and could be measured from the
interference pattern resulting from the reconstruction.

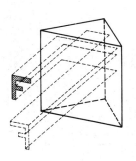 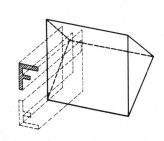 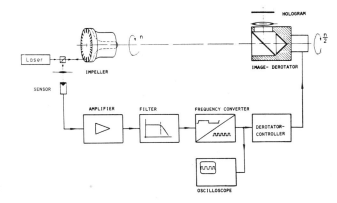

Fig.3. Functional principle of reflection
     type optical image derotator

Fig.4. Control unit of the optical
     image derotator

Besides optimizing as to the mutual arrangement of mechanical and optical components of the holographic interferometer, e.g. in order to achieve a sufficient illumination of the object surface or measuring sensitivity as well as stability against disturbance from the environment, great importance has to be attached to the mutual alignment of object, image derotator and interferometer.

Results

With the holographic measuring arrangement described, vibration interferograms of a rotating impeller, see fig.6 and /4/,/5/, of a radial compressor were carried out under operating conditions, i.e. outside the laboratory.

With open compressor inlet measurements were carried out up to a rotation speed of approx.8500 rpm. With closed inlet holographic measurements of vibration modes at rotation speeds of up to 20500 rpm have been achieved so far. For this purpose, an alignable window and an alignable deflection mirror to illuminate and observe the object were constructed and installed in the inlet tube, see fig.5.

Figures 7 - 10 exemplarily show vibration interferograms for these operating conditions.

With open compressor inlet and a rotation speed of 3000 rpm the blades were excited in the first bending mode, see fig.7. The excitement itself was carried out by the driving wheel teeth of the compressor gear. The reasons for the different vibration amplitudes in the holographic interferogram are the reciprocal coupling effects of the blades with one another. This method to excite vibration proved to be the strongest exciting source of the compressor in the entire range of rotation speed.
An additional resonance vibration in the first bending mode occured at n=7950 rpm, see fig.8. The blades vibrate very uniformly with again recognizable vibration amplitudes resulting from the reciprocal coupling effects.
The interferogram in fig.9 shows a resonance vibration in the second mode at 10940 rpm. No longer vibration modes with only one nodal line (first resonance frequency) appear here, but vibration modes that consist of the first and other modes. This vibration behaviour is mainly a result of the coupling effect between the individual blades.
Above n=16000 rpm there were vibration modes that cannot only be explained by resonance vibrations or their coupling.
The interferogram illustrated in fig.10 shows vibration modes of the blades at n=20450 rpm that are different from one another. The vibration modes consist of the first, second and other vibration modes.
Experiments carried out up to now have shown that holographic vibration analysis by means of a reflection image derotator can be used at a rotating object under real operating conditions up to high rotation speeds of approx.20500 rpm.

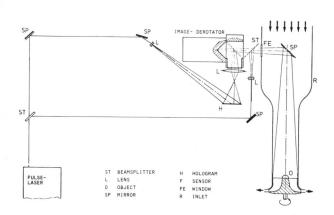

Fig.5. Holographic measuring arrangement to analyse vibrations at rotating objects

Fig.6. Impeller (external diameter 290 mm) of a radial compressor

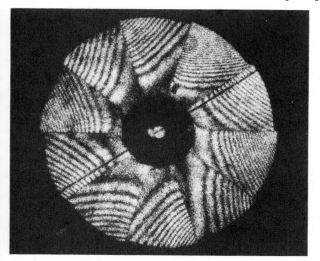

Fig.7. Vibration interferogram of impeller rotating at 3000 rpm.

Fig.8. Vibration interferogram of impeller rotating at 7950 rpm.

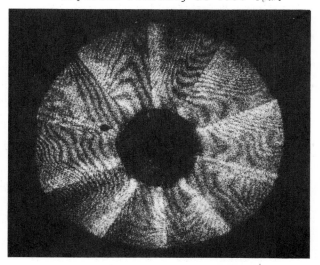

Fig.9. Vibration interferogram of impeller rotating at 10940 rpm.

Fig.10. Vibration interferogram of impeller rotating at 20450 rpm.

2.3 Method with laser triggering related to object location and phase of vibration

Holographic storage of vibration interferograms of rotating objects is possible without any additional mechanical or optical auxiliary means, if the two laser pulses, needed for the double exposure of the holographic plate, illuminate the object in exactly the same spatial location and if the difference of relating vibration phases to be compared holographically do not exceed the holographic measuring sensitivity.

Special features of this method are as follows:

- The holographic arrangement does not differ from the one used to analyse vibration at stationary objects, i.e. additional mechanical and optical components as in the method presented in 2.2 are not necessary
- Image plane holograms as in the method presented in 2.2 are not needed
- Furthermore, no rotation axes are to be aligned collinear, for instance, required in the image derotator method, see 2.2, and in the method with rotating holographic interferometer, see 2.4.
- In contrast to the other methods, it is possible to arrange in the path of the objectwave elements of low optical quality, e.g.as to parallelism, smoothness or optical density that are stationary with regard to the object rotation.
- Another essential advantage compared with the other methods is that rotating objects cannot only be observed in axial direction of the rotating object but in any other desired direction.

The application of this method to vibration modes changing as to time is limited since the time interval of the exposure pulse is to be at least a full revolution period of the object. With high object rotation speed, however, revolution time can be short compared with the period of the vibration mode, so that this disadvantage is then of minor or no importance.

Substantial premises for a successful application of this holographic measuring method are:

- For triggering the laser pulses the chosen object position has to be recorded exactly, that means up to fractions of one um. This may be accomplished by a photodiode gathering the periodic signal of a coding disk fixed onto the object. The signal is put into a counter, that feeds the electronic devices controlling the flashlamp pulses and the quality switch of the resonator, see fig.11 and /1/.
- To achieve the aimed object illumination, after triggering the first laser pulse, the vibration phase of the object after each total revolution has to be compared with the vibration phase during the time instant of the first laser pulse. The second laser pulse is then triggered, if the phase difference

-- lies within a "window", whose upper and lower bounds x and y are determined by the maximum and minimum permissible interference fringe densities,
-- takes a predetermined phase amplitude within this window.

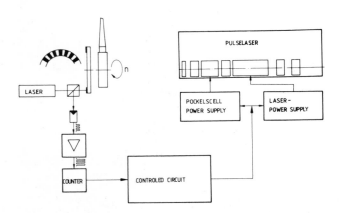

Fig.11. Control of laser triggering related to object location and phase of vibration

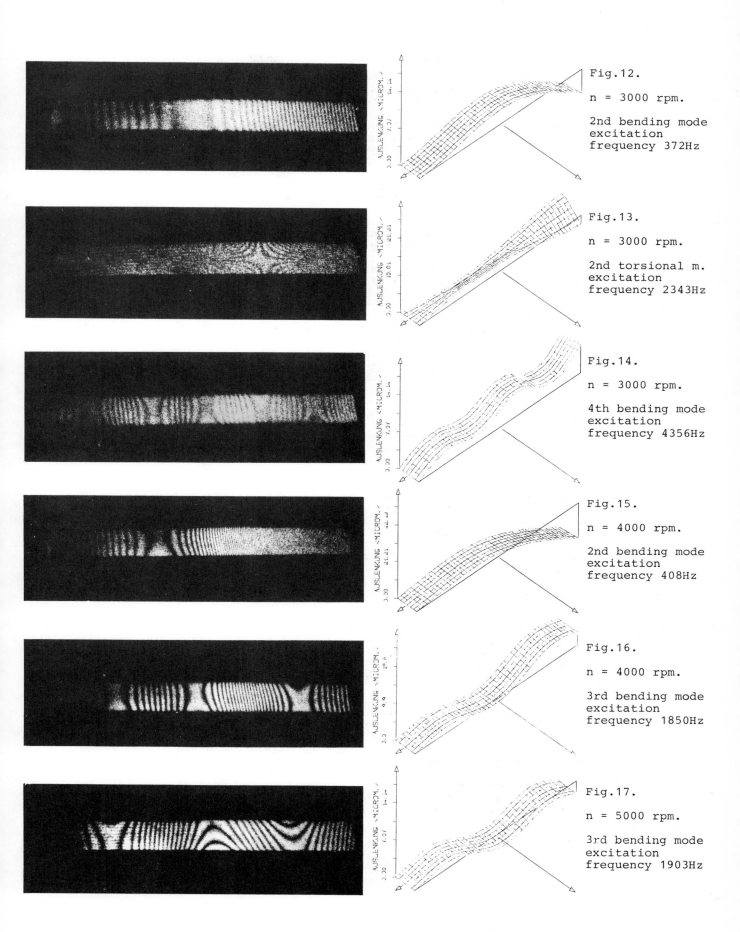

Fig.12.

n = 3000 rpm.

2nd bending mode
excitation
frequency 372Hz

Fig.13.

n = 3000 rpm.

2nd torsional m.
excitation
frequency 2343Hz

Fig.14.

n = 3000 rpm.

4th bending mode
excitation
frequency 4356Hz

Fig.15.

n = 4000 rpm.

2nd bending mode
excitation
frequency 408Hz

Fig.16.

n = 4000 rpm.

3rd bending mode
excitation
frequency 1850Hz

Fig.17.

n = 5000 rpm.

3rd bending mode
excitation
frequency 1903Hz

Vibration interferograms of rotating blade models in a test stand containing a vacuum chamber could be produced by this method under real operating conditions.
In this test stand, the magnetically excited object could rotate within a rotation speed range of up to 5000 rpm and at low pressure of a few Torr in the measurement chamber. The blade position in the measurement chamber could be observed through a plexiglas window of low optical quality as to smoothness and parallelism of the window surface. For the illumination of the rotating object, a ruby-giant-pulse-laser, stabilized as to axial modes and with a power output of approx.20 mJ was used.
Figures 12 to 17 show examples of vibration interferograms which were performed under the conditions mentioned above and observed in axial direction. The quantitative evaluations of these interferograms are shown in the right part of the figures.

2.4 Method with holographic interferometer rotating synchronously to the object

This method for axial observation and measuring vibration modes of rotating objects has already been proposed in /2/ and was initially tested successfully for a laboratory system. Synchronization of both rotations was achieved by a rigid mechanical V-belt-shaft between the rotating object and the rotating holographic system.
A disadvantage of this mechanical coupling is that it is only realizable in exception cases and that particularly at higher rotation speeds the test result is affected by the mechanical behaviour of the shaft, e.g. as a result of torsional vibration.
Therefore, a rotating holographic measurement system has been developed and tested where synchronization of rotation is achieved via an electronical control, see fig.18.

This system has the following advantages:

- The rotating holographic interferometer head can be arranged in any desired distance to the object so that a holographic vibration analysis can be carried out even at larger rotating technical objects of up to more than 1 m^2.
- As image-plane holograms not need to be produced, aligning of the arrangement is simplified, optical aberrations do not occur, and the object appears in full size after reconstruction.

To produce vibration interferograms the rotating object is illuminated through the axis of the rotating holographic interferometer. Lens L in the rotation axis produces a divergent illumination wave. Two stationary reference beams, which are related to both rotating holographic plates H, are produced via two beamsplitters ST, mirrors and expanding lenses.
Both holograms can either be recorded simultaneously or one after another controlled via a shutter in one light path.

Synchronization between object rotating speed and rotating speed of the holographic interferometer is achieved

- by comparing the signal derived via photo diodes by means of encoder marks at the rotating holographic interferometer and the object and by adjusting the speed of the stepping motor which performs rotation of the holographic interferometer by the differential signal, see fig.19, or
- by detecting rotation frequency of the object and direct controlling of the stepping motor.

The rotating holographic interferometer was tested at a rotating impeller model of a radial compressor. Using this arrangement with a CW-laser, interferograms of rotating object in a rotation range of up to 1100 rpm could be produced without any affection of the interference patterns resulting from the object rotation, see fig.20 and 21. Instead of eigen modes fluttering modes could occur in this rotation range, see fig.21..

Developing this holographic method by constructional improvement of the interferometer for even higher rotation speeds and employing it at technical objects by making use of the mentioned advantages will be aimed at in future experiments.

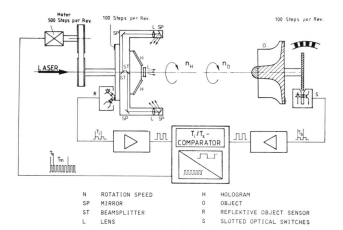

Fig.18. Controlling the rotating
      interferometer

N	ROTATION SPEED	H	HOLOGRAM
SP	MIRROR	O	OBJECT
ST	BEAMSPLITTER	R	REFLEKTIVE OBJECT SENSOR
L	LENS	S	SLOTTED OPTICAL SWITCHES

Fig.19. Holographic arrangement with
      interferometer rotating
      synchronously to the object

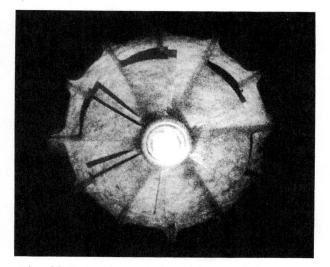

Fig.20.Hologram of impeller model
      at n = 1100 rpm.

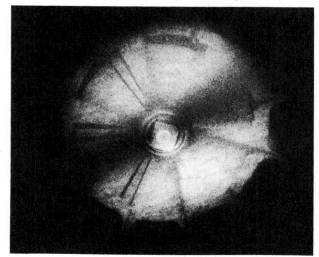

Fig.21.Interferogram of impeller model
      at n = 1050 rpm.

3.Conclusions

In this paper four methods for holographic vibration analysis at rotating objects are presented.
The method with optimized holographic arrangement and the method using an optical image derotator are already employed successfully in engineering.
The method with rotating holographic interferometer and the method using laser triggering controlled by object location have been improved with respect to higher rotanional speed of objects and tested successfully at technical objects under real operating conditions.

The experimental results confirmed the applicability of the proposed method in practice and pointed out the limitations of each method still existing.

Acknowledgements

The authors greatfully acknowledge the financial support provided by the German Research Association (DFG). The authors also thanks Prof.Rautenberg and Dipl.-Ing. Haupt (Institute for Turbomachinery, University Hannover, West Germany) for good cooperation in performing the experiments. Furthermore thanks are given to Dr.Dirr and Dipl.-Ing. Vogt (Institute for Mechanics, University Hannover, West Germany) for providing the experimental facilities and support in performing some of the experiments.

References

/1/   Kreitlow,H.,Geldmacher,J.,Jüptner,W., Entwicklung neuer Verfahren zur holografischen Schwingungsmessung an rotierenden Bauteilen, Proceedings Laser `81, Springer Verlag, 120-126

/2/   Beeck,M.-A.,Kreitlow,H., Conditions and examinations of vibration analysis of rotating blades with help of holographic interferometry, Proceedings Laser `77, IPC-Science and Technology Press

/3/   Fagan,W.F.,Beeck,M.-A.,Kreitlow,H., The holographic vibration analysis of rotating objects using a reflective image derotator, Optics and Lasers in Engineering, Applied Science Publishers Ltd, England, 1981, 21-32

/4/   Haupt,U.,Kreitlow,H.,Rautenberg,M., Blade vibration measurements by means of telemetry and holographic Interferometry on an radial impeller with thin blades, The fourth international conference for mechanical power engineering, Cairo University, Oct.1982,

/5/   Haupt,U.,Rautenberg,M., Blade vibration measurements on centrifugal compressors by means of telemetry and holographic interferometry, Phoenix, USA, 1982

Hologram interferometric measuring system for industry

Z.Füzessy, A.Ádám, I.Bogár, F.Gyimesi, G.Szarvas

Institute of Physics, Technical University Budapest
H-1521, Hungary

Abstract

A portable hologram interferometric system has been developed measuring 3-D displacement due to static/dynamic load in factory environment. The evaluation of interferograms is computer aided. The system is shown both statically and functionally. Two measurements (lathe bed deformation and vibration analysis of the arm of a radial drilling machine) are reported.

Introduction

Hologram interferometry is a versatile tool for measuring deformations of industrial objects where the theory is incapable of living. Nevertheless its acceptance in measuring engineering has been quite slow taking the real capabilities into consideration. There are three areas of technical development required if the full potential of hologram interferometry is to be realized. First, purposeful development of methods and techniques in order to use them as the base for technical development. Second, development of equipments capable to work in factory. Third, development of interferogram- and data processing algorithms and devices. The paper presents a hologram interferometric system for measuring 3D-displacement which demonstrates our ambition to fulfil the requirements mentioned above.

Theory and methods as the basis of the system

It is obvious that determination of all the three components of the displacement vector with approximately equal accuracy is possible when the interferometer is sensitive to the three orthogonal directions in the same way. That optimal situation can be approached by a system, where the sensitivity vectors are realized by one illumination and more observation directions. In the technique proposed determining optical path length difference between the light source and observer for a given surface point results in the value of displacement component along the sensitivity vector, only. Determination of that component is founded upon measurements of fringe order numbers[1].

Using the fundamental equation of the technique in the case of interferograms with three different sensitivity vectors but recorded simultaneously

$$\underline{\underline{S}}\underline{L} = \underline{N}\lambda \qquad (1)$$

(where $\underline{\underline{S}}$ is the sensitivity matrix determined by geometry of the system, $\underline{L}$ is the displacement vector to be determined, $\underline{N}$ is the fringe order vector, λ is the wavelength of the light) the displacement vector can not be determined unambiguously. On the other hand data stored in four interferograms are sufficient for determination of the moduli and line of the influence of the vector[2].

In data collection procedure a difficulty arises consisting of the identification of object surface points on different holographic images. A way of overcoming the problem can be found in[3].

Description of the measuring system

The hologram interferometric measuring system developed at our laboratory consists of an interferometer, a data collecting and processing unit.

In the following the interferometer will be presented. It is portable and can operate right after its transportation to the scene of the measurement. The interferometer is a compact equipment for producing interferograms needed for 3D-deformation measurement. The general side view of the equipment is shown on Figure 1. The interferometer is mounted on a tripod. It can be lifted up hydraulically and rotated along vertical and horizontal axis.

The main components are:
- the frame (ribs strung on pipes);
- arms for setting the measuring holocameras;
- four holocameras (recording is possible in daily light, too);
- the light source built together with the interferometer (it is a double pulse ruby

laser of energy of about 300 mJ);
   - the optical system producing five reference beams and one object beam;
   - the adjusting system;
   - the control holocamera (interferograms are recorded on a thermoplastic film in order to study the displacement field in some ten seconds after exposures).

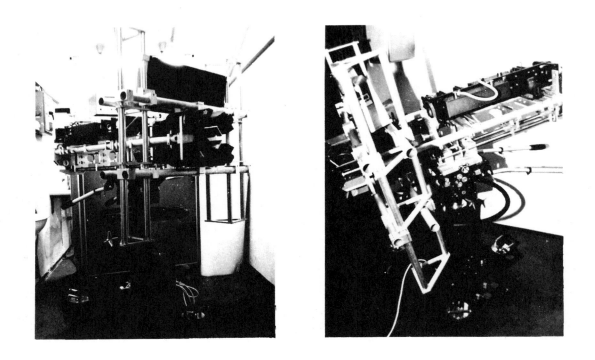

Figure 1. Side views of the interferometer

The block diagram of the system is shown on Figure 2.

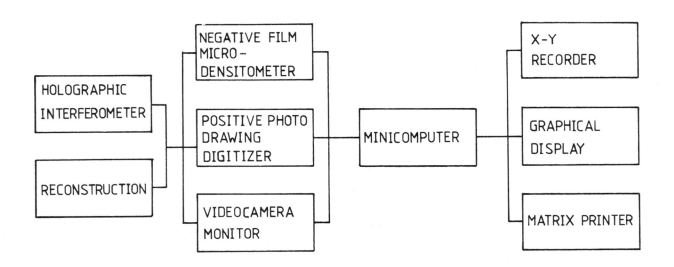

Figure 2. The block diagram of the system

   Functionally the system can be devided into 4 subsystems. In the first column on Figure 2 there are the interferometer introduced previously and equipment for reconstruction of interferograms. Interferograms enter the data collection part of the system. It works alternatively.

Using the microdensitometer the resolution is up to 300-500 fringes on the image of the surface investigated. The number of grey levels (10-100) is determined by the appearance of nonlinearity of photographic material used. The geometrical accuracy is high. Determination of the nonlinearity curve increases the time needed for evaluation. Reliability of data collection is in close connection with visibility of interference pattern. Marking identifying points on the object surface the automation of evaluation of interferograms meets with difficulties. The accuracy of fringe order estimation is better than 0,5 fringe at the edges of interferogram and 0,1 fringe elsewhere (it can approach even 0,05 fringe).

The drawing digitizer is the most popular data collection equipment in our practice in spite of the fact that the data collection is semiautomatic. The resolution is satisfactory: 100 fringes on the image of the surface investigated. Data can quickly be fed into the computer. Drawing digitizer can be used for data collection in the case even if identifying marks are painted on the object surface. Centres of interferometric fringes can be estimated to a satisfactory accuracy on an interferogram where the visibility of interference pattern is low. Perpendicularly to fringes the accuracy is about 0,1-0,2 fringe, but it can be low if the density of interference fringes decreases.

The data collection system based on a videocamera works with interferogram directly on a hologram plate. The light intensity distribution of an interferogram is converted into video signals. The lach of mechanically moving parts is a big advantage of the system. It can be used even if the object surface is covered with marking points, too, but the procedure now is semiautomatic. It can also operate in the case of low visibility of interference pattern. Nevertheless, the resolution is low (30-40 fringes on the image of the surface investigated) and the geometrical distortion is high.

The third block on Figure 2 consists of the computer which is a CAMAC system at present. The software was developed at our laboratory.

The fourth column consisting of X-Y recorder, graphical display and matrix printer.

Evaluating interferograms to determine all the three components of displacement vector two problems must be solved: measurement of elements of $\underline{S}$ and $\underline{N}$ in the equation (1).

The main steps determining the elements of $\underline{S}$:

- sticking identifying points to the surface;
- measurement of coordinates of the identifying points as well as some points on the interferometer by theodolite;
- feeding the coordinates of all that points measured on positive photos (4 ones) to the computer;
- to find an algorithm for identification of the points between the identifying ones and points measured on photos and determination their position is coordinate system assigned to the object.

The main steps determining $\underline{N}$:

- feeding the coordinates of fringes to the computer by drawing digitizer (microdensito-meter or video camera);
- fitting a function to the intensity distribution represented by the fringes on the interferogram and fed into the computer in the form of polinomials.

In data processing determining the components of $\underline{L}$ the equation is solved using computer programs; the actual fringe order even its fraction can be found as a value of the fitted function mentioned above at the object surface point considered.

The system in measurement

As a typical application of the system first the deformation measurement of a lathe bed will be shown. The deformation of the bed has occured due to imitated static cutting force of 2775N. One of the interferogram used for quantitative interpretation is shown on Figure 3. The line diagram of the investigated part of the lathe bed with marking points is shown on Figure 4. The components of the displacement vector of the points lying on a rail between points 1 and 10 are reported on Figure 5. The point 5 is taken for an unmoved one, the values of displacement are given in micrometres. The field of deformation was calculated theoretically by finite element method and in some points the y-component of motion was measured by dial indicator.

Figure 3. Interference pattern on the lathe
bed

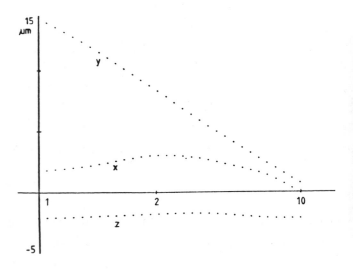

Figure 5. The displacement components of
points between markers 1 and 10.

Figure 4. Line diagram of a part of the
lathe bed

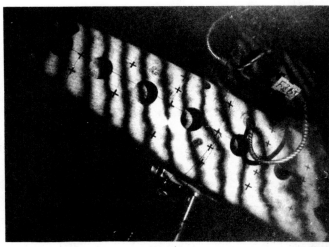

Figure 6. Interference pattern on the arm
of a drilling machine

Another example of application of the system in factory environment is the measurement of
vibration of a radial drilling machine. The horizontal arm was loaded dinamically by
no-load-speed of the main spindle. One of the interferograms used for quantitative evaluation is shown on Figure 6. The line diagram of the arm with marking points is presented on
Figure 7. The interferograms were evaluated along straight lines between points, the vibration amplitude on the surface was determined relatively to the point 24. The components
of the displacement vector of the points lying on the line between points 25 and 3 are
reported on Figure 8. The values of displacement components are given in micrometres. It
can be seen, that as a result of rotation of the main spindle a significant deflection can
be observed at the end of the arm in both y and z directions. The moving support has been
situated around the marking point 15. Because the vibration system has a large mass at that
area, strength of the arm here is higher than at another parts of the arm. This is demon-
stratively reflected by behaviour of x and y components of displacement vector at that
points.

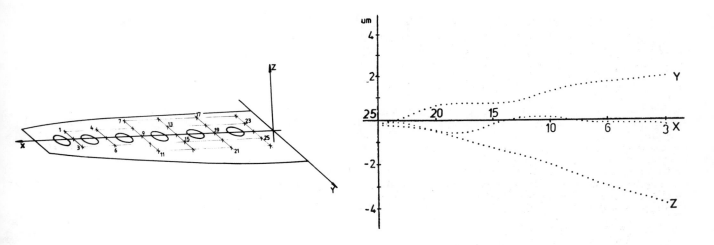

Figure 7. Line diagram of the arm of a
　　　　　drilling machine

Figure 8. The displacement components of
　　　　　points between markers 25 and 3

A piezoelectrical head was cemented at marking point number 14. When frequency of
vibration was 24 Hz the value of vibration amplitude measured was 1,4 μm. The z component
of vibration amplitude at that point calculated from interferograms has been equal to 1,234
μm.

Conclusions

The affinity of industry to any new measuring technique is greater if the measurements
can be carried out by ready-operate devices. According to that a portable hologram interferometric measuring system has been developed. The measurements in factories have shown
that hologram interferometry coming out from laboratories is capable to solve real industrial measurement problems.

Acknowledgements

The work is a part of the National Middle-Period Program of Research and Development and
is supported by National Board of Technical Development, Machine Tool Factory of Csepel
Works and Machine Tool Works.

References

1.　Ennos,A.E., Measurement of in-plane surface strain by hologram interferometry,
J.Sci.Instrum (J. Phys. E) Vol.1. Ser.2. pp. 731-734, 1968.
2.　Bogár,I., "Calculation method to eliminate the sign problem of fringe order in
holographic interferometry" Per. Polytechnica ME, Vol.24, pp. 279-284, 1982.
3.　Füzessy, Z., Abramson, N., "Measurement of 3-D displacement: sandwich holography and
regulated path length interferometry", Appl. Optics, Vol.21, pp. 260-264, 1982.

Industrial Application of Instant Holography

J. Schörner, H. Rottenkolber

Rottenkolber Holo-System GmbH
Erhardtstr. 12, 8000 München 5

Abstract

The development of holographic test methods has lead to a wide industrial application. Today serial tests of aircraft tyres with holography have become routine. A large chemical company is testing their plastic materials like tanks, wheels and fans with the method of holographic interferometry. In the power engine industry turbine blades are tested holographically to find an optimal shape and to test the vibration behaviour. The automotive industry is using holographic methods for construction optimization. The economic application of these test methods was possible by using instant holography. The principle of a new hologramrecorder is presented. The application of this equipment is shown in examples of testing materials, optimizing constructions and vibration analysis.

Introduction

Twenty years ago, Leith and Upatnieks produced the first holograms by means of laser light. Then, it was possible to produce holograms only at night and down in the cellar because the structures responded to vibrations with a high sensitivity and as the laser units had a low output; nowadays, holographic testing equipment is installed in manufacturing plants, even adjacent to machining equipment. This change was possible due to an intensive development of the laser and the holographic recording equipment.

Today, holography in push-button mode has become a reality. Such automation has resulted in an economic application of holographic testing methods on an industrial scale. Holographic interferometry has become generally accepted as a non-destructive testing method in quality control, optimization of construction and vibration analysis.

HOLO-Recorder using thermoplast film

This successful propagation is largely a result of instant holography, which means that the hologram is available for inspection immediately upon registration. In this method, the hologram is stored as a phase information on a thermoplast film[1], and 3 seconds after exposure the test result is displayed on a video monitor. The HF-85 thermoplast film, which is usually used in this method, is a panchromatic material with a high sensitivity and a high diffraction efficiency (Fig. 1). The optimum exposure energy ranges at 10 erg/cm², and even with 2 erg/cm² good results can be obtained. Hence, the sensitivity of this thermoplast material comes under the same range as that of the usually employed holographic silver film materials (e.g. Agfa Holotest 10 E 75 has 20 erg/cm²). There is no other thermoplast recording material with the same sensitivity. Moreover, the photographic thermoplast material allows a wide range in exposure time. Remarkable variations of the diffraction efficiency were not established in the range from 0.1 to 20 concerning the intensity ratio between the object beam and the reference beam.

Hitherto, the use of thermoplast film was restricted to recording of objects having small dimensions as an optimum diffraction efficiency could be obtained only within an angular area of about 20 degrees between the object beam and the reference beam. Such limitation was eliminated due to the HRC-110 hologram recorder of Rottenkolber Holo-System. At the same time, this recorder is a unit which includes both the optical system required for illumination and the reference beam generation in hologram production as well as the instant development on a thermoplast film. When a hologram is shot, the interim object image produced by an image forming objective is recorded

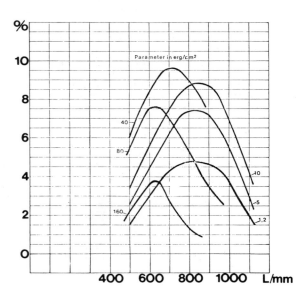

Figure 1. Diffraction efficiency as a function of spatial frequency with exposure energy as a parameter

Figure 2. Schematic set-up of HOLO-Recorder HRC-110,
I main beam, II illumination beam,
III reference beam, ㉖ first image
㉔ hologram

Figure 3. HOLO-Recorder HRC-110 for automatic instant
holography
left side: camera head
right side: control unit with video screen

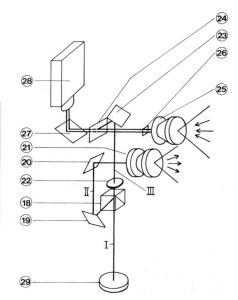

on the thermoplast film. By these means the information to be stored is restricted to the area of optimum diffraction efficiency. Thus, object angle areas as wide as 110° can be covered and yet instant holography is possible. Figure 2 shows the schematic of the structure of this HRC-110 HOLO Recorder. In this hologram-recorder it is only necessary to introduce through reflectors a laser beam expanded to 10 mm. The optical illumination system is a component of this unit. Simple change of the objectives or the use of a zoom objective allows adaption of the field of view to the respective requirements in each case.

The developed hologram is registered by a TV camera and displayed on a monitor. A video recorder may be used for documentation purposes, or the image may be transferred onto a 35 mm film via an additional optical system. Figure 3 shows how this hologram recorder is constituted by various units. This hologram recorder is universal as far as its application is concerned. It is suited for use with both continuous-wave laser and pulsed-laser holography. It allows simple realization of all holographic methods such as double exposure, real-time and time average methods. The packaged design allows the HOLO-Recorder to be used also for testing in areas which involve three-dimensional problems. This hologram camera is an inseparable component of all of our industrial holographic testing equipment.

HOLO-Analyzer

Another requirement for realization of holographic methods with continuous-wave lasers in an industrial environment is the provision of suitable anti-vibration isolation. Rottenkolber Holo-System has developed special active, pneumatic air cushions which are characterized by their high capacity and their very low resonant frequency (Figs. 4 and 5). At an operating pressure of 8 bars, the load may be as high as 1000 kilograms per shock absorber. The natural resonance in the vertical direction is about 1.3 cps, and about 3.2 cps in the horizontal direction. Particularly when holography is used in mechanical engin-

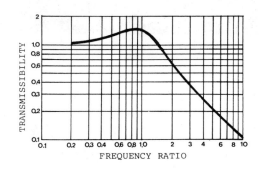

Figure 4. Active pneumatic vibration
isolator HSI-35

Figure 5. Damping characteristic of vibration
isolator HSI-35; resonance frequence
1,3 cps vertical, 3,2 cps horizontal

eering, where cast-iron T-grooved disks are used to clamp the heavy test bars, this pneumatic suspension system must be disigned for high loads. The HT-300 HOLO-Analyzer with its Argon-Ion laser and the HOLO-Recorder (Fig. 6) is a complete measuring equipment on this basis.

Typical applications are deformation or strain studies for design optimization, quality control of products of series fabrication and vibration analyses according to the time-average method. Fig. 7 (BASF, Ludwigshafen) shows the application of the HOLO-Analyzer to optimize the design of a top plate for a copying machine. Fig. 8 (BASF) shows a strain study with a window frame. Depending on the shape and location of the notch in the corner of the frame, different deformations are produced when a spreading or expanding force is applied. Moreover, temperature variations take analogous effects after mounting of the window frame. The interference hologram of Fig. 9 (BASF) produces information about the absorption of forces acting between two intermeshing gears. A holographic disign optimization for the bucket seat of Fig. 10 (BASF) was realized with the consequence that the number of alternations of the failure load was increased from 30.000 to more than 8.000.000.

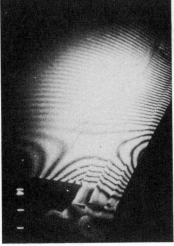
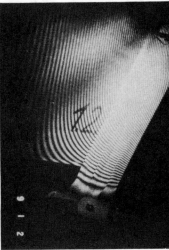

Figure 6. HOLO-Analyzer HT-300 for non destructive testing with the method of instant holography

Figure 7. Design optimization of a top plate of a copying machine with HOLO-Analyzer HT-300, BASF left: before, right: after optimization

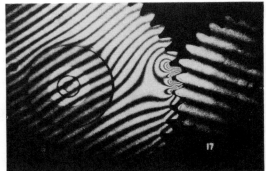

Figure 8. Strain analysis of a window frame (BASF)

Figure 9. Interference hologram showing the absorption of forces acting between two intermeshing gears (BASF)

A great advantage of instant holography is achieved in application of the real-time method. The hologram recorded on the thermoplast film is fixed on the development plate by electrostatic charge so that there is an exact positioning in view of the shot. Repositioning, which is required for the photo-plates developed in the wet method, is omitted. Another advantage resides in the fact that only 30 seconds later a new reference hologram can be shot and developed, which is impossible when photo-plates are used. Roeder et al. describe an example of such use of the holographic instant camera. [3] This method was employed in the study of the action of cement hardening on artificial hip joint components. Real-time holography with Holo-instant camera was employed by Feiertag in his studies of the deformation

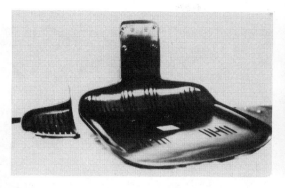

Figure 10. Design optimization of a bucket seat with HOLO-Analyzer HT-300

of electronic pc-boards. [4] Depending on the electrical load acting on the individual components, different expansion values are noted in different board areas. (Fig. 11) At the same time, this method can be used to demonstrate excessive thermal loads on electronic components.

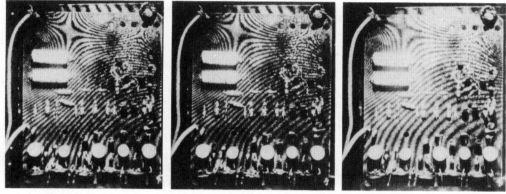

Figure 11. Deformation of electronic pc-boards depending on the electrical load shown by using 'real time' instant holography (Prof. Feiertag)

HOLO-Tyre Testing Unit

Holographic series testing has been realized with the HRT-220 HOLO-tyre testing unit. This equipment is constituted by the essential elements of the HOLO-Analyzer like shock-absorbing system, hologram recorder, and Argon-Ion continuous-wave laser, which are arranged in a testing chamber. A great deal of experience collected in industrial application has been incorporated in the latest development of this HOLO-tyre tester.

A prominent feature is the degree of automation in this system which can be used to inspect tyres for passenger cars, trucks and airplanes for separations, damage and structural weakness. The tyre size to be tested can be set at the controller, and the complete holographic test of both the side walls and the tread can be started by pushing a button, without it being necessary to turn the tyre again. Fig. 12 shows the general arrangement of the system. A pressurised chamber is used as the test station, which is subjected to differences in pressure between 20 and 100 mbar for registration of the interference hologram. Moreover, the pressurised chamber offers protection of the test set-up against airborne sound which exists often due to fans particularly in large manufacturing halls. The tyre is placed onto a turn-table supported on a frame structure in a gymbal mount to be protected against vibration. The HOLO-Recorder is arranged in the centre of the turn-table. The recorder is automatically moved to the respectively desired height, depending on the tyre size. The tyre periphery is checked in four sectors with the illumination and registration objectives scanning an angle of 100°, approximately, each. A 4 W argon-ion laser is used to produce the hologram (i.e. a holographically output ranging between 500 and 1000 mW). One exposure at normal pressure and one exposure after pressure reduction in the chamber are realized per tyre segment. As has been explained in the description of the HOLO-Recorder, the hologram is instantly developed and is immediately available for display at the monitor. The test result can then be documented on a video tape. Here too, the ideal feature is the possibility of real-time inspection which is not possible when the hologram is recorded on conventional photographic material. Fig. 13 shows an interference hologram of a tyre tread. The arrow indicates a flaw.

Figure 12.   HOLO-Tyre Test Unit HRT-220 for automatic holographic instant test of passenger-car, truck- and air-craft tyres

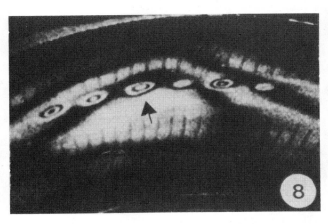

Figure 13.   Tyre hologram showing seperations within the tyre

HOLO-Vibration Analyzer

The vibration analysis both according to the time-average method and using the double-pulse holography is a vast field of industrial application of instant holography. Both methods are employed by leading companies in turbine and automobile industry with great success, with application focussed on development and design.

Time-average method:

Let me give you some examples to demonstrate this point. Fig. 14 (KWU, Muelheim/Ruhr) shows the vibration of a turbine blade at 2686 cps. The representation of the interference lines distinctly shows the nodal lines and the locations of maximum vibration amplitude. The time-average method furnishes a very good representation of the vibrational behaviour with the respective resonance frequencies of a structural element or component, as well as the localization where the disign has to be changed. The knowledge of such vibration diagrams is extremely important in order to determine the localization of maximum sound emission. Daimler, for instance, utilize one of our systems for vibration analysis with an 8-cylinder motor block. [5] Fig. 15 illustrates the study with a metal sheet (Volkswagenwerk). Apart from this time-average method employing a measuring station protected against vibration, such as our HOLO-Analyzer HT-300, the double-puls holography gains an ever-increasing importance.

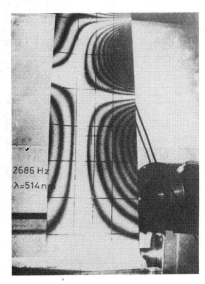

Figure 14.   Vibration analysis at a turbine blade using time-average holography (KWU Muelheim)

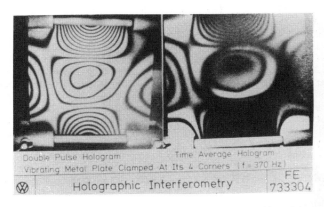

Figure 15.   Holographic test of a vibrating metal plate clamped at its four corners (Volkswagenwerk)

Double-Pulse-Holography:

This method offers the advantage of vibration analysis being carried out under operational conditions. First tests of this kind were made by the Volkswagenwerk. In the meantime, BMW Daimler, Ford, KWU and MAN have adopted the double-pulse holographic method in Germany as well. Fig. 16 shows the camera system which has been developed to meet these requirements. It comprises a camera carriage supporting a scissor-type table which carries the pulse laser and the HOLO-Recorder. An He-Ne-laser beam is directed through mirros into the system for adjustment of the beam path and for hologram reconstruction. The camera carriage is linked up with the mobile controller through a multiple line. The controller cabinet accomodates the laser power supply, a vibration-processing system, a specific trigger system, the HOLO Recorder controller and a TV monitor for reproduction of the interference hologram. Fig. 17 illustrates an example of a holographic vibration analysis at the bottom of an automobile body. The left part of the illustration shows the vibration against time as well as the spacing between the laser pulses. This camera system can be combined with a derotator to study rotating objects (Fig. 18). This system was applied to analyse the vibrations at a turbine periphery at 10.000 revolutions per minute (Fig. 19, BIAS a.TU Hannover).

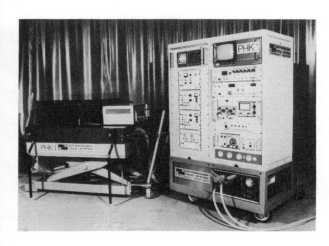

Figure 16.   HOLO-Vibro-Analyzer PHK-100

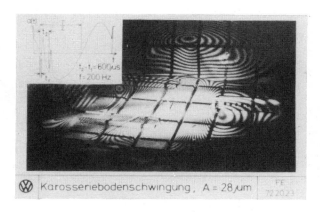

Figure 17.   Double pulse holographic investigation of a car body sheet (VW)

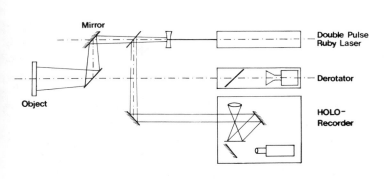

Figure 18.   Schematic set up for combining a derotator with the HOLO-Vibro-Analyzer PHK-100

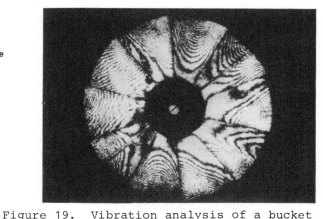

Figure 19.   Vibration analysis of a bucket wheel of a radial compressor (10.000 rpm) using a derotator and pulse holography (BIAS and TU Hannover)

Conclusions

The development and production of holographic test equipment with a high degree of automatization and with instant display of the test result has brought a successful and effective progress in the application of technical holography to the industry. The instant holography was a pressumption for this davelopment. CW-laser- and double pulse laser holography deliver valuable results for non destructivetesting, quality control, optimization of construction and vibration analysis, as well. A further increase in these application can be achieved by developing automatic analysis systems.

Acknowledgements

The authors would like to thank the mentioned industrial companys and research institutes for permanent discussions, which are stimulating the progress of our work.

References

1. Moraw, R., "Photothermoplastische Aufzeichungsmaterialien für die Holographie", 4. Internat. Kongreß für Reprographie und Information 1975 Hannover 13.-17.4.1975,

    Rottenkolber Holo-System GmbH, Produkt-Information HOLO-INTERNATIONAL Holo-Instant-Film HF-85, Oct. 1980.
2. Schörner, J., Rottenkolber, H., Instant Holography - an important step towards automation of holographic test technique "Optoelectronics in Engineering, Proc. Laser 81, Springer Verlag 1982
3. Röder, U., Niess, N., Plitz, W., "Action of cement hardening on artificial hip joint components", SPIE-Conf., Utrecht, Sept. 1980, Vol. 236
4. Feiertag, R.M., "Real-time holography of printed circuits", SPIE-Conf., Utrecht, Sept. 1980, Vol. 236
5. Leis, H., "Vibration analysis of an 8-cylinder V-engine by time-average holographic interferometry" SPIE-Conf. Geneva, April 1983
6. Felske, A., Happe, A., "Schwingungsuntersuchungen an Karosserien und Aggregaten mit Hilfe der holographischen Interferometrie" ATZ Automobiltechnische Zeitschrift, Vol. 75, No. 3/1973 Franckh'sche Verlagshandlung Stuttgart
7. Geldmacher, J., Kreitlow, H., Beek, M.-A., Fagan, W.F., "Schwingungsmessungen an rotierenden Bauteilen unter Betriebsbedingungen mit einem Bild-Derotator", Optoelectronics in Engineering, Proc. Laser 81, Springer Verlag 1982

NON DESTRUCTIVE TESTING USING REAL TIME HOLOGRAPHIC INTERFEROMETRY IN B.S.O. CRYSTALS

J.P. Herriau, A. Delboulbé and J.P. Huignard

Thomson-CSF, Domaine de Corbeville, B.P. n° 10
91401 Orsay (France)

Abstract

Electrooptic and photoconductive Bismuth Silicon Oxyde (B.S.O.) crystals allows the recording of phase volume holograms through the photorefractive effect. We demonstrate applications to real time non destructive testing using holographic interferometry by double-exposure, time average and speckle technics. We also report in this paper, the possibility of using optical fibers for object and reference wavefronts generation. In fine, we show that it is possible to induce energy transfert between reference and object beams. Application to image amplification and mode pattern visualisation of 3-D diffuse object are demonstrated.

Introduction

Photoconductive electrooptic $Bi_{12} Si O_{20}$ crystals (BSO) allow the recording of phase volume holograms by space charge modulation in the transverse EO configuration. Recording and erasure energy S^{-1} at argon laser lines have been found nearly equivalent to high-resolution photographic plates. (λ = 488 nm : BSO crystals $S^{-1} \sim$ 100-300 $\mu J.cm^{-2}$), and no limitation to spatial carrier frequency is observed. The material is biased with a tranverse electric field E_o in the 110 direction and illuminated with holographic fringes. (Incident beams in the $\bar{1}$10 direction). A photoinduced space-charge density due to the difference between the distribution of trapped electrons and trapped holes is so generated. The created resultant space-charge field modulates the refractive index through linear electrooptical effect (phase volume hologram recording). [1]

Among other characteristics, a precise knowledge of the spatial frequency response of the recording parameters in BSO constitutes an important feature for imaging applications. In this paper, we show that it is also possible to record holograms of three-dimensional diffuse objects in BSO crystals utilizing the FWM configuration and to visualize in real-time the mode pattern of a vibrating structure.

Hologram formation in the crystal volume is generally accompanied by an energy redistribution between the two interfering light waves. This intensity transfer (beam coupling), initially observed in photorefractive $LiNbO_3$ crystals by Staebler and Amodei [2], is due to a permanent phase mismatch ψ between the holographic grating and the fringe pattern. It has been shown that the maximum energy transfer is obtained when the incident fringe pattern and the photoinduced index change are shifted by $\Pi/2$. We demonstrate in the following that an efficient intensity redistribution can be obtained with the BSO in the "drift" recording mode when the crystal or the fringes are moving at a constant speed. As analysed in the references [3-4], the induced additional phase shift enhances the amplitude of the stationary $\Pi/2$ shifted component of the grating which causes the beam coupling. This new effect in BSO materials is applied to image amplification and vibrations analysis.

Phase conjugation in four wave mixing configuration (FWM) and applications

Analysis of wavefront reflectivity versus the fringe spacing in the FWM configuration is important for imaging purposes. The experimental results are interpreted on the basis of the Kukhtarev's theory [5], which applies to photorefractive crystals with long transport lengths of the photocarriers. This theory describes in a dynamic way any possible changes of the fringe pattern along the crystal length that are due to the intensity distribution between the two writing beams (beam coupling) for both the initial and the steady-state stages of the hologram formation.

According to Kukhtarev et al. an expression for the reflectivity ρ has been derived as a function of the fringe spacing Λ ; [6]. For low index modulation we derived the following expression :

$$\rho = R \frac{E_o^2 + E_T^2}{1 + \frac{E_T^2}{E_q} + \frac{E_o^2}{E_q^2}}$$

The diffusion field E_T and the maximum space charge field E_q are given by :

$$E_T = \frac{2\pi}{e}\ \frac{kT}{\Lambda} \quad ; \quad E_q = \frac{e}{2\pi\ \varepsilon_0\ \varepsilon_r}\ N\Lambda$$

where N is the concentration of trapping centers.

For high spatial frequencies, the fields E_0, E_q and E_T are of the same order of magnitude, and the experimental curve tends to be asymptotic to that provided by a pure diffusion process. This is shown in Fig.1, in which the amplitude-modulation transfer function ($\sqrt{\rho}$ versus Λ^{-1}) of the coherent four-wave mixing process has been plotted.

From the results presented, conclusions can be drawn for practical applications. BSO or equivalent BGO provides the interesting property that the modulation transfer function can be controlled with externally applied field E_0. Recording by drift with E_0 = 6 kV cm^{-1} provides low-pass spatial filtering. Through a pure diffusion process, E_0 = 0, high-pass filtering is achieved and E_0 = 2 kV cm^{-1}, permitting a flat spatial-frequency response. These specific properties should be important for application of the crystal to coherent image processing.

Holographic interferometry of 3-D diffuse objects is commonly met in the field of industrial non destructive testing of mechanical structures. Now, due to the high sensitivity BSO materials, real-time analysis becomes possible.

The optical configuration used to record the dynamic hologram and to generate the phase conjugate wavefront is given in fig. 2 and more precisely described in ref. [7]. The object structure is illuminated by a diffusing screen. The plane wave reference beam transmitted by the crystal is retroreflected by a mirror and so diffracts in real-time the incident object wavefront. Beam splitter B.S. projects the real diffracted image on an instant camera or on a vidicon tube.

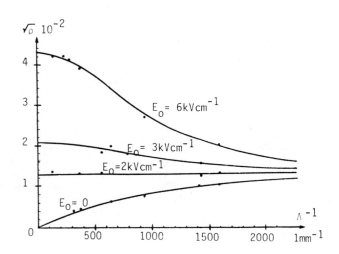

Fig. 1 - Amplitude modulation transfer function : dots, experimental curves

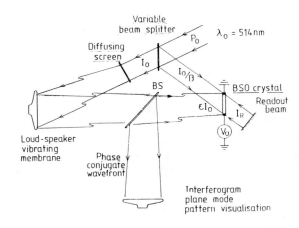

Fig. 2 - Optical set-up dynamic holography and phase conjugation of a diffuse object structure

The object so considered scatters a low power density on the recording medium and consequently it results a very large intensity beam ratio of the interfering waves on the crystal. Since we use a dynamic material, the time constant for hologram build-up and available optical power in the image plane can be now adjusted in real time. When the object structure is vibrating at frequency F, we obtain a diffracted image (Id) on which are superimposed dark fringes corresponding to the zero of bessel function J_0^2 [$4\pi\delta/\lambda_0$]. Following Powell and Stetson theory [8], this arises when the time constant required to build-up the hologram at saturation is long compared to vibration period T = 1/F ($\tau \gg T$), for time average intensity recording in the crystal volume.

$$< Id\ (x,y) > = Id\ (x,y)\ J_0^2\ (4\pi\delta/\lambda_0)$$

δ : local vibration amplitude, Id diffracted image intensity when no excitation on the object structure.

For each excitation frequency of the membrane, the so described method, permits in a dynamic way, a mode pattern visualisation of the vibrating structure. Corresponding experimental results are given in figure 3. Permanent applied voltage to the crystal is 8 kV cm^{-1}. The observed optimum time constant is τ = 500 ms, to build-up the interferograms.

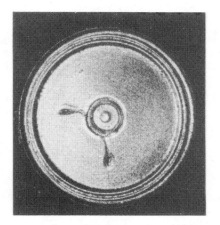 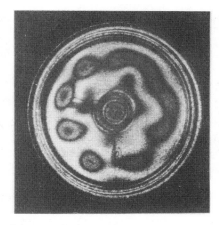

Fig. 3 - Time-average interferogram of a diffuse vibrating object : Mode pattern visualisation of a loudspeaker membrane excited at different frequencies.

Speckle-free time-average interferograms

Imaging with coherent wave front generation usually results in a noisy image because of the use of laser light illumination. The noisy image arises unavoidably from de complex interference pattern of scattered light resulting from dust particles, surface defects, or inhomogeneities of the recording medium. The corresponding speckle texture superimposed on the image plane provides a reduction of spatial resolution and severely degrades the signal to noise ratio of any coherent imaging process.

The proposed method for speckle-noise suppression is based on a time integration in the detector plane of N coherent images having independente speckle noise pattern. Under such conditions, the signal-to-noise ratio in the image plane is improved by $\sqrt{N}$ [9-10]. In this paper we demonstrate that holographic speckle free imaging is possible also when a four-wave mixing configuration is used. The experimental set up is the same of the one shown in fig. 2. In this experiment the object is a vibrating reflecting membrane excited by a transducer at frequency Ω_o. In this case, the diffusing screen is randomly moved step by step. The random speckle texture superimposed upon the phase-conjugate wave front is seen in figure 4a. A large improvement of signal noise ratio is thus achieved by integration on the camera located in the interferogram plane of N exposures of the membrane vibrating at frequency Ω_o. Experimental results are shown in figure 4b for N = 9 and figure 4c for N = 144.

The same type of experiment could be performed by spinning the diffusing screen at a constant speed.

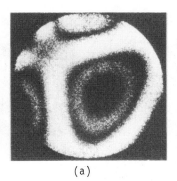 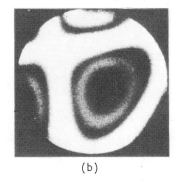 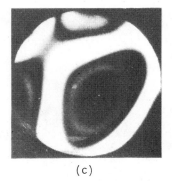

        (a)   (b)   (c)

Fig. 4 - Speckle noise suppression with incoherent superpositions of N coherent interferograms

    (a) N = 1   (b) N = 9   (c) N = 144

Real time holography using optical fiber illuminators

A basic experimental problem in holography is to achieve a positional stability of the components such as mirrors, beam splitter... do not exceed λ/2. Mechanical vibrations, thermal gradients and air turbulence may introduce deleterious effects in the recording step. The use of optical fibers in the recording and readout processes lead to a drastic reduction in the number of components to be adjusted and eases the environmental requirements [11]. The experimental set-up is shown on figure 5. This experiment use a classic four-wave mixing configuration with B.S.O. conjugate mirror. The characteristics of optical cables are the following :

. TH-CSF prototype single mode fibers

. core diameter ∿ 4.5 μm

. numerical aperture ∿ 0,1

. lenght ∿ 6 m

. coupling rate > 70 % at λ = 514.5 nm

The input beam splitter is a uncoated plate of glass, which allows a hight optical intensity (500 mW) in the object single mode fiber and only 20 mW in the reference single mode fiber. As can be appreciated in figure 6, this technic allows a good quality of real time interferograms visualized with a polaroïd camera or a vidicon tube.

The avantages of using the single mode optical fibers are the following :

- the possibility of locating the object to be analyzed at a distance from the laser source and the viewing equipment (camera and T.V. monitor)

- the capability of installing the link between object and the viewing equipment in a highly electromagnetically perturbing environment.

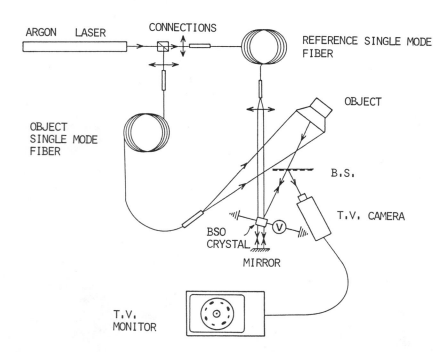

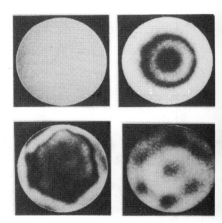

Fig. 5 - Experimental set-up of a degenerated four-wave mixing configuration in BSO crystal, using single mode optical fiber illuminators, for real-time projection of the interferogram on a T.V. camera.
object : piezoelectric transducer

Fig. 6 - Real-time mode pattern visualisation of a diffuse piezotransductor (∅ = 3 cm) using single mode optical fiber illuminators

Energy transfer in two-wave mixing TWM and applications

In photorefractive media, a $\pi/2$ phase shift exists when the photoinduced index change occurs by diffusion of the photocarriers (no "internal" or externally applied electric field). As a consequence of this phase shift, a permanent or transient energy transfer and coherent image amplification have been obtained in photorefractive $LiNbO_3$ crystals [12-13]. However, these materials have a rather low sensitivity and long recording-erasure cycles.

In BSO crystals, the non-local response of the material is induced by moving the fringes at a constant speed. Such a displacement enhances the amplitude the $\pi/2$ shifted component of the index modulation and consequently greatly improves the energy transfer. According to the formalism of the reference, 14 the intensity of the signal wave emerging from the crystal is given by the following expression :

$$I_S = I_{S_0} \exp\left[(\Gamma - \alpha)\ell\right] \quad ; \quad I_0 \ll I_{R_0}$$

where I_{S_0} is the incident signal-beam intensity, I_{R_0} is the reference-beam intensity, α is the absorption coefficient at recording wavelength λ, Γ is the exponential gain coefficient, and 1 is the crystal thickness.

For the recording conditions usually used with the BSO (applied electric field, $E_0 \sim 6\text{-}8$ kV.cm^{-1}; fringe spacing, $\Lambda > 3$ µm), the exponential gain has been expressed as a function of the fringe-displacement speed and the applied field E_0 : [14]

$$\Gamma = \frac{4\pi}{\cos\Theta} \, CE_0 \, \frac{kV\tau}{1 + k^2V^2\tau^2}$$

where 2Θ is the angle between the two recording beams inside the crystal, τ is the time constant to build up the holographic grating, $K = (2\pi/\Lambda)$ is the grating wave vector, and Λ is the fringe spacing. C is a constant whose value depends on intrinsic parameters of the material. A resonance on the energy transfer is reached when the fringe-displacement speed v_0 satisfies the relation :

$$K v_0 \tau = 1$$

The measured value of Γ is 2.4 cm^{-1} at $\lambda = 514$ nm ($E_0 = 10$ kV cm^{-1}). This corresponds to an increase of the incident object wave intensity of about 50 % after passing through the crystal (1 = 10 mm). In this case, amplification is reached and the conditions $\Gamma > \alpha$ is fulfilled.

The optical configuration used for image amplification is shown in fig. 7.

COHERENT IMAGE AMPLIFICATION

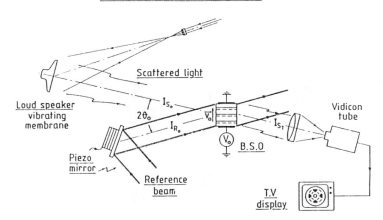

Fig. 7 - Two wave mixing experiment with photorefractive BSO crystals. Recording wavelength, $\lambda = 514$ nm. Crystal size, 10 mm x 10 mm. Thickness, 10 mm ; beam angle, $2\Theta = 10°$. Applied electric field in the 001 direction, $E_0 = 10$ kV.cm^{-1}. Incident intensities $I_{R_0} = 10$ mW.cm^{-2} and $I_{S_0} = 5$ µW.cm^{-2}. Fringe displacement speed, $V_0 = 3.2$ µm.s^{-1}. Solid lines, interference fringes = dashed lines, index modulation.

The three-dimensional (3-D) object to be used is a white painted loudspeaker membrane that scatters a low intensity on the crystal : $I_{S_0} \sim 5 \ \mu W.cm^{-2}$. The object diameter is $\emptyset = 7cm$, and its distance to the crystal is 70 cm. The signal wave interferes with an off-axis reference beam whose intensity is $I_{R_0} = 10 \ mW.cm^{-2}$. The resulting fringe-modulation level incident upon the crystal is $m = 4.5 \times 10^{-2}$, and the corresponding time constant for hologram recording at saturation is τ 150 msec. Since the build up time of the hologram at saturation is inversely proportional to incident light intensity, a shorter time constant could be obtained at higher pump-beam intensity I_{R_0}. The mean value of the angle between the two recording beams is $2 \ \Theta_0 = 10°$, and fringe-displacement speed v is controlled by a piezomirror driven by a ramp generator.

Behind the BSO a lens is placed on the signal-beam path in order to project a real image on the detector (vidicon tube or instant camera). The experimental results are shown in Fig. 8. Figure 8(a) shows the image intensity I_T transmitted through the crystal in the absence of the pump beam : $I_T = I_{S_0} \exp(-\alpha l)$. Figure 8(b) shows the amplified image that is due to energy transfer from the pump beam to the signal : $I_S = I_{S_0} \exp(\Gamma - \alpha)l$ (10 x amplification).

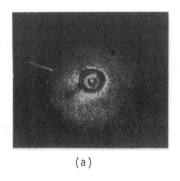
(a)

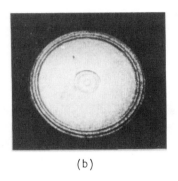
(b)

Fig. 8 - (a) Image transmitted through the crystal (loudspeaker membrane)

(b) Stationary 10X - amplified image that is the self- diffraction of the pump reference beam in the shifted phase-volume hologram.

If the object is now vibrating at frequency $\Omega/2\Pi$, the image intensity emerging from the crystal is given by the relation : [15].

$$I_S = I_{So} \exp \left\{ \left[\frac{2 \ \Pi \ CE_0}{\lambda \cos\Theta} \ J_0 \left| \frac{4 \ \Pi \delta}{\lambda} \right| - \alpha \right] l \right\}$$

If the vibrating object structure has a complex mode pattern, a spatial modulation of the energy transfer is induced. The dark fringes appearing in the image plane (Fig.9) correspond to the zero of the Bessel function $J_0 \ (4 \ \Pi \delta/\lambda) = 0$. With this two-wave mixing configuration, the intensity in the diffracted mode pattern (see relation [6]) is different from that obtained with the usual formalism of time-averaged holography, which does not consider the beam -coupling effects. This approach derives an image intensity of the vibrating structure proportional to $J_0^2 \ (4 \ \Pi \delta/\lambda)$ [8].

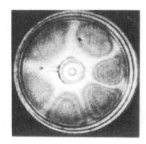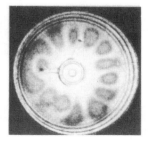

Fig. 9 - Application of the energy transfer and image amplification to the pattern visualisation of the vibrating loudspeaker membrane. (vibration frequencies, f = 2.2 kHz and f = 1.6 kHz)

Conclusion

Dynamic holography and real-time non-destructive testing has been achieved utilizing two different optical configurations with BSO crystals. Appreciable image quality and relatively fast response (T $\sim$ 500 ms) have been obtained with 3-D diffuse object, in the four wave mixing configuration. However, the possibility to use energy transfer in two-wave mixing experiments is an alternative experimental configuration since it allows image amplification and shorter response time (T $\sim$ 150 ms).

Potential applications of this experiments are in the field of the industrial real-time non-destructive testing with the following advantages :

- reusable materials (B.S.O. crystals)
- real-time process
- speckle free imaging
- demonstration of optical fiber illuminators.

References

[1] - J.P. Huignard and F. Micheron, Appl. Phys. Lett. 29, 591 (1976).

[2] - D.L. Staebler and J.J Amodei, J. Appl. Phys. 43 (1972) 1042.

[3] - V.L. Vinetskii, N.V. Kukhtarev, S.G. Odulov and M.S. Soskin, Soviet Phys. Usp. 22 (1979) 742.

[4] - J. Feinberg, D. Heiman, A.R. Tanguay and R.W. Hellwarth, J. Appl. Phys. 51 (1980) 1297.

[5] - N. Kukhtarev, V. Markov, S. Odulov, M. Soskin and V. Vinetskii, Ferroelectrics 22 (1979) 949, 961.

[6] - J.P. Huignard, J.P. Herriau, G. Rivet and P. Gunter, Opt. Lett. 5 (1980) 102.

[7] - J.P. Huignard, J.P. Herriau, P. Aubourg and E. Spitz, Optics Lett. 4 (1979).

[8] - R.L. Powell and K.A. Stetson, Opt. Soc. Am. 55 (1965) 55.

[9] - A. Kozma and C.R. Christensen, J. Opt. Soc. Am. 66, 1257 (1976).

[10] - J.P. Huignard, J.P. Herriau, L. Pichon and A. Marrachki, Opt. Lett. 5, 436 (1980)

[11] - A.M.P.P. LEITE, Opt. Comm. 28, 3 (1979).

[12] - V. Markov, S. Odulov and M. Soskin, Opt. and Laser Tech. April 1979.

[13] - N. Kukhtarev, V. Markov and S. Odulov, Optics. Comm. 23 (1977) 338.

[14] - J.P. Huignard and A. Marrakchi, Opt. Comm. 38, 249 (1981) Opt. Lett. 6, 622 (1982).

New holographic means to exactly determine coefficients of elasticity

Gerhard Schönebeck

Kraftwerk Union AG, D-4330 Mülheim/Ruhr

Abstract

Within this report the practical uses of holography for more exact definition of the module of elasticity E for specific materials will be given. In conjunction with this theme a new holographic method "The Step Hologram" with which the measurement range of the holographic interferometry can be widered to approximatly factor 10...20 times will be discussed. The Step Hologram allows separation of the elastic-linear lines from elastic-non-linear lines respectively and from plastic deformation.

As an application excample, the exact measurement of the E-module for a brittle material such as Porcelin will be given. This material has been specifically chosen for its elastic properties. Porcelin is a white plastic material which, like Gypsum presents practical usable results. Porcelin however unlike Gypsum is not so succeptable to breakage.

Introduction

First one should be familiar with the concept of the new development, the Step Hologram[1], to measure exactly the elastic module E. This development can be utilized in other areas as well.
With a double exposure hologram deformations from subject surfaces can be perceived only to within about 10 μm, because of the closeness of the lines of interferences to one another. The new Step Hologram extends this area of perception to a factor of 10 to approximately 20 so that deformations to one-tenth of a millimeter can be measured. Naturally one can achieve a similar effect through the adoption of separate double exposure holograms with stepped weighting.

By taking 10 to 20 double exposure holograms during the stepwise enlarged load one could achieve this effect as well. But this would result in uncertainty about the exact deformation due to the after effects of deformation during the intervals of two successive double exposures. Because these temporary after effects can not be known, they can also not be holographically defined. The Step Hologram avoids these disadvantages. It is especially suitable to show elastic, thermic or plastic characteristics.
It is possible to show non elastic effects with interferometric exactitude by combination of stepwise applied load and exposures without loadchange but with timed intervals.
On the other hand the elastic module at nearly no load can be measured by this method as well so that over a wide range of load the module of elasticity can be determined to an exactness never reached before.

The necessity of determining the module of elasticity E for various materials (as Gypsum, Porcelin, plastics, synthetic rubber etc.) became obvious when these materials were introduced as model materials for the holographic investigation of stationary or instationary loaded component parts.[2] The Step Hologram can also be used to indisputably clarify the Hooke's elasticity principle.

Up to now the customary strength calculations of the elastic module have had relatively minute meaning as regards the exactness of measurement. The middle value of the E-module will aid in revealing this problem. This value is permissible and also within reason so long as the calculation of the deformation or stress demands only an approximate level of exactitude, i.e. as more or less an understanding of the exact assessment.
When and if with further development of calculation methods are found these can also be holographically controlled. These more exact methods of calculation should then be applied with the more exact values of the E-module. It will only then be possible to definitely assess the validity of the new process.
The new Step Hologram is therefore a very useful tool which has provided a great amount of information in the past few years. It is not to be confused with the partial lighting principle[3] which provides totally different information. Both methods do complement each other in the endeavor to measure linear and non linear phenomenon caused by deformation but those portions of the measurement which were here-to-for unmeasurable can and have been measured through the use of the Step Hologram.

The principle of the Step Hologram

The Step Hologram is a result of overlapped double exposure holograms which are exposed with a special arrangement (motordiven exposure slot).
With a usual double exposure hologram, the hologramplate is exposed completely by the first und second exposures. The Step Hologram is however different. Here only a strip of the width of the slot in a metal foil is opened to exposure. During the first exposure, only a strip of width d is exposed (Zero hologram at slot position 1) after this first exposure, the slot moves by a distance d/2 (slot position 2) and the load and the resulting deformation of the experimental material is changed (Figure 1). During this second exposure a strip with the width d/2 is double exposed. In this way one receives a small double exposure hologram with d/2 overlapped exposures (positions 1 and 2). At the same time a new strip of the width d/2 is now first exposed (new Step Hologram).

When the slot is once more advanced by d/2 (slot position 3) this strip will be exposed once more to produce a double exposure hologram (positions 2 and 3). This process repeats itself until the entire hologramplate has been exposed. The number of repetitions is dependent upon the expected results.

By use of this process we receive a row of double exposed holograms on which all are adjacent to one another providing an ungapped grouping of the deformations of the tested materials in regard to the applied stress. It is therefore possible to extend the amount of deformation normally found through use of holography up to the factor of 10 to 20 times.

The picture in Figure 2 is a view of the slot breach. A single foil with a slot width of d is driven by a motor across the hologramplate exposing at each step a width of plate equal to d/2. By the use of a contact plate and a sliding contact the movement is constrained to d/2 after applying voltage to the motor. The brass foil is fed by the motor from a roll on one side and taken up by a roll on the other side.

The experiment

In order to measure the elastic module E for the regime of small loads prismatic bars with a cross section of H = 12 mm x B = 48 mm and a free length of l = 300 mm clamped at one side were used. The original model was a steel bar made from X22CrMoV121. This was taken as a master model to produce molds out of silicon rubber wherin bars from Gypsum and Porcelin respectively were cast. By that way bars were practically of the same geometric values. The bars were horizontally clamped at one end and on the other end weights were hanged step by step and friction free (Figure 3). A mirror was mounted over the bar at an angle of 45° (Figure 4).

The stress deflection $|\bar{n}|$ on the free end will now be calculated holographically. (See reference 4 and 5.)

The formular is:

$$|\bar{n}| = \frac{\lambda}{2} \frac{1}{\cos \alpha/2} \cdot N \qquad (1)$$

With λ being the wave length of the Laser; α is the angle between direction of illumination and direction of observation; N is the order of the white interference strip until the free end of the bar. The deflection $|\bar{n}|$ on the free end of the bar of a length l is:

$$\delta = |\bar{n}| \qquad (2)$$

The results are calculated for the deflection δ for the clamped prismatic bar with constant cross-section and single load on the free end.

The formular for the deflection δ is:

$$\delta = \frac{1}{3} \frac{F \cdot l^3}{E \cdot I} \qquad (3)$$

Where F = force; l = free length and $I = \frac{B \cdot H^3}{12}$ the areal moment of inertia.

From (1), (2) and (3) results the E-module

$$E = \frac{2}{3} \cdot \frac{F \cdot l^3}{I} \cdot \frac{\cos \alpha/2}{\lambda} \cdot \frac{1}{N} \qquad (4)$$

Results

In the diagrams (Figure 5, 6 and 7) the force F is listed over the deflection δ. As one can see from this information, the calculation lines of Gypsum and Porcelin are proportional to the line for steel in the area of the origin. In this regime there seems to be no hysteresis loop. Gypsum and Porcelin are therefore in agreement with Hooke's principle. That infers that Gypsum and Porcelin can serve as very good model materials for elastic experimentations regarding the properties of steel. The found relation of the E-modules of Porcelin and steel are:

$$\frac{E_{Por}}{E_{st}} \approx \frac{1}{10}$$

This means that under equal stresses the deformations to the Porcelin models will be 10 times that of the steel models. Therefore, more exact calculations can be made at the lower ranges by using Porcelin. This can be used to great advantage in holographic experimentation. The use of Porcelin also saves expense and time over the use of steel.

Table 1 gives an overview of the E-module with the tolerance zone (at 20°C), the rupture stress and the specific gravity, to destinate the relations between the model materials steel, Porcelin and Gypsum for steady and unsteady experiments.

Table 1

Material	E-Module	Rupture stress	Specific gravity
	$N \cdot mm^{-2}$	$N \cdot mm^{-2}$	$kg \cdot dm^{-3}$
Gypsum	5 140 ± 20	2,89	1,08
Porcelin	21 050 ± 70	11,17	1,86
Steel X22CrMoV121	210 000 ± 700	(950)	7,75

References

1. Schönebeck, G.: Das Stufenhologramm, ein neues Verfahren. VDI-Berichte Nr. 480, 1983.

2. Schönebeck, G.: Die Anwendung von Modellwerkstoffen in der Holografie. VDI-Bericht Nr. 366, 1980, s. 79 - 82.

3. Neumann, W., Krause, F.: Das holografische Partialbelichtungsverfahren. Techn. Messen, 42. Jahrgang, 1982, Heft 3, S. 99 - 103.

4. Schönebeck, G.: Eine allgemeine holografische Methode zur Bestimmung räumlicher Verschiebungsfelder. VDI-Berichte Nr. 313, 1978, S. 155 - 162.

5. Schönebeck, G.: Eine allgemeine holografische Methode zur Bestimmung räumlicher Verschiebungen. Diss. TU-München, Febr. 1979.

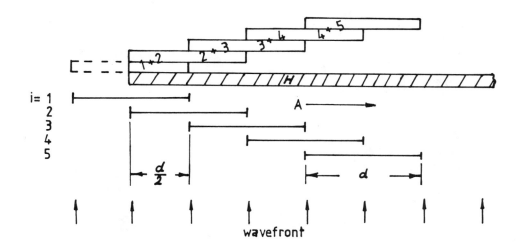

Figure 1. Principle of the step hologram
   Hologram: H; Aperture slot width observation: d, Aperture step width: $\frac{d}{2}$,
   Exposure number: i = 1,2,3...1, Double exposure hologramm: 1+2, 2+3, 3+4, 4+5 etc.
   direction of movement of the slot d: A.

Figure 2. Motor driven Aperture slot for
   the Exposure of step holograms .

Figure 3. Experimental set up with experimental bar
   clamped to the right and miror at 45°.

Figure 4. Porceline bar, clamped to the right with displayed interference lines depicted by a step hologram.

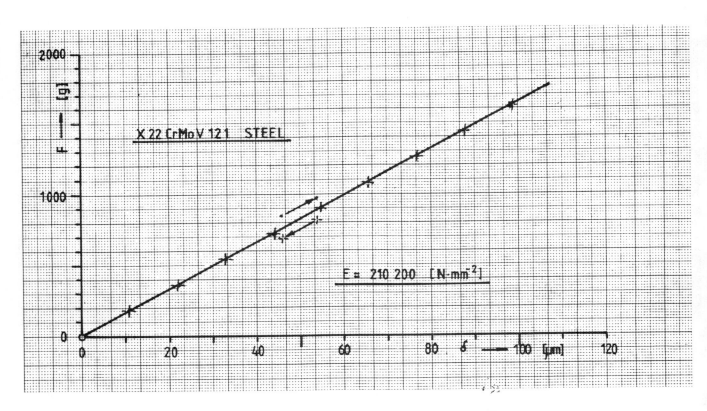

Figure 5. Deflection δ and force F on a steel bar of X22 Cr Mo V121 at 20 °C (small deflection).

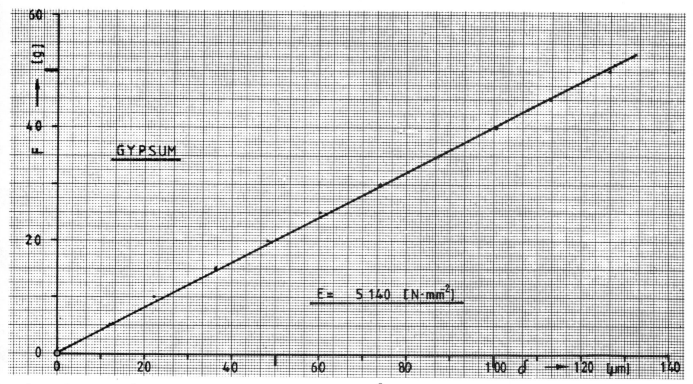

Figure 6a. Deflection δ and force F on a Gypsum bar at 20 °C (deflection 0....125 μm).

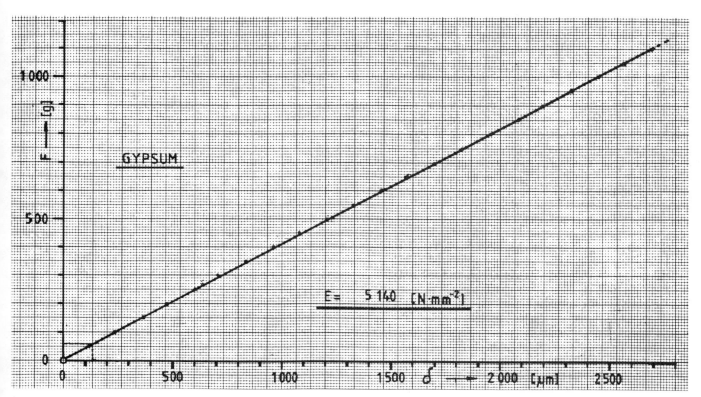

Figure 6b. Deflection δ and Force F on a Gypsum bar until breakage occurs (0....2500 μm).

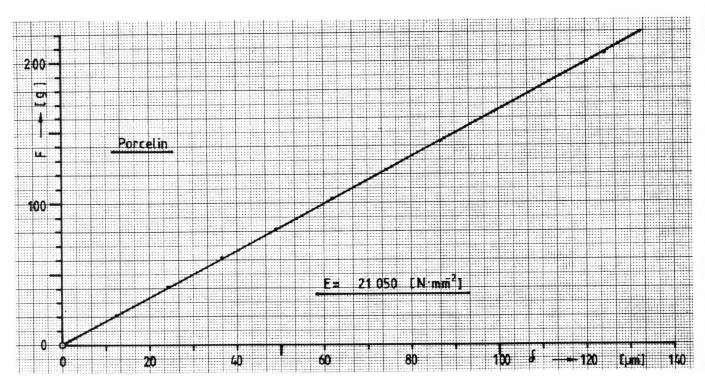

Figure 7a. Deflection δ and force F on a Porceline bar at 20 °C (deflection 0...125 μm).

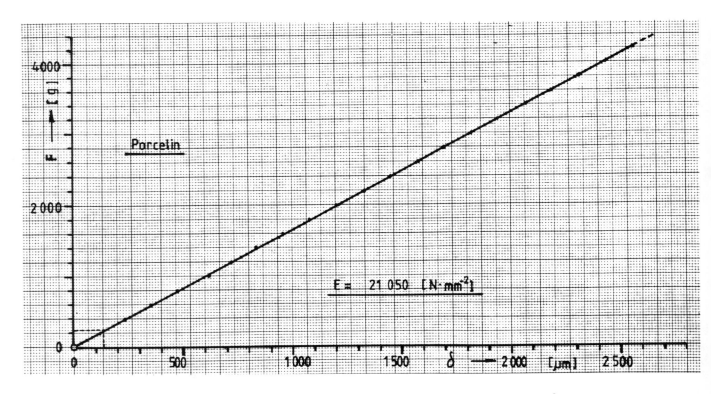

Figure 7b. Deflection δ and force F on a Porceline bar until breakage occurs at 20 °C (0...2500 μm).

Session 3

Holography, Speckle, and Moiré Techniques I

Chairmen
H. Kreitlow
Bremen Institute for Applied Radiation Techniques, West Germany
P. Paulet
Centre Technique CITROEN, France

Application of Denisyuk pulsed holography to material testing

R.L. van Renesse and J.W. Burgmeijer

Institute of Applied Physics TNO, P.O. Box 155, 2600 AD Delft,

the Netherlands

Abstract

When holography is applied outside the laboratory, some well known problems are experienced: vibrations, rigid body motion, stray daylight. Pulsed holography can overcome the difficulties with vibrations but the other problems are less easily solved. When the object area to be holographically tested is small, a very simple and convenient method may be employed, which was earlier described by Neumann and Penn[8]; they fixed the hologram holder rigidly on the object under test, thus avoiding rigid body motion of the object with respect to the hologram. In a similar configuration Denisyuk reflection holograms are made without any necessity of darkening the environment. The authors believe that the main reason that this technique is not widely used, is due to difficulties generally encountered in processing the Denisyuk hologram to good quality. A simple processing technique is described resulting in high quality reflection holograms which may be analysed by microscope up to interference fringe densities of about 30 fringes/mm. As examples the results of a projectile impact study and the study of early fatigue crack detection in a critical aeroplane structure will be presented.

1. Introduction

Surprisingly the history of wavefront reconstruction by recording standing waves in a volume-medium, appears to date back as far as 1810, when Goethe's extensive work on chromatics is published.[1] In this work we read the findings of T.J. Seebeck that silver-chloride under circumstances tends to adopt the color of the light it is exposed to. Along this principle the first color photographs were inadvertently made by E. Beckerel about 1848, who was unaware yet of the underlying principle.

After W. Zenker's explanation of the phenomena through the theory of standing light waves in 1868 it was the celebrated M.G. Lippmann who in 1891 succeeded in making brilliant color-photographs of excellent keeping quality that have stood the tooth of time and can still be found reconstructed in some musea.[2] It was soon recognized that color photography was to be developed along more practical lines and thus the fascinating technique of Lippmann photography became of only academic interest. It must be emphasized however that at the turn of the century the fabrication of simple volume reflection holograms was practically possible and was in fact accomplished by H.E. Ives in 1908[3] who actually made holographic optical elements using the green mercury line while applying a photographic bleach technique. In 1892 Lippmann had already demonstrated dichromated gelatin as an excellent medium for volume color photography.[4] The real potential of the technique, apart from color photography, was not realized however and practical applications had to wait until concrete form was given to Gabor's ideas (1948) on wavefront reconstruction. It was Yu.N. Denisyuk[5] who in 1962 revived the old Lippmann-process and demonstrated its suitability to wavefront reconstruction. While the simplicity of the optical principle has promoted its application to holographic interferometry, contrary to this the need for extremely fine-grained emulsions and the difficulties connected with their processing have restricted the general application of the technique in this field.

It is the purpose of this paper to discuss these difficulties and their convenient solution as well as the practical applicability of the technique to holographic interferometry.

2. An NDI-application

2.1 Crack-detection

An investigation presently carried out by our institute will serve as an introduction to the principles of the technique. We are developing a non-destructive inspection (NDI) method that will detect the initiation and growth of fatigue cracks at fastener holes in the wing of an aircraft. Using a periodical inspection procedure, fatigue crack growth has to be detected before the crack size has reached its critical value. The demand for reliable inspection methods comes from the need to fly some aircraft well beyond their initial design lifetime.

For the most current inspection methods, the detection of cracks under installed fasteners forms an only partly solved problem. The inspection procedure must be appropriate for use in the aircraft hangar without the need to dismantle the wing or dismantle fasteners. This last restriction is understood by taking into consideration that badly installing of fasteners is one of the principal causes of crack initiation in fastener holes.

One is most interested in the inspection of three fastener holes in a highly stressed area, the so called 'critical area', where a higher probability of fatigue crack initiation exists. This critical area has a flat, diffuse metal surface and is approximately circular with a 60 mm diameter. It is situated in the lower wing skin on a place not easily accessible for current NDI-apparature. Holographic techniques are known for the detection of fatigue crack growth[6,7], but many of them have serious drawbacks for our application. For the detection of these flaws not only a suitable loading mechanism has to be found, but also some extra conditions inherent to the application outside the optical laboratory must be satisfied. We will now consider six principal items connected with the technique.

1. A loading method has to be found causing a detectable anomaly in the fringe pattern, which is a map of surface deformation, in the neighbourhood of holes were cracks have reached a length much smaller than the critical length. At the tip of the flaw a plastic deformation zone will be formed, which will reach the surface far earlier than the crack opening. Most NDI-methods are based on crack detection and not on detection of this plastic deformation zone. It is assumed that this zone will induce an anomaly in the deformation during loading. Several loading methods can be considered, for instance direct mechanical stressing, vibrational excitation, and impulse loading. For the latter two methods a double pulse laser would be necessary. In a preliminary experiment in a test facility for aircraft model constructions, we have demonstrated that a mechanical stressing method is successful. For an inspection procedure in a hangar however, this may be unpractical and therefore other loading methods are still searched for.

2. During hologram exposure vibrations and motions must not cause translations between object and holographic plate more than a fraction of the wavelength. The method applied by most holographers is to shorten the exposure time, such that translations during this time-interval are restricted. To get sufficient exposure energy however a high power pulsed laser is needed. We have used a ruby laser having a pulse width of 25 ns and a pulse energy of 30 mJ.

3. Rigid body motions of the object relative to the holographic plate between the two exposures of the hologram will give an unwanted extra number of fringes during reconstruction. There are many solutions for this problem, which all have their own drawbacks. These fall into 3 categories:
a. Using a double pulse laser, the pulse interval can be as short as 1 µs. The object translation in this interval will be small enough to prevent most of the formation of unwanted fringes. A drawback then is the restriction to dynamical loading techniques, for instance vibration excitation and impulse loading.
b. A compensation after the exposure of the interferogram is possible, e.g. using fringe control, sandwich-holography, holographic-moiré. At the present time these techniques are still to sophisticated for a routine-based inspection by untrained personnel.
c. During the exposure of the hologram, rigid body motions can be compensated for by a technique introduced by Neumann and Penn.[8] They clamped the holographic plate holder to the object with the holographic plate as near to the object as possible, and used a collimated laserbeam as to optimally compensate for rigid body motion. (Figure 1). The holographic technique applied is Denisyuk reflection holography.[5] Although Denisyuk reflection holography has the advantage of simplicity, the impossibility to adjust the intensity-ratio between object and reference beam limits its applications. Generally objects holographed this way must possess a high reflectance or otherwise be coated with a suitable metallic paint, to render a sufficient intensity-ratio.[8]

We have used the Denisyuk technique without any surface preparation of the object. Therefore it is demanded that the holographic processing performs optimally with respect to diffraction efficiency and signal-to-noise ratio. This subject will be discussed in section 3. A drawback of the technique may be that the size of the inspected area is restricted to the size of the holographic plate. For our application however the size of the area to be inspected was small enough to be covered by a 4x5" plate. The technique offers a very efficient compensation for machine vibrations and motions. Some small vibrations between object and plate cannot be compensated for, so that the use of a pulsed laser is still needed for their complete elimination.

4. A surplus of daylight exposure on the holographic plate will seriously decrease the contrast of the interference fringes, and along with it the brightness of the reconstructed image. The technique of Denisyuk pulsed holography however offers a good opportunity to expose holograms in a daylight environment.

A plateholder was constructed which is shielded on one side by the object and on the other side by a mechanical shutter with a large aperture. During the opening of the shutter (.1 s) the pulsed laser is fired and the exposure to daylight will generally be negligible.

5. Simplicity of the set-up and its operation is demanded in case of operation by unskilled personnel. The need for a simple set-up is the more urgent since the location of the area to be inspected is not easily accessible. As is illustrated by figure 1 the Denisyuk holographic set-up is very simple and easy to adjust. The plateholder can be manufactured compact so that it can be conveniently installed at all kinds of locations.

6. Interpretation of the fringe pattern will be most easy if it can remain qualitative. In that case, the abrupt changes in the curvature of the fringes that appear near the plastic deformation zone, allow the qualitative discovery of a crack before it can be detected by the usual means. If the anomaly in the fringe pattern is only slight, or if the crack-size has to be evaluated quantitatively, automatic fringe readout and processing becomes necessary.

2.2 The effect of shrinkage

Due to the processing of the hologram the emulsion will generally shrink, which affects reconstruction angle and wavelength. This section evaluates these reconstruction parameters while the causes of this shrinkage will be discussed in section 3.

The holographic plate is exposed by a coherent plane wave of wavelength λ_e under an angle α_e with the normal (Figure 2). Part of this wave is transmitted through the emulsion and subsequently diffusely reflected by the object. Each object point will radiate a nearly spherical wave towards the plate, which will interfere with the original plane wave. The period of the resulting intereference pattern is nearly $\lambda_e/2n$ in a medium of refractive index n. If the angle α_e is negligible, the interference pattern consists of planes practically parallel to the surface of the holographic plate. In practice a small angle α_e is adjusted as to separate the zero-order from the reconstructed image.
After processing the fringe period is given by:

$$d = s \frac{\lambda_e}{2n} , \tag{1}$$

where s denotes the shrinkage factor.
For the reconstruction of the hologram a wavelength λ_r is used, which is not necessarily equal to λ_e. According to Bragg's law constructive interference of the light reflected by successive planes is obtained by:

$$2nd \cos \alpha_r = \lambda_r \tag{2}$$

Equations (1) and (2) may be considered in two different ways:
1. If the reconstruction is made under the same angle as the exposure ($\alpha_r = \alpha_e \simeq 0$), an optimally bright reconstruction is obtained by selecting the proper wavelength λ_r of the reconstruction light source.
From eq.(1) and (2) then follows:

$$\lambda_r = s \lambda_e \tag{3}$$

Due to the shrinkage of the emulsion the reconstructed wavelength is shifted by a factor s with respect to the original wavelength λ_e. Thus the shrinkage factor s may be experimentally determined by measuring λ_r. The results of these experiments for different processing parameters will be discussed in section 3.

2. For optimal resolution of the image details in the reconstruction a monochromatic light source is needed. The application of a He-Ne laser for this purpose is most convenient. The reconstructed wavelength λ_r is then fixed so that we have to optimize the angle α_r. It follows again from eq.(1) and (2) that:

$$\cos \alpha_r = \frac{\lambda_r}{s \lambda_e} \tag{4}$$

Next to the ruby laser ($\lambda_e = 694$ nm), we use a He-Ne laser ($\lambda_r = 633$ nm) for reconstruction of the holograms, so that according to eq. (3) we find $\lambda_r/\lambda_e = 0.91$. Therefore, if the shrinkage is more than 9% (s < 0.91) there is no solution for eq. (4) and reconstruction becomes impossible unless a laser of shorter wavelength is applied. Without shrinkage of the emulsion (s = 1) the angle of reconstruction $\alpha_r = 24^O$. Due to refraction this angle will become approximately 37° outside the hologram.

Generally, in a conventional set-up for transmission holograms the object can only be observed under limited angles, practically normal to the object surface. This is due to the rather large object to plate distance with respect to plate size. As a result the sensitivity of a suchlike set-up for object deformations is practically normal to the object surface so that mainly out-of-plane deformations are detected. Contrary to this the Denisyuk holographic set-up allows minimal object to plate distances, so that the object may be observed under rather large angles with the normal. Therefore sensitivity to in-plane deformations is obtained as well.

Considering the foregoing, it appears necessary to find a holographic process performing well enough to allow the making of Denisyuk holograms without surface preparation of the object. Moreover, this process should allow sufficient adjustment of the shrinkage factor s. In the next section the relevant processing parameters will be discussed.

3. Direct bleach

3.1 Efficiency

The conversion of the developed image silver of an amplitude transmission hologram into a transparent silver compound image, after removal of the residual halide by the fixer, creates a phase hologram which diffraction efficiency is considerably increased. A suchlike conversion, which is generally referred to as "direct bleach", results in a silver halide, a silver ferrocyanide, or a silver mercurochloride image. For a long time there has been a deeply rooted belief of many holographers that the conversion of the image silver into a compound with a refractive index substantially greater than that of the gelatin medium was demanded in order to prepare efficient holograms. The cause of the original misunderstanding that the refractive index modulation should increase along with increasing refractive index of the silver compound grains, is probably attributable to the implicit supposition that we deal with grains with large respect to the wavelength of the light in the medium. In that case the geometrical-optics formulation gives the resulting refractive index as the mean, weighted by volume, of the refractive indices of the respective components. However this supposition is obviously invalid in the case of the extremely fine-grained holographic emulsions where the grain diameter amounts a few tens of nanometers only. Such grains may be regarded as Rayleigh scatterers and the refractive index of the composition is given by the Lorentz-Lorenz equation, which predicts an approximate linear relation between refractive index of the composition and the polarizability of the composite substances. Efficient phase transmission holograms therefore may be obtained independent of the bleaching agent applied, as long as differences in concentrations of polarizable silver compound molecules are optimized by adjusting the exposure energy. This explains why in the past decades innumerable chemical formulations have been proposed, a research-effort reminiscent of the alchemic quest of the philosophers' stone, not resulting however in anything like a "final" formulation.

In earlier work[9] is was set forth that not the refractive index, but the molecular polarizability and the molecular volume of the silver compound concerned are the crucial parameters governing the phase variations resulting from the variations in prebleach optical density. The -at the first sight paradoxal- conclusion that follows from this proposition is, that efficient phase transmission holograms may be obtained by conversion of the image silver into a silver compound of refractive index equal to that of the gelatin. This was experimentally confirmed by conversion of the image silver into silver ferro-cyanide which refractive index virtually equals that of the gelatin.

In view of photographic sensitivity the only and minor consideration that follows from the foregoing is that we might wish to apply the bleaching agent that converts the least possible density variations into sufficient phase variations. Evidently this is accomplished by the conversion of the image silver into the compound with the highest possible polarizability, silver mercurochloride.

It must be emphasized that the foregoing specifically applies to transmission holograms where, beyond spatial frequencies of some hundreds of lines/mm, the gelatin surface will no longer show thickness variations in accord with variations in concentration of the silver or its compounds (ref. 9 section 4). The contrary is true for reflection type volume holo-grams, where Bragg-planes are oriented perpendicularly with respect to transmission type holograms.
In this latter case, even at very high spatial frequencies the gelatin is able to follow such variations in concentration to a large extent, therefore the foregoing considerations only apply to transmission holography.

3.2 Scatter

The one-sidedness of the previous considerations becomes evident once the scattering of the silver compound grains is taken into account. A bleaching agent that, for the smallest possible optical density modulation, renders optimal diffraction efficiency, may well result in an objectionably high noise-level. As it appears, the intensity of the light scattered by a particle, small with respect to the wavelength, is proportional to the square of its total polarizability.[10] This observation shows already that diffraction efficiency is indissolubly linked to light scatter. The situation becomes more complex however because the grain may in general not be regarded as a scatterer in a homogeneous gelatin medium. On the one hand the gelatin medium may be tanned through cross-linking of the gelatin molecules in the vicinity of the grain by the oxidation products of the developer, on the other hand the gelatin may be locally compressed through the forces exerted by the grain which increases in size when being bleached. These processes may change the mass density of the gelatin near the grain, thereby causing a change in the refractive index of the gelatin enveloping the grain. These considerations led to the proposition of the model of a scatterer as depicted schematically in figure 3.

The spherical core of radius $f \times r$ consists of silver compound of refractive index n_1 and is enveloped by a spherical gelatin shell of refractive index n_2; thus composing a scatterer of radius r, immersed in the regular gelatin medium of refractive index n_3.[11] The total polarizability α of such a composite particle, small with respect to the wave-length, is given by Van de Hulst.[12]

The validity of the shell-model was confirmed experimentally by scattering measurements as well as phase measurements, the results of both approaches being in close agreement.[11] It is a remarkable coincidence that the proposition of this shell-model was accompanied by complementary experimental evidence collected by Joly e.a.[13], who applied developers of different tanning activity and showed by means of electron photographs that the average size of the bleached grains is independent of the type of developer. Joly concluded therefore that the observed differences in noise-level must be attributed to gelatin distorsions within as well as at the surface of the emulsion. As mentioned before the growth of the grains may be expected to exert radial forces on the adjacent gelatin, forces that may compress the shell of tanned gelatin, thus increasing its molecular density and with it the total polarizibility of the scatterer. Alternatively however these forces may indeed further enhance the amplitude of the surface irregularities coincident with the surface grains. A simple comparative experiment with a developer of low tanning activity reveals indeed that it is not the collaps of the gelatin layer that accompanies the fixing step, but the bleaching that causes excess "reticulation" of the gelatin surface. Joly has shown that surface irregularities decrease with increasing tanning activity of the developer. From the foregoing it follows that a suchlike suppression of the surface scatter goes at the expense of an increased intra-emulsion scatter, since the decrease in surface irregularities is compensated by increased molecular density in the shells. Therefore, depending on the nature of processing the balance between intra-emulsion scatter and surface scatter may deflect between extremes. These observations sustain our conclusion that the intensity of scattered radiation may vary widely with the type of processing applied and that it is therefore not adequate to relate results of various bleaching agents without investigating the influence of processing parameters on these separate contributions to the total emulsion scatter.

This may be accomplished by utilizing an index-matching technique. Evidently, application of the direct bleach process will only yield optimal results if the total polarizibility of the scatterers is minimized through a subtle processing that prohibits shell forming by allowing the emulsion to swell completely, and if the surface is index-matched as to eliminate surface scatter.

3.3 Gelatin collaps

The intricacy of the direct bleach process is further increased by the fact that the removal of the bulk of residual silver halide by the fixer causes the gelatin layer to collaps. For volume reflection holograms this involves a decrease in distance between Bragg planes, resulting in a generally unacceptable shift to shorter reconstructed wavelengths, an inconvenience that demands the application of a swelling agent like triethanolamine.[14] This is illustrated by a series of experiments with holographic volume gratings exposed on Agfa Gevaert 8E75 HD emulsions at 633 nm wavelength. Two plane waves of about equal intensity impinge perpendicularly on either side of the plate thus forming an interference pattern of period $\lambda_e/2n$ parallel to the surface. The holograms were developed in pyrogallol developer of the following composition:

A pyrogallol	20 g	B sodium carbonate sicc.	60 g
sodium sulfite sicc.	nihil	destilled water to	1 l
destilled water to	1 l		

The solutions A and B are mixed in equal proportions immediately prior to development. As the mixture is liable to fast oxidation it should only be used once. Without sodium sulfite solution A has a limited shelf life.

Holograms are developed for 4 min. (20°C-cont.agit.), washed for 2 min. in running water, subsequently fixed in a 200 g/l sodium thiosulfate neutral fixer (10 min.- interm.agit), washed 10 min. in running water and finally bleached in formulations 2 or 4 as to convert the image silver into AgBr or $Ag_4Fe(CN)_6$ respectively.[9] By means of a spectrometer the reconstructed wavelength λ_r was measured at each stage of the processing, as a function of optical density. The results of these experiments are given in the figures 4 and 5. Immediately after development a decrease in reconstructed wavelength nearly proportional to optical density is established. This is due to the first gelatin collaps after removal of bromine through reduction of AgBr by the action of the developer. As will be shown in the next section, the hardening activity of the pyrogallol developer has a limiting effect on the gelatin collaps. Subsequent removal of the residual AgBr by the neutral fixer drastically converses this picture. As the most of the bulk halide is removed at low optical density, the gelatin collaps is the most severe in this region as is indicated by the reverse in slope. A total shift in reconstructed wavelength of about 100 nm is encountered for optical densities between 2 and 3. The final conversion of the image silver into a transparent silver compound results in a considerable growth of the grains, in accordance with the increase in molecular volume, and consequently causes the gelatin layer to swell again. As the molecular volume of $Ag_4Fe(CN)_6$ exceeds that of AgBr by a factor of 2.4, the slope of the graph is expectedly higher in the former case.[9]

The large amount of alternate shrinking and swelling of the gelatin appears to be disadvantageous since it is only at the bleaching stage of the processing that serious surface scatter is induced. A considerable band-broadening is observed as well (up to 60 nm), indicating that with these large variations in gelatin thickness and grain size the smoothness and coherence of the Bragg-planes diminishes at the expense of diffraction efficiency (max. observed 10%).

Considering the difficulties apparently inherent to the rather circumstantial develop-halide solvent-rehalogenate procedure, an alternative processing that avoids these problems should be very useful. A more direct approach to be considered therefore is the develop-silver solvent process that modulates the concentration of the residual silver bromide.

4. Reversal bleach

4.1 General

If after development of an amplitude hologram, omitting the fixing step, a suitable silver solvent is applied that does not significantly remove any of the residual silver-halide, a phase hologram is obtained as well. In this case, which is generally referred to as "reversal bleach", the concentration of the silver bromide that was originally deposited in the emulsion, is modulated. This reversal process owns a few advantages over the direct process:
- The omission of the fixing-step reduces the process to essentially two steps, which considerably curtails processing time.
- The residual AgBr grains have not suffered any changes in size, neither has the gelatin adjacent to the grains undergone any differential tanning. Therefore these grains are not enveloped by any shell-structure, neither have they been the cause of any surface irregularities, and thus may be expected to render the least scattering.
Conversely it has been argued by Phillips[15], that removing irregular chunks of silver will leave an irregular vacancy in the gelatin host, and the scatter may thus be as severe as if the image were fixed and rehalogenated. For the same reason however one might argue that the fixing step, which removes the AgBr grains, must be expected to bring about a suchlike effect, which as experiments show, does not appear. If indeed the silver grain removal would leave marked voids in the bulk gelatin, a considerable background scatter should be observed at higher prebleach densities. Our experiments however have shown invariably that the reverse is the case: scatter decreases with prebleach density and becomes negligible at densities beyond 2 or 3, in the case of reversal bleach. Evidently, the bulk of residual halide, which is inversely proportional with optical density, is the main source of scatter. As will be shown in section 3.2 the application of a strong hardening developer prohibits further shrinking of the gelatin layer on removal of the silver grains. The creation of vacuoles in this case seems inevitable therefore, so that on second thoughts the observed decrease in scatter with optical density is somewhat surprising. Evidently the concept of bubble-like voids remaining in the gelatin after removal of the silver grain seems erroneous. It is difficult however to imagine what alternatively may take place in the emulsion, as collaps of the voids seems excluded by the absence of any wavelength shift.

4.2 Experimental

We have conducted a series of experiments, similar to those described in section 2.3, with the reversal bleach technique to the purpose of disclosing the influence or processing parameters on the wavelength shift.

The straight pyrogallol developer was selected again because of its strong hardening activity which was expected to minimize gelatin collaps and thus to minimize wavelength shift. This is illustrated by a series of comparative experiments with a metol developer, the formulations of which is obtained by substituting metol for pyrogallol in solution A (section 2.3) and adding 100 g/l sodium sulfite.

Metol as a developing agent has a low tanning activity, while the high sulfite content further eliminates any remaining hardening activity by reducing the primary oxidation products of the developing agent so that these can no longer crosslink the gelatin molecules. The result of these comparative experiments is illustrated in figure 6. After development in the pyrogallol developer a shift in reconstructed wavelength is observed, virtually equal to that of the previous experiments, due to the liberation of bromine. Subsequent dissolution of the image silver by a 4 g/l potassium dichromate bleach with low pH (4 cc H2SO4/l) surprisingly does not induce any further collaps of the gelatin, as was already mentioned in section 3.1. This is probably due to the fact that the hardening activity of the developers' oxidation products has a limited diffusion lifetime so that the gelatin is cross-linked in the direct vicinity of the developed grains. Gelatin collaps by removal of these grains by the silver solvent is therefore prohibited by the rigidity of the locally hardened gelatin. Completely contrary to this, the removal of the undeveloped residual AgBr, consequently contained by unhardened gelatin, will cause a strong collaps of the gelatin layer as was demonstrated in the figures 4 and 5. The results with the non-hardening metol-sulfite developer shown in figure 6, sustain this explanation. Not only does the virtual absence of any hardening effect allow a considerably larger gelatin collaps during development, subsequent dissolution of the silver by the dichromatic bleach now brings about an expected further gelatin collaps. It may be noted that the largest shrinkage is rendered by removal of the bromine that has an atomic volume almost twice as large as that of silver.

These experiments show that phase holograms may be obtained by means of a simple reversal bleach procedure without the large and disastrous shrinking and swelling movements of the gelatin that occur during the cumbrous direct-bleach procedure. Shrinking occurs though, but is limited to a few tens of nanometers by the hardening activity of the pyrogallol. These experiments further suggest that the amount of the wavelength shift may be manipulated to a large extent by adjusting the hardening activity of the developer formulations by the sulfite content.
Figure 7 shows the results of a series of experiments demonstrating this possibility of reconstructed wavelength manipulation by increasing the sulfite content from nihil to 10% by weight. The wavelength shift increases linearly with about 10 nm/% sulfite up to a 3% concentration, beyond which the effect gradually saturates. At a concentration of 10% the shrinkage appears to have reached maximum values as no more difference between a pyrogallol-sulfite and the metol-sulfite developer can be observed. For prebleach optical density between 2 and 3.5 the color of He-Ne exposed holograms may thus be varied between a deep orange-red and a neutral-green.

Apart from the apparent advantages of the reversal bleach procedure combined with a tanning developer over the direct bleach procedure, the resulting diffraction efficiency of the holographic Bragg-gratings thus obtained, is of great interest. Diffraction efficiencies were measured from the spectrographs and are presented in figure 8. Optimal efficiencies typically amount 20% and are obtained for prebleach optical densities between 2.5 and 3.5. Bandwidths typically amount 20 nm.

Because the hardening pyrogallol development is accompanied by the formation of a considerable brownish stain of the gelatin, the transparency of the hologram suffers a decrease proportional to prebleach density. Evidently the resulting absorption will cause a decrease in diffraction efficiency; in view of the simplicity of the related processing however, this disadvantage seems of little interest for NDI-holography. It may be mentioned however that this gelatin stain can be removed for instance by a 1-1½ min. bath in a 0.05% potassium·permanganate/0.4% sulfonic acid solution in water. As it appears that gelatin stain and gelatin hardening are independent to a high degree[16], the resulting gelatin collaps is only slight (10-15 nm). This stain removal however is accompanied by an increased background scatter, which is to be expected because this stain effectively absorps multiple scattering by the halide grains. A further processing step, which consists of dying the gelatin with a suitable dye, is therefore demanded to suppress this scatter while preserving the gain in diffraction efficiency, which is increased to typically 30%.

5. Results

The excellent performance of the reversal bleach processing described, rendered the possibility to make Denisyuk holograms of low reflectance diffuse objects. Although the diffraction efficiency considerably decreases with the reflectance of the object, reconstructions remained sufficiently bright to allow for their convenient evaluation. As an example a few results are presented of our investigation of the fatigue crack growth discussed in section 2.1. The area of interest has a diffuse reflectance of only 10% for the ruby wavelength of 694 nm. The holograms were made in an environment with machine vibrations and stray daylight. Figure 9 shows the double-exposure interferograms of the critical area for an increasing number of stress cycles. The interference pattern given in figure 9b significantly indicates the development of a crack well before it could be visually detected.

An important advantage of the Denisyuk technique moreover is that the image is reconstructed in the immediate vicinity of the plate, so that it is generally accessible for low magnification microscopes. Extremely fine details of the interference pattern may therefore microscopically be evaluated without difficulty. This advantage was utilized in a recent study of impact phenomena.[17] Figure 10a gives an example of an interferogram of a ceramic target 1 μs after impact. Figure 10b gives an enlarged portion of this interferogram showing the development of cracks after impact. Microscopic observation of such interferograms showed fringe densities up to 30 lines per mm. Fringe visibility in these areas was still sufficient to allow evaluation of the interferograms.

6. Discussion

A simple processing has been developed that allows the making of low-noise, high-efficiency Denisyuk phase reflection holograms. Even of objects of relatively low diffuse reflectance, holograms are obtained of a quality sufficient for their convenient interferometric evaluation. Further advantages of the processing described are the low emulsion shrinkage as well as the possibility of shrinkage manipulation, and so, reconstructed-color manipulation, through the sulfite content of the developer.
Application of this processing allows the practical and easy utilization of Denisyuk holography outside the laboratory. The latter technique readily solves the problems connected with rigid body motion if the plate holder is fixed to the object.[8] Moreover the reconstruction of the object in the immediate vicinity of the plate allows microscopic evaluation of the fringe pattern up to high spatial frequencies. Finally the object may be observed under widely different angles so that sensitivity to in-plane as well as out-of-plane deformation is obtained.

7. Acknowledgement

The authors wish to thank the National Aerospace Laboratory in Emmeloord for their collaboration in the experiments on aircraft wing specimen. The research work, discribed in this paper, has been supported by the National Defence Research Organization in the Netherlands.

8. References

1. Goethe, J.W., Naturwissenschaftliche Schriften, Bnd. V, 156-159, Ausg. R. Steiner Verlag, Dornach (1982).
2. Lippmann, M.G., La photographie des couleurs, Compt.Rend. 112, 274-277 (1891).
3. Ives, H.E., An experimental study of the Lippmann color photograph, Astrophys. J.27, 325-352 (1908).
4. Valenta, E., Die Photographie in natürlichen Farben, Encyclopädie der Photographie(2), W. Knapp (ed.), Halle a. S. (1894).
5. Denisyuk, Yu.N., On the reproduction of the optical properties of an object by the wavefield of its scattered radiation, Opt. and Spectr. 15, 279-284 (1963).
6. Erf, R.K. (ed.), Holographic Nondestructive Testing, Ac.Press, New York (1974).
7. Vest, C.M., Holographic Interferometry, John Wiley & Sons, New York (1979).
8. Neumann, D.B. and Penn, R.C., Off-table Holography, Exp. Mech. 15, 241-244 (1975).
9. Renesse, R.L. van, and Bouts, F.A.J., Efficiency of Bleaching Agents for Holography, Optik 38, 156-168 (1973).
10. Kerker, M.,The scattering of light, p. 39f., Academic Press, New York/London (1969).
11. Renesse, R.L. van, Scattering properties of fine-grained bleached emulsions, Phot.Sci. Eng., 24, 2, 114-119 (1980).
12. Hulst, H.C. van den (ed.), Light scattering by small particles, p73f., John Wiley & Sons, New York (1957).
13. Joly, L. Vanhorebeek, R., Development effects in white-light holography, Phot.Sci.Eng. 24, 2, 108-113 (1980).
14. Rowley, D.M. A holographic interference camera, J.Phys.E: Sci.Instrum.12, 971-975 (1979).
15. Phillips, N.J., Ward, A.A., Cullen, R., Porter D., Advances in holographic bleaches, Phot.Sc. and Eng., 24, 2, 120-124 (1980).
16. Tull, A.G. Tanning development and its application to dye transfer images, J.Photogr. Sci., 11, 1-26 (1963).

17. Kolkert, W.J., Amelsfort, R.J.M., Renesse, R.L. van, On the early response of a sintered n-Al$_2$O$_3$ ceramic target plate impacted by a low velocity lead projectile, 6th int. symp. on ballistics, 27-29 Oct. 1981, Orlando, Florida.

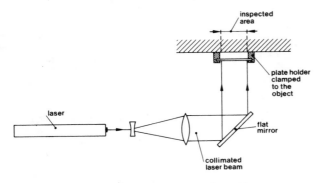

figure 1   arrangement for the exposure of reflection holograms

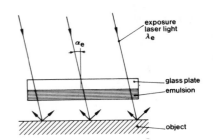

figure 2   formation of interference planes during exposure

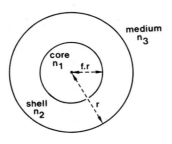

figure 3   model of a silver-compound scatterer enveloped by a tanned gelatin shell in a non-tanned gelatin medium

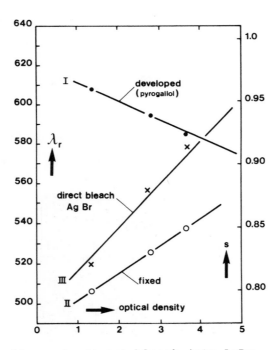

figure 4   direct bleach into AgBr

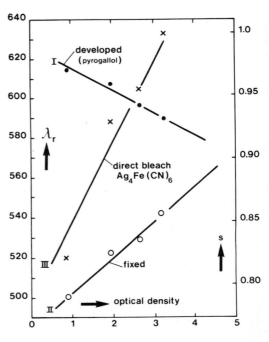

figure 5   direct bleach into Ag$_4$Fe(CN)$_6$

Reconstructed wavelength λ_r and shrinkage factor s as a function of optical density.

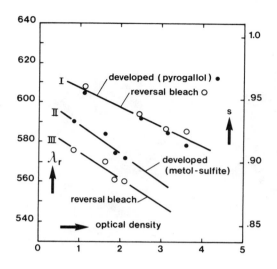

figure 6   comparison of a non-hardening
            metol developer and a
            hardening pyrogallol developer

figure 7   influence of sulfite content
            ● pyrogallol - 10% sulfite
            · metol   - 10% sulfite

reversal bleach: reconstruction wavelength λ_r and shrinkage factor s as a
                 function of optical density.

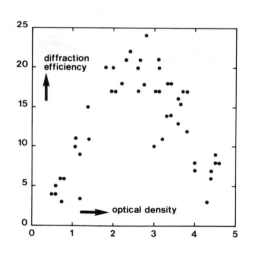

figure 8   reversal bleach:
            diffraction efficiency (%) as a
            function of optical density.

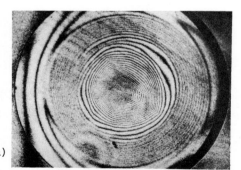

a)

b)

figure 10 a) interferogram of ceramic target 1 μs after impact (1.5 x enlarged)
          b) detail of a), showing crack development (13 x enlarged)

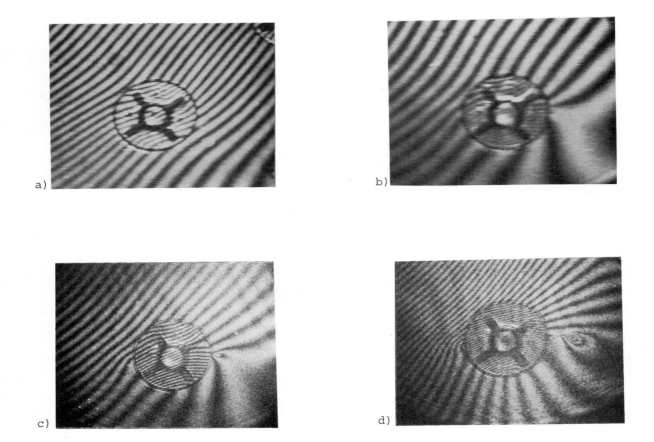

figure 9   Double exposure interferograms of the critical fastener in a specimen
           of the wing. The mechanical load on the specimen is increased between
           the two exposures.
           The interferograms are made after an increasing number of stress cycles:
           a) no anomaly in the pattern,
           b) indication of the development of a crack on the right side of the fastener,
           c) the crack on the right side reaches the surface; crack development on the
              left side is also indicated in the pattern,
           d) a few cycles before complete fracture of the specimen.

Holographic interferometry applied to external osteosynthesis :
comparative analysis of the performances of external fixation prototypes

Pierre Jacquot, Pramod K. Rastogi, Leopold Pflug

Laboratory of Stress Analysis, Swiss Federal Institute of Technology,
CH-1015 Lausanne, Switzerland

Abstract

The use of external osteosynthesis in the healing and the management of fractured bones is in rapid progression. The method employs an external rigid frame which is mounted to keep the fractured bones in state of immobilisation by means of percutaneous transfixing pins traversing the bones. In this paper, holographic interferometry is used to investigate the mechanical behaviour of the ball-joint - a central element in the fixation frame - subjected to realistic loads. Besides, the work has permitted to compare several models of this piece (of comparable handiness) as to their characteristics of rigidity and of resistance to slipping.

Introduction

The aim of osteosynthesis is to promote the healing of fractures and to endeavour to restore, during treatment, the intact mechanical properties of the bone. In external osteosynthesis, a number of pins interlinked by a frame go right through the fractured member (figure 1). Although the first use of an external fixation dates back to 1902, it was only in the early seventies that this surgical technique really progressed[1], judging by the number of publications devoted to the biomechanics of external fixation[2] or by the multiplicity of available fixator models on the market since then. The reasons for this progression are to be found in the advantages of external osteosynthesis, as amputation is less and less frequently resorted to and as, unfortunately, the number of cases of severe injury increases, particularly in motorcycle accidents.

The advantages[1,3,4] include the relative simplicity of the surgical intervention and its adaptability to a great number of cases, the fact that the reduction of fractures can be effected a posteriori and that a three-plane correction remains possible after the application

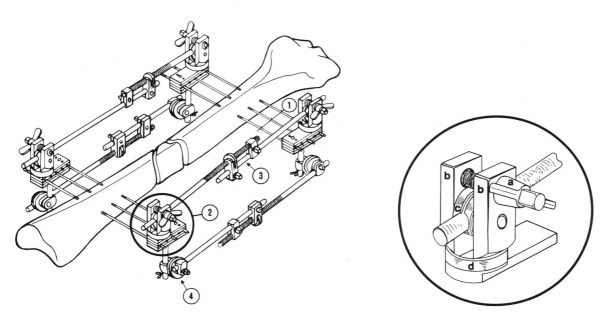

Figure 1. Schematic of the Hoffmann-Vidal external fixation frame holding a fractured bone. The four components making up the frame are (1) transfixing pins (2) universal ball joint with rod (3) adjustable connecting rods and (4) articulation couplings. In the medal is shown a close-up view of the ball joint which is composed of (a) clamping screw (b) cheeks (c) clips and (d) bowl.

of the external fixation. The adjustable connecting rods permit functioning in conditions of compression (healing), distraction (bone lengthening) or neutralization (loss of osseous substance, infected pseudarthroses). The frame is sufficiently well ventilated so as not to darken x-ray radiographs or angiograms and leaves the operative field free for complementary surgical interventions such as grafts and treatment for loss of osseous substance. The risk of infection is often found to be smaller than in the case of internal osteosynthesis. The external fixation facilitates the patient's early mobilization.

However, an external fixation must meet a number of strict requirements if it is really to present all these advantages. A long term rigidity and stability, for a minimal weight and number of elements, are required of the frame which must remain very manageable.

In this context, a distinction is drawn between the seeking of a primary healing and of a secondary healing. The former, which promotes the direct joining of the fracture surfaces by autogenous union, only tolerates movements of less than 5-10 μm on the fracture region[4]. The latter, which aims at the formation of a periostic callus, seems, on the contrary, to be stimulated by well controlled micro-movements about the fracture region.

Intolerance reactions from the tissues as well as the necessity for sterilizing at high temperatures limit the choice of materials. The frame geometry is imposed by purely medical considerations. Thus, according to the type of fracture, certain regions for the grafting of pins and certain pin spacings are excluded so as not to impair the blood vessels and nerves in the region.

It follows that there is no rational plan leading to the complete optimization of these devices. The attempts to reduce fractured bone-frame behaviour to a global model, based on the use of the finite elements method and carried out by Burny[3], in particular, have not yet succeeded. On the contrary, there is a series of rules which recommend, according to the injured member and the type of fracture, the number of pins to be used, their diameter, their spacing, the distance to the fracture seat as well as the distance between the connecting rods. On the other hand, the conflicts between the requirements of engineers and those of surgeons are empirically settled on the basis of the progressive improvement of devices which takes the accumulated clinical experience into account and which benefits from the feed-back of the medical profession and patients. This situation opens a vast field of investigation to experimental methods. Research on the fixator has been carried out essentially by means of gauges, in particular by Chao and others[5]. Holographic interferometry, considering the success it has achieved in neighbouring branches of biomechanics[6] or internal osteosynthesis[7], has an important role to play in external osteosynthesis.

Definition of the problem

Three examples of the application of holographic interferometry to the mechanical analysis of Hoffmann's external fixation are reported herein. According to Burny[3], the main problems relate to :

- the mechanical characteristics of the bone-pin interface and their evolution in time,

- the behaviour of the link between the pin and the adjustable connecting rod. This link is provided by a key element of the device : the ball joint which not only provides the connection of the pins and rods, but also permits the blocking of the four degrees-of-freedom on the rod through tightening of a single nut,

- the frame architecture as a function of the mechanical characteristics of the fracture.

Owing to the experimental difficulties raised by the first problem and not disposing of a series of graded bones, we have tackled the second problem and the three examples refer exclusively to the quality of the functioning of the ball joint.

It is a question of (i) defining the behaviour of the ball joint under the effect of tightening and of differently orientated forces applied to the connecting rod; (ii) comparing the respective rigidity of two different bowls; (iii) comparing the respective resistance of two similar ball joint models to a rotation of the rod around its axis.

Use was made of two classical holographic interferometers in the laboratory. The first yields the lines of equal out of plane displacement, w, with half optical wavelength sensitivity for illumination and observation normal to the object surface :

$$w = \frac{n\lambda}{2} \tag{1}$$

The second, called holographic moiré, provides the lines of equal in-plane displacement, u or v, according to the relations :

$$u = \frac{n\lambda}{2\sin\vartheta_x} \quad ; \quad v = \frac{n\lambda}{2\sin\vartheta_y} \tag{2}$$

where ϑ_x and ϑ_y are the half angles of the beams illuminating the object symmetrically with respect to its normal and contained in the horizontal or vertical planes.

These two optical arrangements are exploited in real time, either by using an immersion plate holder, or an instant holocamera with thermoplastic material. In both cases, the interferograms are displayed on a monitor screen. The real time version of the second optical arrangement is extensively described in reference 8. In the first optical arrangement, a set of mirrors makes it possible to examine simultaneously the views of three ball joint faces lit and observed in the same geometry.

Characterization of the mechanical behaviour of the ball joint

The ball joint under examination has been in use for several years and has the reputation of giving full satisfaction. Nevertheless, considering the inevitable part of empiricism which led to its shape, it is desirable (i) to make sure that it properly performs the functions for which it was conceived; (ii) to define objectively its real mechanical behaviour in conditions of realistic loads. The response of the ball joint to these loads will be quantified by means of simplifying parameters, while avoiding the risk of disregarding a global trait of behaviour. Holographic interferometry is quite apt to accomplish this kind of task[9],[10] and is to be preferred to the gauge methods whose simplicity is more and more questioned since the advent of instant holocameras.

The parameters thus obtained will be used as reference data in the expectation of a future evolution of the design of the element.

Tightening of the ball joint

Figure 2 represents the out of plane displacement fringes as a function of an increment in the tightening torque of the clamping screw. This illustration has been taken from a series which includes the torque increments (9 - 9.2 Nm; 9 - 9.4 ; ... 9 - 10 Nm) and (13 - 13.2 Nm ; ... 13 - 14 Nm). It is typical inasmuch as the fringes keep the same general aspect and only their densities change. In practice, surgeons apply tightening torques ranging from 7 to 11 Nm.

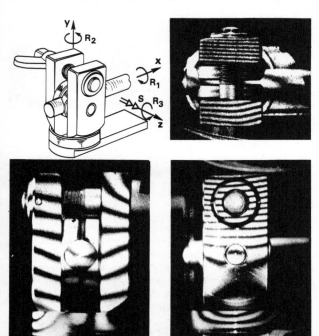

The front-view fringes are characteristic of a flexion of the cheek, the most strained area being that between the axis of the connecting rod and the bowl. The maximum deflection is at the top part of the element. The gradient of the normal displacement at the top part of the element was chosen as significant parameter of this loading case. On an average, over all the loading steps analysed and per unit of tightening torque increment, this parameter, directly deducted from the fringe spacing, has the value:

$$\langle\overline{\beta}\rangle = 400 \ \mu rd/Nm \tag{3}$$

The fringes to be found on the kneecap, which receives the tightening screw, are slightly displaced with respect to the others, indicating a sliding of the kneecap in its seating, which is precisely its function.

Figure 2. The out of plane displacement fringes relative to a tightening torque increment of 9-9.2 Nm.

On the top view, the fringes are nearly parallel and equidistant and display therefore a rigid body rotation R_1 of the upper side around x axis. This rotation is linked with the flexural deformation. However, the fringe density clearly varies from one cheek to another. There is, therefore, a differential cheek rotation around x axis. Similarly, there is, according to the side view, a differential cheek rotation R_3 around z axis. From all the loading steps applied, it is apparent that these micro-rotations, which also take place around y axis, vary within the range (0 - 150 μrad) per Nm tightening torque increment. These micro-rotations are likely to put the connecting rods in slight prestress merely by the action of tightening.

It may also be noticed from the front and side views that the bowl becomes ovalized in a regular and progressive way.

<u>Axial force applied to the connecting rod</u>

Five loading increments were exerted always starting from an initial axial force of 50 N. Figure 3a shows the out of plane displacement fringes for the maximum increment of 50 to 250 N.

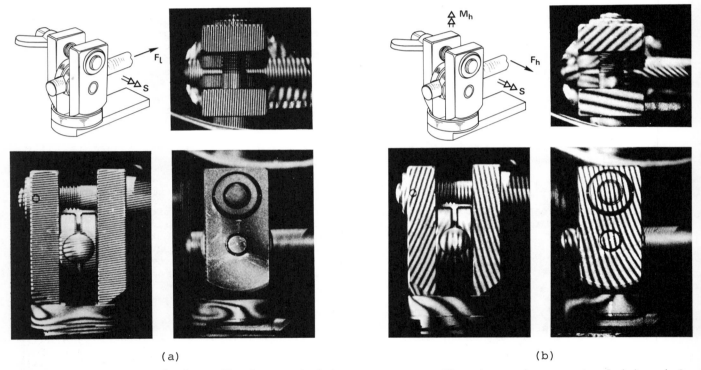

(a)   (b)

Figure 3.   The out of plane displacement fringes corresponding to an increment of (a) axial force between 50 and 250 N, (b) horizontal force between 0 and 21 N, at 0.125 m from the y axis. In both the cases the tightening torque is maintained at 9 Nm.

The front view is fringeless : if the axial force is pure, there are no parasitic rotations around the x or y axes.

The side view of the two cheeks shows fringes in equal numbers, of closely related orientation, rectilinear, parallel and equidistant. Consequently, the cheeks undergo a rigid body rotation R_3 around z axis. The relationship between the value of this rotation and the axial force increments is linear over the series of loading examined. The reversion to the flat tint on the three views is noticed when the axial force is brought down to its initial value of 50 N. This is observed rapidly in the real time holographic arrangement.

The parameter chosen to characterize the response to axial loading is, therefore, rotation R_3 converted to the unit axial force increment which has the value :

$$\overline{R}_{3A} = 2.5 \ \mu rd/N \tag{4}$$

The top view fully confirms the preceding observations.

Subject to a pure axial force and over the range examined, the ball joint has an elastic behaviour, like a monolithic block, even though the device involves three connections through friction between four groups of independent elements.

Horizontal force applied to the connecting rod

Five steps of loading were examined, ranging between (0 - 10 N) and (0 - 32 N) for a horizontal force applied at 0.125 m from y axis. From the out of plane displacement fringes, of which figure 3b is a sample, the following information may be directly inferred :

The front view demonstrates that the cheek turns preponderantly around y axis (R_2). However, as the fringes are more dense on top than at the bottom, the presence of a torsional deformation around y axis is also noticed. Moreover, the slight obliquity of the fringes means that rotation R_2 is not pure. The redundant profile and top views make it possible to isolate component R_2, which is used for establishing the significant parameter of this loading case. On an average and converted to the unit torque increment exerted by the horizontal force, applied at 0.125 m from y axis, this parameter has the value :

$$\langle \overline{R}_{2M} \rangle = 75 \ \mu rd/Nm \tag{5}$$

The side and top views confirm the existence of micro-rotations R_1 and R_3 around x and z axes. These micro-rotations are found to vary in the range (0 - 125 μrd) for the five steps of loading increments. On the profile view of figure 3b, the two cheeks turn in a remarkably symmetrical way.

As in the case of the axial force, the behaviour of the ball joint is reversible, i.e. the deformations disappear when the horizontal force is brought down to its initial level 0 N.

One of the loading steps was exerted for some twenty hours, without there being any noticeable evolution in the fringes. This illustrates the long term stability of the locking device.

Vertical force applied to the connecting rod

Five loading increments ranging from (0 - 10 N) to (0 - 32 N) were exerted at 0.125 m from

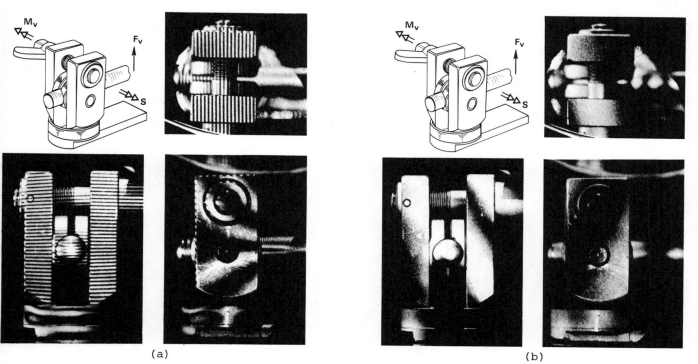

(a)   (b)

Figure 4.   (a) The out of plane displacement fringes corresponding to a vertical force increment of 0 - 26.5 N. (b) The residual fringes obtained when the vertical force is brought down to its initial level of 0 N after having been mounted to 32 N. (S = 9 Nm).

y axis. The out of plane displacement fringes (figure 4) present a remarkable similitude with the case of the application of the axial force, in that the device responds with a pure rigid body rotation R₃ around z axis, proportional to the force exerted. This rotation, converted to the unit torque due to the vertical force constitutes the significant parameter of this case and has the value :

$$\overline{R}_{3V} = 75 \ \mu rd/Nm \tag{6}$$

Here again and in the range of the exerted forces, the device has an elastic behaviour, like a monolithic block. This characteristic is illustrated on figure 4b, where a single fringe subsists when the force is brought back to its initial level 0 N.

Remarks

Holographic interferometry in real time has permitted to expose, in a simple and rapid manner, the real behaviour of the ball joint. This behaviour is defined both qualitatively and quantitatively by means of parameters designed to play a reference role. Apart from the case of screw tightening, the observations relating to the displacement-deformation reversibility when the loadings are suppressed and to long term stability, provide a rational basis for determining the right conception of the device. These characteristics are essential owing to their direct impact upon the behaviour of the fractured surfaces and, consequently, upon the rapidity and the solidity of the osseous healing.

Comparison between two bowl designs

Having accepted that the conception of the model examined is sound, it is conceivable to endeavour to perform some modifications with a view to improving it. The lightening of the frames is one of the permanent concerns of manufacturers. As a matter of fact, the patient must carry, for several months, frames often comprising more than six ball joints. His comfort depends on the reduction in weight, even it is small.

Two ball joint bowls of analogous conception are compared as to their respective rigidities. One is made of a particular type of steel, the other of a special aluminium alloy. The latter is 35% lighter than the former despite an increase in volume that is necessary to compensate differences in the modulus of elasticity of the two materials. Besides, in the case of the aluminium model, slight modifications pertaining to the dimension figures and the vertex of the re-entrant angles have also been deliberately introduced. However both bowls may otherwise take the same elements.

Figure 5 shows the out of plane displacement fringes relating to a tightening increment from 9 to 9.8 Nm for the two models. The two bowls were identically sectioned as shown by the shaded regions in Figure 5. The plane cross-section so arising facilitates the interpretation of the fringes and gives information about deformations at the core of the bowls. The ovalization of the bowls, which results in a rotation of the sections around y axis is quite perceptible. In order to characterize the importance of the ovalization, a parameter, $\langle n_w \rangle$, was defined in the following way :

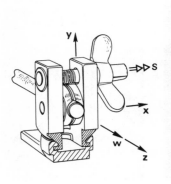
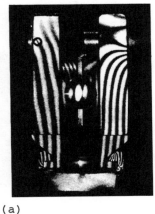
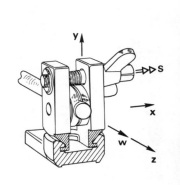

(a)   (b)

Figure 5.   The out of plane displacement fringes relative to a tightening torque increment of 9 – 9.8 Nm for the bowls made in (a) steel and (b) aluminium alloy.

$<n_w>$ is the average number of fringes, over a series of tightening torque increments varying between 9 – 9.2 Nm and 9 – 9.8 Nm, converted to the unit tightening torque increment and appearing on the lateral edges of the bowl. The two parameters, $<n_{ws}>$ and $<n_{wa}>$ for steel and aluminium, respectively, have the value :

$$<n_{ws}> = 9 \text{ fringes / Nm}$$

$$<n_{wa}> = 3 \text{ fringes / Nm} \tag{7}$$

The same test was carried out for in-plane displacements (u) in the direction of x axis, with the help of a real time holographic moiré set up. It is noticed (figure 6) that the lateral edges of the steel and aluminium bowls tend to turn in the opposite direction in their plane following a rotation R_3 around z axis.

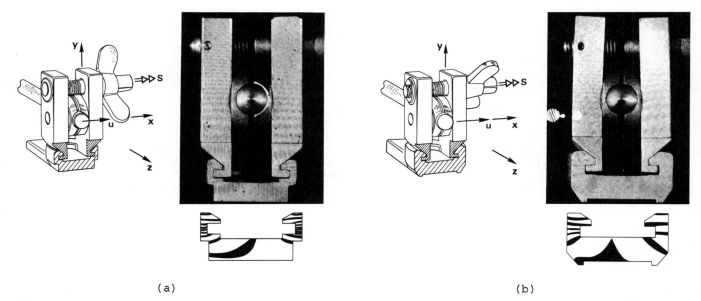

(a)   (b)

Figure 6.   The in-plane displacement fringes relative to a tightening torque increment of 9 – 9.6 Nm for the bowls made in (a) steel and (b) aluminium alloy.

As before, we define a fringe number $<n_u>$ which measures, over several cases of tightening varying between (9 – 9.2 Nm and 9 – 9.8 Nm) and per unit tightening torque increment, the importance of the in-plane rotation.

These two parameters $<n_{us}>$ and $<n_{ua}>$ for steel and aluminium respectively, have the value :

$$<n_{us}> = 15 \text{ fringes / Nm}$$

$$<n_{ua}> = 7 \text{ fringes / Nm} \tag{8}$$

In both cases, therefore, an increased resistance of the aluminium bowl to in-plane and out of plane deformations is observed, despite the reduction in weight mentioned. The light alloy element presents a greater rigidity in the proportion given by parameters in equations (7) and (8).

Resistance of the connecting rod to the rotation around its axis - a comparison

Both in the case of a primary and a secondary healing and even for an ideal frame, the long term stability is impaired by the bone creep at the pin-bone interface. This makes it necessary to renew the compression periodically and to check the locking of the ball joints frequently. In certain healing cases, three-plane corrections can be effected through modification of the frame geometry.

Throughout treatment and at the time of these interventions, resistance to rotation of the cheeks in their bowl, of the clips in the cheeks or of the connecting rods in the clips, is of primary importance, to the extent of determining the superiority of a prototype over another. The buckling of the frame being a major drawback, we must concern ourselves with the analysis of the resistance of the rod to rotation around its axis with a view to determining which of two ball joint models is more satisfactory.

Externally the two models being compared are quite similar. The two models (of which one is shown in figure 2) differ from one another essentially in the conception of the clips: in the "new" model one half of the clip in the form of a "C" clasps the other half between its jaws, an arrangement which causes the two clip halves to rotate together and facilitate the introduction of the connecting rod.

The resistance of the rod to the rotation around its axis is globally examined, with no attempt to isolate the phenomena which exclusively concern either the friction or the geometry or the deformation.

A tightening torque (S) is first applied to the locking screw. Increasing torques (Cb) are then exerted on the rod, step by step. At each loading step, the connecting rod undergoes a rotation which can be separated additively in an elastic rotation and an irreversible micro-rotation due to the slipping of the rod in the clips. In order to isolate the irreversible component, one proceeds as follows : Before each torque increment ΔCb, the torque Cb is brought to zero. The associated value of the residual micro-rotation of the rod is recorded and added to the micro-rotation (R) caused by preceding accumulated torque increments applied to the rod. At each stage, the rod is acted upon at the same place. Weights are hooked on to a solid lever placed at about ten centimeters from the clips in order to exert torque Cb. The torque is thus maintained constant during the time necessary to attain the stable equilibrium of the rod rotating around its axis.

On the whole, a graph like the one in figure 7a is obtained, in which the cumulated rotations of the rod (R) are ordinates and the successive torques (Cb) are abscissae. For equal tightening torques (S), the same torque ascending series {Cb} is applied to the two ball joints, although within the interval considered, pairs of practically overlapping curves are obtained for distinct series of torques. The dispersion of the values around a curve R=f(Cb) is experimentally found to be very small.

The measurement of residual rotations is carried out using a holographic interferometry arrangement in real time comprising a holocamera with thermoplastic material. A thin

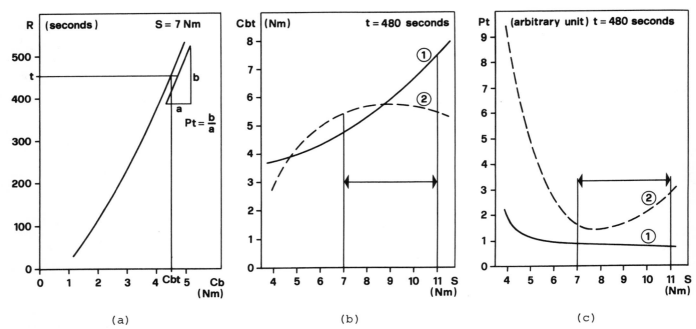

Figure 7. (a) Variation of the connecting rod residual rotation R in function of the torsional couple (Cb) applied to the rod for a fixed tightening torque (S) acting at the clamping nut. Cbt and Pt represent respectively the abscissa and the slope of the curve at any predetermined rotation value (t). These curves are drawn for different values of S and allow one to draw the curves (b) Cbt=g(S) and (c) Pt=h(S) for the two ball-joints being compared.

elongated plate in sandblast aluminium is fixed on to the connection rod, in a vertical position, at the clip outlet. The reference hologram records the plate in the state of zero torque (Cb). At each loading step, the residual rotation fringes appearing on this plate are counted, after a return to zero torque (Cb). Taking into account the length of the plate and relation (1), the sensitivity is found to be lower than one second per fringe.

Moreover, an autocollimator, with a 2 second-sensitivity is used (i) to control the elastic rotation of the rod; (ii) to determine the time taken by the rod to find a position of stable equilibrium, after which it is possible to go back to zero torque (Cb). The mirror aimed at by the autocollimator - which has an excellent flatness and a large diameter and is, therefore, heavy and cumbersome - is placed on the solid lever used to exert torque Cb.

The preference given to holographic interferometry for the measuring of the residual rotations is justified by the difficulty of placing such a mirror near the clips and by the accessible measuring range : it was possible to cover the interval (0 - 600 seconds) with the help of a single hologram by performing some ten consecutive fringe compensations till the completion of each torque series {Cb}.

Moreover, holographic interferometry allows a global control of the operations. The hologram of the aluminium plate covers simultaneously the ball joint and its support, as well as a guidance support located near the lever. It is always noticed that the measurements effected are not tainted with an eventual creep of these supports.

The curve in figure 7a is summarized by a set of three values (S, Cbt, Pt) where Cbt is the torque corresponding to a predetermined threshold (t) of rotations (R) and Pt is the slope of the curve R=f(Cb) at the determined threshold. The process is repeated for the series of tightening torques S (4-5-6-7-8-9-10-11 Nm) and for the two implements. The ensemble of the values (S, Cbt, Pt) permit to draw for each implement (1) and (2) the curves Cbt=g(S) and Pt=h(S) (figure 7b and c).

For a well designed implement, the permissible torque (Cbt) is expected to increase with the tightening torque (S). The slope Pt should stabilize at as low a value as possible for a high tightening torque, indicating that an abrupt disastrous sliding is impossible.

From figure 7b and c it may be inferred that model (1) is correctly designed, unlike model (2). In particular and for the latter model, the change in tendency of the slopes (Pt) to an upward rise soon after the tightening torque (S) enters the service load domain (7 - 11 Nm) is an obvious indication of its bad design.

Although not the aim of our analysis, we were able to partly understand this unexpected behaviour of the second implement. At a certain torque (S) stage, the cheeks become so deformed that one of the extremities of the clip-envelope "C" touches the opposite cheek. All additional tightening has the effect of deforming this "C", more and more compressed between the cheeks, resulting in a progressive loosening of the connecting rod.

Conclusion

The three examples reported herein show, in our opinion, that holographic interferometry is a powerful tool for defining the behaviour of mechanical parts, for comparing different models thereof, for judging the superiority of a manufacture to another, when the production designer is confronted with a series of strict and varied requirements. Holographic interferometry brings the possible choices to light and becomes thus a genuine conceiving aid.

The recent advent of instant holocameras, combined with video cameras and monitor screens, amounts to a considerable progess, likely to supplant the gauge methods in many cases. Finally, it is fitting to underline that, in the three examples given, the interpretation of the fringes is immediate, does not require complicated processing and involves extremely simplified computations.

Acknowledgements

The authors wish to thank Mr. M. Wagenknecht of Jaquet Orthopédie SA for his many helpful comments and advice throughout the course of this work. The authors are equally grateful to him for having generously provided them with the relevant materials.

References

1. Vidal, J., "Historique et perspectives d'avenir de la fixation externe", pp. 5-14, <u>Proceedings of the 7th International Conference on Hoffmann</u> ^{TM} <u>External Fixation</u>, Montpellier, January 1980, Diffinco, Genève.

2. <u>All Type External Fixation Bibliography 1852-1982</u>, Hoffmann External Fixation Information Center, Jaquet Orthopédie SA, Genève.

3. Burny, F., "Biomechanics of external fixation. A general review", <u>Proceedings of the 9th International Conference on Hoffmann</u> ^R <u>External Fixation</u>, Genève, September 1982.

4. Mears, D.C., "History of External Fixation", pp. 3-10, <u>External Fixation, The Current State of the Art</u>, edited by Brooker A.F. and Edwards C.C., Williams and Wilkins Co, Baltimore, 1979.

5. Chao, E.Y.S., Briggs, B.T., Mc Coy, M.T., "Theoretical and experimental analysis of Hoffmann-Vidal external fixation system", pp. 345-370, Ibid. 4.

6. <u>Holography in Medicine and Biology</u>, edited by von Bally, G., Springer-Verlag, 1979.

7. Hansen, U., "Quantitative evaluation of holographic deformation investigations in experimental orthopedics", pp. 27-33, Ibid. 6.

8. Sciammarella, C.A., Rastogi, P.K., Jacquot, P., Narayanan, R., "Holographic moiré in real time, <u>Experimental Mechanics</u>, 22, pp. 52-63, 1982.

9. <u>Holographic Nondestructive Testing</u>, edited by R.K. Erf, Academic Press, 1974.

10. Pflug, L., Jacquot, P., "Application de l'holographie à l'étude d'un outil de coupe", <u>Wear</u>, 62, pp. 21-36, 1980.

Biomechanical Research of Joint II
Experimental Research of Primate's Femur Facies of Articulation of Knee

Zhang Ren-xiang   Lu Ming   Lan Zu-yun   Zhang Xong-zi

Beijing University of Iron and Steel Technology, 100083, Beijing
CHINA

The movement of animal articulation depends on the drive of the muscle. It sustains changeable load. It proceeds movement of the rule of unsingle movement. It has the action of rolling motion and sliding motion itself and the ability of self growth. Many articulation performs complicated movement. The animal articulation is different remarkably from the moving pair of ordinary engineering. Therefore it is necessary to make special investigation.

There is strict and mutual relationship between the mode articular movement and the shape of the surface of the articulations. We apply the moiré technique on animal. We have given the moire contour fringes of teleost, amphibian, reptile, bird and the mammal class. We have studied some articular facies of vertebrates.[1-4]

Application of moire contour fringes is the optical system which can be easily used. It is a new content of biomechanics research. It is valuable in the observation to distinguish the biological and paleotological types and follow their developments, as to construct the structures of bio-engineering.

The great majority of Primates are full tree perch's animals, it uses four limbs to walk, run and jump. The acticity of its limbs and foots are very strong. The construction of the joint in four limbs ensures greater ability of rotation. During movement, it bears pulsation load produced by weight. The monkey sustains body weight with four limbs. Movements are very nimble. Its bones are small.

We have taken the moiré fringes of Presbytis phayrei, Macaca assamensis M'celland and Hylobates concolor ieucongeuys. From these moiré fringes we may arrive at the following conclusions.

1. There are different curved surface equations of knee. We know that the Macaca assamensis M' celland and Presbytis phyrei are close. They are different with Hylobates concolor ieucongenys. If the Hylobates concolor ieucongenys's left and right condylu's curve equation gradient are grad z_1, grad z_2, than $|\text{grad } z_1|$ and $|\text{grad } z_2|$ are smallest than Ceropithecidae.

2. Hylobates concolor ieucongeuys's right femur facies of articulation of knee fore curve equation's grident are smallest.

3. From the moiré fringes of femur facies of knee , we know that the golden-haired's moiré contour lines are close with the Macaca assamensis M'celland and Presbytis phyrei and different with the Hylobates concolor ieucongeuys.

To the Department of Animal Lab. of Beijing University, Department of Anatomy of Beijing Medical College and Department of Vertebrate Classification in the Institute of Animal Biology of the Academia Sinica for suppling the skeleton thank is due.

Reference

1. Zhang Ren-xiang, Lan Zu-yun, The application of moire contour lines to study articulation of vertebrates, Proc. JSMC, pp. 18-19. 1982.
2. Zhang Ren-xiang, Lan Zu-yun, Zhang Xong-zi, Lu Ming, Biomechenical research of joint 1, Acta Mechanica Sinica, 3, pp. 308-310. 1982.
3. Zhang Ren-xiang Lu Ming, Lan Zu-yun, Zhang Xong-zi, Stereometric analysis of the human body and articular surfaces, Proc. SPIE, Vol. 361, pp. 284-286 1982.
4. Zhang Ren-xiang, Lan Zu-yun, Zhang Xong-zi, Application of moire contourography to the human body, Proc. SPIE, Vol. 389, pp. 59-61 1983.

NEW MEAN FOR THE ASSESSMENT OF FLATNESS BY HOLOGRAPHY

Prof. Dr. A. F. Rashed (*) Vice-Dean, Faculty of Engineering
Prof. Dr. S. M. El-Fayoumi (**) Assistant Prof. Production Engineering Department
Eng. M. A. Younis, Assistant Lecturer, Production Engineering Department
Faculty of Engineering, Alexandria University, Egypt.

Abstract

The objective of this work is to find a new mean for macro and micro assessment of flat surface. The holography technique is used to get three dimensional measurement of flat surface under consideration. The holographic contour generation using two reference waves is applied. From the hologram mathematical model for the suggested assessment is established.

Seven flat specimens with different machining conditions were prepared for the application of the proposed technique. The assessment parameters that are established to distinguish between the various surfaces are:-

1. The coefficient of correlation of the surface points fitted to a flat surface.
2. The mean deviation of the perpendicular distance between surface points and least square plane.
3. The true bearing area diagram (not the bearing length diagram).
4. The ratio between the volume surface topography above the reference plane and the area of the surface.

Introduction

The quantitative assessment of surface flatness of relatively rough surface, produced by milling, shaping or turning operations is very difficult using stylus probe contact method or convenional interferometric contactless method. The aim of this work is to suggest a new mean for the assessment of flatness using holography techniques. The holography technique applied is contour generation using two reference waves.

Contour Holography

The flat surface under consideration is exposed twice to a collimated laser wave at two different angles and double exposure hologram is recorded for the surface. The hologram recorded set up is shown in Figure 1.

After correct processing the hologram results in a contour image for the flat surface under study. The large angle between the reference and the object waves makes white light reconstruction possible. Figure 2 shows the contour image photography set up, where a slide projector supplies the reconstruction wave. The contour spacing of the obtained image is given by the following equations:-

$$\Delta z = \frac{\lambda \cos (I)}{2 \sin (\theta_a - I) \sin \dfrac{\Delta\theta_R}{2}} \qquad (1)$$

$$\theta_a = \frac{\theta_{R1} + \theta_{R2}}{2} \qquad (2)$$

$$\Delta\theta_R = \theta_{R1} - \theta_{R2} \qquad (3)$$

where

   z is the contour spacing of the resulting contour image
   λ is the wave length of laser source in μm
   I is the view angle. The angle between the line of sight and the normal to hologramplate

* Visiting Professor King Abdel Aziz University, Jedah, Saudi Arabia Kingdom.
** Guest worker National Physical Laboratory, England, United Kingdom.

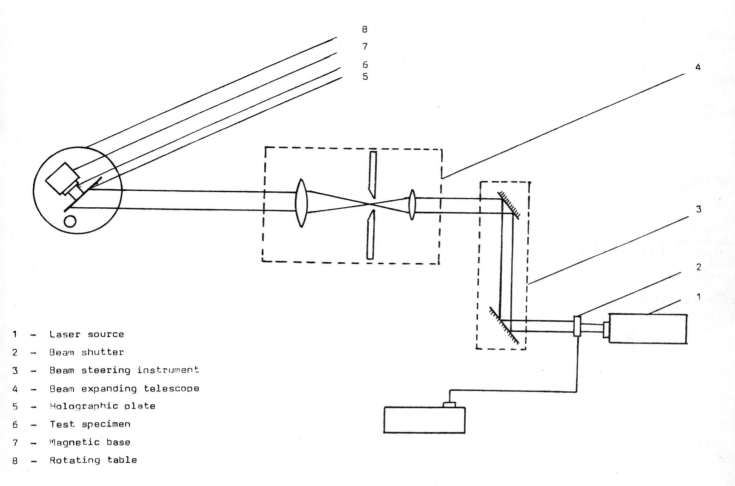

1 — Laser source

2 — Beam shutter

3 — Beam steering instrument

4 — Beam expanding telescope

5 — Holographic plate

6 — Test specimen

7 — Magnetic base

8 — Rotating table

Fig. 1.   Schematic diagram of the experimental set-up

θ_a is the average incidence angle.

θ_{R1}, θ_{R2} are the first and second incidence angles.

The incidence angle is the angle between the direction of propagation of wave and the normal to the hologram plate.
$\Delta\theta_R$ is the difference in magnitude between two incidence angle.

From equation (1) it is obvious that the contour maping depends on the value of viewing angle (I). This means that the tested specimen can be scanned at different contour spacing through the same holograph which is the major advantage of the technique.

Experiments

Seven square flat specimens of 20 mm side length were considered in the study. A hologram for each specimen was prepared to get the contour image of the corresponding surface. Figure 3 shows the contour image of one of the flat surface under investigation. The surface contour image was divided into a square grid such as given in Figure 4i-vii with spacing of 2 mm. The grid is drawn by dividing the image contour map to a number of equally spaced lines equal to the side of square (20 mm) divided by 2 mm. The spacing of the grid determines the nature of the study whether it is macro or micro inspection of geometrical form. In other words the larger the grid spacing the more the tendency to study macrogeometrical form and visa versa. The surface height at each node was determined applying equation (1). Meanwhile, the position of the grid nodes and the nodal height are the three co-ordinates of the surface points.

Mathematical Modeling of Flatness

The measurements data are fitted to a plane surface as given in equation (4) which is the

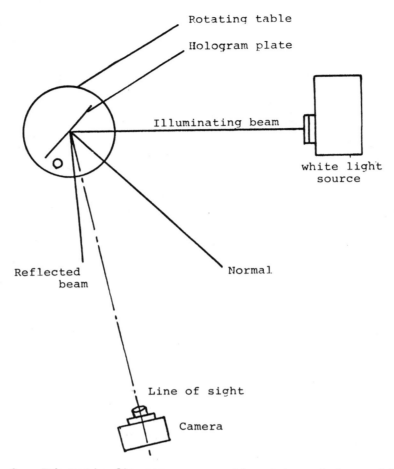

Fig. 2. Schematic diagram representing image photographing least square plane.

$$Z = aX + bY + c \tag{4}$$

where

X, Y, Z are the coordinates of surface points
a, b, c constants of the plane surface

This plane is an excellent reference to the total geometry of surface under investigation. Moreover, a parallel plane enables a good reference to flatness study.

The perpendicular distance between surface points and the least square plane is given by the following equation.

$$P = \frac{aX + bY + C - Z}{\sqrt{a^2 + b^2 + 1}} \tag{5}$$

where

P is the perpendicular distance between the surface point and the least square plane

The characteristics of flatness is dependent mainly on the perpendicular distance between surface points and the least square plane, or a specified plane parallel to that plane.

Flatness Assessment

Flatness assessment depend on spacious measurements. Four methods were applied to categorise the surface flatness fluently and evaluate the surface geometry.

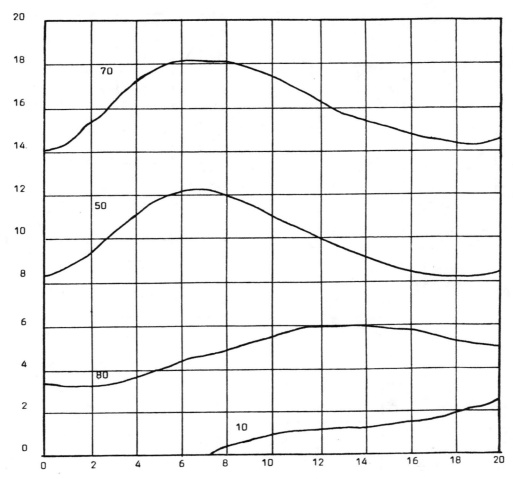

Fig. 3. Schematic diagram representing the square grid

1. Coefficient of correlation

The coefficient of correlation of the least square fitted plane to measurements data, is a good indication of how far the data points are belonging to a flat surface.

2. Average and variability deviation

The average deviation from the least square plane which is the average perpendicular distance from surface points (nodes) and least square plane. The standard deviation of these perpendicular distance indicate the surface variability. These two parameters give a clear idea about surface flatness.

3. True bearing area diagram

The true bearing area diagram considers the three dimensional measurements and not the bearing length diagram as plane measurements. The reference plane is a plane parallel to the least square plane and contain the lowest surface point. The coordinates of the measurement data are adjusted in such a way that the reference plane is the X-Y plane Figure 5.

The maximum z coordinate is the top of the bearing area diagram. The increment used to draw the true bearing area diagram is dependent on its range of height and the required accuracy. The bearing area at the X-Y plane is the full area of the solid material. The area at each increment is the sum of triangles areas of each adjacent three nodes having average z coordinate below the height of considered plane. The true bearing area diagram give a good picture for the surface texture.

4. Volume surface ratio

The ratio between the volume of solid above the reference plane X-Y and surface area of

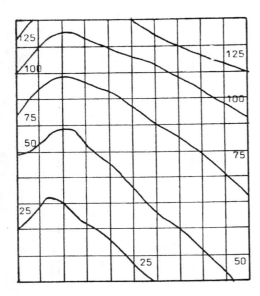

i   Specimen number 1

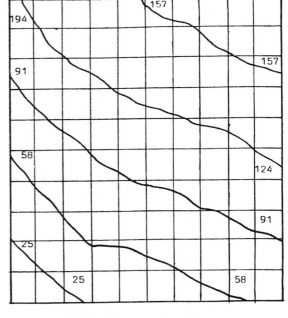

ii Specimen number 2

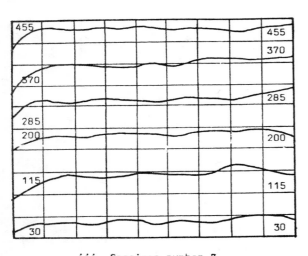

iii Specimen number 3

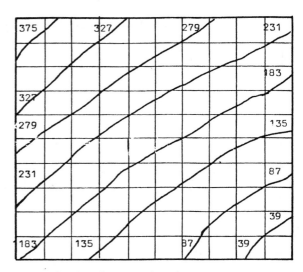

iv Specimen number 4

Fig. 4.   Contour image of the specimens (i-iv)

the texture is an ideal quantitative evaluation of flatness. The element of volume is the product of average nodal height of three adjacent point by the base area of these noeds (at reference plane). The total solid volume above the reference plane is the sum of elements of volume. The element of texture area for each adjacent three nodes is given by the following equation:

$$A^2 = \frac{1}{4} \begin{vmatrix} Y1 & z_1 & 1 \\ Y2 & z_2 & 1 \\ Y3 & z_3 & 1 \end{vmatrix}^2 + \frac{1}{4} \begin{vmatrix} z_1 & x_1 & 1 \\ z_2 & x_2 & 1 \\ z_3 & x_3 & 1 \end{vmatrix}^2 + \frac{1}{4} \begin{vmatrix} x_1 & y_1 & 1 \\ x_2 & y_2 & 1 \\ x_3 & y_3 & 1 \end{vmatrix}^2 \tag{6}$$

where

A is the element texture area
x, y, z are coordinates of nodes

164

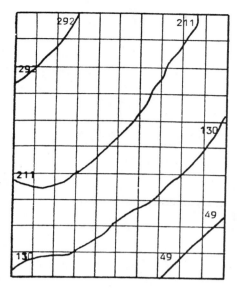

v.   Specimen Number 5

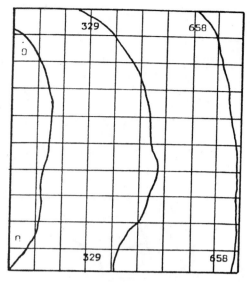

vi.   Specimen Number 6

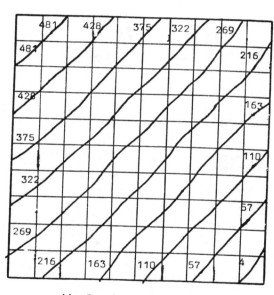

vii   Specimen Number 7

Fig. 4.   Contour image of the specimens   (v-vii)

The sum of elements area give the total texture area.   Thus the volume surface ratio is given by the following equation:

$$VS = \frac{VT}{AT} \qquad (7)$$

where

VS is the volume surface ratio
VT is the solid volume above the reference plane
AT is the texture area

Computer Program

A computer program was constructed to calculate the proposal assessment parameters.   The program considered the specimens to be of square shape.   The length of the square side and

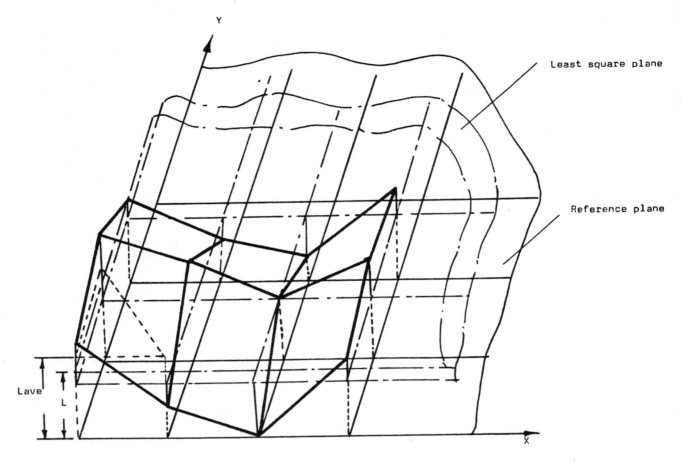

Fig. 5. Schematic diagram representing reference plane and
least square relative to surface points

the grid increment can be adapted to any dimensions. The program has been executed on PDP
11/70 time sharing system.

Results and Discussion

The results of assessment of surface texture by previous mentioned parameters are given
in table 1 and Figure 6. The data given in table 1 showed that the specimens number 2, 5,
6 have small flatness error. Their coefficient of correlation are 0.99111, 0.97292 which
are height and indicate proper surface flatness. While the volume to surface ratio are the
smallest values 180, 92, 69 μm. Specimen number 1 possess the maximum flatness error its
coefficient of correlation is too poor 0.17267 and its volume surface ratio is the height
value among the specimen 448 μm.

The three dimensional (true) bearing area diagram Figure 6 indicated that the specimens
of the best true bearing area diagram is number 6 while the one of poorest bearing area is
number 1.

It is obvious from these results that the coefficient of correlation, volume to surface
ratio and the three dimensional (true) bearing area diagram are compatible. These parameters are good means for spacious evaluation of surface flatness. The average and the
variability of deviation can only distinguish between a surface of a very poor flatness
such as specimen 1 and surfaces of proper surface flatness as specimens 5, 6. These
parameters are not sensitive enough to distinguish between the flatness of various surfaces
unless a huge amount measurements at small increment is determined.

Conclusions

The contour holography technique is a good tool for the three dimensional assessment of
flatness. The contour image that may be obtained can be varied by changing the average
incidence angles. The major advantage of the present technique is the possibility of varying the contour spacing allowing for the surface at different height intervals to be

Table 1. Values of flatness assessment by various parameters

Specimen No.	Machining process	Coefficient of correlation	Average deviation μm	Variability deviation (standard deviation μm)	Volume to surface ratio μm
1	Grinding	0.17267	214	126	443
2	Grinding	0.99111	8	6	180
3	Milling	0.85118	5	3	188
4	Milling	0.28806	2	12	252
5	Milling	0.99264	4	3	92
6	Milling	0.97292	7	5	69
7	Polishing	0.76709	4	3	262

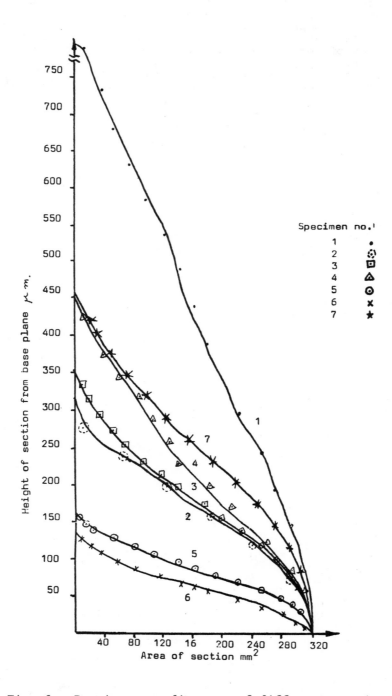

Fig. 6. Bearing area diagrams of different specimens

detected; only by changing the viewing angle.

The proposed methods of three dimensional assessment of flatness are satisfactory. These give a complete picture and precise description of the surface. The assessment methods can help in studying surface phenomena such as friction, wear, plastic and elastic deformation. Moreover, it is a helpful tool in determining surface qualities of machining processes. The suggested methods of assessment of surface texture is a reliable mean for spacious evaluation of surface topography.

References

1.   Smith, Howard M., "Principle of Holography" John Wiley (1975).
2.   Henshaw, PhD and Ezekial, S., "High Resolution holographic contour generation with white light reconstruction". Optic communication, Vol. 12 n.1. (Sept. 1974).
3.   Haines, K. and Hildbrand, B. P., "Contour generation by wave front reconstruction" Physics letter Vol. 19 No. 1 (Sept. 1965).
4.   Yu, F. T. S. and Tai Arthong, "Modulation detection holographic image contouring" Proceedings ICO Conference of Optical Methods in Science and Measurements Tokyo 1974.
5.   Tsuruta, T. and Shiotalse, N., "Holographic generation of contour map of diffusely reflecting surface by using immersion method. "Japanese Journal of Applied Physics (1967).
6.   Prows, D. B., "The calculation of the mean plane of a surface plate" Microtecnic Vol. XXI n.6 (1967).

Measurements of microdisplacements by holographic digital image processing

Ri. Peralta-Fabi

Lab. Micromecánica, Instituto de Ingeniería, Universidad Nacional
Autónoma de México (UNAM), Coyoacán,04510, México, D.F.

Abstract

This paper is concerned with an automatic method designed to produce a series of holographic interferograms that contain the history of surface displacements of a material undergoing a deformation process.

The purpose of automatic control of the hologram formation stage, is to insure that every interferogram displays similar fringe patterns, which can then be analyzed by another automatic setup for fringe counting based on digital processing of the holographic images. The formation parameters are supervised by a microcomputer that effects a decision of the exposure time. With this approach, it is possible to follow a dynamic phenomena and to produce measurements of statistical significance which, when expressed in terms of distributions, can help characterize the response behaviour of complex structured materials.

Introduction

Holographic interferometry has been utilized for the last fifteen years for the evaluation of 3-D surface displacements. However, it is only recently that an effort to relay the tedious and repetitive aspect of the technique has been bestowed to computer assisted devices. Besides the now classic paper of Aleksandrov and Bonch-Bruevich[1] in which several quantitative details were brought to light, one of the first fruitful attempts to provide for some basis of automatic evaluation was presented by Fossati Bellani and Sona[2]. Their scanning procedure, as well as the fibre optic read-out system were both subject of servo control, thus facilitating automatization. Several workers adopted this general approach[3,4,5] and more recently, fully automatic analysis techniques have been published[6,7] by other groups.

In this study, the object under investigation is a clay soil submitted to a constant compression load. The response of the material is a steady deformation process, of which we are interested in quantifying the 3-D components of displacement, as a function of time. Various tests were conducted to establish the main parameters which affected fringe geometry. Once a set of rather parallel fringes with discernible density was obtained, an interferometric analysis was conducted, to evaluate the exact rigid body movement of the instrumented loading device. By reproducing the axial displacement within a given value, all interferograms present similar fringes from different perspectives. In our case, the interference pattern is located in space, between the object and the observer, thus, upon a change of the observer's position the pattern moves relative to the object's surface. Another option to this process is to project the image onto a screen by means of a conjugate beam. By scanning the beam across the holographic plate, when the position of the screen is the same as the original object's surface, the image remains fixed, whilst fringes sweep across the object.

The method here presented utilizes image processing techniques to effect the counting of fringes in many points simultaneously, therefore allowing a fast and accurate assessment of displacements.

Experimental arrangement and procedure

Experimental tests are carried out in a vibration isolated table, weighing about 4 t, that is held in suspension over a 5 cm air cushion. Each corner rests on a piston-bellow container connected to a small surge tank. For low frequency excitations there is an air interchange between both parts, which is self-damping. However, in the event of higher frequencies, the connection between the container and the tank introduces a capillary resistance, enough to impede air circulation, thus limiting the response of the system to the container only. A full study of the effectiveness of vibration isolation was carried out, the natural frequency being about 1.2Hz, and demonstrating its ability to reproduce our design and construction criteria. Holograms could have long (2 min) exposure times, and interferograms may take 10 minutes in between multiple exposures. However, the approximate value of actual parameters in our tests, is 6 ms for exposures (effectively freezing the deformation process) and up to three min in between first and second exposures.

An important device in this setup is the loading apparatus. It is composed of a small magnetic table base that holds two guides, vertical and parallel. In these guides runs a cart with high displacement stability, as evaluated by interferometric tests. A relevant aspect of actual tests is that interferograms always included the loading device on the exposure, thus allowing for a self evaluation of true one-dimensional loading, essential for our mechanical tests. The device is instrumented by a set of displacement transducers consisting of well known LVDT's together with fotonic sensors for fine measurements. In our particular case, new exposures are taken when the cart is displaced 17 µm approximately (in view of the appropriate fringe density obtained) it is therefore important to resolve fine movement. In the case of load, it is monitored by a load cell to permit an accurate control of this extremely important parameter. Fig 1 shows a photograph of the loading device.

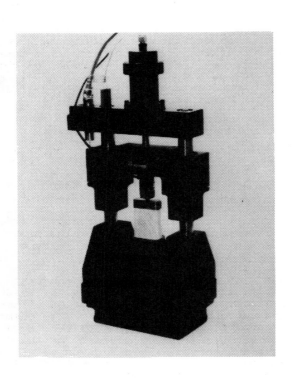

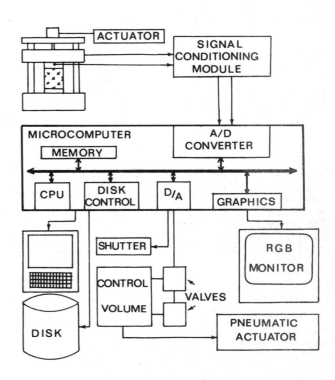

Figure 1. Photograph of sample loading device

Figure 2. Control system for automatic interferogram production

In order to control the exposure sequence, both transducers feed their output to a computer, that has adaptive-control programs. First, the signal is fed to conditioning preamplifiers that raise the voltage from mV to volts, filter all high frequency noise, and send the signal away (8 m) to the central microcomputer of the laboratory. In this device the signal is converted to digital form with a 12 bit resolution and at high speed. The signal is further amplified to allow for 5 V TTL logic to take over. At this point, it is entered as data into the control program which compares actual to desired values, and if necessary, takes corrective action. Simultaneously a data file created for this particular test, is being augmented with all the experimental parameters as a function of real time.

Control over the load is based on a continuous supervision of actual load values. When there exists a deviation beyond a threshold, a signal is sent to either of two solenoid valves, presided by fixed regulator valves, that permit the exit or entrance of air onto a control volume, in turn connected to a pneumatic actuator. Because the transducer is located between the cart and the sample, measurements reflect the instantaneous value of the load.

With respect to displacement, there is no correction necessary but a continuous supervision of displacement is needed to produce the opening of an electronic shutter that will allow for exposures to realize.

The closing of the shutter is effected by an electronic device that integrates light flux up to a predetermined value, given by the emulsion's sensitivity.

With this system, holographic interferograms are produced sequentially and on a predictable basis; the general block diagram for the control system is shown in fig 2.

During the experiment, it is necessary that the parameters most important to the test are displayed in real time inside of the laboratory. A limitation in this case is the dark room environment needed for exposures. However, the plates utilized are insensitive to red ; thus, a color RGB monitor displays black and red graphics of load and displacement. The information is presented in two forms, analog and digital since at times a glance is sufficient to judge the state of the test, but there are cases in which numerical data is more useful to human interpretation, (see fig 3). It is important to realize that the cost and availability of holographic plates is an aspect we cannot ignore, and that due to the sequential nature of these tests an error is usually expensive in terms of plates. Besides the above mentioned automatic control in the formation stage, that will adjourn a test upon a deviation from threshold values, and the real time display that allows more intelligence to enter the control process, an effort to save plates (that may be 40-50 for a given run) is done by exposing only sectors of a plate, by running a shade in front of any of the plates, allowing for multiple double exposures on a single plate. A schematic diagram of hologram formation is shown in fig 4.

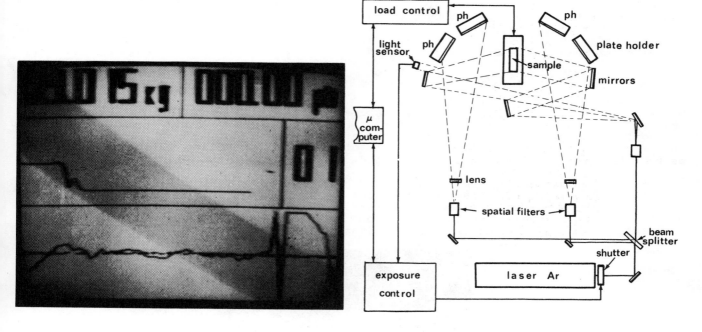

Figure 3. Photograph of T.V. screen showing control graphics display

Figure 4. Schematic diagrams of hologram formation

Sequential exposures

From the last figure, it can be observed that on each side of the object there are two plate holders. Every time the shutter opens all plates are illuminated; however, considering only one side of the sample, that is, only two of the holders on any side, the exposure sequence is as follows: calling the plate holders ph1 and ph2, ph1 is first exposed (1) whilst covering ph2, when the second exposure (2) is automatically set, ph2 is uncovered, thus ph1 receives its second exposure and ph2 its first; before the next exposure (3) we replace ph1 with a virgin plate and expose both plates. Ph2 is thus exposed for a second time whilst the new plate in ph1 is exposed for a first time. At this point the plate replaced is that from ph2 since it has accumulated two exposures. The process continues in this manner until the test is complete. With this procedure a continuous coverage of a dynamic phenomena is obtained. Provided that plates can be switched in time (enough for vibrations to damp) this extension of conventional holographic interferometry can permit a wide variety of tests to be designed for the evaluation of complex quasi-dynamic problems.

Interferometric analysis

Once the plates have been exposed and properly processed, an analysis to extract displacement information can be carried out.

This analysis implies replacing the plate onto its original position and illuminating it with a conjugate reference beam. The image is then projected towards a translucent screen, and the reconstruction beam swept across in three predetermined directions. Scanning the beam will cause the movement of interference fringes across the image since these are not located at the surface. During a scan, the beam is arrested every millimeter and the image is captured by a television camera whose signal is fed to digitization circuitry and stored. A special program will follow the gray levels of any set of fixed points throughout the sequence of stored images and evaluate the passage of fringes in many points simultaneously. A full description of this technique has been presented recently and can be read in Ref 8.

Discussion of results

The results obtained with the automated control of hologram formation, provide a set of interference patterns that reflect the overall behaviour of a complex material undergoing a deformation process. Whilst it has been possible to achieve similar results before for a given interval of time, this technique allows for continuous phenomena to be recorded, and later quantitatively evaluated . By ensuring that each double exposure produces discernible fringes and of similar appearance it is readily possible to count them by automatic methods as they sweep across the object. The general matrix equation that relates fringe numbers to actual displacement measurements is adopted from Ref 2 and we may represent it as follows:

$$U_i = A_{ij}^{-1} N_i \lambda \tag{1}$$

where U_i is the 3-D displacement we are interested in and A_{ij}^{-1} is the geometrical matrix that relates any point of evaluation at the surface of the original sample, with the position of the holographic plate, the reference beam and the object plane; N_i is the number of fringes that has passed in front of any fixed point on the object's surface as the plate is swept in three independent directions; λ is the wave length of the laser.

Detailed discussions on the properties of this well known relation can be found in refs 2,4,5&6 together with the error analysis included in our evaluation program. With respect to experimental accuracy, our main parameters, load and displacement, are controlled to 0.05% of the test load and 0.03% for displacement. Both allow high quality holograms to be obtained, given that the holographic illumination ratios are properly managed (see fig 5).

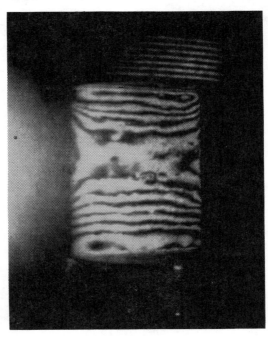

Figure 5. Photograph of interferogram showing rather parallel fringes

On the other hand, one of the main virtues of automated interferogram production and interpretation is the single scan technique, in which all data for 3-D displacement evaluation is obtained, during a single scanning of the three directions. In addition, this saves a considerable amount of time.

Numerical results from these tests have several important aspects: first, that from every plate we obtain measurements of up to 100 points simultaneously. These points are distributed over the entire surface of the sample in the form of an imaginary mesh of 5x5 mm.

Microdisplacements can be expressed in terms of statistical distributions, with the mean approximating a macroscopic measurement, but indicating the degree of randomness of the structure through the deviation. Also, since these measurements are obtained as a function of time, in view of the sequential exposures, the intensity of the transition probabilities of displacements can be evaluated readily. This information is unique to particular materials, it is thus possible to characterize material's behaviour in this precise manner.

Conclusions

The method presented in this paper combines several techniques of recently published work. It is believed that it introduces various advantages over conventional approaches with respect to controlled interferogram production and interpretation, when investigating materials with complex structures. It also permits considerable time savings whilst maintaining a high accuracy in the quantification of displacements. The above conveniences allow for the application of holographic interferometry to practical problems in soil mechanics, as it is our case, and to many other difficult materials that require extensive testing and accurate assessment of response behaviour. By combining automatic control in hologram exposures and image processing in fringe interpretation we relay the tedious and repetitive aspects, to a microcomputer; idoneous device for this task.

Acknowledgements

The author would like to thank G Hernández and E Vicente for the writing of the software, and some of the hardware work.

References

1.   Akeksandrov, E. B., Bonch-Bruevich, A. M., "Investigation of Surface Strains by the Hologram Technique," Soviet Physics Tech. Phy. Vol. 12, N° 2, Aug 1967
2.   Fossati Bellani, V., Sona, A., "Measurements of Three-Dimensional Displacements by Scanning a Double-Exposure Hologram," Appl. Opt.,Vol. 13, N° 6, June 1974
3.   Kreis, T. M., Kreitlow, H., "Digital Processing of Holographic Interference Patterns," Top.Meeting on Holog. Interf. Speckle Metrol. OSA-SESA, P. TUB2-1, Cape Cod, Mass. June 1980
4.   Ek, L., Biedermann , K., "Implementation of Hologram Interferometry with a Continuously Scanning Reconstruction Beam," Appl. Opt., Vol. 17, N° 11, June 1978
5.   Peralta-Fabi, Ri. "Experimental Investigation of Creep Behaviour of Bond Paper," Ph.D. Thesis, McGill U., Montreal 1978
6.   Axelrad, D. R., Rezai, K. "Determination of Surface Displacements by Holographic Electrooptical Processing," Appl. Opt. , Vol. 13, N° 11, June 1982
7.   Ek, L., Majlöf, L. "Evaluation of 3-D Displacement by Holography and Computer," Top. Meeting Holog. Interf. Speckle Metrol. OSA-SESA , P. TUB3-1, Cape Cod, Mass., June 1980
8.   Peralta-Fabi, Ri. "Opto-Electronic System for Automatic Holographic Fringe Counting," 10th Int. Opt. Computing Conf. IOCC-ICO., MIT., Mass. April 1983

Application of holographic interferometry to shock wave research

K. Takayama

Institute of High Speed Mechanics, Tohoku University
2-1-1 Katahira, Sendai, JAPAN

Abstract

Paper reports a successful application of holographic interferometry to the shock wave research. Four topics are discussed; i) transonic flow over an aerofoil, ii) shock wave propagation and diffraction past a circular cross-sectional 90° bend and two-dimensional straight or curved wedges, iii) stability of converging cylindrical shock waves and iv) propagation and focusing of underwater shock waves. Experiments were conducted on shock tubes equipped with a double exposure holographic interferometer. In each case isopycnics around shock waves were determined and three-dimensional shock wave interactions were also observed. Results are not only bringing forth new interesting findings to the shock wave research but also showing a further potentiality of holographic interferometry to the high speed gasdynamic study.

Introduction

It is known that in the flow visualization study a holographic interferometry has a wide versatility[1] which was impossible by the conventional flow visualization techniques, such as a shadowgraph, a schlieren method or a classical Mach-Zehnder interferometry. A shock tube is a device which can very easily produce a high temperature and high speed flow necessary to the high speed gasdynamic study[2]. Today the shock tube is used not only in the gasdynamic study but also in other branches of science. However, the working duration of the uniform flow in the shock tube is limited from a few milliseconds to several microseconds depending on the condition, so that it is crucial to use a short time measuring system, such as a holographic interferomtry with a Q-switched ruby laser as a light source.

In the Institute of High Speed Mechanics, Tohoku University, a double exposure holographic interferometry has been developed and applied to the shock wave research by using the shock tube. The present paper reports a recent result of successful application of holographic interferometry to shock wave phenomena in gases and liquids. The following topics are discussed here; i) transonic flow over a NACA 0010 aerofoil, ii) shock wave propagation and diffraction past a circular cross-sectional 90° bend and shock transition over straight or curved wedges, iii) stability of converging cylindrical shock waves and iv) propagation and focusing of shock waves produced by underwater microexplosions.

In the first three examples, experiments were conducted on a trasonic shock tube and conventional shock tubes. In the fourth example, underwater shock waves were produced by detonating lead azide pellets of 4 mg with a pulsed laser. In each case, isopycnics around shock waves were determined and by using three-dimensional holographic interferometry the complicated shock wave interaction was also observed.

Transonic flow over an aerofoil

Shock tube

In a shock tube a uniform transonic flow exists behind an incident shock wave. Therefore, the shock tube can serve the purpose of a simple short duration transonic wind tunnel[3]. As shown in Fig. 1, a transonic shock tube of the Institute of High Speed Mechanics, Tohoku University, consists of a 1.5 m long high pressure chamber and a 8.0 m long low pressure channel of 60 mm x 150 mm cross section. The test gas was dried air at 0.3 - 1.01 bar and the driving gas was helium at 4.0 - 8.0 bar. Mylar films of 100 - 250 μm were used as diaphragms separating the high pressure driving gas from the low pressure test gas. Shock speed was measured by two pressure transducers (Kistler 606L) placed 250 mm apart ahead of the test section. Shock Mach number realized was from 1.5 to 1.9, i.e., the uniform flow Mach number was from 0.6 to 0.9 and the Reynolds number to the aerofoil chord length was $0.8 - 1.6 \times 10^6$.

A NACA 0010 aerofoil of 60 mm chord length was placed in the test section. The test section has plastic glass windows of diameter 130 mm and 25 mm thickness. The aerofoil was fixed in the grooves on the plastic glass windows and the angle of attack of the aerofoil was variable from 0° to ± 7°.

Holographic interferometer [4]

A schematic of the experimental arrangement is shown in Figure 1 and is self-explanatory. The present optical system consists of a set of paraboloidal mirrors of 300 mm daiameter and 3,000 mm focal length, auxiliary mirrors and lenses, and a double pulse holographic ruby laser, Aplollo Inc., 22HD, 2J/pulse.

In order to obtain finite fringe holograms, a collimating lens L_4 of 150 mm diameter and 1,000 mm focal length to the reference beam is shifted before the event as shown in Figure 2. In this case the first exposure was done manually and the second exposure was separately triggered by the event. The fringe interval was determined for the given optical arrangement by changing the displacement of the axis of the lens L_4.

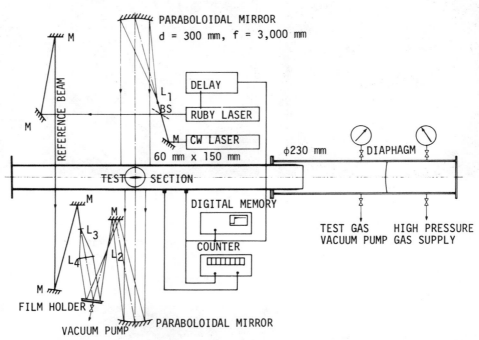

Figure 1. Schematic of experimental arrangement.

The angle between the refernece beam and the object beam was about 20°. The path difference of these beams was less than 10 mm. The magnification of the object was about 0.6. Films, Agfa 10E75 100 mm x 125 mm sheet film, were hold on a film holder evacuated by a vacuum cleaner and were processed by Kodak D-19 developer at 293 K 4.5 minutes for a double exposure and 6.5 minutes for a single exposure.

Reconstruction of holograms was carried out by using an argon-ion laser (Toshiba LAI-104-A, 1W for 514.5 nm wave length) and Neopan SS sheet films.

Results. Figure 3 shows the example of transonic flow over a NACA 0010 aerofoil for the uniform flow Mach number $M_\infty = 0.90 \pm 0.01$, the Reynolds number $R_e = 0.8 \times 10^6$ and the angle of incidence $\alpha = 4°$ - a;a finite fringe double exposure interferogram, b;an infinite fringe double exposure interferogram and c; a triple exposure interferogram.

Combining a finite fringe interferogram with one without the event, i.e.,an interferogram of parallel lines at equal distances, we can obtain an infinite fringe interferogram, since interactions

Figure 2. Collimating lens L_4.

of the finite fringe with the parallel lines at equal distances correspond to the integral multiples of the fringe shift. This process is possible by constructing a triple exposure hologram with the first exposure done by shifting the lens L_4 in Figure 1 before the event, the second exposure without shifting the lens L_4 again before the event and the third exposure triggered by the event without shifting the lens L_4.

Although in Figure 3 a,b,c, each interferogram is of different flows to the same initial conditions the flow patterns are almost identical. Isopycnics are immediately determined in Figure 3b and the density distribution over the aerofoil can be obtained from Figure 3b and 3c. If the isentropic relation is assumed over the aerofoil the pressure coefficient or the lift and drag coefficients of the aerofoil are immediately determined

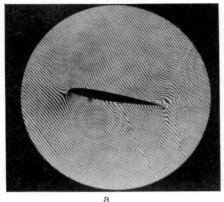

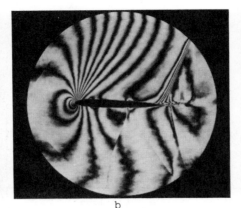

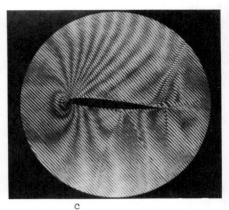

a   b   c

Figure 3. Transonic flow over a NACA 0010 aerofoil.

Shock diffraction past a circular cross sectional bend and wedges

The holographic interferometry can be applied to various shock tube flow problems, which might otherwise be impossible to be substantiated especially by conventional flow visualization techniques. In this section firstly diffraction and transmission of shock waves past a circular cross sectional 90° bend and secondly reflection and diffraction of oblique shock waves over straight , convex and concave wedges are discussed.

Three-dimensional bend flow[5]

When an initially planar shock wave is propagating past a circular cross sectional 90° bend, the movement of the transmitting shock wave and the flow behind it are found significantly different from those in the two-dimensional case. However, in order to make quantitative measurement of three-dimensional shock wave interactions past a circular cross sectional 90° bend, the test section should have a property that a collimated incident light ray can traverse the circular cross sectional inner bore parallel and emerge parallel as shown in Figure 4. The test section was made of plastic glass and manufactured at the workshop of the Institute of High Speed Mechanic Tohoku University. Tracing a light ray as shown in Figure 4b, we can immediately determine the outer aspheric configuration as follows ,

a

$$\frac{s}{s_0} = \frac{n \cos\theta + \cos\psi}{(n+1)\cos\psi} \quad , \quad \sin\psi = n \sin\theta \quad (1)$$

Figure 4. Potograph of the test section and aspheric lens.

where s, s_0 and n are wall thickness, the maximum wall thickness and the index of refraction which is experimentally determined to be 1.4915 ± 0.0005 , respectively. θ is the angle defined in Figure 4c.

Figure 5a shows a calibration of the test section where the first exposure was done with the test section evacuated down to 100 Pa and the second exposure was done with the section filled by atmospheric air. Since the path length of each parallel light ray is different across the inner bore, therefore the change of fringe intervals indicates the accuracy of the test section. Consequently from Figure 5a, the effective angle is found to be 75°. Optical defects of plastic glass which are fatal in a Mach-Zehnder interferometry can be fortunately cancelled out in the present double exposure holographic interferometry.

The test section was connected 80 mm behind a circular cross sectional sharp 90° bend. In Figure 5b,c reconstructions of a three-dimensional hologram for an incident shock Mach number $M_s = 1.28$ are shown. The two photographs were reconstructed with the different view angles from the same hologram. It is clear that coexistence of regular and Mach reflections can be seen and a horse shoe shaped locus of triple point is observable on the transmitting shock wave. Figure 5d shows a reconstruction of an image hologram for M_s = 1.28. With the knowledge from Figure 5b,5c we can evaluate the flow.

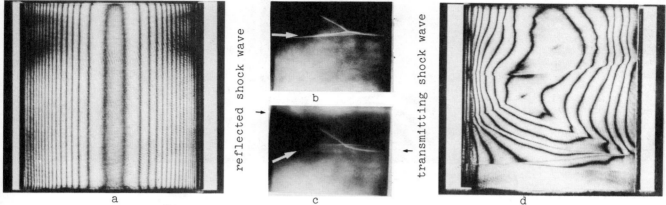

Figure 5. Reconstructions of holograms.

Shock diffraction over straight, convex and concave wedges. The classical problem of reflection of an oblique shock wave over a planar or curved wall is an important uncompleted research task [6]. The shock tube is a convenient tool to investigate this problem. The previous optical observation was done by mostly a shadowgraph or a schlieren method and a Mach-Zehnder interferometry. The former two methods can provide informations of shock wave configurations. A Mach-Zehnder interferometry is the most reliable method, however, the range of its application is limited, less flexible than the present holographic interferometry.

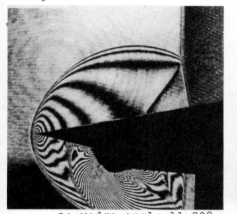 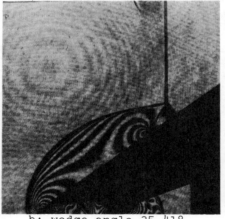 

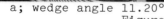
a; wedge angle 11.20°   b; wedge angle 35.41°   c; wedge angle 46.36°
Figure 6. Shock reflection over straight wedges.

Keeping M_s constant and increasing the wedge angle, the Mach reflection terminates at the critical wedge angle beyond which only the regular reflection can exist. Figure 6 shows typical Mach reflection for M_s = 2.0 in CO_2, a; simple Mach reflection, b; complex Mach reflection, and c; double Mach reflection. Infinite fringe interferograms can provide not only the

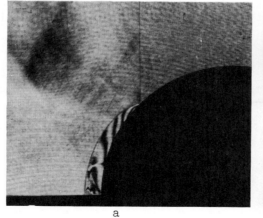 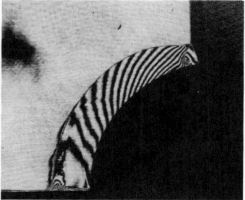
a   b
Figure 7. Shock reflection over a convex wall.

information on shock wave shape but also reveal flow field itself behind the reflected shock wave. We see how isopycnics are changing with increasing the wedge angle.

When a shock wave encounters a curved wedge where the wedge angle is not constant any more, two types of shock transitions are possible, that is, the transition from regular to Mach reflection over a convex wall as shown in Figure 7 and the transition from Mach to reguler reflection over a concave wall shown in Figure 8. These belong to truly non-stationary flow phenomena, whereas the shock diffraction over straight wedges belongs to pseudostationary flow. In an

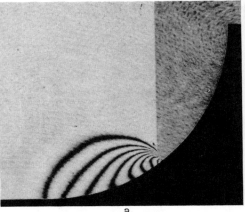
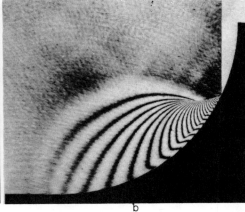

a   b

Figure 8. Shock diffraction over a concave wall, M_s = 1.5.

experimental work, it is important to obtain the isopycnics behind the reflected shock wave so that use of infinite fringe holographic interferometry is particularly valuable. A convex wall of 40 mm radius and a concave wall of 60 mm radius were placed in a 40 mm x 80 mm shock tube. Figure 7 is for M_s = 2.50 in air, a; regular reflection, b; Mach reflection and Figure 8 is for M_s = 1.5 in air, a; Mach reflection, b; regular reflection

Stability of converging cylindrical shock waves

Converging cylindrical and spherical shock waves can produce very high temperature and pressure at the centre of convergence. Many theoretical and experimental investigations were conducted on the propagation and stability of converging cylindrical and spherical shock waves , however the experimental investigation is still far from being completed. Here the experiment we have done is described briefly.

A converging cylindrical shock wave was produced in a 230 mm diameter annular shock tube in which a planar incident shock wave is turned 90° through an axisymmetric 90° bend as shown in Figure 9. A 50 mm diameter shock tube produced an incident shock wave which was transformed into a ring shaped transmitting shock wave in the annular section consisting of a 210 mm diameter inner tube and a 230 mm diameter outside tube. The annular section is long enough to stabilize the transmitting shock wave. The inner tube is supported by struts of 12

mm diameter whose effects on stability of the transmitting shock wave are found negligible. The gap distance of the axisymmetric converging section is 8 mm. As seen in Figure 9, the diameter of parallel observation section is 130 mm whose inner and outside walls are a metal mirror and a glass window, respectively.

Nominal shock Mach numbers were measured by pressure transduces (Kistler 603A) and are ranging from 1.10 to 2.10 in air and CO_2. A very good repeatability of the shock Mach numbers was obtained.

A double exposure holographic interferometry was used whose optical arrangement is similar to the Twyman-Greene interferometry and this is probably the first time for holographic interferometry to be applied to converging shock wave observation.

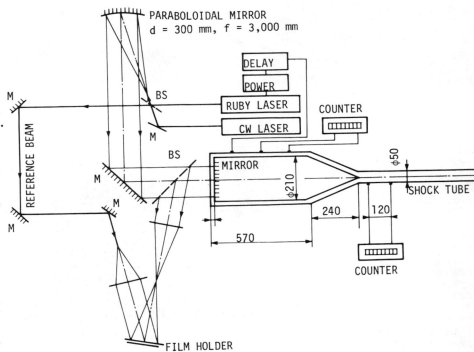

Figure 9. Schematic of converging shock wave experiment.

Figure 10 shows infinite fringe interferograms for nominal shock Mach number $M_S = 1.37$ in CO_2 at the initial pressure of 40 KPa, a; behind a converging cylindrical shock, arrowed, of diameter d = 6.4 mm at this instant, isopycnics show the instability of mode m = 4 is about to start, b; d = 2 mm and the density non-uniformity is remarkable showing the deformation of the shock front is more striking, c; showing the beginning of reflection at the centre of convergence and d; a reflected shock wave, arrowed, i.e., a diverging cylindrical shock wave is found stable, however at the centre of convergence, there is a remain of the four vortices which indicate that once a very strong instability did exist there.

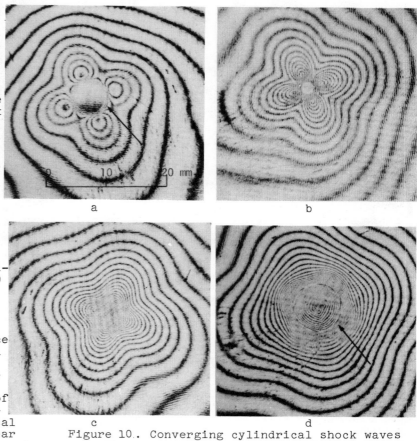

Figure 10. Converging cylindrical shock waves near the converging centre, $M_S = 1.37$ in CO_2.

In previous works, a schlieren method was used for the flow visualization. However as seen in Figure 10 the density change behind a converging shock wave is too delicate to be detected by the conventional method. It is important to examine the effect of the initial disturbance on the shock instability. Artificial shock front disturbances, twelve circular rods of 14 mm diameter were placed at the exit of axisymmetric bend. Surprisingly the instability of mode m = 4 still existed. It is concluded that the converging cylindrical shock wave is always unstable and near the converging centre the instability of mode m = 4 exists.

Propagation and focusing of underwater shock waves

The advantage of using holographic interferometry in the flow visualization study is best exhibited in an underwater microexplosion study. In this case the non-uniformity of the medium is also ignored so far as using a double exposure holographic interferometry. In the Institute of High Speed Mechanics, Tohoku University, a technique was developed to produce undisturbed spherical shock waves in liquids. A pellet of lead azide, PbN_6 4.0 ± 0.5 mg and about 1.0 mm diameter, was suspended by a fine thread in distilled water and ignited by a Q-switched ruby laser pulse, 2.0 mm beam diameter, 20 nsec pulse duration and 100 mJ.

The energy addition from this laser ignition was found negligible if compared with the amount of the released energy of 6J from 4.0 mg of lead azide. A perfectly spherical shock wave was obtained in good reproducibility. The optical arrangement was the same as in Figure 1 and the pressure was also measured by a pressure transducer, Kistler 603A, Figure 11 shows an underwater shock wave, a; finite fringe interferogram and b; infinite fringe interferogram. In this axisymmetric flow fringe shifts do not directly indicate the isopycnics but a simple calculation

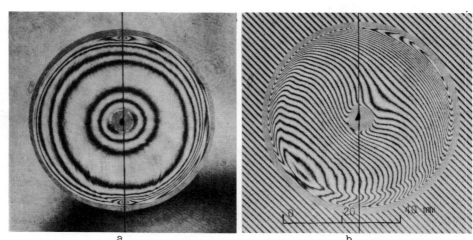

Figure 11. Underwater microexplosion of PbN_6.

179

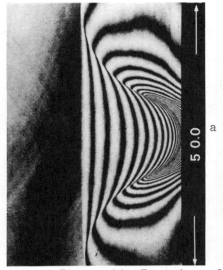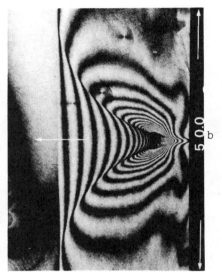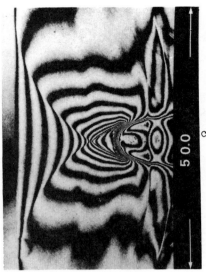

Figure 12. Focusing of underwater shock waves from an ellipsoidal reflector.

is needed for evaluation. The thus evaluated density and pressure jump at the shock front agrees very well with the pressure measurement. Figure 11a shows a shock wave of $M_s = 1.02$.

As an application, the underwater shock wave focusing was conducted by using a 50 mm diameter reflector whose configuration is a part of ellipsoid. A lead azide pellet was placed at the first focus point and the shock wave focusing was expected to occur at the second focus point 7.5 mm from the reflector edge. Figure 12 shows the focusing of different shock waves with the same initial condition, a; 4 μsec before the focusing, b; very close to the focusing and c; about 4 μsec after the focusing. A shock front concave to the direction of propagation and the catching up of an expansion wave from the reflector edge can be clearly seen.

Evaluating fringe shifts in Figure 12b, we can determine the pressure profile in Figure 13, abscissa; radial distance r mm, ordinate; the pressure p bar and parameter X mm; axial distance. The maximum pressure of 880 bar does exist at X = 5.83 mm. In this case about one per cent of the total released energy seems to be focused. However, if the energy source is placed at the second focus point inside the reflector and then the shock wave forcusing occurs at the first focus point, the pressure amplification will be enormous.

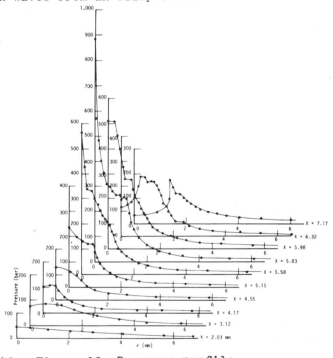

Figure 13. Pressure profile.

Acknowledgements

The author wishes to express his gratitude to Professor M. Honda, Mr. O. Onodera Mr. O. Ojima and Mr. S. Hayasaka of the Institute of High Speed Mechanics, Tohoku University, for encouragements and helpful discussions.

References

1. Caulfield, H. J., Handbook of Optical Holography , Academic Press 1979.
2. Glass, I.I. and Hall, J. G., Handbook of Supersonic Aerodynamics, Section 18, Shock Tube NAVORD Report 1488 1958. 3. Cook, W. J.," Test Section Configuration for Aerodynamic Testing in Shock Tubes," Proc. 12th International Symp. on Shock Tube and Waves, pp. 127-136. 1979.
4. Takayama, K. and Watanabe, W., "Shock Wave Reflection over Concave Walls," Mem. Inst. High Speed Mech., Vol. 45, pp. 1-33. 1980. 5. Takayama, K.,"Shock Propagation past a Circular Cross Sectional 90° Bend," 9th Symp. on Shock Dymanics, pp. 164-169. 1982.
6. Ben-Dor, G., "Domains of Shock Wave Reflexion ," J. Fluid Mech., Vol.92, pp.459-496. 1979.

HOLOGRAPHIC RECONSTRUCTION OF A SYNTHETISED SUBPICOSECOND PULSE

G. Tribillon, R. Salazar

Laboratoire d'Optique, L.A. 214, Faculté des Sciences et des Techniques
Université de Franche-Comté, 25030 Besançon Cedex, France

Abstract

Subpicosecond pulses have been synthetized with a dye laser. The 8000 GHz broadband source corresponds to a temporal resolution of $\Delta t = 1/\Delta\nu = 0,12$ ps. The holographic recording set-up proposed by N. Abramson has been used to visualize the light in flight, but the wavelength dependence of the hologram introduces a dispersion in the reconstructed image. Compensation techniques have been applied and an other recording system has been developed. Experimental results of the reconstructed holographic wavefront are presented. The interesting possibility to obtain a reconstruction in white light is compared with the coherent case. The signal-to-noise ratio of the image is discussed.

Introduction

A new holographic technique to record a light wavefront through space has been proposed by N. Abramson.[1-2] The light source can be an argon laser without etalon or a mode locked ion-laser. The spatial width for a 10 picosecond pulse wavefront is about 3 mm. These pulses advantageous in some applications - studies on ultra-short phenomena but also studies on physical phenomena requiring a high temporal resolution can be synthetised by a dye laser tuned in frequency.[3]

Holographic visualization by the Abramson's method

The broadband source is a tunable dye laser. The tuning over a wavelength range of 0,58 to 0,59 μm is obtained by means of an electro-optical cell or by means of a Lyot filter mechanically rotated.

In a first time, the recording set-up proposed by N. Abramson (see figure 1.a) has been tried. A painted plane surface is illuminated at a large incidence angle and the reference wavefront sweeps across the photographic plate. After the beam-splitter LS, the optical lengths of the object beam (LS, O_2, H_2) and the reference beam (LS, M, H_2) are identical. The scattered light coming from the different points O_1, O_2, O_3 gives with the reference beam an interference phenomenon in H_1, H_2, H_3 respectively because the coherence of the source is very close. To check the spectral band of the radiation an hologram has been recorded with the dye laser without tuning. In this experiment, the electro-optic cell was placed in the cavity of the dye laser. The virtual holographic image is presented the figure 1.b. Longitudinal modes of the cavity can be observed. The width of the modes is about 3,3 mm that represents a line width of 90 GHz in good agreement with the caracteristics of the modulator. On account of the longitudinal modes produced by the windows of the electro-optic

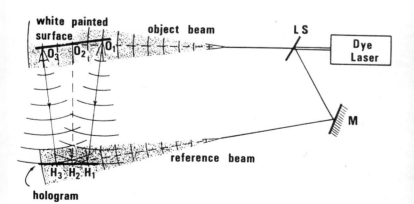

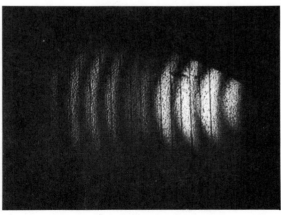

Figure 1.a - Holographic recording
set up

Figure 1.b - Longitudinal modes of the dye laser
supplied with the electro-optic cell

cell, we have worked with a dye laser tuned by a Lyot filter.

By using the same experimental set-up a tuning of 450 GHz is performed during the hologram recording. The width of the reconstructed wavefront (Figure 2.b) can be compared with the holographic image obtained without tuning (Figure 2.a). A synthetised picosecond wavefront is shown (Figure 2.c) when the spectral band is about 1300 GHz.

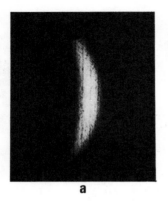
a

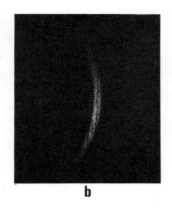
b

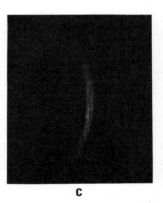
c

Figure 2. Reconstructed wavefronts obtained with the
dye laser supplied with the Lyot filter

Between the cases (b) and (c) the width of the wavefront is not decreasing. On these results, two remarks are needful with regard to the influency of the tuning amplitude.
- the contrast of the holographic image decreases,
- the sharpness of the synthetised picosecond pulse reachs a maximum.

The explanation is to discover in the experimental set-up that occurs a diminution of the signal to noise ratio and that introduces chromatic aberrations in the holographic reconstruction. These aberrations can be measured by using the properties of the reconstructed holographic images.[4] To explain the value of signal-to-noise ratio we must show in what manner the interference fringes are created on the holographic plate. Each point of the object scatters the light through space, particularly in the direction of the photographic plate. If N elements are resolved in the image, we know that the signal-to-noise ratio is decreasing as 1/N. As the spectral band of the laser source is increasing with the tuning, the fringe visibility is vanishing. This point has been outlined by H.O. Bartelt and al[5] who propose one coding technique of the time variable.

Holographic recording of subpicosecond pulses

Methods[6-7-8] have been developed to eliminate the effects of the partial coherence of the source on the sharpness of the reconstructed image. These methods are not matched to our case. An other simple experimental set up has been used (Figure 3). Reference and object beam have been inverted. The path lengths of the two beams must be approximatly equal to ensure mutual coherence. Two advantages appear : the image is reconstructed over the hologram and the image can be observed in white light. This white light reconstruction increases the signal to noise ratio as discussed by P. Chavel.[9-10]

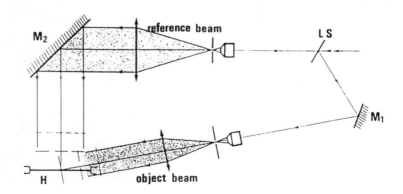

Figure 3 - Second experimental set up

Experimental results

The dye laser is tuned by means of the Lyot filter. The tuning determined the temporal resolution. These pulses are represented on the holographic plate by a bright fringe. The 37 μm wide line is the actual 0,12 picosecond wavefront. The pulse reconstructed in coherent light is shown in figure 4.a. White light reconstruction is reported in figure 4.b.

Some metrological possibilities can be outlined. For exemple a double exposure hologram has been recorded. We have set a 6,5 mm thick glas plate on the object beam and we have tilted this plate between the exposures. This experiment has been made with a 4000 GHz tuning. The tilt is about 32 degrees. The 0,36 mm path length difference is represented by the space between the bright fringes (see figure 5.a.). The result obtained with a 8000 GHz tuning is shown (see figure 5.b.).

Figure 4.a. 0,24 picosecond wavefront reconstructed in coherent light

Figure 4.b. 0,24 picosecond wavefront reconstructed in white light

Figure 5.a. Wavefronts obtained by double exposure for a 4000 GHz tuning

Figure 5.b. Wavefronts by double exposure for a 8000 GHz tuning

We begin to apply this method to studies such as fiber optics dispersion and such as the time of flight difference between various modes.

Conclusion

Dye laser tuned in frequency can be a good tool to synthetise subpicosecond pulses. A matched holographic recording can give the true image of these wavefronts. The signal-to-noise ratio does not decrease although the recording source corresponds to a partial coherence source.

References

1 - Abramson N. , "Light in Flight Recording by Holography", Opt. Lett., Vol. 3, pp. 121-123, 1978.

2 - Abramson N. , "Holographic High Speed Motion Picture of a Light Pulse Focused by a Lens", I.C.O. XII, Graz, 1981.

3 - Barthelemy A., Piasecki J., "Subpicosecond Temporal Display of Fiber Mode Pattern by Coherent Optical Filtering", Opt. Lett., Vol. 6, pp. 269-271, 1981.

4 - Meier R. W. , "Magnification and Third Order Aberrations in Holography" J.O.S.A. , Vol. 55, pp. 987-992, 1965.

5 - Bartelt H. O. , Case S. K. , Lohman A. W., "Visualization of Light Propagation", Opt. Comm. ,Vol. 30, pp. 13-19, 1979.

6 - Leith E. N. , Upatnieks J. , "Holographic with Achromatic-Fringe-Systems" J.O.S.A. , Vol. 57, pp. 975-980, 1967.

7 - Kathyl R. H. , "Compensating Optical Systems : Generation of Holograms with Broadband Light", App. Opt. , pp. 1248-1254, 1972.

8 - Leith E. N. Chang B. J. "Space-Invariant Holography with quasi-Coherent Source", App. Opt. , Vol. 12, pp. 1957-1967, 1973.

9 - Chavel P. , Lowenthal S. , "Noise and Coherence in Optical Image Processing", J.O.S.A. , Vol. 68, pp. 721-732, 1978.

10 - Chavel P. , "Optical Noise and Temporel Coherence" , J.O.S.A. , Vol. 70 , pp. 935-943, 1980.

Study of holographic gratings formed by ultrafine silver particles

Pavel Fiala, Gojko Lončar, Jiří Růžek, Tomas Jerie

Department of Physical Electronics, Faculty of Nuclear Science and Physical Engineering
Czech Technical University of Prague
Břehová 7, 115 19 Prague 1, Czechoslovakia

Abstract

The theoretical properties of diffraction gratings formed by ultrafine grained silver particles are studied. The effectivity index of gratings is introduced and the dependence of diffraction efficiency on grain size, polydispersity, grating thickness and modulation damping in thickness is described. Some of our theoretical results have been verified on laboratory-prepared photographic emulsions, developed in a special semi-physical developer. Finally, the influence of some technological factors is treated and analysed.

Introduction

The silver halide emulsions (SHE) still belong to the most frequently used recording materials in practical holography. The reasons for this fact are the following:

- high sensitivity in whole visible spectrum, the highest of all known recording materials,
- relatively high diffraction efficiency and acceptable noise,
- the possibility of producing larger formats,
- reasonable price.

There are several ways of processing these materials. In particular let us mention those which create the volume and phase, or absorption-phase character of the hologram:

1) Bleaching procedures, i.e. converting metal silver with high light absorption into silver salt with a low absorption and refractive index different (higher) from that of gelatine. These holograms have a relatively large diffraction efficiency, however, as result of the relatively large size and irregular shape of the silver salt grains also considerable scatter. A further disadvantage is that the created silver complex is partially sensitive to light and consequently, these holograms do not have long-term light stability.

2) Procedure 1) can be modified by "reversal bleaching", the grating being formed by the originally unexposed desensitized halide and the metal silver is removed from the hologram. These holograms have a slightly lower scatter than types described sub 1), however, the disadvantage of residual sensitivity remains.

3) If ultrafine grained, so-called Lippmann emulsions, with relatively little silver content, are used, fine grain or semi-physical developers can be employed, which create gratings whose particle dimensions do not, as a rule, exceed the grain sizes of the original halide. The absorption of these particles is then small and the holographic grating has a mostly phase charakter. The holograms have low scatter, are light stable and also have a relatively high diffraction efficiency (30 - 40%) in yellow-red part of the spectrum.

Our paper deals with the problems of the third group, i.e. with the questions of behaviour and properties of holographic gratings created in ultrafine grained SHE's processed by semi-physical developers. Holograms of this type have found to be particularly useful in the field of pictorial holography, in museology, the arts, medical dokumentation, the pedagogical process, etc. However, they can also be used in holographic interferometry, because these materials display low emulsion distortion.

Theoretical description

Diffraction from a volume grating formed by a set of particles

Assume a set of small particles with the volume concetration $N(y)$ which is modulated in a thin layer dz with periodicity Λ :

$$N(y) = N_0 + N_1 \cos(\frac{2\pi}{\Lambda}y) \tag{1}$$

Assume the polydispersity of the set to be characterized by a normed distribution function of particle size frequency $f(a)$ with radii a , and assume a_0 to be the radius with most frequent occurrence.

If $2a_0 < 0.1\lambda$, $dz < \Lambda$, we are able to formulate the transmittance function[1]:

$$T = \exp\left[-ik(n_0 + n_1\cos\frac{2\pi y}{\Lambda}) - (\alpha_0 + \alpha_1\cos\frac{2\pi y}{\Lambda})\right]dz \tag{2}$$

where $k = \frac{2\pi}{\lambda}$, λ is the wavelength, n_0 and n_1 the mean and differential refractive indices, and α_0, α_1 the mean and differential absorption coefficient.
The appropriate coefficients for the particle set are :

$$\alpha_0 = \frac{2\pi N_0}{k^2}\ \text{Re}\int_0^\infty f(a)\ S(0,0,a)\ da \tag{3}$$

$$\alpha_1 = \frac{2\pi N_1}{k^2}\ \text{Re}\int_0^\infty f(a)\ S(\vartheta,\frac{\pi}{2},a)\ da \tag{4}$$

$$n_0 = n_g + \frac{2\pi N_0}{k^3}\text{Im}\int_0^\infty f(a)\ S(0,0,a)\ da \tag{5}$$

$$n_1 = \frac{2\pi N_1}{k^3}\ \text{Im}\int_0^\infty f(a)\ S(\vartheta,\frac{\pi}{2},a)\ da \tag{6}$$

n_g is the refractive index of gelatine, Re and Im denote the real and imaginary parts respectively.

Function $S(\vartheta,\frac{\pi}{2},a)$ is the amplitude function[1] of particle which depends on the particle parameters: the size, shape, relative refractive indices, wavelength and angle of scatter. The relations for S can be found in reference[1].

Equations (3) – (6) describe the equivalent parameters of plane grating. However, we are interested in the volume grating. From Alferness's papers[3,4] it is evident that exists some kind of equivalence in the approaches to solving this problem: the diffracted field of the volume grating can be obtained either by solving the equation of the electromagnetic field for coupled waves[5], or by means of the limiting process of successive diffraction on elementary plane gratings[3,4]. We may thus extrapolate Eqs (3) – (6) to coverage volume structures, provided we insert the angle of diffraction for the angle ϑ in functions $S(\vartheta,\pi/2,a)$.

Effectivity index of grating

One of the important objectives of our deliberations is determining the limiting values of the diffraction efficiency for given sets of particles and the conditions under which this maximum can be reached.

We shall start with Kogelnik's coupled wave theory[5]. Our type of grating represents roughly the case of a phase grating with absorption, i.e.:

$$\alpha_1 \approx 0, \quad \alpha_0 \neq 0, \quad \varkappa = \frac{\pi n_1}{\lambda} \quad (\varkappa \text{ is coupling parametr}) \tag{7}$$

By calculating[2] we can prove that the condition for maximum diffraction efficiency is satisfied, if:

in transmission gratings
   - the grating planes are perpendicular to the hologram surface,
   - the grating has an optimum thickness:

$$d = \frac{\cos\Theta}{\varkappa}\ \text{arctg}\ \frac{\varkappa}{\alpha_0} \tag{8}$$

Θ is angle of incidence of the reference wave.
Then:

$$\eta_{max} = \sin^2(\text{arctg}\ \frac{\varkappa}{\alpha_0})\cdot\exp(-\frac{2\alpha_0}{\varkappa}\ \text{arctg}\ \frac{\varkappa}{\alpha_0}) \tag{9}$$

in reflection gratings
   - the grating planes are parallel with the hologram surface,
   - the grating thickness $d \longrightarrow \infty$
Then:

$$\eta_{max} = \left(\frac{\varkappa}{\alpha_0}\right)^2 \frac{1}{\left[1 + \sqrt{1 + (\varkappa/\alpha_0)^2}\right]^2} \tag{10}$$

Clearly, in both cases the maximum diffraction efficiency is only a function of the factor $\varkappa/\alpha_o$ which we have introduced as the effectivity index of the grating Q :

$$Q = \frac{\varkappa}{\alpha_o} = \frac{N_1}{2N_o} \frac{\text{Im} \int_o^\infty f(a) \, S(\vartheta, \pi/2, a) \, da}{\text{Re} \int_o^\infty f(a) \, S(0,0,a) \, da}$$ (11)

Functions (9) and (10) can be seen in Figure 1. In the next section of this paper, we shall give numerically calculated dependences of the index Q on the individual parameters.

Volume-inhomogeneous grating

In practice it was found that the case in which the grating is not homogeneous throughout its thickness, e.g., as a result of non-uniform developing throughout the grating volume, is important. For the sake of simplicity, let us assume that:

$$N_o(z) = N_o \, g(z) \, , \qquad N_1'(z) = N_1 \, g(z)$$ (12)

In the case of Bragg's angle of diffraction, we may then obtain an analytical solution[2] of the diffracted wave. Calculations have shown that the relations are completely identical to the original Kogelnik's relations for a homogeneous gratings, provided we substitute the mean values for functions $\alpha_o(z)$ and $n_1(z)$, or if we substitute the mechanical thickness of the grating d by its "effective" value:

$$d_{ef} = \int_o^d g(z) \, dz$$ (13)

It does not, therefore depend on the actual behaviour of the parameters $\alpha_o(z)$, $n_1(z)$, but only on their mean values. However, the actual behaviour may be manifest in studying the spectral or angular relations.

Numerical results and theoretical conclusion

For purposes of numerical calculation of the amplitude functions, we assumed silver particles to be of spherical shape with a dimension not exceeding 60 nm. Using Mie's theory and the procedure described, e.g. by Hulst[1], we have formulated relations for S and parameters α_o , α_1 , n_o , n_1 which are suitable for numerical calculation[2]. The total amount of silver was assumed to be the same (0.1 g/cm³) when we varied the particle radius (a_o = 5, 10 and 15 nm). We used the values of the complex refractive index of silver[1]:

$$\lambda = 633 \text{ nm} \qquad n_{Ag} = 0.38 - 4.68i$$
$$514 \text{ nm} \qquad\qquad 0.27 - 3.36i$$

and the refractive index of gelatine n_g = 1.52 .
The gratings was oriented so that one of beams was always perpendicular to the hologram and another incident at an angle of 45° (in air); the fringe visibility was maximum ($N_1 = N_o$). The calculated results are shown in Figures 2 - 6.

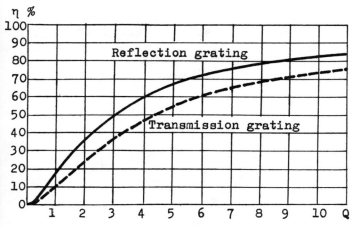

Figure 1. Diffraction efficiency of the volume grating as a function of the effectivity index Q.

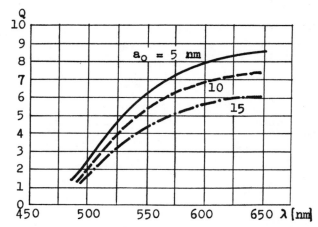

Figure 4. Spectral properties of the effectivity index Q; degree of polydispersity p/a_o = 0.2 .

The distribution of silver halide particle size is roughly lognormal. We assume the same also for the silver particle set; halfwidth p of the size distribution is defined : $f(a_0 + p) = 0.5 \; f(a_0)$; parameter p/a_0 is degree of polydispersity.

The numerical results enable us to draw the following theoretical conclusions:

1) Only the effectivity index Q determines the maximum diffraction efficiency of transmission and reflection gratings (Figure 1). Q decreases with increasing size of particles and with the degree of their polydispersity (Figures 2 and 3). Q also decreases considerably with decreasing wavelength (Figure 4).

2) There is no sense in endeavouring to achieve extremely small silver particles and consequently, also silver halide particles; decrease of radii below 5 nm have very little effect on the diffraction efficiency, but the sensitivity of the halide decreases very substantially. (The statement concerning the effect of the dimensions of silver particles on diffraction efficiency is only valid if the refractive index of silver does not change substantially.)

3) The effect of moderate polydispersity of particles on efficiency (or on Q) in particle size below 20 nm is small (Figures 2 and 3); on the contrary, moderate polydispersity may be advantageous in that the optimum emulsion thickness is displaced to lower values under small decrease of diffraction efficiency (Figures 5 and 6).

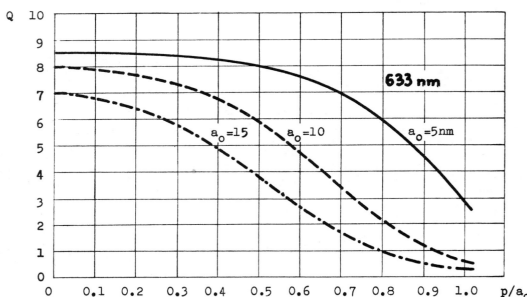

Figure 2. Effectivity index Q as a function of internal grating parameters. λ = 633 nm, a_0 is radius of particles with most frequent occurrence.

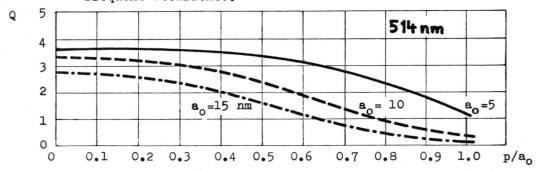

Figure 3. Effectivity index Q as a function of internal grating parameters (radius of particles, polydispersity). λ = 514 nm.

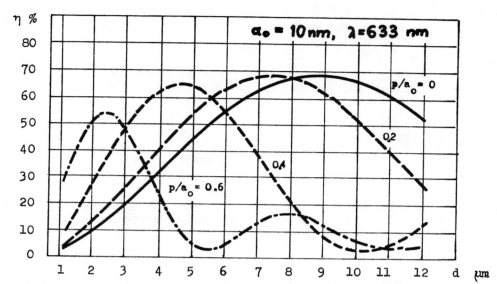

Figure 5. Diffraction efficiency of the transmission grating as a
function of its thickness; wavelength 633 nm, a_o = 10 nm

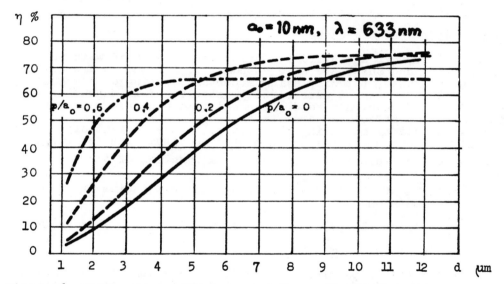

Figure 6. Diffraction efficiency of the reflection grating as a
function of its thickness; wavelength 633 nm, a_o = 10 nm

4) As regards transmission holograms in the green spectral range it is more expedient to
use thinner emulsions, up to 4 μm; the standart thickness of 7 μm is satisfactory in
the red region.

5) The limiting values of the diffraction efficiency of gratings formed by silver particles are shown in Table 1.

<u>Experimental results</u>

The above theoretical analysis was carried out primarily with the purpose of determining the fundamental dependences and possibilities of real gratings in relation to the
process of synthesis and processing of laboratory emulsions. The primary criteria were the
diffraction efficiency and the scatter, as well as small settling of emulsion by processing and long-term stability. The result of our efforts is emulsion ZE-3[δ] with possibilities of band or panchromatic sensitization and with an optimized method of processing.

Before comparing the experimental results with theoretical limits, we should mention some of the phenomena we encountered in the course of research and which slightly complicate the interpretation of the theory.

Non-linearity recording of the interference field

This involves the problem of recording linearity of the interference field (fringe visibility), i.e. investigating the deviation from the ideal dependence :

$$\sqrt{\eta} \sim m \tag{14}$$

where m is the modulation coefficient (fringe visibility); m is given by the beam ratio $P = I_1/I_2$:

$$m = \frac{2\sqrt{P}}{P + 1} \tag{15}$$

The linearity of dependence (14) in whole range m is rarely satisfied in practice; if we adopt the efficiency value for m = 1 as initial, we can formulate the "amplification factor" A at low degrees of modulation:

$$A = \frac{\eta(m)}{m\,\eta(1)} \tag{16}$$

where $\eta(m)$ and $\eta(1)$ are the diffraction efficiency when recording with modulation **m** and m = 1, respectively. In our case the value of A varied from 1.3 to 2.8 for m = 0.57 (P = 10). This phenomenon is clearly the result of increasing polydispersity of the set at higher values of m when a coarser grained structure is created in the exposure minima of the interference field than in the field maxima. Thus, higher values of m lead to a decrease of the effectivity index of the grating Q.

Internal inhomogeneity of the grating

The complicated chemical and diffusive processes associated with semi-physical developing of ultra fine grained SHE's, are also responsible for the volume grating not being of quite homogeneous thickness, but displaying a certain damping of the grating parameters. The hypotheses for this statement were already established in measuring the angular dependences[8] of the diffracted wave, when good agreement with Kogelnik's theory was only achieved on introducing so-called effective parameters n_{1ef}, d_{ef}.
The hypothesis was proved by measuring the absorption coefficient of a uniformly exposed wedge with variable thickness of the emulsion. Let us assume the folowing:

a) the absorption coefficient of the layer is a function of the distance from the grating surface only;

b) the value of the most frequently occurring particle radius a_0 does not change throughout the emulsion.

As a result of relations (1),(3),(6) and for m = 1 ($N_0 = N_1$) it holds approximately that:

$$n_1 \sim N_1 \sim N_0 \sim \alpha_0 \tag{17}$$

Under these assumptions it is possible to substitute the functional variable of the thick emulsion z by the parameter of the wedge thickness d .

Measuring the transmitance of this wedge proved the existence of a kind of subsurface concetration barrier (Figure 7) which apparently also contributes most to the diffraction efficiency; away from the grating surface the particle concentration diminishes and, therefore, also the value of the differential refractive index n_1 decreases.

As regards emulsion thicknesses of less than 3 μm, assumption a) and, consequently also function $\alpha(z)$ in this region is disputable, nevertheless, the damping can be qualitatively proved in the interval above 5 μm.

Non-equivalence of reconstruction both sides of a reflection grating

If GP-type semi-physical developers are used[6,8] , the diffraction efficiency of reflection gratings is not the same on both sides; it is about 20 - 50 % lower on the emulsion side than on the glass side. However, the conclusions drawn from the theory of inhomogeneous grating discussed above, indicate that the damping cannot be responsible for this phenomenon. It can only be explained under the assumption that there is a thin absorption film on the surface of the grating (emulsion),which weakens the reconstruction on the emulsion side, but does not affect the reconstruction on the opposite side.
In measuring the actual parameters of the emulsion we encountered another phenomenon[8]: the volume concentration of the developed silver amounts to about 80 % of the total volume

concentration of silver in the undeveloped emulsion,whereas, with a view to the actual transmittance of the developed layer and the assumed particle diameter, the grating should only contain 40 – 50 % of this total value[6]. These facts can only be reconciled by assuming that part of the silver in the emulsion is in a colloidal and most freely dispersed form (the absorption of such particles is small), and therefore it does not take immediate part in diffraction, the decrement of overall transmittance being relatively small. The deposition of this silver particularly at the boundary of the emulsion may be responsible for the generation of the said absorption film and thus provide an explanation for this phenomenon. If fine grain chemical developers in which silver complexes in the solution are not formed and released, are used, this phenomenon is negligible. Thus, this fact also substantiates the hypothesis of generation of colloidal silver in the emulsion.

Comparison of results and conclusion

In the experiments we used the laboratory prepared,ultrafine-grained SHE ZE – 3 with a particle size of around 20 nm and a layer thicknes of 7 μm. The developing process was at 20°C for 15 minutes in a GP – 2 developer. Some experimental results are shown in Table 1.

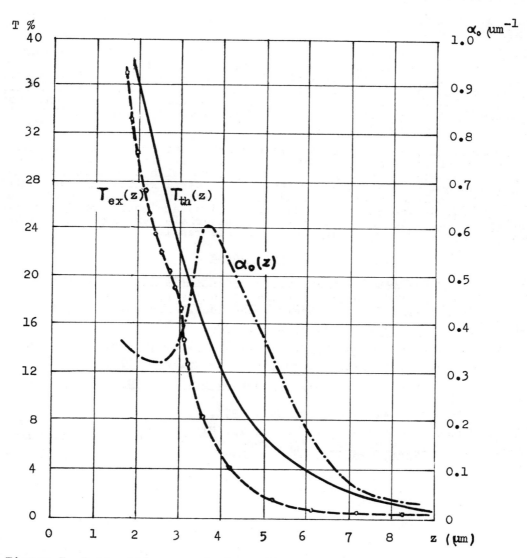

Figure 7. Transmittance and absorption coefficient of the developed
emulsion as a function of the emulsion thickness (of the
distance from the emulsion surface respectively). λ = 514 nm
T_{th} – theoretical transmittance function of the homogeneous
layer;
T_{ex} – experimental values of the real layer;
$\alpha_0(z)$-absorption coefficient calculated from the T_{ex} .

Table 1. Theoretical and experimental results

Type of grating		transmission			reflection		
Wavelength (nm)		530	568	633	530	568	633
Exposure energy ($\mu J/cm^2$)		630	400	130	630	1000	200
Diffraction efficiency %	Theoretical maximum	52	65	70	65	74	80
	Experimental maximum m = 1	12	28	32	23	30	41
	*Experim.value linear extrapolated from m = 0.57 to m = 1	30	80	68	34	40	77
Amplification faktor A(m), m = 0,57 (see relation (16))		2.5	2.8	2.1	1.5	1.3	1.9

* This value is obtained by linear extrapolation from experimental efficiency $\eta(m)$ (m = 0.57) to m = 1 : $\eta(m)/m^2$

It is evident that for line 633 nm and a modulation of 0,57 (P = 10) very good agreement with theory has been achieved. Greater discrepancies at shorter wavelength can be explained by higher degree of non-linearity of recording material in that spectral range. By increasing the emulsion thickness to more than 10 μm the diffraction efficiency is mostly decreased, which represents a certain difference from the theoretical model, however, this is apparently the results of imperfect developing in volume and distortion of the grating.

The presented analysis of ultrafine grained SHE´s has proved theoretically and experimentally the good properties of ultrafine grained materials processed by semi-physical developer, particularly if the region of the spectrum with longer wavelengths is employed. However, they are quite unsuitable for the blue spectral region. By decreasing the dimension of the silver halide, however, we do not achieve our objective also because the sensitivity of the halide decreases considerable, and we are consequently deprived of the main advantages of SHE. Neverthless, the given materials were applied to a two-coloured reflection hologram (red, green) with a reasonable result.

References

1. Hulst, H. C. van de, Light Scattering by Small Particles, New York 1957.

2. Jerie, T., Study of Diffraction on Volume Structures, Diploma work, Faculty of Nuclear Science and Physical Engineering Tech.University, Prague 1982 (in Czech).

3. Alferness, R., Appl. Phys., Vol. 7, p.29, 1975.

4. Alferness, R., Optics Communications, Vol. 15, p. 209, 1975.

5. Kogelnik, H., Bell Syst. Tech. Journ., Vol. 48, p. 2909, 1969.

6. Usanov, Ju. E., Kosobokova, N. L., Zh. nauch. prikl. Fotogr. Kinem., Vol. 22, p. 449, 1977.

7. Technical Information Agfa-Gevaert, July 1971, p. 7.

8. Fiala, P., Růžek, J., Lončar, G., Jerie, T., Ultrafine Grained Halide Silver Emulsions for Holographic Recording, research rep., Czech Technical University of Prague, FJFI, Prague 1982, (in Czech).

Industrial Applications of Image Derotation

W. F. Fagan* and P. Waddell**

* AOL-Dr. Schuster GmbH, Abelegasse 8, A-1160 Vienna, Austria
** Department of Mechanical Engineering, University of Strathclyde, Glasgow, Scotland

Abstract

The technique of image derotation allows a variety of measurements to be made on rotating components by optically compensating for their rotational motion. A description is given of a novel technique required to achieve the high level of alignment accuracy necessary for the successful implementation of holographic interferometry and laser doppler velocimentry in conjunction with the image derotator. Several recent industrial applications in the gas turbine and automobile industries are described.

Introduction

Image derotation is an optical technique that compensates for the rotational motion of engineering components[1,2,3,4,5]. Many important measurements can then be made on the component as if it was stationary. Holographic Interferometry and Laser Doppler Velocimetry can be employed with derotation to record the travelling wave vibrations of rotating objects. The principle of the method of image derotation is shown in figure 1. Light from the rotating object passes through the beamsplitter onto the rotating right-angled prism. If the prism rotates at exactly half the speed of the object and in the same direction then the object's image reflected in the prism will be non-rotating. The image is then observed reflected in the beamsplitter located in front of the prism. The exact 2/1 speed synchronization ratio between the object and the prism is achieved by means of an encoder unit attached to the rotating shaft of the object. This provides a series of electrical impulses, the spacing of which is directly proportional to the object's rotational speed. These impulses are fed into a control console that employs special electronic circuitry to drive the servomotor, which is attached to the rotating prism, at exactly half the speed of the object. The rotational axis of the object must coincide with the axis of the rotating prism if a clear, non-rotating image of the object is to be obtained. Reference 6 describes the optical effects of a misalignment between the two axis. The image of the object, although non-rotating, exhibits an out-of-plane oscillation the amplitude of which is governed by the degree of angular misalignment and the radius of the object. This optical oscillation is undesirable as it leads to false measurements being recorded when either holographic interferometry or laser doppler anemometry are employed for vibration studies on a rotating object. Figure 1 shows the optical setup for image derotated holographic interferometry. The hologram plate is located at the viewing aperture of the derotator where an objective lens images the object onto the plate. The interference fringes of the hologram are formed by making two pulsed laser exposures of the derotated vibrating object. The fringes are a contour map of the vibration displacement of the component in an axial direction during the time elapsed between the two laser pulses. An additional optical effect is observed, in the case of holographic interferometry, when the object beam does not symmetrically illuminate the object with respect to the object axis of rotation. Reference 6 also describes how error fringes are produced by this effect in the recording of a double-exposed holographic interferogram. A new method of aligning the object's rotational axis and the central ray of the object beam's illuminating cone has been devised to minimise these misalignment effects.

Alignment Procedure

Figure 1 also illustrates the optical system required for alignment. Stetson[7] employed the method of observing laser speckle, through the derotator, reflected from the rotating object's surface to align the derotator prism's axis with the object's axis. A strong contrast speckle pattern is only observed when the two axis are collinear. Retroreflective tape was attached to the area around the object's rotational axis. This material consists of thousands of small glass beads coated on the rear surface with a reflecting paint. When light falls on this material the beads reflect the light back along the incident direction with a narrow beam divergence. In the case of coherent light a diffraction halo, figure 2, is observed in the back reflected light. Figure 1 shows how light from a small He-Ne laser is incident on the object's surface that has been coated with retro-refelctive material. The diffraction halo is reflected in the derotator prism to a beamsplitter in front of the prism and finally to the observer. If the object is now rotated slowly the halo will move round in a circle if the central ray of the illuminating cone is not collinear with the object's rotational axis. The direction of the incident beam is altered by means of two tiltable mirrors until the halo is observed to be stationary. The derotator is now switched off and with the object still rotating slowly the speckle pattern reflected from the surface will appear to rotate in a circle. The derotator axis is now moved until the centre of speckle rotation coincides with the centre of object rotation. Upon switching on the derotator the speckle effect can now be seen on the object. Fine adjustments of the derotator axis are now made to maximise speckle contrast and together with the stationary halo the optical system is now aligned for holographic interferometry. It is important to ensure that the optical path of the ruby laser, used for double-pulsed holography, follows the direction of the He-Ne laser beam. This can be achieved by use of a moveable mirror that can deflect the He-Ne laser beam into the optical layout, when present, and allow the Ruby laser beam to enter when absent. Figure 3 shows a double exposed hologram of a rotating automobile ventilator cooling fan which is undergoing fluttering vibration. This hologram was recorded through the derotator which was properly aligned to the fan's axis. The fringes seen on the hub of the fan are due to rigid body tilting of the fan axis with respect to the derotator.

Industrial Applications

Several interesting applications of derotation have appeared recently, applied to aero engine design. Storey, reference 8, of the Rolls Royce Company has reported the use of an image derotator combined with holographic interferometry to study the fluttering vibrations of a fan up to speeds of 10,000 RPM. Similar work has been reported by Stange and MacBain[9] of the US Air Force Aero Propulsion Laboratory, who investigated dual mode vibrations phenomena in a mistuned bladed disk and also by Bearden and Clarady[10] of the Pratt and Whitney Aircraft Group who investigated the resonant response of a gas turbine engine bladed disk assembly up to speeds of 7,500 RPM. Haupt and Rautenberg of the Institute of Turbomachinery Hannover, Germany, have studied[11] the blade vibration of radial impellers up to speeds of 13,000 RPM. Other groups active in this research area include the KWU Company of Mülheim in Germany, the IHI Co. of Tokyo and an Institute of Turbine Research in the People's Republic of China.

Another major application area of image derotation is in automobile research and development. Volkswagen of Germany have used an image derotator combined with a laser doppler system to study rotating fan vibrations. A laser doppler beam is directed through the image derotator onto the rotating object, the beam constantly monitoring the same area as it is scanned around with the object by the rotating prism. Although measurement is made only at one point at a time on the surface the back scattered light which passes through the derotator to the photodetector of the laser doppler unit contains information about the frequency spectrum of the vibration. Tiziani[12] of the University of Stuttgart has used such a system to measure the vibrations of a rotating tyre. This same study also included the use of pulsed holographic interferometry and derotation to study the vibration amplitude distributions on a rotating tyre. Figure 4 shows a photograph of a hologram made by Tiziani. This was recorded by a combination of image derotation and a double pulsed ruby laser and shows the out-of-plane vibrations of an automobile tyre rotating in contact with a simulated rough surface. This work provides important information about tyre noise generation. Davis and Morris of the Ford Motor Co. Dearborn[13] have reported the initiation of a programme of study of the application of image derotation to the study of tyres, wheels, brakes and gear trains using holographic interferometry and photoelasticity. The latter technique involved the construction of an automobile tyre in photoelastic material which was studied, while rotating, using an image derotator, photoelastic, bench. Real time vibration analysis of a rotating disk has been realised by Harding and Harris of the University of Dayton Research Institute Ohio, by means of a Projection Moiré Interferometer and an image derotator. This work, yet unpublished, uses a laser interferometer to project Moiré fringes onto a rotating disk with a stroboscopic Argon laser. Interference takes place between a reference grid and the stroboscopically frozen fringes projected onto the disk. The resulting moiré interferograms recorded by a video camera contain the vibration amplitude distribution of the disk's surface.

Two papers in this seminar also report on image derotator applications. Dr. Preater of the City University, London, will talk about the use of an image derotator combined with electronic speckle pattern interferometry and a pulsed laser to study the strain distribution on rotating structures and Mr. J. Geldmacher, et al., will compare image derotation with other methods employing holography to study the vibrations of rotating objects.

Finally a most recent application of image derotation has been that of disk brake testing. Figure 5 shows the optical layout of the measurement system. Alignment requirements are much less restrictive here as infra-red imaging is used to measure the brake performance. As no interferometric technique is used with this method the resolution requirements of the system are greatly relaxed. An infra-red camera is used to record the heat distributions on the surface of the disk during braking, the derotator optics providing a non-rotating, thermal, image. Valuable measurements can be made on brake wear, noise and fading mechanisms by means of this novel technique.

Conclusion

Image Derotation has shown itself to be an extremely useful technique in the research and design of modern rotating components. Future areas of application will include ball bearing technology, grinding wheels and propellors. Further research will be directed towards

1. measuring stresses directly on rotating objects using the derotator and the SPATE camera technique[14] which records the stresses of dynamically excited objects directly by measuring the heat generated by different areas of the surface subjected to varying degrees of stress, and
2. crack detection and measurement on rotating components using a combination of derotation and laser scanning methods.

References

1. Waddell, P.   The real time non-stroboscopic examination of centrifugal stress on rotating photoelastic disks utilising an optical image derotation technique, Conference on Optical Methods in Scientific and Industrial Measurements, Tokyo, August 1974.

2. Fagan, W. F., and Beeck, M. A.   The laser applications of a reflective image derotator related to the study of the dynamic behaviour of rotating machinery and fluids, Laser 79 Opto-Electronics Conference, Munich, July 1979, published by IPC Press Guildford, Surrey, England.

3. Beeck, M. A., and Fagan, W. F.   Study of the dynamic behaviour of rotating automobile cooling fans using image derotated holographic interferometry, Topical Meeting on Hologram Interferometry and Speckle Metrology, June 2 - 4, 1980, Cape Cod, Massachusetts, published by the Optical Society of America.

4. Fagan, W. F., Beeck, M. A. and Kreitlow, H.   The holographic vibration analysis of rotating objects using a reflective image derotator, Optics and Lasers in Engineering, 2 (1981) 21-32, published by Applied Science Publishers Ltd, Barking, England.

5. Fagan, W. F.   A motion compensated image derotated, holographic interferometer, Laser 81, Opto-Electronics Conference, Munich, July 1981, published by Springer Verlag, Berlin, Germany, 1982.

6. Fagan, W. F., Beeck, M. A. and Kreitlow, H.   The Practical Application of image derotated holographic interferometry to the vibration analysis of rotating components, Proc. of the 1980 European Conference on Optical Systems and Applications, Utrecht, The Netherlands, September 23 - 25, published by SPIE, Bellingsham, Washington, USA, 1980.

7. Stetson, K. A. and Elkins, J. N.   Optical system for dynamic analysis of rotating structures, Air Force Aero Propulsion Lab. contract F 33615-75-C-2013, AFAPL-TR-77-51 (Oct. 77).

8. Storey, P. A.   Holographic Vibration Measurements of a rotating fluttering fan, 18th Joint Propulsion Conference of AIAA, paper 82-1271, June 1982, Cleveland, Ohio. to be published by AIAA.

9. Stange, W. A. and MacBain, J. C.   An investigation of dual mode phenomena in a mistuned bladed disk, Design Engineering Technical Conference, September 20 - 23, 1981, Hartford, Connecticut, published by ASME, 1981.

10. Bearden, J. L., and Clarady, J. F.   Spin pit application of image derotated holographic interferometry, United Technologies Corp., Pratt and Whitney Aircraft Group Florida, United States Air Force report number AFWAL-TR-80-2083, September 1980.

11. Haupt, U., and Rautenberg, M.   Investigation of blade vibration of radial impellers by means of telemetry and holographic interferometry, 27th Internaional Gas Turbine Conference, London, England, April 18 - 22, 1982, to be published by ASME.

12. Tiziani, H. J.   Real time measurements in optical metrology, Laser 81, Opto-Electronics Conference, Munich, July 1981, published by Springer Verlag, Berlin, Germany 1982.

13. Davis, C. W., and Morris, C. J.   Holographic vibration analysis of rotating objects, Sound and Vibration, April, 1981, ISSN 0038-1810, USA.

14. Oliver, D. E., Razdan, D., and White, M. T.   Structural Design assessment using thermo-elastic stress analysis (TSA). Proceedings of the State of the Art in Measurement Techniques. Joint BSSM/R. Ae. Soc. meeting, Univ. of Surrey, England, 6-9 Sept., 1982.

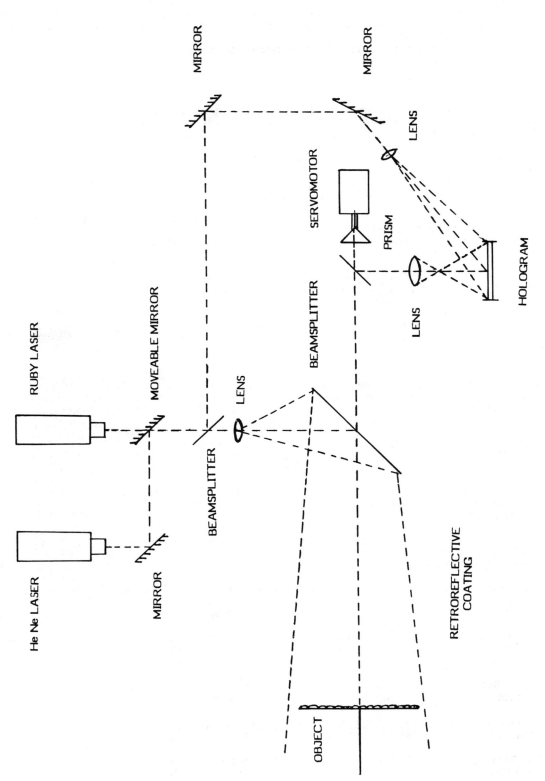

Figure 1

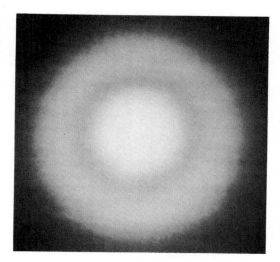

Figure 2

Figure 3

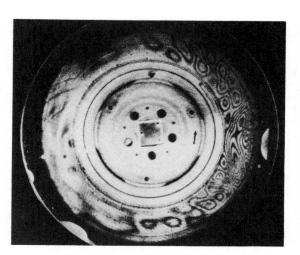

Figure 4

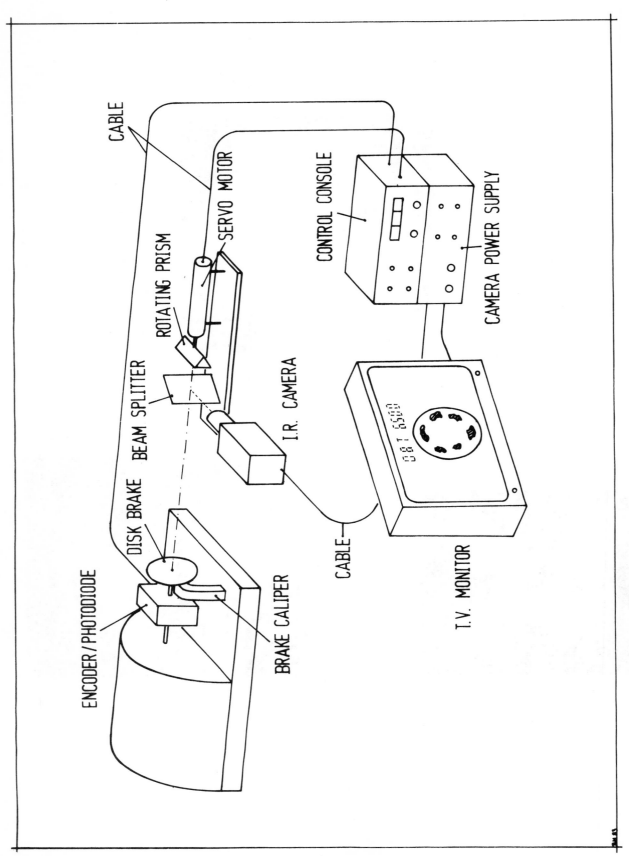

Figure 5

198

INDUSTRIAL APPLICATIONS OF LASER TECHNOLOGY

Volume 398

Session 4

Holography, Speckle, and Moiré Techniques II

Chairmen
Z. Füzessy
University of Budapest, Hungary
J. McKelvie
Centre for Industrial Innovation, Scotland

Laser moiré-schlieren system of variable sensitivity

P. Waddell

Department of Fluids and Thermodynamics, Mechanical Engineering
Group, University of Strathclyde, Glasgow, Scotland

Abstract

A simple inexpensive, wide field schlieren system has been developed to visualise the
fluid flow patterns around heated components. The system has variable sensitivity, ranging
from a high sensitivity to visualise small heat flows around the human body to a low
sensitivity to visualise the large heat flows around a bunsen flame. The variable
sensitivity is achieved by adjusting the distances between the optics. The wide field of
view is obtained by using a novel flexible membrane mirror, providing a two foot diameter
collimated light beam for inspecting the disturbed fluids.

Introduction

Fluid flow visualisation is commonly carried out by shadowgraph, schlieren and Mach
Zehnder apparatus, in which a usually collimated light beam is passed through the fluid to
be investigated. Disturbances in the fluid will produce refractive-index and density
changes, which if large enough will deflect the light rays in the collimated beam. The beam
deflection is utilised in different ways in the above apparatus. The general principles of
the three types of apparatus can be illustrated, reference Figure 1. A light ray passes
through a fluid, for no fluid refractive index changes the ray strikes a screen at P_1, a
fluid refractive index change will deflect the ray to strike the screen at P_2. The
deflection distance P_1-P_2 is extremely small, Figure 1 is drawn at a greatly enlarged scale.
A longer time is required to reach P_2, let ΔT be the path length time difference.
Observations can be made of ΔQ, $\Delta \theta$ and ΔT, in practice ΔQ is observed in the shadowgraph, $\Delta \theta$ in the schlieren and ΔT in the Mach-Zehnder interferometer.

In the shadowgraph a collimated non coherent light beam is passed through the disturbed
fluid. Assume n is the refractive index of the non disturbed fluid and X is a direction
in the fluid normal to the beam direction, the beam is observed on a screen placed after the
fluid. For $\frac{\partial^2 n}{\partial x^2} = 0$ the collimated rays stay parallel and no intensity change is observed
on the screen. For $\frac{\partial^2 n}{\partial x^2} > 0$ the rays spread out towards the screen and hence there will be
a localised decrease in intensity for this fluid image area on the screen. For $\frac{\partial^2 n}{\partial x^2} < 0$
the rays converge towards the screen and an increase in intensity is observed on some parts
of the screen. Since fluid refractive index and density are related, one also observes the
rate of change of density gradient. The shadow graph is useful when abrupt changes in
density occur, an example are the disturbances around shock waves. The shadow graph is very
insensitive to slow or continuous variation of refractive index in planes normal to the
light beam direction.

The Mach - Zehnder interferometer uses a collimated partially or fully coherent light
beam (mercury vapour or laser). The beam is optically split before the disturbed fluid, one
beam traverses the depth of the disturbed fluid and the second beam traverses a similar
depth of the same non disturbed fluid. The two beams are optically recombined after the
fluid by a semi reflecting mirror and observed on a screen. The disturbed fluid deflects
the light rays of that beam passing through, thus producing path length changes or time
differences between corresponding rays in the two beams. On the screen the two beams are
now in and out of phase and produce interference fringes, indicating the degree of
refractive index or density changes for different fluid points.

The schlieren normally uses collimated non coherent light, however schlieren
interferometers can be built which require partially or fully coherent light. The normal
schlieren light is a line source, usually obtained by placing a thin rectangular slit over
the light bulb, the slit width will be limited by diffraction effects. In general the
schlieren sensitivity increases as the slit width decreases, the minimum width being
limited, as mentioned previously, by diffraction effects. Light from the slit is collected
and collimated through the fluid by either a lens or a mirror. A second lens or mirror is
placed after the fluid and is used to focus the slit light on to a straight opaque edge,
which lies parallel to the length of the slit. The edge is adjusted across the parallel,
focussed rectangle of light, in order to partially reduce the intensity of light transmitted
past the edge to a screen placed after the edge. On the screen the slit image is out of
focus and not seen, but the working area of the fluid is in focus and sharply imaged. Let
X be a direction in the fluid which is normal to the slit source length, and let Y be
orthogonal to X. For the slit and edge in the Y direction then fluid disturbances in the X
direction are seen and vice versa, reference Figure 2 with slit and edge in the Y direction.
Consider the rays deflected by an angle θ_x in the X direction as shown, it can be seen
that more light passes over the edge to reach the screen. θ x in the opposite direction will

decrease the light intensity on the screen, the light not passing over the edge. It can be shown that [1].

$$\theta_x = \int \frac{1}{n} \frac{\partial n}{\partial x} \cdot dZ \quad , \quad \theta_y = \int \frac{1}{n} \frac{\partial n}{\partial y} \cdot dZ$$

The deflections are an integrated effect over the depth of fluid 'Z'. It can be shown that the screen contrast is given by

$$\frac{\Delta I}{I} = \frac{f2}{a} \cdot \int \frac{1}{n} \frac{\partial n}{\partial x} \cdot dZ$$

For large sensitivity i.e. a large contrast change for a small change of refractive index then 'f_2' the focal length of the second lens should be large and 'a' the width of the focussed slit light on the edge should be small. The depth of fluid Z is a parameter, large depths increase θx and θy and vice versa. Typical schlieren mirrors are 8 to 10 inch diameter and f 12 to f 15 i.e. long focal lengths. The range over which θx or θy are proportional to the screen contrast is given by

$$\theta_{X\ MAX} = \frac{a}{f_2}$$

If air is the fluid used, density c and refractive index n are linked by the Gladstone - Dale coefficient K_{G-D}. It can be shown that [1].

$$\frac{\Delta I}{I} = \frac{f2}{a} \cdot \frac{\partial c}{\partial x} K_{G-D} \cdot Z$$

For air at atmospheric temperature and pressure, n is 1.0. Knowing c for the undisturbed fluid points together with $\frac{\partial c}{\partial x}$ and $\frac{\partial c}{\partial y}$ enables one to estimate the density, pressure and temperature variations in the disturbed fluid. The system sensitivity can be doubled by retroreflecting the light beam back through the same fluid points. It is possible to increase the sensitivity even further by placing semi-reflecting mirrors on either side of the fluid and repeatedly reflecting the light through the fluid, light losses are obviously large [2]. A bad feature of all the systems mentioned, is the integrating effect on the light of the different refractive index changes throughout the fluid depth. It is extremely difficult to assess the effect of any one 2D plane in a 3D disturbed fluid. Obviously a thin 2D plane would mean a very small 'Z' value and a greatly decreased sensitivity. The schlieren has been successfully converted to a sharp focussing system, [3,4]. to isolate 2D slices in a 3D volume of disturbed fluid, by using a multiple slit light source and corresponding multiple straight edges. Such multiple slits and edges are made by using straight line grids having equal width opaque and clear lines. One grid is placed between the light bulb and the first lens, the lines of the second edge grid are aligned parallel to those of the first grid and placed in the focussed plane K after the lens L2 reference Figure No 3. Only the disturbances in plane D are seen on the screen, [3,4]. the position of plane D can be altered by moving the optics.

It is obvious in all the above systems that in order to examine large areas of fluid one requires lenses or mirrors at least as large as the area of fluid under investigation. In general for large diameter optics it is better and cheaper to use front surfaced mirrors and not use lenses. If lenses are used they will be heavy, the light source will have to be monochrometric in order to prevent absorption and chromatic aberration. Mirrors must also be used with care, for on and off - axis operation in order to avoid coma, astigmatism and spherical aberration. The above effects and how to avoid them, so as to produce initially uniform intensity on the screen before any fluid disturbances are described [5].

Non sharp focussing moiré-schlieren systems

Such systems have been described [6,7], reference Figure 4. An illuminated grid of equal line to space ratio G1 is imaged by a lens L_3, placed after the fluid, on to a similar grid G2. G2 is usually a photographed negative of grid G1, placed back exactly in the position where it was first recorded. The opaque lines on G2 act as the straight opaque edges for the slit light sources from the clear spaces on G1. Perfect superposition of G1 on G2 results in a uniform intensity on a screen placed after G2. G1 is larger than the flow area, grids of 60 L.P.1. were used resulting in several hundred L.P.1. in G2, presenting great problems in camera stability, in order to record G2 as a negative. System optical sensitivity is inversely proportional to the width of the spaces on G2 and proportional to the focal length of lens L_3. Any disturbance in the fluid distorts the image of G1 and results in a moiré pattern appearing on the screen. In a 1982 paper the author reversed the optical layout of Figure 4 [1]. A small slide grid G1 of 500 L.P.1. was projected from a slide projector and passed through a camera zoom lens to be imaged through the fluid on to

201

a vertical grid of 60 L.P.1. The grid was made from inexpensive Letratone Ltd straight line grid sheet. The grid G2 was adjusted until the imaged black lines of G1 lay exactly in the clear spaces of G2 creating a zero fringe condition on G2. Any further movement of G2 results in a series of straight line moiré fringes appearing on G2. Any fluid disturbance will now create a moiré pattern from either the zero fringe condition or distort the initial moiré straight line pattern on G2. System sensitivity can be increased by changing the slide grid G1 to one of say 1000·LP1 and or increasing the distance of G1 from G2. A retroreflective screen [8.] was placed behind G2 and used to double the sensitivity of the system by sending the light back through the fluid. Projecting grids of 1000L.P.1. and greater presents problems of diffraction even for good quality diffraction limited optics. A new system has been evolved and is illustrated in Figure 6. A laser beam is expanded by a microscope objective and pinhole arrangement to be reflected through the fluid as a collimated beam by means of a membrane mirror drum. The mirror is created by stretching an aluminised membrane over the hoop of an orchestral drum, the drum skin having been removed from the hoop. Application of a vacuum to the drum interior produces a concave surface on the membrane, the focal length of which can be adjusted by changing the vacuum pressure, f numbers down to 1.0 have been achieved. An off axis arrangement is readily attained by suitable adjustment of the drum skin tensioning screws. The two identical straight line grids G1 and G2 are now placed after the fluid with the lines of the two grids either parallel in the horizontal or vertical directions. The grids were manufactured by photographing inexpensive Letratone Ltd 65 L.P.1. grids, these grids are mounted on semi opaque plastic backing sheets and are of no use for transmission work. From the negative the two positive grids G1 and G2 are produced on thin clear plastic. The two grids are mounted with their lines parallel and placed vertically upright with a space between them, the collimated laser beam now casts the shadow of G1 onto the lines of G2. For a non disturbed fluid the G1 shadow is a perfect fit on G2 and a zero fringe condition is easily obtained. Slight adjustment of G2, usually a small rotation of α , will produce straight line moiré fringes across G2, such fringes are usually arranged to be vertical or horizontal. Any fluid disturbance will now either create a moiré pattern in the zero fringe condition on G2, or distort the moiré pattern created between the slightly misaligned grids G1 and G2. Either moiré pattern can be used to quantitatively examine the density changes in the disturbed fluid, reference Figure 7. G1 is imaged on to G2 with a small angle α between the grids, vertical moiré fringes appear at the intersections of the G1 and G2 lines at $P-P^1$. Density changes in the fluid result in a^1 being shifted up or down by an amount ΔX, the new grid intersection is P^{11}, the moiré fringes have thus moved horizontally. The distortion of the grids is proportional to $\frac{\partial n}{\partial x}$ and $\frac{\partial n}{\partial y}$ for the grid lines horizontal, $\frac{\partial n}{\partial x}$ is the cause of the vertical moiré moving horizontally, for vertical grid lines $\frac{\partial n}{\partial y}$ causes the horizontal moiré to move vertically. The localised displacement are measured and plotted as $\frac{\partial n}{\partial x}$, X., or $\frac{\partial n}{\partial y}$, Y graphs. One can also plot $\frac{\partial^2 n}{\partial x^2}$, X, $\frac{\partial^2 n}{\partial y^2}$. Y graphs. System sensitivity is varied by the depth of fluid used, distance of the fluid centre from G1, the gap width between G1 and G2 and the lines per inch on G1 and G2. Figure 8 illustrates the membrane mirror drum.

Test Results

Figures 9 and 10 illustrate the zero fringe background disturbed by the heat around a very low pressure, flickering, bunsen flame and the heat around two matchsticks in the authors hand.

Figure 11 illustrates the distortion of the horizontal moiré background pattern between G1 and G2, the heat is from two matchsticks.

Figures 12 and 13 illustrate the distortion of the vertical moiré background fringes between G1 and G2, the heat is from a low pressure bunsen flame.

Conclusions

A simple wide field inexpensive moiré-schlieren system has been presented which possesses a large range of sensitivity. Although results with laser illumination have been presented, the system also works well with filtered mercury vapour illumination. A novel inexpensive large diameter membrane mirror of variable focus has been presented.

References

1. Waddell, P.,"A Large Field Retroreflective Moiré-Schlieren System". Electro Optics - Laser International 82 U.K., Butterworth Scientific,London, pp. 74-82. 1982.
2. Hall, L.S., "High Sensitivity Schlieren Technique". Rev-Sci. Instrum., Vol. 37, pp. 1735-36. 1966.
3. Buggard, R.D., "Description of a Three Dimensional Schlieren System", 8th Intl. Congress High Speed Photography, J. Wiley, pp. 335-340, 1968.

4. Dixon Lewis, G., Isles, G.L., "Sharp Focussing Schlieren Systems for Studies of Flat Flames". J. Sci. Instrum., Vol. 39, pp. 148-151, 1962.

5. Hasch, J.W., Walters, J.P., "High Spatial Resolution Schlieren Photography". Applied Optics, Vol. 16, pp. 473-482, 1977.

6. Mortenson, T.A., "An Improved Schlieren Apparatus employing Multiple Slit Gratings". Rev. Sci. Instrum, Vol. 21, pp. 3-6, 1950.

7. Burton, R.A. "A Modified Schlieren Apparatus for Large Areas of Field". Jnl. Opt. Soc. Am., Vol. 39, pp. 907-908, 1949.

8. Waddell, P., Fagan, W.F. "The Role of Retro-Reflective Materials in Holography". Jnl. Phot-Science, Vol. 23, No 2, March/April 1975.

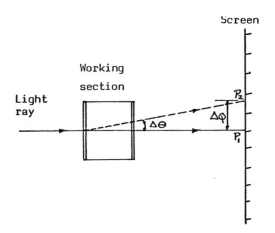

Figure 1. Light ray deflection by refractive index change.

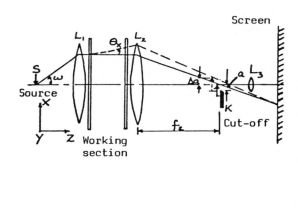

Figure 2a. Transmission schlieren using lenses.

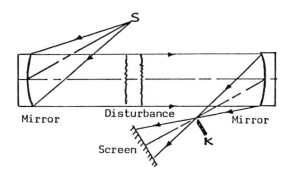

Figure 2b. Transmission schlieren using mirrors.

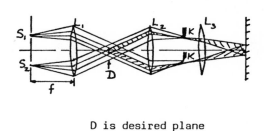

D is desired plane

Figure 3. Sharp focussing schlieren system.

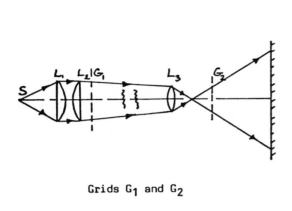

Grids G$_1$ and G$_2$

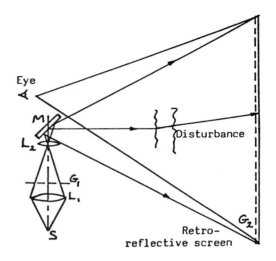

Figure 4. Moiré-schlieren system.

Figure 5. Retroreflective moiré-schlieren system.

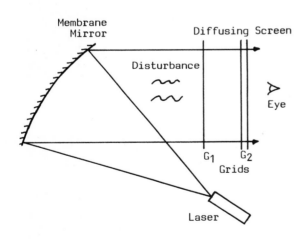

Figure 6. Laser moiré-schlieren

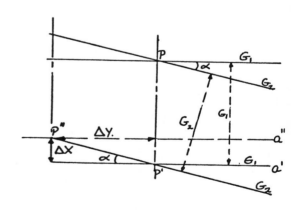

Figure 7. Moiré distortion caused by fluid density disturbance.

Figure 8. Membrane mirror drum

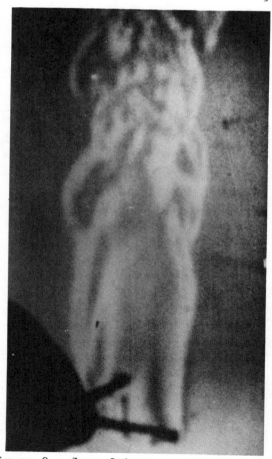

Figure 9. Zero fringe - matchstick flames.

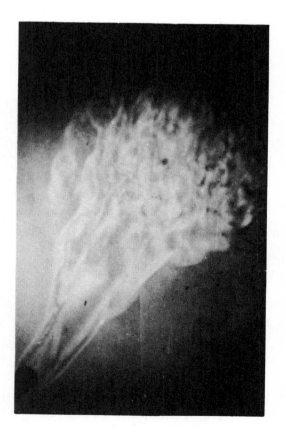

Figure 10. Zero fringe - bunsen flame.

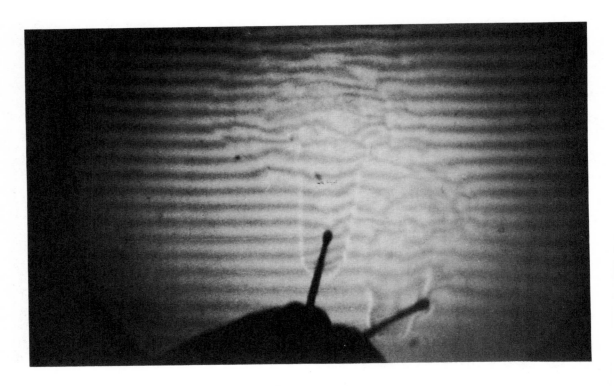

Figure 11.   Moiré distortion by matchstick flame

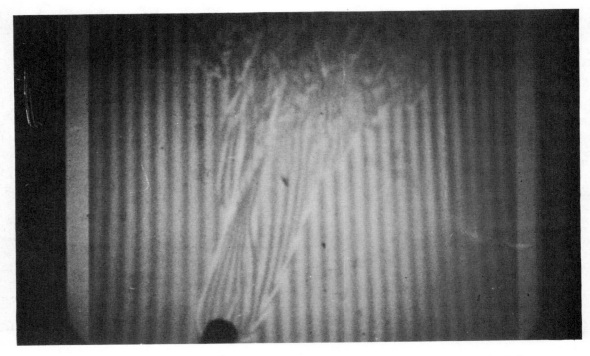

Figure 12a.   Moiré distortion by bunsen flame, vertical fringes

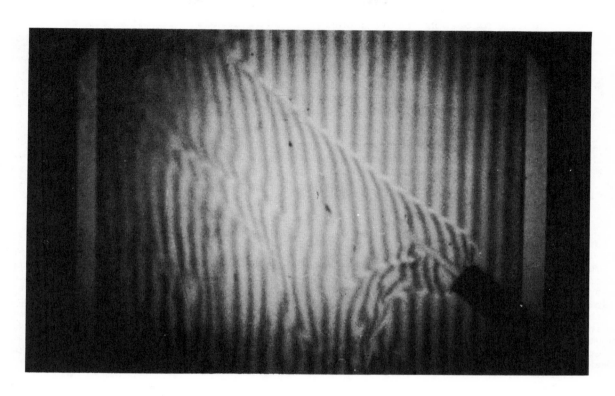

Figure 12b. Moiré distortion by bunsen flame, vertical fringes

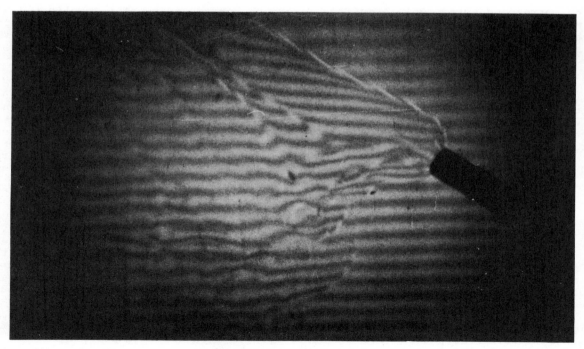

Figure 13. Moiré distortion by bunsen flame, horizontal fringes

Moiré Topography using a charge coupled device television camera

Andreas Nordbryhn

Central Institute for Industrial Research
Oslo, Norway

Abstract

A simple and straightforward way to make projection Moiré topograms is described. A grating pattern is projected on the object, which is imaged by an interline transfer (ILT) charge coupled device (CCD) television camera. High contrast Moiré fringes that can be equal to topographical contours are then obtained. Experiments showing examples of this technique are presented and discussed.

Introduction

Moiré methods for topographical contouring of 3-dimensional objects constitute a well-established field for various applications.[1-3] A range of different ways to make Moiré contourgrams exist, they can mainly be grouped into two: Contact and projection methods. In contact methods, a plane grating is mounted very close to the object to be contoured, and a Moiré pattern between the grating itself and the shadow of the grating on the object constitutes the desired contour pattern. In projection methods the object is illuminated by projecting a grating pattern on it from an angle. The object is viewed through another grating from a different angle. The Moiré beating between the two gratings will, if the geometry is properly arranged, show topographical contour lines of the object. The projection method is the most versatile of the two. This paper is concerned with one way to make projection Moiré.

Most often, photographic methods have been used for Moiré contour recording. This gives a very high spatial resolution, but no real time imaging. Another problem, just like in ordinary topographical maps, is deciding whether a fringe set corresponds to an uphill or a downhill slope.

A proposed way to overcome both these problems is to use a TV camera to view the illuminated object, and using a digital image processor to create an artificial reference grating.[4] This is an elegant approach that gives real time contours as well as a way to determine the sign of the slopes. In addition the shape of the reference grating can be made to correspond to a nonplane surface, e.g. the correct shape of a complicated machine part, to control the shape of such parts under production. The need for an image processor, however, makes this a costly approach. This paper will describe a different, and very much simplified method for TV pickup of Moiré contour patterns, using a CCD camera.

Description of the method

Figure 1 shows the setup for CCD camera Moiré contouring. The object is illuminated from one angle by an ordinary slide projector, projecting a fine grid on the object. The object is viewed from another angle with an interline transfer charge coupled device TV camera. In the experiments, a Fairchild 380 x 488 element ILT CCD (CCD 222) sensor has been used. (Probably at present the only commercial sensor usable for just this purpose.) By adjusting the distance between object and camera, a Moiré beating is obtained between the projected grating and the grid of detector elements in the camera. This Moiré beating will be correct topographic contour lines, if the geometry is properly arranged.

In figure 2 is shown the geometry of the detector elements of the Fairchild ILT sensor. The important detail of the structure is the vertical stripes of detector elements, with a separation between the stripes of approximately the same size as the width of the detector elements. The vertical detector element stripes constitute a light sensitive grating, and when arranged parallell to, and with the same periodicity as the illumination grating when focused onto the sensor's focal plane, very high contrast fringes appear. In addition, this sensor's very high light sensitivity, makes possible Moiré contouring of low reflectivity objects.

Figure 3 illustrates how this way of projecting a grid and viewing with a sensor array creates a contour line pattern through Moiré fringes. In order to achieve the geometry shown in figure 3, the projector and camera would have to be very far away from the object. Otherwise, the difference in projection and viewing angle across the object would create cylindrical, not plane contour surfaces at the object. Alternatively, the projector and camera could be modified so that their aperture planes and focal planes were all parallell. The latter method will give plane contour surfaces, but slightly non-equidistant.

In the experiments performed, both ways to avoid cylinder shaped contour surfaces were impractical. Therefore the results presented are slightly nonperfect in this respect.

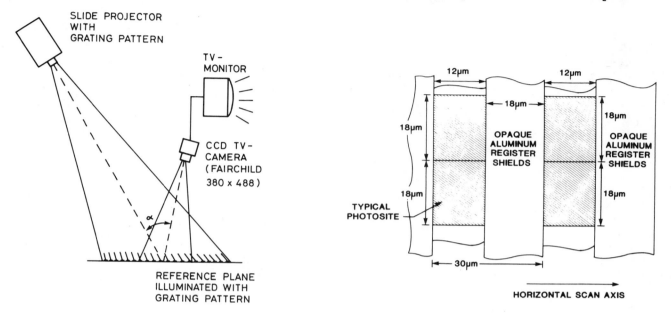

Figure 1. Schematic setup for Moiré topography with a CCD TV camera

Figure 2. Photosite dimensions of CCD 222 (From Fairchild catalog)

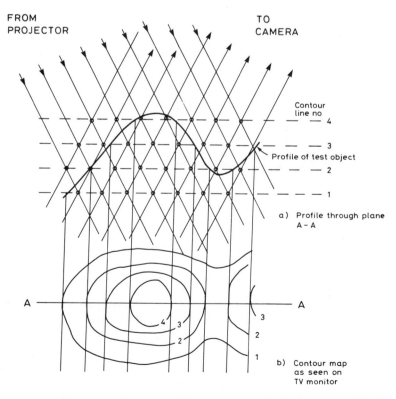

Figure 3. The generation of Moiré contours of a 3-dimensional object using the described method.

Moiré patterns and sampling theory

An appealing way to deal quantitatively with Moiré fringes in CCD imagers is through the use of sampled data theory. The pick-up of an image using an array of discrete detector elements is equivalent to a sampling of the image in the spatial dimension. A Moiré pattern can be considered as the aliasing resulting when undersampling an image having a periodic pattern. This occurs if the input pattern contains frequencies higher than half the sampling frequency (the so-called sampling limit or Nyquist frequency).

In solid state imagers aliasing is most certain to take place. This is especially so with interline transfer CCD imagers.[5]

In figure 4 is shown the frequency response for horizontal spatial frequencies in an ILT CCD, with detector element widths half the element-to-center spacing. The frequency response of one element is the Fourier transform of a rectangle function, i.e. the frequency response is a sinx/x (drawn in a solid line). Because of the repeated sampling, the transfer function is also repeated. Because of the spacing between samples and the width of the frequency response of one element, a strong overlap occurs between the repeated versions of the frequency response. This overlap is the cause of aliasing patterns.

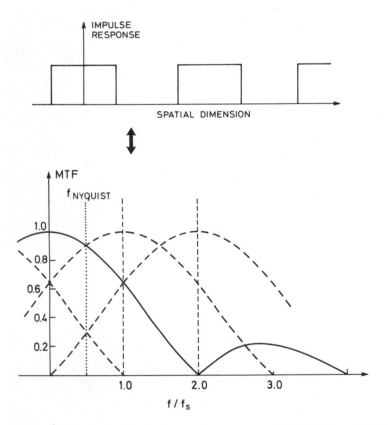

Figure 4. Spatial impulse response of ILT CCD and its MTF, which is the real parts of the Fourier transform of the impulse response.

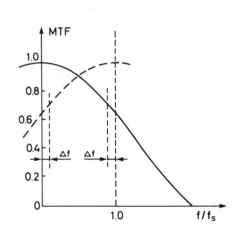

Figure 5. Moiré pattern at a frequency Δf interpreted as a kind of aliasing of a pattern at a frequency $fs - \Delta f$.

Figure 5 shows what happens if a periodic input pattern is focused on the detector, with a frequency being Δf different from the sampling frequency. This frequency component cause strong aliasing, and will be folded down to a frequency Δf, that is, a very low frequency pattern.

This is a Moiré pattern, and can be utilized for topographic contouring. From figure 5 one can easily find the contrast of the Moiré pattern. Moreover, it is easy to assess different contrast reducing effects in a given setup, using the MTF's of the different components and processes.

Experiments

A number of different types of objects have been imaged in the described setup, in order to illustrate the method. A f=50 mm lens was used with the CCD camera and the aperture was altered between F/1.4, and F/8 for the various objects. Some examples are shown in figures 6-12. From these, one can see that a large variety of reflectances and surface textures can be successfully contoured. All white objects, like the elephant in figure 6 and the folded paper in figure 7 give very high contrast contour lines. All black, but diffusely reflecting objects also give reasonably good fringes (bowler hat in figure 8). The most difficult type of objects are specularly reflecting ones, like metal parts and glossy painted objects. Most often reasonably good fringes are nevertheless obtained, but care must be taken to avoid blooming due to specular reflexes. An example showing how specular reflexes and blooming affects the result is shown in figure 9. Figures 10 and 11 illustrates the increase in depth sensitivity when increasing the angle between illumination and viewing. Larger angle means better sensitivity and more fringes. In figure 12 is shown a typical fringe contrast on human skin.

In the experiment performed the depth resoltuion (fringe to fringe distance) was varied between about 1 mm and 10 mm.

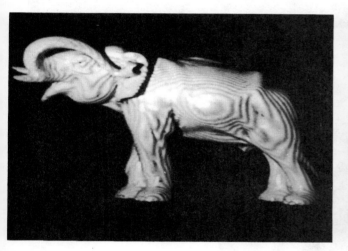

Figure 6. Moiré pattern of white porcelain elephant.

Figure 7. Paper that has been folded and then flattened.

Figure 8. Black bowler hat.

Figure 9. Glossy gas bottle top.

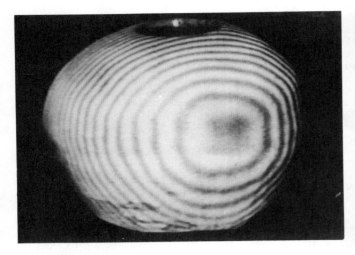

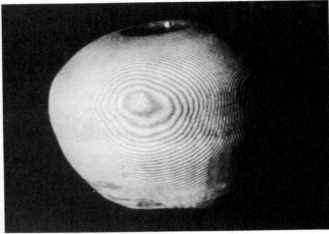

Figure 10. 15 cm diameter red clay vase, seen with a narrow angle between light source and camera axis.

Figure 11. Same as fig.10, but with a wider angle between light source and camera axis.

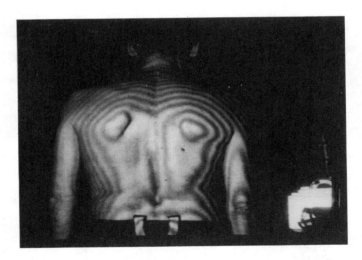

Figure 12. Moiré fringes of a human back.

Discussion

As illustrated above, the described method is a suitable one for making Moiré contours of 3-dimensional objects.

One important limitation of the method is a trade-off between depth resolution and image field size. The depth resolution can be seen from figure 3 to be dependent on the angle between illumination and viewing, and on the projected line density. Usually the depth increment between adjacent fringe will be of the same order of magnitude as the line density on the object. The total number of projected lines in the field of view, using the Fairchild CCD 222, is 380. This means that a fringe-to-fringe distance of about 1 mm gives a field of view of about 38 cm. In most cases it is possible to interpolate between fringes, to increase the resolution.

The described technique would be applicable in all cases where an object is to be mapped in 3 dimensions. It would, however, be especially interesting in situations where a dynamic behaviour of a 3-dimensional object is to be mapped. During casting of different materials, the settling of a free surface during cooling can be of interest, and the described Moiré technique could be used. To check for hidden faults in construction materials, e.g. wooden fiber plates one can stress the materials, and look for deviations from a normal in the strain pattern.

In our setup an ordinary slide projector with a 250 W lamp was used. It is fully possible to increase the power of the light source much, in order to map larger structures, like vehicles or buildings.

Acknowledgement

The author wishes to thank Mr. Bakke of Datamatik in Oslo, Norway and Mr. Eckersand of Fairchild Semiconductor AB in Stockholm, Sweden for making available a CCD camera for the experiments reported.

References

1. Takasaki, H.: Appl. Optics 12, 845 (1973)
2. Idesawa, M. & al: Appl. Optics 16, 2152 (1977)
3. Yatagai, T. & al: AIP Conference Proc. No.65, 579 (1981)
4. Gåsvik, K.J.: Tekn. Ukeblad, 129, 34 (1982)
5. Barbe, D.F. & al: NATO ASI, Ser E-No.16, 623 (1976)

Contouring of cutting tools inserts in real time using moiré of interference

J.-M. Burry

Laboratory of Stress Analysis, Swiss Federal Institute of Technology
CH-1015 Lausanne, Switzerland

Abstract

The method presented in the paper permits a global observation of the contours of any diffusing object, in real time, with both a varying sensitivity able to reach 10 μm or even less, and a possible choice of the orientation of the reference plane in the same measurement.

It is related to multiplicative moiré methods but does not need the use of a filtering grid, nor the preliminary registration of a phase or amplitude image, but involves only interference phenomena, due to a particular use of Sagnac interferometer.

Because the interchangeability of analysed objects may be carried out without any delay, and contouring of larger objects by scanning is very easy, this method is well adapted to routine controls of the topology of any small diffusing surfaces, for example rake faces of cemented carbide tool inserts.

Theoretical explanations, experimental results and also theoretical and practical performances and limitations of the method are presented in this paper, so that its field of application is well determined.

Introduction

In using current contouring moiré methods (observation through an undistorted reference grid pattern of a grid of the same pitch projected onto a distorted surface - therefore this is also deformed)[1-2] one of the two grids at least is material, as is in all cases the reference grid pattern. Only the distorted grid is occasionally obtained by an interferential phenomenon or by shadows[3].

The presence of these material systems often limits the performance of the methods, and does not permit the variation of the sensitivity without modifying the observation apparatus.

The speckle interferometry[4-5] and holographic methods[6-7] remain very satisfactory, but are bounded during use - either with an immersion tank, which is not always possible, depending on the analysed material, or with the use of a tunable laser. In addition, effective observation, even in real time, is only made after a necessary delay in the registration of a first hologram or speckle image, a more or less long delay, depending if one uses a photosensitive emulsion, a photo-thermoplastic film or an electro-optical crystal. As regards to stereo-photogrammetry[8], it requires a complicated apparatus, and does not allow the real time

The method presented is related to projection moiré, but in which both distorted and reference grid patterns are obtained with interferential phenomena.

Principle

The image of a surface, onto which a distorted grid pattern is projected interferometrically is observed through a modified Sagnac's interferometer, permitting the superimposition on this image of an interferogram formed of quasi rectilinear fringes, which gives place to a particular phenomenon of multiplicative moiré.

Study of the interferometer

Without going back into detail to the functioning of Sagnac's interferometer (Fig.1) used as much for holography[9], image processing[10], laser gyroscope[11] as for that concerning us : shearing interferometry[12-13-14], we recall that it allows the lateral shift of two beams emerging from a same point light source, introducing no path difference between them. Therefore one can use a spatially incoherent source, and obtain interference fringes localised at infinity, which materialise in the focal image plane of a lens L2 placed at the exit of the interferometer. The use of a monochromatic source permits the obtention of a quasi rectilinear and unit contrast fringes interferogram.

Besides, if the extended source lies at infinity, or in the focal object plane of a lens L1 placed at the entrance of the interferometer, we shall have the interferogram localised in the image plane of the source.

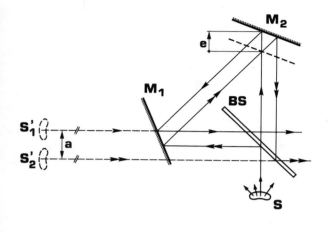

Figure 1. Sagnac interferometer. BS beam splitter. M_1, M_2 flat mirrors. S light source. S_1', S_2' laterally shifted source images. a : lateral shift. e : mirror M_2 translation.

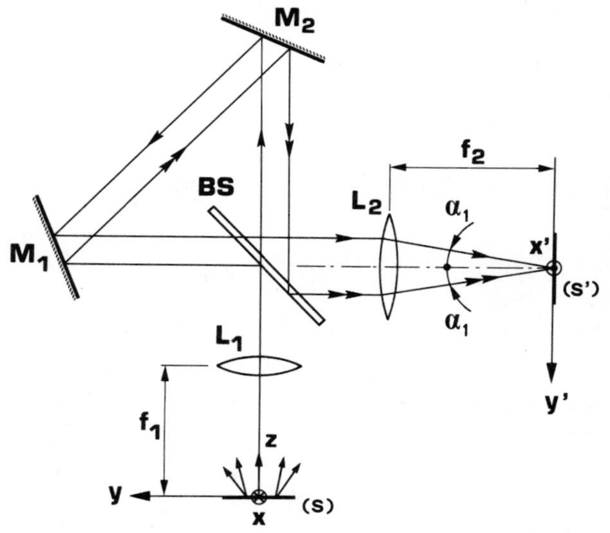

Figure 2. Modified Sagnac interferometer. L1, L2 lenses. F_1, F_2 focal lengths. M_1, M_2 flat mirrors. BS beam splitter. (S) light source. (S') light source image. α_1 half angular shift in the image space of L'2.

The characteristics of the interferometer presented in figure 2 are as follows:

$$a = e \sqrt{2} , \tag{1}$$

where e is the translation of the M2 mirror along the z axis, from the zero shift position. The magnification G is :

$$G = \frac{f_2}{f_1} . \tag{2}$$

The angular shift $2\alpha_1$ in the image space of the lens L2 is:

$$2\alpha_1 = \frac{a}{f_2} . \tag{3}$$

If one expresses $A_0(X'/G , Y'/G)$ the amplitude in the object plane, and $h(X',Y')$ the spread response of the interferometer in coherent light, the image amplitude $A_i(X',Y')$ yields:

$$A_i(X',Y') = \left[\frac{A_0}{2} (\frac{X'}{G},\frac{Y'}{G}) \otimes h(X',Y') \right] e^{ik\alpha_1 Y'} + \left[\frac{A_0}{2} (\frac{X'}{G},\frac{Y'}{G}) \otimes h(X',Y') \right] e^{-ik\alpha_1 Y'} . \tag{4}$$

This immediately gives

$$A_i(X',Y') = \left[A_0(\frac{X'}{G},\frac{Y'}{G}) \otimes h(X',Y') \right] \cos k\alpha_1 Y' . \tag{5}$$

The amplitude $A_0(X'/G , Y'/G)$ is characteristic of the mean reflectivity and at the same time, the speckle on the object. In our experiment, the speckle is only a noise phenomenon, consequently the value interesting us is the expression of the intensity in the image plane obtained after adequately averaging the speckle : $\langle I_i(X',Y') \rangle$:

$$<I_i(X',Y')> = <|A_0(\frac{X'}{G},\frac{Y'}{G}) \otimes h(X',Y')|^2> \cos^2 k\alpha_1 Y' . \tag{6}$$

This case has been studied precisely[15], finding that:

$$<I_i(X',Y')> = \left[<|A_0(X',Y')|^2> \otimes |h(X',Y')|^2\right] \cos^2 k\alpha_1 Y' . \tag{7}$$

Knowing that the spread response in incoherent light of the interferometer, $H(X',Y')$ is:

$$H(X',Y') = |h(X',Y')|^2 , \tag{8}$$

and that the object plane illuminance $I_0(\frac{X'}{G},\frac{Y'}{G})$ is

$$I_0(\frac{X'}{G},\frac{Y'}{G}) = |A_0(\frac{X'}{G},\frac{Y'}{G})|^2 , \tag{9}$$

gives the results

$$<I_i(X',Y')> = \left[<I_0(\frac{X'}{G},\frac{Y'}{G})> \otimes H(X',Y')\right] \cos^2 k\alpha_1 Y' \tag{10}$$

$$= \frac{1}{2}\left[<I_0(\frac{X'}{G},\frac{Y'}{G})> \otimes H(X',Y')\right](1+\cos 2k\alpha_1 Y') . \tag{11}$$

The final mean image is therefore modulated by a system of interferences of unit contrast whose step p_1' in the image plane center is given by equations (3) and (11):

$$p_1' = \frac{2\pi}{2k\alpha_1} = \frac{f_2}{a}\lambda . \tag{12}$$

Obtention of the moiré of interference

The source is an unlevel diffusing surface lighted by two collimated beams perfectly coherent, one inclined in relation to the other, and intersecting at surface level, making an angle of respectively $\vartheta+\alpha$ and $\vartheta-\alpha$ with the object plane (Fig.3). The surface is then lighted by a system of delocalised rectilinear fringes whose step p_0 in the object plane is:

$$p_0 = \frac{\lambda}{2\sin\alpha \cos\vartheta} . \tag{13}$$

Reduced to the image dimensions, this step becomes:

$$p_0' = G p_0 . \tag{14}$$

In the case of an unlevel object, figure 3 shows that a variation ΔZ of the altitude provokes a translation v in the fringe system:

$$v = p_0 \frac{p}{\cdot} = \Delta Z \, tg\,\vartheta . \tag{15}$$

The expression of the amplitude re-emitted by the object is then:

$$A_0(\frac{X'}{G},\frac{Y'}{G}) = \bar{A}_0(\frac{X'}{G},\frac{Y'}{G}) \cos\pi \frac{\frac{Y'}{G}-v}{p_0} , \tag{16}$$

which gives an object illumination in the form (Eq.(9)):

$$I_0(\frac{X'}{G},\frac{Y'}{G}) = \bar{I}_0(\frac{X'}{G},\frac{Y'}{G})(1+\cos 2\pi\frac{\frac{Y'}{G}-v}{p_0}) , \tag{17}$$

216

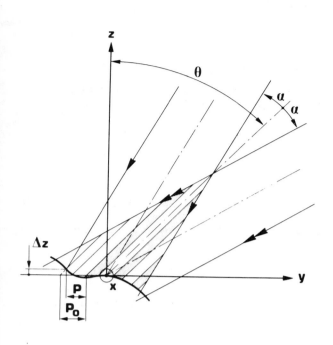

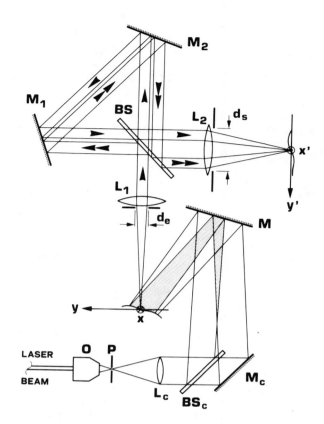

Figure 3. Illumination pattern. (OX,OY) reference plane. $\vartheta \pm \alpha$ illumination angles. p_0 ordinary step. p real step. ΔZ height variation.

Figure 4. Experimental set-up. O microscope objective. P pinhole. L_C, L_1, L_2 lenses. BS_C, BS beam splitters. M, M_C, M_1, M_2 mirrors. d_e entrance pupil diameter. d_s exit pupil diameter.

where $I_0(X'/G , Y'/G)$ is a value taking the speckle on the object into account, which needs to be averaged, thus finding:

$$<I_0(\frac{X'}{G},\frac{Y'}{G})> \ = \ <\overline{I}_0(\frac{X'}{G},\frac{Y'}{G})> \ \left[1 + \cos 2\pi \ (\frac{Y'}{Gp_0} - \frac{v}{p_0})\right] \ . \tag{18}$$

Equations (11) and (12) induce the expression of the average intensity in the image plane:

$$<I_i(X',Y')> \ = \ \left[\left\{<\overline{I}_0(\frac{X'}{G},\frac{Y'}{G})> \left(1 + \cos 2\pi \ (\frac{Y'}{Gp_0} - \frac{v}{p_0})\right)\right\} \otimes H(X',Y')\right] (1 + \cos 2\pi \ \frac{Y'}{p_1'}) \ . \tag{19}$$

The quantity $<\overline{I}_0(X'/G , Y'/G)>$ having a very narrow spectrum is almost not affected by the convolution, the equation (19) may be written :

$$<I_i(X',Y')> \ = \ <\overline{I}_i(X',Y')> \ \left[1 + V \cos 2\pi (\frac{Y'}{Gp_0} - \frac{v}{p_0})\right] \left[1 + \cos 2\pi \ \frac{Y'}{p_1'}\right] \ . \tag{20}$$

V being the value of the modulation transfer function of the interferometer, $\mathcal{H}(\rho)$, for the spatial object frequency, $\rho = 1/p_0$, one can write:

$$V = \mathcal{H}(\frac{1}{p_0}) \ . \tag{21}$$

And if the adjustments are such that

$$p_1' = Gp_0 = p_0' ,$$ (22)

then

$$<I_i(X',Y')> = <\overline{I}_i(X',Y')> \left[1 + V \cos 2\pi \left(\frac{Y'}{p_0'} - \frac{v}{p_0}\right)\right] \left[1 + \cos 2\pi \frac{Y'}{p_0'}\right] ,$$ (23)

a characteristic equation of a multiplicative moiré, which shows up, after development, a surmodulation phenomenon of the image with visibility $V/2$, in the form:

$$S = \left(1 + \frac{V}{2} \cos 2\pi \frac{v}{p_0}\right)$$ (24)

and taking the equation (16) into account:

$$S = \left(1 + \frac{V}{2} \cos 2\pi \frac{\Delta z \, tg \, \vartheta}{p_0}\right) ,$$ (25)

dark moiré fringes appear on the explored image surface in the places where this presents variations in the height such as:

$$\Delta Z = \frac{1}{tg\vartheta} \left(\frac{p_0}{2} + mp_0\right) ; \quad m \in Z ,$$ (26)

which allows visualisation of the curves at equidistant levels with a sensitivity Δh of:

$$\Delta h = \frac{p_0}{tg\vartheta} .$$ (27)

In the experimental set-up schematized by the figure 4, the orientation of the mirror M simultaneously with the displacement of M2 permits varying together p_1' and p_0', and thus we have a variable sensitivity, but above all, a choice of reference plane different from the focused plane by reorientation of the fringes on the object. One can thus artificially create inclined sections, and above all focus on the nearest plane to the surface to be con- toured, which permits an optical transmission of the object grid pattern, and thus we have a depth of analysis superior to the depth of field.

Theoretical performance and limitations of the method

Modulation transfer function of the interferometer. Fringes visibility

In the best cases where there are neither pupils decorrelation nor aberrations introduced by any element of the set-up, we can consider that the interferometer modulation transfer function $\mathcal{H}(\rho)$ is that of objective Ll. The expression of $\mathcal{H}(\rho)$ is given by J.W. Goodmann[16]

$$\mathcal{H}(\rho) = \begin{cases} \frac{2}{\pi} \left[\text{Arccos} \frac{\rho}{2\rho_0} - \frac{\rho}{2\rho_0} \sqrt{1 - \left(\frac{\rho}{2\rho_0}\right)^2}\right] & \rho \leqslant 2\rho_0 \\ 0 \text{ anywhere else} \end{cases}$$ (28)

where the quantity ρ_0 is the cut-off frequency in coherent light :

$$\rho_0 = \frac{d_e}{2\lambda f_1} = \frac{1}{2m\lambda}$$ (29)

the parameter m being the aperture number of lens Ll. From equations (21), (28) and (29), it is possible to express the value of the imaged grid contrast V,

$$V = \mathcal{H}\left(\frac{1}{p_0}\right) = \frac{2}{\pi} \left[\text{Arccos} \frac{\lambda m}{p_0} - \frac{\lambda m}{p_0} \sqrt{1 - \frac{\lambda^2 m^2}{p_0^2}}\right]$$ (30)

If we take into account equation (27), we can express equation (30) using a more practical parameter than p_0, $\Delta h \, tg \, \vartheta$:

$$V = \frac{2}{\pi} \left[\text{Arccos} \frac{\lambda m}{\Delta h \, tg \, \vartheta} - \frac{\lambda m}{\Delta h \, tg \, \vartheta} \sqrt{1 - \left(\frac{\lambda m}{\Delta h \, tg \, \vartheta} \right)^2} \right] \tag{31}$$

Depth of the field

Because the method has to be applied to unlevel surfaces, the depth of the field appears as a fundamental limitation for the method.

In the case there is a small misfocusing, the distance X_1 between lens L1 and the observed point of the surface can be written:

$$X_1 = F_1 - d \quad , \quad \text{with } d \ll F_1 \tag{32}$$

Such a misfocusing introduces an optical path difference w at the edge of the pupil of lens L1, whose diameter is d_e. If we call ε_d the quantity

$$\varepsilon_d = \frac{1}{X_1} - \frac{1}{F_1} \tag{33}$$

then the formula expressing w is given by J.W. Goodmann[16]

$$w = \frac{d_e^2}{8} \varepsilon_d \tag{34}$$

Equations (32) and (33) permit us to write

$$\varepsilon_d = \frac{d}{F_1 X_1} \simeq \frac{d}{F_1^2} \tag{35}$$

Substituting this last equation in (34), one can write the final formula giving w :

$$w = \frac{1}{8} \frac{d_e^2}{F_1^2} d = \frac{d}{8m^2} \tag{36}$$

To find the maximum admissible misfocusing distance , d_{max} , on both sides of the focal object plane of lens L1, we shall apply the Rayleigh's quarter wavelength rule, admitting that the modulation transfer function $\mathcal{K}$ is only slightly affected by misfocusing until the optical path difference w remains smaller than $w_{max} = \lambda/4$. From equation (36), we find:

$$d_{max} = \pm \, 8 \, m^2 \, w_{max} = \pm \, 2 \, m^2 \lambda \tag{37}$$

Then the maximum number of resolved planes without appreciable contrast change n_{max} , is:

$$n_{max} = \frac{2 |d_{max}|}{\Delta h} = \frac{4 \lambda m^2}{\Delta h} \tag{38}$$

Limitation due to speckle

As it has been seen earlier, speckle is only a noise phenomenon. We verify in practice that it becomes awkward when the reference or object grid pitch dimension becomes lower than about twice the mean speckle grain value , p_s . This value is known to be:

$$p_s = 1,22 \, \lambda \frac{f_1}{d_e} = 1,22 \, \lambda m \tag{39}$$

We deduce from equations (27) and (39) that, if the speckle is not eliminated, the limit atteinable value of the sensitivity, Δh_s , is:

$$\Delta h_s = 2 p_s / tg \vartheta = 2,44 \, m\lambda / tg \, \vartheta \tag{40}$$

The limit value of the visibility, V_s , under which speckle is very awkward, is given by equation (31), for $p_0 \leqslant 2p_s$:

$$V_s = \Re \left(\frac{1}{2p_s} \right) \simeq 0,5 \tag{41}$$

General graph

We shall represent on the same diagram two families of parametric curves giving the number of resolved planes n_{max} function of the inverse of the sensitivity $1/\Delta h$ (i.e. number of resolved planes per millimeter), with constant visibilities V for the first family, with constant aperture numbers m for the second one. Besides, because the parameters λ and $tg\vartheta$ have to be taken into account, we shall express the equations of the curves in the form:

$$\frac{n_{max}}{tg^2\vartheta} = C(V) \cdot F \left(\frac{1}{\Delta h(mm)} \cdot \frac{\lambda(\mu m)}{0,5} \right) \text{, for the first ones} \tag{42}$$

$$\frac{n_{max}}{tg^2\vartheta} = G(m) \cdot \ell \left(\frac{1}{\Delta h(mm)} \cdot \frac{\lambda(\mu m)}{0,5} \right) \text{, for the second ones} \tag{43}$$

C, F, G, ℓ being independent functions.

Constant visibilities parametric curves. Equation (31) shows that the image grid contrast is a function of the quotient $\lambda m/\Delta h\, tg\,\vartheta$. The conditions to obtain constant image grid visibility are then given by:

$$\frac{\lambda m}{\Delta h\, tg\,\vartheta} = constant = c(V) \tag{44}$$

Table 1 gives the computed values of c(V), for different values of V, and V/2 which is the moiré fringes visibility (see equation (24)).

Table 1							
Reference grid visibility	0	0,2	0,4	$V_s=0,5$	0,6	0,8	1
Moiré fringes visibility V/2	0	0,1	0,2	0,25	0,3	0,4	0,5
c(V)	1	0,69	0,49	0,4	0,32	0,16	0

Equations (38) and (44) permit us to express equation (42) in the form:

$$\frac{n_{max}}{tg^2\vartheta} = 8000 \, [c(V)]^2 \cdot [\frac{1}{\Delta h(mm)} \cdot \frac{\lambda(\mu m)}{0,5}]^{-1} \tag{45}$$

Constant aperture numbers parametric curves. Besides, we deduce immediately the expression of equation (43) from equation (39):

$$\frac{n_{max}}{tg^2\vartheta} = 2.10^{-3} \, (\frac{m}{tg\vartheta})^2 \, [\frac{1}{\Delta h(mm)} \cdot \frac{\lambda(\mu m)}{0,5}] \tag{46}$$

Figure 5 shows the drawing of this graph, whose examination allows us to define precisely the field of application of the method. Once the wavelength and the value of $tg\vartheta$ are chosen (depending respectively on the laser used and on the morphology of the studied surface), the scales are determined and the graph gives:
- either the maximum attainable visibility and the aperture to be used for lens L1, when the sensitivity and the number of resolved planes have been chosen, and vice versa;
- either the visibility and the maximum number of resolved planes when the aperture of lens L1 and the sensitivity have been chosen, and vice versa;
- either the maximum number of resolved planes and the aperture of lens L1 to be used when the sensitivity and the maximum attainable visibility have been chosen, and vice versa;
- the effects of the speckle phenomenon (according to the approximative admitted criterium).

Examples: for $\lambda=0,5$ μm and $tg\vartheta=1$, if we impose the sensitivity to be 100 mm^{-1} and the number of resolved planes to be 10, then the maximum visibility for moiré fringes will be about 0.3 and the aperture of lens L1 between 8 and 5.6 (point A on figure 5).

If the needed sensitivity had been 300 mm^{-1} for the same number of resolved planes, then the maximum visibility would have been about 0,15 , provided that speckle is eliminated, and the aperture of lens Ll about 4 (point B on figure 5).

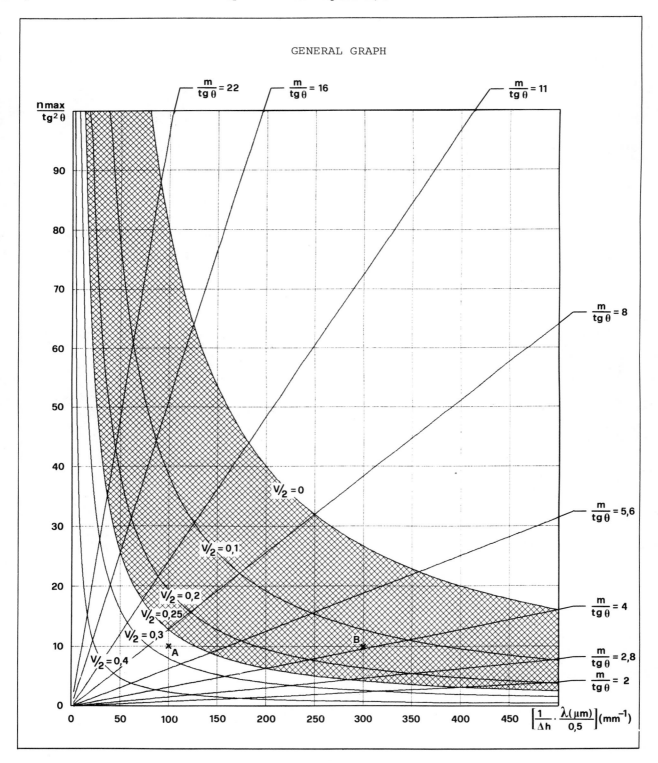

Figure 5. General graph linking the main parameters of the method. Hachured surface corresponds to the points for which speckle must be eliminated (according to the approximative admitted criterion).

This method was applied to the study of the topography of cutting tools, whose morphology in the vicinity of the cutting tips and edges is primordial. Argon laser light at wavelength 514,5 nm was used, object illumination inclination being about 45°.

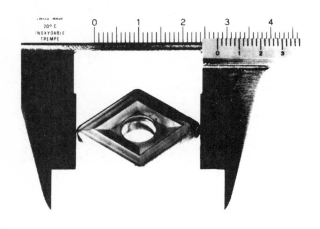

Figure 6.   Example of carbide tool insert, showing its size.

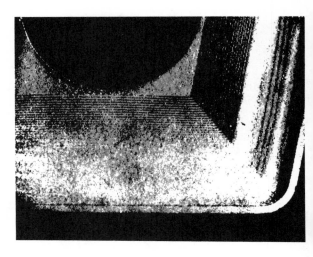

Figure 7a.   Lower level contours, Δh =
22,5 μm.

Figure 7b.   Intermediate contours, Δh =
22,5 μm.

Figure 7c.   Upper level contours, Δh =
22,5 μm.

Figure 6 shows an example of cemented carbide tool insert, showing its size.

Figure 7 shows the contouring of the insert shown on figure 6 with Δh = 22,5 μm. The photographs a), b), c) correspond to the necessary focusings when the chosen reference plane (i.e. plane surrounding the internal diameter) is parallel to the focused plane, bringing about depth of field problems.

On figure 8, focusing is now made on the nearest plane of the surface by slightly inclining the insert. There is no more problem with the depth of field, but the reference plane does not appear explicitly.

On figure 9, the focusing is accomplished in the same manner as previously, but a slight modification of the orientation of the mirror M_C modifies the object grid orientation in such a way to produce a fictive movement compensating the tilt that had been applied to the analysed surface, and permits the choice of the desired reference plane.

Figure 10 shows the profile estimation by tilting the insert. Most of the fringes disappear, due to the depth of field problems. Δh = 21,5 μm.

Figure 11 shows the profile estimation by modifying the reference plane. Fringes cover the whole edge of the insert. Δh = 21,5 μm.

Figure 8.   Focusing on the nearest plane
            of the surface by tilting the
            insert. Δh = 22,5 μm.

Figure 9.   Focusing on the nearest plane
            of the surface with arti-
            ficial tilt compensation. Δh = 22,5 μm.

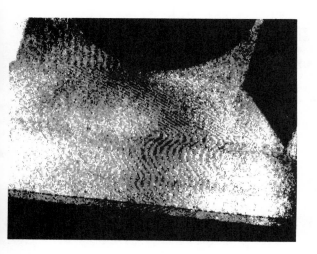

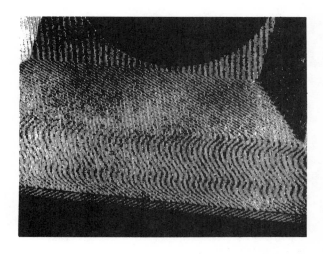

Figure 10.   Profile estimation by tilting
             the insert.

Figure 11.   Profile estimation by modi-
             fying artificially the reference plane.

Practical limitations

We shall not make a detailed study of the other limitations of this method in the present paper. Nevertheless, we must know that pupil decorrelation due to the finite dimensions of the devices composing the set up limits the interference field - and thus the object field. We can also demonstrate that some systematical error is attached to the method, due to a very slight distorsion of the reference grid. Besides, as in classical moiré experiments, problems due to lighting inclination can appear for objects presenting large declivities. It is obvious that all the calculations presented in this paper are essentially available for diffraction limited devices; therefore, all the optical elements of the set up must be of excellent quality. The lighting elements L_C , BS_C , M_C , M must produce perfectly rectilinear fringes in the object plane. The objectives L1 and L2 must have a very good M.T.F. and an extremely weak distorsion, and the element BS, M_1 , M_2 composing the interferometer must not affect in any way the image transmission nor the image grid formation.

Conclusion

This method offers several fundamental advantages with regard to classical methods of contouring. With respect to shadow and projection moiré methods, better sensitivities can be reached because no material grids are used; therefore, awkward phenomena due to the diffraction disappear. Besides, the continuously varying sensitivity and the possible choice of any reference plane within the same measurement permit analysis of objects of varied morphology. With respect to the methods related to speckle interferometry and holography, it does not need any preliminary registration of a phase or amplitude image; therefore, interchangeability of analysed objects is carried out without any delay, and routine controls can be done efficiently.

Nevertheless, the method is limited by the speckle, and requires relatively costly devices.

References

1. TAKASAKI, H., Applied Optics 9, p. 1467. 1970.
2. THEOCARIS, P.S., Moiré Fringes in Strain Analysis, Pergamon Press. 1969.
3. PIRODDA, L., Rivista di Ingegnera 12, p. 913. 1969.
4. YAU HUNG, Y., Speckle Metrology, ed. Robert K. Erf, Academic Press, pp.68-70. 1978.
5. BUTTERS, J.N., JONES, R., WYKES, C., Speckle Metrology, ed. Robert K. Erf, Academic Press, pp. 146-155. 1978.
6. VARNER, J.R., Handbook of Optical Holography, H.J. Caulfield, Academic Press, pp.595-600. 1979.
7. KÜCHEL, F.M., TIZIANI, H.J., Optics Communications 38, p. 17. 1981.
8. Handbook of Non Topographic Photogrammetry, H.M. Karara. 1979.
9. COCHRAN, G., Journal of the Optical Society of America 56, p. 1513. 1966.
10. GRIMES, N.D., Applied Optics 11, p. 914. 1972.
11. ROSENTHAL, A.H., Journal of the Optical Society of America 52, p. 914. 1962.
12. HARIHARAN, P., SEN, D., Journal of the Optical Society of America 49, p. 1105. 1959.
13. HARIHARAN, P., SEN, D., Proceedings of the Physical Society of London, 75, p. 434. 1960.
14. MONTGOMERY, A.J., Journal of the Optical Society of America 57, p. 1121. 1967.
15. LÖWENTHAL, S., ARSENAULT, H., Journal of the Optical Society of America 60, p. 1478. 1970.
16. GOODMANN, J.W., Introduction à l'Optique de Fourier, Masson Editeurs, Paris, pp. 106-115. 1972.

Electronic Speckle Pattern Interferometry for Rotating Structures using a Pulsed Laser

R.W.T. Preater

Department of Mechanical Engineering, The City University
Northampton Square, London EC1V OHB., U.K.

Abstract

The development of Electronic Speckle Pattern Interferometry (ESPI) for the analysis of in-plane strain measurement on rotating structures has been proceeding at The City University. ESPI is a technique which not only possesses the accuracy of holographic interferometry but with the ease of electronic processing of television pictures avoids the delays involved in normal photographic techniques. Using a pulsed laser to overcome the rigorous stability requirements of conventional holography gives this method the potential of an experimental technique which may be used under service environmental conditions, with little or no shut-down of costly production-line plant.

The experimental results achieved so far show reasonable fringe contrast for in-plane displacements over a wide range of tangential velocities up to nearly 5 ms^{-1}. Both rotational speed dependent strains and strains independent of speed have been measured in the laboratory with repeatability.

Introduction

Experimental confirmation of mechanical engineering design procedure by analysis of the prototype under service conditions is particularly attractive if the necessity for costly shut-down of production-line plant is eliminated. A non-contact method of analysis which requires the minimum of surface preparation, spraying with matt white paint, has considerable advantages over conventional experimental methods. Holographic techniques possess the required accuracy and sensitivity to permit measurement of small out-of-plane and in-plane displacements on components subjected to the low stress levels under fatigue conditions in service.

ESPI pioneered at Loughborough University by Butters and Leendertz [1] for the measurement of static in-plane strains in the laboratory is now being developed to examine rotating structures. The use of a J.K. System 2000 pulsed ruby laser in place of the c/w laser removes the rigorous stability requirements of conventional holography and the pulse width of 40 - 70 ns freezes the component movement making measurement under service operating conditions possible. Closed circuit television incorporating video disc and tape storage facilities together with electronic processing for subtraction of single tv-frame recordings, produces clear interference fringes for in-plane displacements.

The television camera used is sensitive to low light levels so that relatively large areas up to 600 mm in diameter may be covered on engineering structures. Use in the field would also require the equipment to be portable.

Both steady state and transient load conditions may be studied provided an initial no load speckle pattern image may be recorded. The in-plane displacement interference fringes are obtained by electronic subtraction of the live load speckle image from an initial no-load recorded image.

Electronic equipment

The study of rotating or moving components using holographic techniques requires either c/w laser illumination and a high speed camera or pulsed laser illumination with precision timing and a conventional television and recording process. The latter method is more attractive since this enables electronic processing of the speckle images to be used rather than the slower conventional photographic technique. The narrow pulse width of the pulse laser, of 40 - 70 ns, not only freezes the movement but can, with precise triggering, produce two speckle images of the component several revolutions apart which on subtraction displays the relative radial displacement alone.

The actual firing of the laser requires two pulses, the first to fire the laser flash tubes, the second to fire the Q-switch. For optimum laser performance these must be separated by 1.26 ms, whatever the component speed, and the firing of the Q-switch requires locational precision to within 1 speckle or triggering to within 1 µs.

Standard television equipment is used but with minor modifications to provide camera blanking, so that complete single frames may be recorded even though laser Q-switching may not coincide with the start of a tv-frame. The silicon tv-tube fitted here to give sensitivity to low light levels also has sufficient persistence in the absence of the scanning electronic beam to allow recording to await the arrival of the next frame pulse.

The video equipment used is shown schematically in Fig. 1. The initial state high contrast speckle image of the component is recorded on the video disc and continuously replayed on the monitor. Several revolutions later a "live load" speckle image is subtracted from the first and the resulting interference fringe pattern is recorded on the video tape recorder. This tape recorder allows a slow motion replay and a single frame search for the frame of interest as well as a permanent record of the sequence of events.

The latest method of precision triggering has shown a marked improvement over previous techniques. A much wider range of component speeds has been observed with very little evidence of mis-match on image subtraction. Automatic pulse separation for laser triggering has been achieved whatever the component speed. There is a distinct improvement in fringe contrast overall but this has yet to be optimised for each of the test results which have been obtained. The upper limit of tangential velocity producing acceptable fringes has yet to be established but previous evidence [4] and the present results show that the maximum tangential velocity will be in excess of 5 ms^{-1}. This is rather better than the 2 ms^{-1} suggested earlier by Cookson, Butters and Pollard [2].

Experimental results

In-plane displacement measurement using ESPI requires two beam illumination of the component surface. Modification of the optical layout with the use of a plane mirror, shown in Fig. 2 simplifies the system and allows access to areas of the components local to the shaft axis.

The plane containing the direct and reflected illuminating beams defines the direction of resolved displacement that is detected, in this case the horizontal radial displacement of a rotating component. A small rotation of the mirror about an axis normal to the plane of the diagram will simulate differential horizontal displacement for test purposes only and produce a series of vertical fringes. This will occur on both static components and rotating ones also, providing the laser Q-switch triggering is correct. Any mis-match or timing error will result in an additional relative rotation of the images producing horizontal fringes similar to the static test result shown in Fig. 4. Radial plus circumferential displacement due to triggering error produces fringes at some inclination to the vertical as was experienced in earlier results [3].

The results presented here are for two test components tested in the laboratory to assess the feasibility of strain measurement on rotating components. The first is a 360 mm diameter perspex disc with heavy steel segmented rim in the form of a flywheel, producing radial movement with increasing speed. The second, a small tensile specimen with a 6 mm hole, mounted radially on a 360 mm diameter aluminium disc.

Results achieved from the flywheel tests show that mirror tilt produces vertical fringes or apparent differential radial displacement over the surface local to the shaft axis. Figs. 5 and 6 are for the same tilt but for subtractions recorded at tangential velocities of 4.3 ms^{-1} and 2.3 ms^{-1} respectively. The change in fringe spacing between the two recordings indicates the effect of radial contraction of the perspex at the lower speed. The reduction by 4 in the total number of fringes here represents over this area of component a strain of 30 μ-strain. The fact that both sets of fringes are vertical is an indication of the precise laser triggering.

For the tension specimen, on the rotating disc, similar mirror tilt again shows consistent vertical fringes at tangential velocities of 4.3 ms^{-1} and 2.2 ms^{-1} in Figs. 7 and 8 on subtraction.

The fringe pattern in Fig. 9 is again for the tension specimen where an initial no load recording was made at 4.6 ms^{-1}. On stopping the component, tension was then applied before re-starting and performing image subtraction again at 4.6 ms^{-1}. Reducing the speed of the component then to 3.1 ms^{-1} shows no apparent change in the interference pattern in Fig. 10 over this tangential velocity change of 1.5 ms^{-1} indicating an independence to component rotational speed. Both these fringe patterns display a symmetrical shape and distinct similarity to that shown in Fig. 3 for the static test on a tensile specimen some 5 times longer than could be mounted and rotated at speed.

Both the above test components were made of non-metallic materials, in one case perspex, the other epoxy resin. These materials require smaller loads and thus less robust rigs to produce measureable strains but being less stable the fringe contrast deteriorates relatively quickly. The next stage in the test procedure will be to produce more sophisticated rotating test rigs and optimise the fringe contrast in order to establish the upper limit of tangential velocity producing acceptable fringes.

Conclusions

The latest method of precision laser triggering providing locational accuracy to within 1 speckle has produced interference fringe patterns for rotating components which are independent of component tangential velocity. Where a flywheel form of test component has been used, clear evidence of the ability to measure radial strains of the order of 30 μ-strain has been displayed for a change of angular velocity. Although no upper limit of tangential velocity has yet been determined it is likely that ESPI alone will be limited to the lower speed rotating structures.

Some preliminary tests have been done incorporating a derotator system with ESPI, and reported elsewhere [3]. It is possible that at some stage in the future the speed range of in-plane strain measurement may well be extended in this way.

Acknowledgements

This work has been supported by The City University and the Science Research Council. The author would like to thank Mr. J.W. Heaphy of The City University and Mr. T.J. Cookson late of Loughborough University for their help in this project.

References

1. Butters, J.N. and Leendertz, J.A., A double exposure technique for speckle pattern interferometry, J. Phys. E., Sci., Instrum., Vol. 4, pp. 277-279, 1971.

2. Cookson, T.J., Butters, J.N. and Pollard, H.C., Pulsed lasers in electronic speckle pattern interferometry, Optics and Laser Technology, June, pp. 119-124, 1978.

3. Preater, R.W.T., A non-contact method of in-plane strain measurement on rotating structures, SPIE, Vol. 236, pp. 58-61, Utrecht, Sept. 1980.

4. Preater, R.W.T., Further developments in in-plane strain measurement of rotating structures, SPIE, Vol. 369 (to be published) Edinburgh, Sept. 1982.

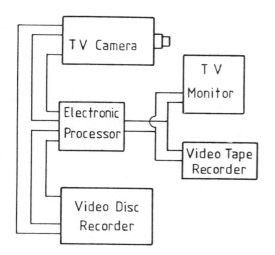

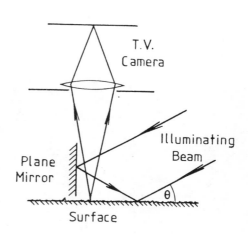

Fig. 1. Electronic equipment - schematic diagram.

Fig. 2. Optical component layout for the measurement of in-plane strain.

Static component tests

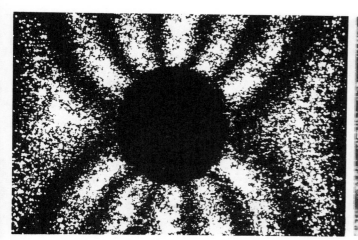

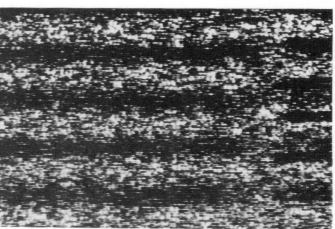

Fig. 3. Interferogram of in-plane displacements around a hole in a bar in tension.

Fig. 4. Interferogram due to a static rotational displacement equivalent to pure shear.

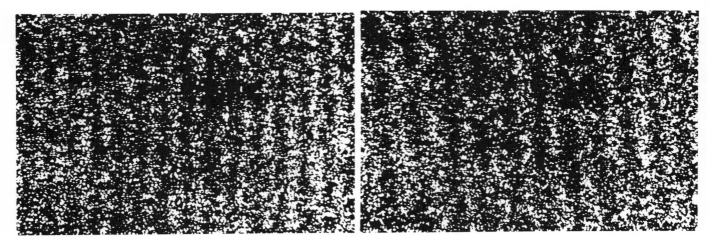

Fig. 5.   4.3 ms⁻¹   Fig. 6.   2.3 ms⁻¹
Interferograms of vertical fringes produced by mirror tilt with subtractions recorded at different
tangential velocities on the flywheel component.

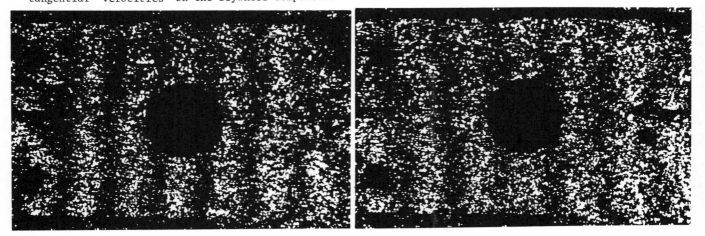

Fig. 7.   4.3 ms⁻¹   Fig. 8.   2.2 ms⁻¹
Interferograms of vertical fringes produced by mirror tilt with subtractions recorded at different
tangential velocities on the tension piece.

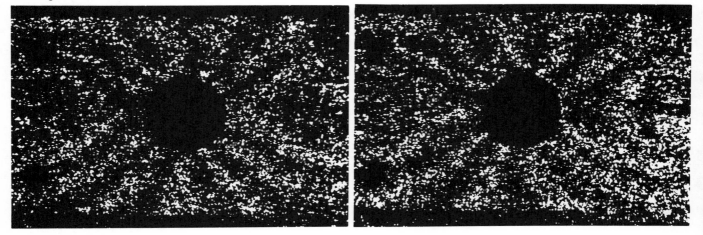

Fig. 9.   4.6 ms⁻¹   Fig. 10.   3.1 ms⁻¹
Interferogram of in-plane displacements around a hole in a tension piece with subtractions recorded at
different tangential velocities.

The Application of Laser-Speckle Photography to ultimate load investigations on Reinforced Concrete Constructions

Klaus-Peter Groß

Institut für Massivbau*, Universität Hannover
3000 Hannover 1, Callinstr. 32

Abstract

When assessing the ultimate load of statically undetermined reinforced concrete beams, an exact knowledge of the bond mechanism and the rotation capability of the plastic hinges is of great importance. The size of these plastic areas is determined conclusively from the efficiency of bond between concrete and reinforcement. Experiments were carried out at the Institut für Massivbau, Hannover University, in order to measure the bond strenght and the extension of the plastic areas on models of reinforced concrete constructions. Laser-Speckle Photography was used for these investigations.

The experimental devices, peculiarities affecting the application of the procedure to reinforced concrete constructions, like optical characteristics of the concrete, fracture development etc., and the results of the test series are reported on.

Introduction

The principles of Laser-Speckle Photography can be described as follows: If an optically rough surface is laser-illuminated, then light is diffusely reflected in all directions of the room. One describes surfaces as being optically rough when their roughness is a multiple of the light wavelength. The interference of the reflected light creates a random, stationary interference field made up of bright and dark spots or speckles on the illuminated surface.[1,2] In case of displacement of the surface the speckle pattern is also displaced. Superimposition of the photographs of the displaced and undisplaced patterns by a double exposure photograph is the optical storage of the entire displacement field.

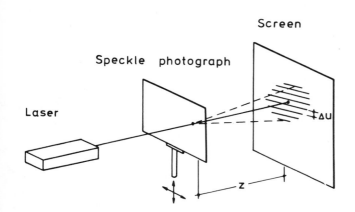

Figure 1. Evaluation of a speckle photograph.

When the processed film is analysed point-by-point, by directing a narrow laser beam throught it, then the Young's fringes as they are called form on a screen positioned behind it (Figure 1). From the distances between the fringes and their direction the extent and direction of displacement of the point in question can be calculated.[3,4]

The application of this method for investigating concrete construction elements it a great advantage, in view of its low tensile strength and the opportunity for measuring crack formation.

The suitability of concrete surfaces for the application of Laser-Speckle Photography

For deformation measurements involving use of Laser-Speckle Photography the surface of the investigated object must fulfil two conditions:
1. It must not adsorb incident light, it must reflect it.
2. It must be optically rough so that a speckle pattern can form, through the interference caused to the diffusely reflected light.

The optical properties of concrete surfaces are determined from their geometry and their constituents, cement matrix and aggregate. For this reason, the geometrical roughness of concrete surfaces and the contrast of the Young's fringes were measured for matrix and aggregate substances. Analysis of these tests led to the following results:
1. Concrete surfaces (grain size up to 32 mm) provide high-contrast speckle photography.
2. Due to the varying composition (matrix, aggregate) the contrast is not uniform.

*Director: Prof. Dr.-Ing. K.W. Bieger

3. The use of a polarization filter can heighten the contrast in many cases (Figure 2).
4. Large pores and holes have a negative effect on the contrast. Good compaction of the concrete is important for this reason.

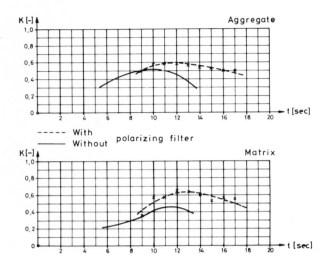

Figure 2. Results of the contrast measurements.

The attainable accuracy of the measurements is of importance for the value of statements of deformation measurements carried out on concrete surfaces. Systematic investigations to this end were also carried out and their results were statistically evaluated. It was revealed that the standard variations achieved were generally small, and the variation coefficients were V < 1 %.
A relation to the demagnification between object and its photograph could not be established.[5]

Demands on the experimental devices

When designing the testing installation it must be borne in mind that all displacements occurring in the investigated object are registered by Laser-Speckle Photography. Apart from displacements within the object, this also includes the rigid body motions. They limit the measureable size of the deformations of interest or falsify them. It is therefore important that the size of the rigid body motions are reduced by an appropriate testing setup. How this can be achieved is shown by way of example through the bonding tests undertaken by our institute. The test setup is shown in Figure 3. It consists of two frame constructions.

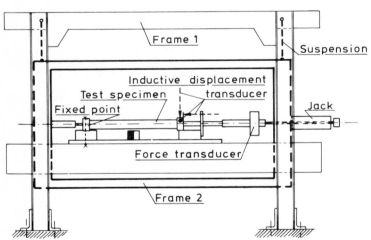

Figure 3. Test setup for the strain tests.

Frame 1 serves mainly as the suspension mount for frame 2 and for supporting the test specimen. Frame 2 is the actual load frame. Together with the test specimen it forms a closed system of forces that is suspended in frame 1 as a pendular unit. The test specimen is mounted in this frame and is connected at one end with frame 1. During loading by the jack the test specimen and frame 2 are deformed, whereas frame 1 and the fixed point with the specimen are not affected. Due to this, there is an area on the test specimen which acts as an immovable pivot, i.e. without rigid body motions.

At the other end of the test specimen the movements of the test specimen are measured triaxially by inductive displacement transducers. This measurement is an additional check on deformations. It also serves the purpose of recording the out-of-plane movements of the specimen. These movements (displacement and tilting) likewise lead to falsification of the optical measurements.[6] In our tests they were measured for this reason and were taken into consideration in the electronic evaluation of the measurements.

Bond tests with Laser-Speckle Photography

Crack formation, crack width and crack spacing substantially determine the bearing capability of reinforced concrete constructions. A series of parameters like e.g. type of reinforcement, tensile strength and bond strenght have a substantial influence on crack formation. As the location of the crack formation is not known beforehand, optical methods enabling areal evaluation of deformations like e.g. Laser-Speckle-Photography, are a great advantage.

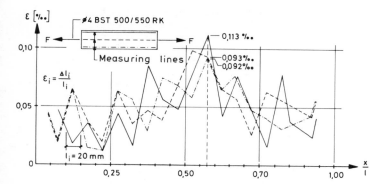

Figure 4. Tensile bar strains in the uncracked state.

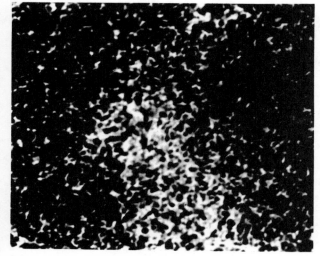

Figure 5. Speckles in a crack tip.

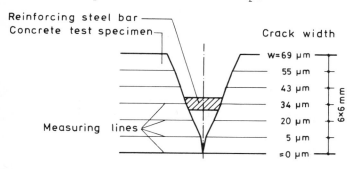

Figure 6. Initial crack width during origination.

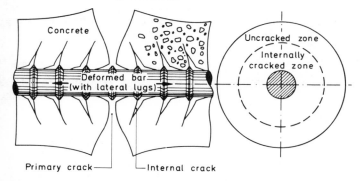

Figure 7. Deformation around the reinforcing bar.[7]

Tests on axially loaded, reinforced tensile specimens were carried out using the testing devices described in the previous section. In this case both the strains and the crack widths were measured, using Laser-Speckle Photography. Prior to the formation of the first cracks, displacements (bond slip) occur between the reinforcing bar and the concrete. The result is a non-uniform state of strain, as shown in Figure 4. The first crack formed at the point of greatest strain, at about the middle of the bar.

In the point-by-point evaluation of a speckle photograph cracks in the tensile bar are shown by an abrupt change in the slope (inclination) of the Young's fringes. The width of the crack can be concluded from the difference in deformation of the two sides of the crack. Within the crack the reflection of light is distorted. Depending on the width of the crack, either no speckles or only a few form. Figure 5 shows the speckle field in the vicinity of a crack tip as seen under the microscope.

The uneven strains of the tensile bar (Figure 4) are also clear in Figure 6 from the line of the crack width, as measured using Laser-Speckle Photography. At the point where a strain peak was present the crack widened most.

If a reinforced concrete construction element is subjected to tensile loading, then apart from the visible (primary) cracks, internal cracks unnoticeable at the surface of the concrete also form around the reinforcing bar (Figure 7). These internal cracks have a great influence on the bond mechanism between steel and concrete. As they are not very wide and also close again after unloading of the tensile bar they can only be verified with great difficulty. A method is described in the literature for reinforced concrete with large aggregate (∅ 16 mm) and reinforcing bars.[7] Internal cracks could be verified for concrete with smaller aggregates (∅ 4 mm) and thin reinforcing bars for reinforced models using the method following at our institute. The bar shown in Figure 4 was loaded almost to the point of failure and then unloaded again. Following this, the original cross section was reduced by grinding to a width of 8 mm (Figure 8), so that the internal cracks extendet up to the new surface. The bar was strained afresh and the deformations resulting from this were measured using Laser-Speckle Photography. The presence of internal cracks could be verified using two methods:

1. Larger cracks were revealed by a quick scanning of the exposures during a sudden change in the inclination of the fringes.

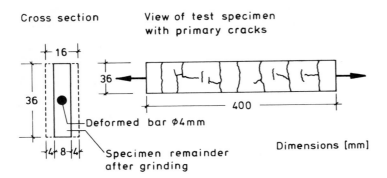

Cross section

View of test specimen
with primary cracks

Deformed bar ∅4mm

Specimen remainder
after grinding

Dimensions [mm]

Figure 8. Treatment of the tensile bar after
the initial loading.

2. A better indication of cracks
emerged from a systematic scanning
of the exposures and the calculation of relative deformations
between the individually stages.
The mean strains measured above
the cracks were several time larger than the ultimate strain of
the concrete (Figure 9).

As the cracks observed in this manner
were present from the lowest loading step
onwards, one can assume therefrom that
was not a case of new formation during
the second loading step. The width of
the internal cracks in the lowest loading
steps was w ≅ 1 - 5 μm, in the higher
loading steps w ≅ 30 μm.

Maximum deviations of the individual measurements from the mean value determined during
these deformation measurements were < 1 %. Figure 9 shows the resultant calculated curve of
the relative deformations. A clear connection between the distance of the lugs and the
cracks cannot be recognized. The sections with negative deformations occur during the unloading of individual areas during reassessment of the crack.

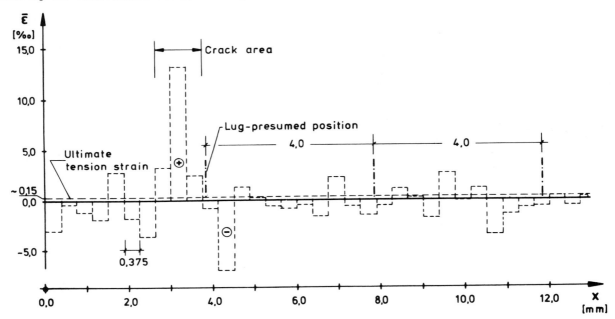

Figure 9. Relative deformations of the tensile bar in the vicinity of the reinforcing bar.

Conclusion

During the use of Laser-Speckle Photography in reinforced concrete construction, the
method has proved to be a valuable new means for crack determination and crack width measurement. The demands of the method on the test setup could be met without difficulty.

Acknowledgement

The author expresses his thanks to Prof. Dr. Bieger for his generous support during this
work.

References

1. Pohl, R.W., Optik und Atomphysik. Springer-Verlag, Berlin-Heidelberg - New York,
12. Auflage, 1967.
2. Ridgen, J.D., Gordon, E.I., The Granularity of Scattered Optical Maser Light.
Proceedings of the IRE 50, 1962.
3. Archbold, E., Burch, J.M., Ennos, A.E., Recording of in-plane Surface Displacement

by Double-Exposure Speckle Photography. Optica Acta, Vol. 17, No. 12, 1970.

4.   Archbold, E., Ennos, A.E., Displacement Measurement from Double-Exposure Laser Photographs. Optica Acta, Vol. 19, No. 4, 1972.

5.   Groß, K.-P., Modelluntersuchungen von Stahlbetonbauteilen mit der Laser-Speckle-Photographie. Werner-Verlag, Düsseldorf 1983.

6.   Archbold, E., Ennos, A.E., Virdee, M.S., Speckle-Photography for Strain Measurement - A critical Assessment. SPIE, Vol. 136, 1st European Congress on Optics. Applied to Metrology, 1977.

7.   Goto, Y., Cracks Formed in Concrete Around Deformed Tension Bars. ACI Journal, April 1971.

Laser measuring techniques in commercial vehicle construction and mechanical engineering

B. Breuckmann

Maschinenfabrik Augsburg-Nürnberg AG
Dachauer Straße 667, 8000 München 50

Abstract

This paper explains the specific advantages and disadvantages of interferometric measuring and testing techniques using the example of holographic interferometry. Applications from commercial vehicle construction and mechanical engineering are given for the most important fields in which holographic measuring techniques are used. Particular attention is thereby paid to considering the techniques from the point of view of the industrial user: to their availability outside special laboratories, to their reliability in day-to-day industrial use and to their economy. The present state of the art and the expected developments in the field of automatic evaluation are also discussed.

Introduction

The invention of the laser has fired man's imagination like few other innovations before it. Although many of the initial hopes and expectations have not been realised, there are today many processes which were not technically possible or economically feasible until the advent of the laser. The industrial applications of the laser can be divided into three main groups:

- telecommunications
- material processing
- measuring and testing techniques.

In the first of these areas, laser diodes are being used more and more often. The demands made on typical laser properties - low spectral width, small divergence of ray - should however been seen here in relation to the corresponding properties of other optical transmitters and are thus in general less strict than they are in measuring technology, for example. In material processing, the CO_2 and Nd-Yag lasers, which have in the meantime become very reliable, are used in order to generate the required high energy densities in a specified, closely restricted space. In measuring technology, various properties of the laser are exploited according to the particular application and the most varied types of laser are accordingly in use. The necessarily higher demands made on the devices for measuring purposes are however in many cases not met. The lack of technical maturity frequently means low reliability and time-consuming preparations, which means that many types of laser can be eliminated as unsuitable for industrial applications from the start. Laser-based measuring and testing techniques also often prove to be unsuitable for the same reason although applications and a demand exist. In view of this, the following sections are devoted to a discussion of the question of the availability of laser-based procedures in measuring techniques specifically from the point of view of the industrial user. The example taken is holography.

Holography in industrial practice

Holography, the application of which is particularly closely linked to the development of the laser, has been the object of great expectations since the mid-1960s. A large number of groups concerned themselves with the new technology and so after only a few years not only the physical principles but also the necessary requirements for lasers, optical structure and recording material were known and had also been realised for most of the present fields of application. Unfortunately, many of the dreams of this time have only been fulfilled in laboratory set-ups, if at all, such as holographic films or even 3-D screens, all of which means that holography has not yet made an impact on everyday life.

In the field of measuring techniques, the situation is similar, although the physical principles, the measuring set-up, the stability requirements and the equations for evaluating have been known for a long time. The possibilities of combining holography with other optical measuring techniques such as speckle technique, moire technique and optics of stress and strain have also been shown. All these procedures have been used and can show some success in solving problems of measurement. However, a large proportion of holographic measuring techniques are still only applied in university institutes or in research and development laboratories of the larger companies. Their use in series production or in quality control has to date been confined to only a few instances such as tyre checks and measuring of optical elements. Using as examples three important applications, the following sections will examine the possibilities offered by holographic measuring and testing techniques in industrial use, the limitations to which they are at present subject and the developments that can be and needs to be carried through if these techniques are to enjoy a more widespread use. Many of the arguments employed here in support of holography can in our opinion also be applied to other interferometric methods of measurement which, like holography, facilitate two-dimensional representation of measured data.

Vibration analysis and acoustics

The essential feature of holographic interferometry is that it makes it possible to recognise vibration patterns even of large and complex objects "at a glance" with high local resolution. It is particularly in the analysis of vibration and noise of machines in operating conditions that holography has clear advantages over other methods of measuring, because often only a few holograms disclose the most important features of the vibrational behaviour. The holographic vibration diagrams for a 6-cylinder 100 kW diesel engine (see Fig. 1a) in Figs. 1 b, c show high vibration amplitudes on the intake pipe and the whole exhaust system, whole body vibration of the injection pump and fuel filter, high vibration amplitudes at both sides of the engine and the vibration patterns excited on the frame.

1 a

1 c

1 b

Figure 1a: 6-cylinder Diesel-engine, 100 KW
Figure 1b: Interferogramm of the intake side
Figure 1c: Interferogramm of the exhaust side

Holographic interferometry has the same advantages in examination of transient processes also. In the analysis of noise development in commercial vehicle brakes it was above all the results of holographic measurement of the squealing brake drum (see Fig. 2 a,b) which made it possible to design a brake system with which it has to date been possible to achieve a sizeable reduction in both the occurance and the sound intensity of brake squeal on the test facility. The holographic method of measurement guarantees a very detailed assessment of the vibration and noise behaviour especially in combination with the methods of modal analysis and sound intensity measuring techniques. There are several reasons why holographic vibration analysis is still seldom used despite these possibilities:

- the investment costs for a powerful pulse holography device are rather high.

- pulse lasers, and particularly their supply installations, are rather large and are thus not as mobile as might be desired.

- because their technology is not yet fully developed, the systems still have to be operated by highly qualified personnel so that a mode-free beam quality is guaranteed even at high power.

- preparation of the conditions necessary for holographic measurements (darkness, laser safety) is a involved procedure.

 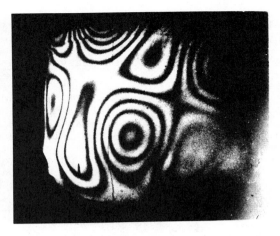

Figure 2a: Double exposure hologramm of a brake drum vibrating at 5.3 KHz

Fig. 26: Vibration pattern of thesquealing brake drum frequency   5.5 KHz

It can nevertheless be assumed that the rate of return on the great expense of installing a holographic vibration laboratory will increase as the demand for efficient machines and vehicles which also do not harm the environment grows, and particularly demands for low-noise design and lightweight construction. It is therefore no coincidence that holographic processes were used first in the car industry, in which the above criteria have been an important market factor for years.

Deformation measuring technique

Here too, representation of measured values in form of a picture is the main advantage of holography. Since the very high sensitivity also makes it possible to take measurements when loads are extremely small, the deformation behaviour even of complex test objects under load can be assessed in qualitative terms very simply. This applies particularly to components on which no information is available. Two important applications of the holographic measuring technique should be mentioned at this point: a) determination of the points at which other recording systems, measuring only at selected points, taking continuous spot measurements, e. g. for long term operation, can be applied to good effect and b) determination of areas and load limits within which viscoelastic and/or plastic deformation occurs. As an example of the use of holography in deformation measurement techniques, let us consider a nozzle casing in a power station; the heated steam flows through this nozzle on to the turbine. The deformation of the nozzle housing as measured by holography is shown in Fig. 3 a-c and the zones of high component stress determined from this can be seen in Fig. 3 d.

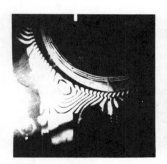   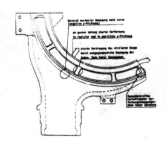

Figure 3 a-c: deformation patterns of a nozzle housing, louded with an internal pressure of about 1 bar.

Figure 3 d:
Zones of high deformation and component stress

The internal pressure load required for these shots amounted to only approx. 1 bar for a wall thickness of some 30 mm. It is possible to evaluate the holographic measurements quantitatively in three dimensions and this was carried out for some of the critical areas. Here too, however, the affort for both the actual measuring set-up for representing all three spatial deformation components and the solution of the evaluation equations is too high; this is particularly true of holographic stress analysis, which almost always requires a precise quantitative evaluation. A further disadvantage - this applies expecially to use of the technique in the production shop - is that, for most applications, either vibration isolation is required or that special optical arrangements, some of them sophisticated, have to be set up with which disruptions can be suppressed.

Non-destructive testing

Holographic testing procedures have a series of advantages which come to the fore in the product development phase and in the establishment of testing criteria. They guarantee evaluation of the whole test object, or at least of large parts of it, which is of great importance if points where defects are particularly liable to occur are not yet known or when defects, e.g. on coated surfaces, can occur anywhere in fairly large areas. Apart from this, the examination can generally be carried out when the load is very small and in relation to the application. The size of the load in such cases can be selected within a very wide range and the development of a defect can therefore be observed as the load is increased up to the point at which the test object is destroyed. It is thus above all possible to assess the defect relevance; defects which are found can be classified and distinguished according to, type, position and size, as is shown by the example of CfK-structures in Fig. 4 a-d.

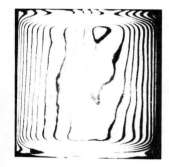 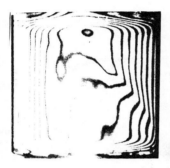 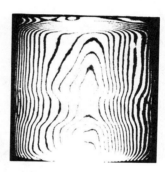 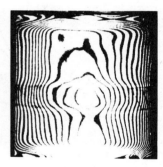

Fig. 4a: default-free, Figure 4 b-d: CfK-structure with various faults CfK-structure

In spite of this, holographic procedures have managed to establish themselves in only a few applications. In our opinion, the reasons for this are as follows. Firstly, the detection limit for recognition of defects is frequently inadequate despite the high absolute accuracy of holographic procedures, as defects can on principle only be detected on the basis of disruptions to the overall deformation pattern. This is particularly true for very stiff components or defects which are a long way from the visible surface. Secondly, a holographic set-up in most cases falls far short of the ideal sought in series testing, which is best signified by the term "black box". In this field, the demands made by holography on vibration isolation, handling and, to some extent, on the evaluation of the measured data are still too high.

To summarise, it can be stated that holographic and other pictorially measuring interferometric methods such as speckle photography are an indispensable part of many areas of research and development. On the other hand the fact should not be disregarded that there is still a series of obstacles to the use of these test methods in series production and quality control. Using the example of holography once again, the following paragraphs will now explore the prospects for such methods in applications which seem possible from today's point of view.

Prospects

In non-destructive testing, systems are conceiveable which to a large extent meet the requirement for a "black box"; these are suitable for testing smaller objects which can be measured adequately with the power of He-Ne lasers. Testing costs of a few $ per holographic picture can be expected, but several test objects may be depicted at the same time. If higher power is needed, no alternative to the inert gas-ion laser is as yet in sight. However, the considerably higher investment costs and the increased expenditure for maintenance mean that testing costs of less than 5 $ can be achieved in only a few cases.

A large number of new applications is definitely to be expected from a considerable
increase in the detection limits. One procedure, which has been proposed by D.B. Neumann[1],
is of interest here: In a first stage, master holograms are produced from a defect-free
object. The real reconstructed images of this then serve in a second stage as illumination
waves for the test object. This principle should make it possible to suppress those
portions of the interference pattern which do not originate from the defect itself, and
thus to show up only the influence of the defect.

Fig. 5 shows a holographic procedure developed by us which follows the basic principles
of Neumann's idea but avoids some of the inherent difficulties of the measuring set-up
he describes. If the hopes which we have for this procedure are realised, and at the moment
there are many indications that they will be, an increase of more than one order of magnitude can be expected in the detection sensitivity.

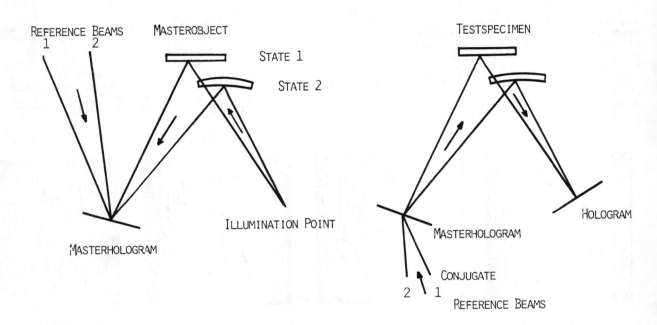

Fig. 5: schematic set up for a comparative holography

A more widespread use of holographic deformation analysis is primarily to be expected
from computer-aided evaluation which should meet the following requirements:

- evaluation of the whole image. Only in this can holography's advantage, namely that
  it measures pictorially, be exploited to the full.

- determination of the interference phases with high accuracy.
  The aim here should be to achieve a phase accuracy of $\lambda/100$ so that expansions can
  also be calculated from the deformation pattern with sufficient reliability.

- 3-dimensional evaluation from one hologram.
  In practice, measuring procedures with several points of illumination/observation
  require too much work to adapt the set-up concerned to the specific object to be tested.
  The use of mirrors is no longer possible, at least not with larger objects.

- Data transfer and compatibility
  Here particularly, the aim should be to produce one issue of the measured data which
  admits of direct comparison with FEM calculations. In addition to this, the possibility
  of combining holographic deformation analysis techniques with speckle holography
  should be borne in mind, as an advanced evaluation process already exists for this.

In the field of computer-aided evaluation of holograms, we have for about a year
now been following up a concept in cooperation with Labor Dr. Steinbichler in which the
interferogram is taken with two separate reference waves. A method has thereby been developed which guarantees the use of the "two reference wave procedure" even when the
wavelengths of the recording and reconstruction lasers are differnt. This makes it possible
to use the evaluating process even when pulse lasers are used, which is particularly

important for vibration analysis. In this case, the two reference waves are switched between the laser pulses by means of synchronised electro-optical components.

In the reconstruction, the phase of the two reference waves and thus of the two reconstructed object waves can be shifted by means of a piezo-electric-driven mirror. Three interference images phase-shifted by 120°, $I_1 \ldots I_3$, are then read in with a television camera. From these images, the interference phase γ modulo π can be determined for all points on the image with high accuracy from the formula[2]

$$\tan \gamma = \frac{(I_3 - I_2) \cos \gamma_1 + (I_1 - I_3) \cos \gamma_2 + (I_2 - I_1) \cos \gamma_3}{(I_3 - I_2) \sin \gamma_1 + (I_1 - I_3) \sin \gamma_2 + (I_2 - I_1) \sin \gamma_3}$$

The orders of the interference fringes are then calculated for the whole image using the interference phase as well as the phase jumps which occur.

A computer-aided evaluation process such as this will be of similar importance for holographic analysis of vibrations and noise. This applies particularly where a series of double-exposure holograms triggered at precise intervals are to be used on running machines to examine not only the spatial distribution of the vibration patterns but also their development over a period of time.

In conclusion, a further note on the insufficient technical standard of many of the devices on the market. The reason for this is certainly to be found in the fact that these devices are still produced in comparatively small numbers. The manufacturers should, however, take into account the fact that the existing market, which is certainly quite a large one, will respond only if the systems are made considerably more reliable, easy to maintain and thus economic than is the case at present. This naturally applies especially to the laser itself, the further development of which will have the greatest influence on the importance of laser measuring procedures in the long term.

References

1. Neumann, D. B., Comparative Holography, Tec. Digest, Topical Meeting on Hologram Interferometry and Speckle Metrology, Opt. Soc. Am., pp. MB 2-1 to MB 2-4. 1980
2. Dändliker, R., Thalmann, R., Willemin, J.-F., Fringe Interpolation by Two-Reference-Beam Holographic Interferometry: Reducing Sensitivity to Hologram Misaligument, Optics Communication, Vol. 42, pp. 301-306. 1982

Difference holographic interferometry

Z. Füzessy and F. Gyimesi

Institute of Physics, Technical University Budapest
H-1521 Budapest Hungary

Introduction

Holographic interferometry in contradiction with conventional interferometry can work with diffusely reflecting objects as well. However, this great advantage is restricted only to the comparison of the states of the same object. The states of two diffuse objects can not be studied by purely optical methods because of their different surface microstructure. To do this, first one has to evaluate their interferograms separately and then compare the evaluated data themselves.

Nevertheless, in holography it is very much possible to compare the simple displacement directly. That can be realized by appropriate changing of the wavefront shape of the object illuminating beam between the two states investigated. Using conventional optical elements the shape of the illuminating beams are really restricted to very simple forms: planes, spheres, cylinders, etc. But an extension of the method towards more complicated illuminating wavefronts is possible with the help of holography. The illuminating wavefront can be the conjugate wavefront of a master object. It can be reconstructed holographically with the conjugate reference beam.

The idea of holographic illumination from preferably specular master objects was published by Denby et al[1] in 1976 for contouring by electronic speckle pattern interferometry. In the present paper a successful attempt is reported where holographic illumination from diffusely reflecting object is used directly in holographic interferometry to compare displacements. The application of the method for contouring and transparent object investigations seems to be straightforward too.

Basic considerations

Figure 1. illustrates the well-known evaluation formula for the holographic interferometry of displacements. A and B are the initial and final position of an object surface, P_A and P_B are the positions of an object surface point before and after loading, $\underline{L}$ is the displacement vector, $\underline{K}_1$ and $\underline{K}_2$ are illumination and observation directions. The $\Delta\Psi$ phase difference has the form of

$$\Delta\Psi = (\underline{K}_2 - \underline{K}_1)\,\underline{L}. \tag{1}$$

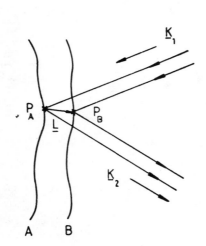

Figure 1. Derivation of fundamental equation

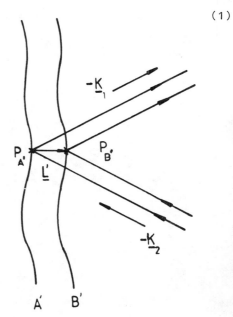

Figure 2. Line diagram for calculating the phase difference at $\underline{L}'$

In proper arrangement the object wavefronts may be holographically reconstructed to travel back separately as new illuminating beams and a second hologram can be taken from the direction of the original illuminating beam after changing the master object for the object to be compared. That is schematically illustrated on Figure 2 where A' and B' are the initial and final positions of an object to be compared with master one; $P_{A'}$ and $P_{B'}$ are the positions of an object surface point before and after the load works; $\underline{L}'$ is the displacement vector; $-\underline{K}_2$ and $-\underline{K}_1$ are reversed illumination and observation directions.

Using the master wavefronts for illumination in the original sequence, the $\Delta\varphi'$ phase difference corresponding to a displacement $\underline{L}'$ of surface point of the object to be compared will show directly the difference of the displacements $\underline{L}$ and $\underline{L}'$. That can be expressed as follows:

$$\Delta\varphi'_- = \left[(-\underline{K}_1) - (-\underline{K}_2)\right] \underline{L}' - \Delta\varphi = (\underline{K}_2 - \underline{K}_1)(\underline{L}' - \underline{L}) .$$ (2)

Using the illuminating master wavefronts in reversed sequence the sum of the displacements will be measured:

$$\Delta\varphi'_+ = \left[(-\underline{K}_1) - (-\underline{K}_2)\right] \underline{L}' + \Delta\varphi = (\underline{K}_2 - \underline{K}_1)(\underline{L}' - \underline{L}) .$$ (3)

The whole train of thought is based on the simple and not perfectly exact evaluation idea that the observable interference fringes are created only by the corresponding points of the object states (P_A and P_B on Figure 1 or $P_{A'}$ and $P_{B'}$ on Figure 2). This idea was developed for smooth illuminating wavefronts (planes, spheres, etc.). In our case where the illuminating wavefronts have a microstructure corresponding to the master object, the straightforward extension of the idea requires that the illuminating beams, too, should belong to corresponding points of the master object (e.g. P_A and P_B). These requirements for illumination and observation can not be met perfectly at the same time because of the difference of the displacements ($\underline{L}$ and $\underline{L}'$).

Otherwise the interference fringes of the master object displacement have to be localized on the surface of the object (if not everywhere) because the object to be compared acts like a screen even if an altering one.

Both requirements lead to the use of restricted illumination and observation angles and therefore enlarge the speckle size and limit the range of the measurable displacements.

Implementation forms

The main problem of the implementation is the creation of the holographic illuminating wavefronts. They have to be recorded in such a way that they could be reconstructed separately but exactly coincidentally. If they are reconstructed simultaneously they have to produce the same interferometric fringe system of the master object as if it were taken by a conventional double exposure hologram.

The conjugate object wavefronts can be reconstructed by the conjugate of the reference beam with the hologram in the original recording position. The simplest way is to choose a plane reference beam and reverse its direction by a plane mirror after passing through the hologram plate. The rough position of the mirror can be checked by the position of the reflected beam after passing through the collimator again[2]. It is important to note that to the perfect alignment the mirror is to be realigned after every hologram plate because the plates are usually wedge shaped with different angles.

The separate recording and reconstruction of the master object wavefronts can be achieved in three different arrangements(called Form I, II and III).

A form I type setup is shown in Figure 3 schematically without the beam expanding elements. The laser light coming from the right hand corner of Figure 3 is devided into two beams. The beam passing through the beamsplitter BS_1 illuminates the master object O_m. The beam reflected by beamsplitter BS_2 is the reference beam R_1. The reference beam R_2 is directed onto hologram plate by beamsplitter BS_3. The adjusting beam is formed by a reflection at beamsplitter BS_4. The beam passing through the beamsplitter BS_4 illuminates the hologram plate H_d as a reference beam during recording of holograms of object to be compared. Reversing mirrors are denoted by M.

The two states of the master object are recorded on the same plate (H_m) and with separate reference beams (R_1, R_2) coming from different directions. Each reference beam will reconstruct its object wave in the original direction and the other object wave in a different direction. To the fine joint alignment of the mirrors reversing the reference beams, an additional spherical wavefront (A) is recorded by both reference beams with their object waves simultaneously. Reconstructing it by both conjugate reference beams at the same time, its fringe free state will mark the correct alignment of the reversing mirrors. This fine joint alignment is very delicate and has to be repeated for every master hologram plate (H_m).

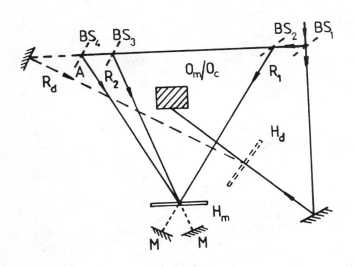

Figure 3. The experimental setup

The Form II type setup is the simplest of all because only one reference beam is used as in the conventional double exposure holography. Reversing the reference beam, both master object wavefronts are reconstructed and used in both illumination. The interference of these four wavefronts produces the required difference interferogram and other distrubing interferograms as well: the sum of the displacements and twice the actual displacement of the object to be compared. If the displacements to be compared are so big that their interferometric fringe system is too dense to observe it in itself, the difference interference fringe system will be the only visible one. In this setup the fine mirror alignment is not needed at all for the conjugate reference beam common for both master wavefronts.

The Form III type setup combines the advantages of the preceeding two implementation forms. It uses one reference beam inspite of its selective reconstruction capability. The master object wavefronts are recorded on the same hologram plate with the same reference beam but they are separated on their way toward the hologram plate by a beamsplitter or some movable optical element so that they arrive on the hologram plate from different directions. At the reconstructions only the appropriate way is open while the other is blocked.

Experimental results

The master object was the bottom of a pressure chamber. Between the two exposures the pressure was increased causing a bulging of the bottom and producing concentric interferometric fringe system. For the sake of simplicity the same chamber was used for the object to be compared as well but in the latter case the chamber position was changed a little to simulate the different microstructure.

The interferogram on Figure 4 was produced in a Form I type setup in several steps with different covering of the object to be compared under the exposures. The aim was to demonstrate the existence of the difference fringes undoubtedly. The upper half of the interfero-

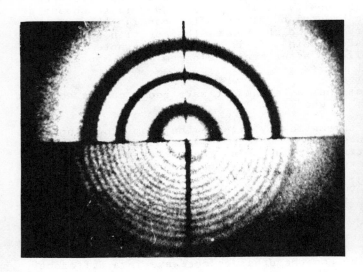

Figure 4. Interference patterns
on the bottom of a pressure chamber

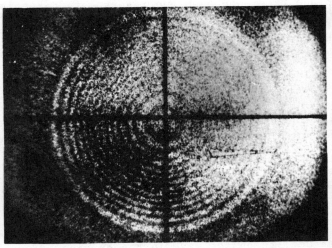

Figure 5. Difference interference pattern on the bottom

gram shows the actual displacement of the object to be compared and the left lower quarter shows the displacement of the master object while the right lower quarter shows their difference. Figure 5 shows only the difference fringe system of approximately 40 times bigger displacements as before so that their own interferometric fringes were already too dense to resolve them by the naked eye.

The visibility of the difference fringe systems is not too good which may be the consequence of the poor quality optical elements used in the arrangement. The interferograms produced in the Form II type setup were approximately of the same quality but in this case this may be caused rather by the disturbing effect of the additional wavefronts. In the Form III type setup only preliminary experiments could be done because of external difficulties. Nevertheless, the results obtained so far give hopes of really good difference interferograms.

Conclusions

Preliminary theoretical and experimental results were presented to prove the possibility of difference holographic interferometry. Further work is needed to determine its limits and accuracy and to extend its application to contouring and transparent object investigations.

References

1. Denby, D., Quintanilla, G.E., and Butters, J.N., Contouring by electronic speckle pattern interferometry, Proceeding of the Strathclyde Conference 1976 (Cambridge University Press), pp. 323-349.
2. Ost, J., and Storck, E., Techniques for generation and adjustment of reference and reconstruction waves in precision holography, Optics Technology, Vol. 1, pp. 251-254. 1969.

Applications of holography in automotive industry

Romuald Pawluczyk

Department of Physical Optics, Central Laboratory of Optics
Kamionkowska 18,   03-805 Warsaw, POLAND.

Abstract

This report provides a brief summary of works concerning the applications of optical holography in automative industry carried out in Central Laboratory of Optics in Warsaw, during the last ten years period. These works cover applications of holographic interferometry for nondestructive testing of engine valves, holographic investigations of pneumatic tires, double-pulse interferometry of driving gear, and holographic registration of oil mist.

Holography is widelly used in different areas of modern technique [1 - 4] and its applications in automotive industry are well known [5 - 7] . These investigations have a great influence on safety aspects of automative systems, so, it is clear that currently they arouse a great interest. As a result, some investigations in this area have been started out in Poland. Partially, these investigations have been carried on in Central Laboratory of Optics in cooperation with different research laboratories. Some results, obtained during last ten years period are presented in this report.

Holographic investigations of engine valves

The analysis of damage causes for different kinds of engines shows that sometimes the only reason is valve breaking. So, it is very valuable to get a method enabling nondestructive testing of these important parts of engines. Preliminary analysis shows that breakings of valves are mostly caused by microcracs and inside material inhomogenities, as well as changes of physical properties of material during engine exploitation. Detection of these failures is still an important problem. It seemed that this problem could be partially solved by means of holography. Double exposure holographic interferometry with mechanical loading of investigated valves has been selected as the investigation method.

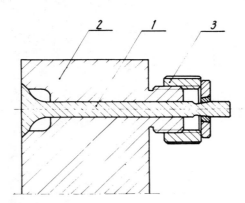

Figure 1. Holder of the holographed valve:
1 -valve, 2 -holder, 3 -tightening nut.

The investigated valves were clamped in special holder as it is shown in fig.1. As it can be seen, the valve head fits in a kind of valve cage and the valve end is tightened by means of a screw-type mechanism. The tightening force can be regulated with a dynamometric wrench. Changes of frontal surface of valve head as the tightening force is changed, are registered by holographic double exposure method. Some examples of obtained interferograms are shown in fig.2. The first of them shows a valve with a small crack near the valve head flange. Next two interferograms show two valves of the same type tightened with the same force at the first and second expositions of interferograms. The only difference is that the first valve was a new one, and the second had worked for some time in normal conditions in engine. In spite of identical conditions during interferogram exposition, deformation of the first valve is much smaller than of the second one. It is clear that the modulus of elascity of the second valve is smaller because of its working in high temperature. Investigations show, that the obtained interference pattern strongly depends on

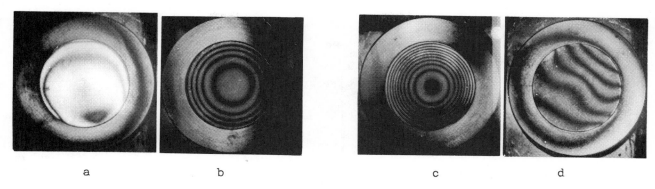

a   b   c   d

Figure 2. Holographic interferograms of the loaded valves.

contact between the flange of the valve and the edge of the cage /fig. 2d/. It seems, that the method can be useful for the control of fitting the valves and cages in real engines as well as the control of modulus of the elascity of the valves material.

Holographic tire analysis

Successful development of holographic method for tire testing[5,6] induced interest in this field in Poland. In result, holographic tire analyser was developed at the end of 1974. The general view of the analyser is presented in fig. 3a and open chamber with

a   b

Figure 3. Holographic tire analyser. a - general view, b - open chamber with tire.

tire in fig. 3b. Functionally it is similar with analysers produced by GCO Inc. in USA i.e. investigated tire and holographic system are located inside air-tight chamber with changable air pressure. Optical scheme of the analyser is shown in fig. 4. As it is shown, holographic camera with holographic plates and focus of illuminating beam are located almost on the axis of the tire, so optical path difference for light rays reaching different points of the tire is relatively small. On the other hand, an optical delay line enabling equalisation of the reference beam path-lenght and the mean path-lenght of the object beam, depending on diameter of the holographed tire, is located in reference beam. Thank to this, a coherence reguirement can be reduced and an ordinary He-Ne laser is applied as a source of coherent light. A glass wedge with one surface at the Brewster's angle to the incident laser beam is used as a beam splitter. As it is well known, reflectivity of such element depends on the polarization direction of the incident beam. The polarization direction can be changed with a half-wave plate what results in changes of the reference beam to the illuminating beam intensity ratio. In such a way this ratio can be optimized. As it was mentioned above, holograms are recorded on holographic plates. Some technical

245

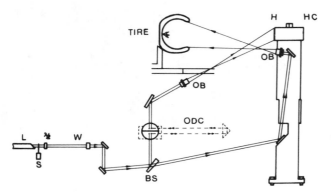

Figure 4. Optical scheme of the holographic tire analyser:L – He–Ne laser, S – shutter, W–window in the chamber wall, BS–Brewser's beam splitter, ODC–optical delay line, OB–objectives, H–hologram, HC–holographic camera

limitations cause that only one-fifth part of the tire perimeter can be holographed on one hologram: as a result, five holograms must be made for holographic inspection of the whole tire. To avoid a mistake, the number of holographed sector is automatically displyed. The air pressure inside the chamber and exposure time are also automatically controled and displayed. The holographed tire is placed on a rotating table and as one interferogram is made, the table with the tire is rotated by one-fifth part of a full turn. Some examples of obtained interferograms of tires with inside separations are shown in fig. 5.

 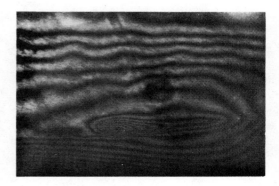

Figure 5. Holographic interferograms of tires with inside separations

The smallest separation which was detected had 7mm in diameter and was placed about 5mm under tire surface.

Holographic investigstions of selected parts of car

Vibration patterns of so complicated an object as a car are a result of complex inter-actions of resonant oscillations of particular parts and vibrations forced by working engine. Holographic interferograms of such an object are troublesome for analysis and, opposite to the method described in report[7]stimulation of vibrations by means of electro-mechanical shaker is applied. It makes possible the vibrations of every car part to be excited with any desirable frequency independent on working conditions of engine.
The resonant frequencies of investigated part of car had been preliminary electroni-cally measured. Next, the investigated part was excited with a shaker at a selected reso-nant frequency and at a fixed phase of vibrations a holographic interferogram was taken by a double-pulse ruby laser. Detailed description of the ruby laser system as well as holographic arrangement are presented in paper[8]. Optical scheme of the ruby laser system is shown in fig.6. As it is shown, the laser system consists of a double-pulse electro-optically Q-switched laser generator with transverse and longitudinal mode selectors and two laser amplifiers. The system provides two laser pulses lasting about 30 nsec each with a time interval between them regulated in range of 0.15-0.25 msec., and energy up to 1 J. Suitable electronic control system secures obtaining of laser pulses at proper phase of investigated vibrations. Time correlation of investigated vibrations with control elec-trical signals and light processes in laser system is shown in fig.7 where t_o is an ac-

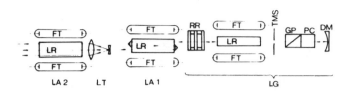

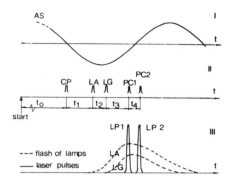

Figure 6. Optical scheme of double-pulse ruby laser for holographic investigations: LG-laser generator; IR-ruby rods, FT-flash tubes, PC-Pockels cell, GP-Glan polarizer, TMS-transverse mode selector, RR-resonant reflector, i.e. longitudinal mode selector, LAI and LA2 - laser amplifiers, LT -telescope.

Figure 7. Time sequences of phenomena in holographic and laser systems: I -analysed signal from piezoelectrical sensor; II-control pulses: CP-first control pulse, LA and LG -triggering pulses of amplifier and generator respectively, PCI and PC2 -control pulses of Pockels cell; III -light processes: LG -in generator, LA -in amplifier, LPI and LP2 -laser pulses.

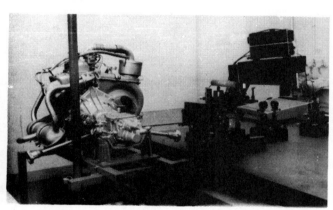

Figure 8. General view of holographic arrangement with driving gear

cidental time interval when the system, after external start pulse, is waiting until the sinusoidal signal from piezoelectical sensor reaches the given /increasing or decreasing/ edge of the designed value /usually zero point/. The time intervals t_2 and t_3 are fixed for optimal conditions of laser work, t_4-time interval between laser pulses /regulated in range of 0.15$_7$0.25 msec/, and t_1-time interval depending on need. The regulated time interval t_1 enables generating of laser pulses at desired time. The general view of the holographic arrangement /without the ruby laser system/ and the investigated object are shown in fig. 8. Some examples of obtained interferograms are presented in fig. 9. The first in-

Figure 9. Holographic interferograms of the driving gear

terferogram shows vibration pattern of original driving gear. The nodal area of vibration is clearly visible as a curved interference fringes in right part of the photo. The second interferogram shows the same part of driving gear after some mechanical modifications in body of the gear box. The same method was applied for vibrations pattern analysis in other parts of car. Obtained results made possible some construction changes in investigated model of car. Follwing inwestigations of redesigned car showed improvements in its vibration and noise characteristics.

Holographic registration of oil-mist

The lubrication method by oil-miast is presently often used in different mechanisms. Considering transporatation and deposition conditions, the optimal diameter of oil-mist drops is about 2 μm. Depending on the generation method their diameter may be enclosed in range of 0.1 to 30 μm. Optimization of oil-mist generation methods demands information about the real diameter of oil drops generated by particular lubrication system. At the same time method used for diameter measurement of oil drops must not disturb neither the oil-mist generation process nor the shape of oil drops. Holographic method may fulfil these requirements, but previously developed method have had unsatisfactorily resolving power. Theoretical analysis of holographic process shows that by using ruby laser light in aberration-free holographic system at appriopriate aperture is possible to obtain imaging of objects with diameter about 1 μm, whereas really obtained results are much worse. Predominantly this is caused by high speckle noise and optical aberrations of holographic system mostly induced by a bad quality of glass plates used as substrates for holographic plates.

The holographic system with side plane reference beam and convergent illumination beam has been used, as it is shown in fig.10. The convergent illuminating beam is focused

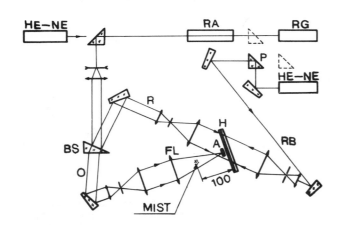

Figure 10. Optical scheme of the holographic arrangement used for holographic registration of oil-mist: RG-ruby laser generator, RA-laser amplifier, HE-NE-He-Ne laser for alignment, BS-Brewster's beam splitter, R-reference beam, O-object beam, RB-reconstructed beam, FL-focusing lense, H-hologram

into small absorbing cone placed close to the hologram surface, so only light diffracted on the oil drops reaches the hologram, whereas nondiffracted part of illuminating beam is absorbed. In such a way it is possible to apply very strong illumination and thereby a strong diffracted object beam with optimal object to reference beam intensity ratio, without anxiety that hologram may be overexposed by a strong nondiffracted part of object beam. Additionally both beams vere filtered by diamond spatial filters. Independently every holographic plate was interferometrically preselected using argon laser light with wavelenght equal to 488 nm, i.e. in spectral region where holographic plates used for ruby laser light, produced by Agfa-Gevaert have lov sensitivity. If an object is placed as near to the hologram as it is possible, high registration aperture can be obtained. As the plane reference beam is used, there are no troubles to obtain conjugate reconstruction beam, and whole the aperture can be utilized in the reconstruction process. When real-time holographic plate holder is used, reconstruction might be carried out in the same holographic system with conjugate reconstruction beam as it is shown in the fig.10. In an opposite case, hologram is fixed in a special 6-degree of freedom chuck enabling a precise hologram orientation with respect to reconstruction beam. This makes it is possible to obtain resolved images of oil drops with diameter almost as small as 1 μm, which are being produced by different lubrication systems.

In reported works besides the autor other investigators took part:
    -in part devoted to holographic investigations of car's parts vibrations:

Mr Zbigniew Kraska of Central Laboratory of Optics in Warsaw and Mr Jan Polański of Research and Development Centre of Automobile Industry in Bielsko Biała;
                    -in part devoted to holographic registration of oil-mist: Dr Bolesław Zachara of The Stanisław Staszic Technical University of Mining and Metalurgy in Cracow and Mr Zbigniew Kraska of Central Laboratory of Optics.

Acknowledgments

The author wishes to express his gratitude to Mrs Małgorzata Sochacka and Mr Zbigniew Kraska for their assistance in preparation of this report.

References

1. Holographic Nondestructive Testing, Erf, R. K., Ed., Academic Press, 1974.
2. Vest, Ch. M., Holographic Interferometry, John Wiley and Sons, 1979.
3. Ostrovskii Yu. I., Butusov, M., M., Ostrovskaya, G. V., Holographic Interferometry, Nauka, 1977, /in Russian/.
4. Holografia Optyczna, Pluta, M., Ed., PWN, 1980, /in Polish/..
5. Cannazzaro, M. J., Hill, F. W. Jr., Holographic Inspection of Tires, SAE Automotive Engineering Congress, Detroit, 1974.
6. Potts, G. R., Csora, T. T., Tire Vibration Studies - The State of the Art, Akron Rubber group Winter Meeting, 1974.
7. Felske, A., Happe, A., Double Pulsed Laser Holography as a Diagnostic Method in the Automotive Industry, The Engineering Uses of Coherent Optics, Robertson, E. R., Ed., Cambridge University Press, 1976.
8. Pawluczyk, R., Kraska, Z., Pawłowski, Z., Holographic Investigations of Skin Vibrations, Appl. Optics, Vol. 21, pp. 759-765, 1982.

INDUSTRIAL APPLICATIONS OF LASER TECHNOLOGY

Volume 398

Session 5

Laser Measurement Techniques I

Chairmen
A. F. Rashed
King Abdul Aziz University, Saudi Arabia
William F. Fagan
AOL-Dr. Schuster GmbH, Austria

Precision Measurement of Surface Form by Laser Autocollimation

A.E. Ennos and M.S. Virdee

National Physical Laboratory, Teddington, TW11 OLW, Great Britain

Abstract

The profile of an optical surface can be measured to nanometre accuracy by integrating slope values obtained by laser autocollimation. An instrument based on this principle for measuring X-ray microscope mirror surfaces of conicoid shape has been developed, and its application to monitoring precision lapping is described. A second system for the measurement of optical flats is also under development.

Introduction

Measurement of optical surface form is usually carried out by some type of interferometry. This yields a fringe contour pattern that subsequently has to be analysed to derive the deviation of the shape from spherical or flat. For high precision, subdivision of fringes is necessary, and this may involve polarization or heterodyne techniques that are complicated and expensive. An alternative method for deriving surface form, based upon the measurement of surface slope, is described in the paper. It is considerably simpler than optical interferometry to implement and has been shown to give very high accuracy, of the order of 1 nanometre, when measuring the profile shape along a line on the surface. In principle the method can be extended to determine the complete shape by combining profiles. It has the advantage of being non-contacting, gives a direct digital output of surface form, can be varied in its sensitivity and spatial resolution on the surface, and is readily automated. While limited so far to the measurement of near-cylindrical or flat surfaces that reflect specularly, it has potential application to the more general case of curved surfaces when very high precision is required.

Principle of Method

A laser beam is directed on to the surface at near normal incidence and the direction of the beam reflected back is accurately measured by autocollimation, using a position sensitive detector at the focus of a lens (Fig 1). If the beam is stepped along a line Ox on the surface while monitoring small changes θ in the reflected beam direction, the profile height h_X relative to the origin is obtained by integration

$$h_X = \int_0^X \frac{\theta}{2} \cdot dx \approx \sum_{n=0}^{n=\frac{X}{d}} \frac{\theta_n}{2} \cdot d \tag{1}$$

Integration by summation of slopes, as indicated by the second near equality, assumes that the reflection is substantially specular and the slope does not vary appreciably across a distance equal to the beam diameter. Thus the profile is 'smoothed' over this distance,

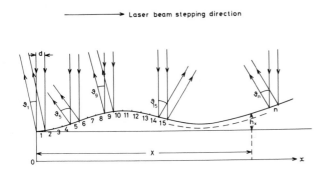

Fig. 1 Principle of surface profile measurement by laser autocollimation

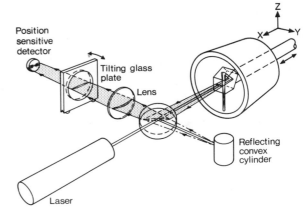

Fig. 2 Laser autocollimation system for measurement of shallow conicoids

and higher frequency undulations cannot be resolved. In practice the laser beam may be stepped along the surface by less than the beam diameter, giving a degree of redundancy when slope measurements are made at each point.

For high accuracy, two important requirements must be fulfilled:-

1) The direction of the laser beam emitted by the laser must remain constant throughout the measurement process.

2) The method of stepping the laser beam along the surface must not introduce variations in its direction.

The first requirement is met by the choice of laser and the method of mounting it. The laser beam direction may also be monitored continuously and appropriate corrections made. For the second requirement a suitable prism system giving constant beam deviation is employed.

Theoretical accuracy of measurement

The accuracy of profile measurement is related to the accuracy with which angular changes in the beam direction can be detected. If the laser beam has a diameter d, and it is focused on to a position sensitive detector by a lens of focal length f, the focused spot diameter σ is given by

$$\sigma \simeq \frac{\lambda}{f} \tag{2}$$

where λ is the wavelength of the light.

If α is the fractional unbalance between signals which can be detected by a given bipartite photocell, then the minimum spot displacement that can be detected will be $\delta\sigma$ where

$$\delta\sigma = \alpha\sigma \tag{3}$$

The smallest detectable angle $\delta\theta$ corresponding to this movement is

$$\delta\theta = \frac{\delta\sigma}{f} \tag{4}$$

so, combining (2), (3) and (4),

$$\delta\theta = \frac{\alpha\lambda}{d} \tag{5}$$

The smallest detectable profile height change, obtained by integration over a distance d, is thus

$$\delta h = \frac{\delta\theta}{2} \cdot d = \frac{\alpha\lambda}{2} \tag{6}$$

δh is therefore independent of the laser beam diameter and is a small fraction of the wavelength of light. For a bipartite type cell and low noise amplifier, α may typically be $\frac{1}{2500}$. Thus $\delta h \sim \frac{\lambda}{5000} \sim 0.1$ nanometre.

Since the measurement process involves integration, errors will accumulate at each step of the measurement, and the final uncertainty will be the resultant of the uncertainties at each stage, combined according to statistical probability. A similar consideration applies when measuring the flatness of surface tables by autocollimator or Talyvel[1]. The net result is that the overall uncertainty in profile shape increases with the number of steps of integration, a small diameter measuring probe giving a greater error than if a large one is used. For example, reducing the beam diameter by 10 times will give improved lateral resolution but is likely to increase the uncertainty of measurement by $\sqrt{10}$ times.

Uniaxial system for profile measurement of near-cylindrical surfaces

A laser autocollimation system was developed in the first instance for measuring the profile of X-ray microscope mirrors[2]. These are diamond-turned and lapped reflecting surfaces

Fig 3.   Laser profilometer with X-ray
         microscope mirror in place

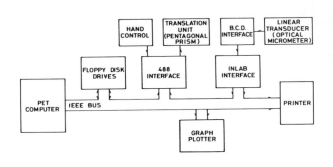

Fig 4.   Block diagram of control system
         for automatic profilometer

of axial symmetry, approximating to shallow cones.   The longitudinal surface generators
are, however, hyperbolae or ellipses, which are so far off axis that the maximum departure
from a straight line is only about half a micrometre in a distance of 15 mm.   The laser
autocollimator system developed for measuring the profiles of these mirrors is shown in
Fig 2.   The beam from a 2 mW helium-neon laser (Hughes model 3222 H-PC) is directed along
the axis of the mirror and reflected on to its surface at near-normal incidence by a pentagonal prism.   The beam reflected back by the quasi-conical surface is focused astigmati-
cally, spreading out into a slightly curved horizontal line on its return through the prism.
A beam-splitter reflects part of the return beam on to a bipartite photodetector (a UDT type
PIN SPOT 4/D quadrant detector with upper and lower adjacent quadrants electrically connected), positioned in the focal plane of a plano-convex lens.   The pentagonal prism, which is
held at the end of an arm mounted on a translation stage, is stepped along the mirror axis.
Any change in slope of the illuminated surface will then result in a vertical deflection of
the line of light focused on the detector.   An optical micrometer (a tiltable glass plate),
introduced between the lens and detector, is used to return the beam to the central position
by nulling the amplified detector signal.   The angle of this glass plate thus gives a sensitive measure of the direction of the reflected beam.   In order to monitor the directional
stability of the beam from the laser it can alternatively be reflected from a fixed convex
cylindrical surface, as shown by the broken lines.   This is effected by means of a shutter
which is operated at intervals throughout a measurement run.

The use of a pentagonal prism for beam tracking ensures that it is always deviated
through 90° in the XZ plane, independently of small prism translations or rotations due to
imperfect stage slideways.   If the prism is rotated slightly about the X or Z axes, the
deviated beam will track along a line on the mirror surface that is not quite straight, but
it will be reflected back along the original direction provided that the beam from the laser
is approximately on the mirror axis.

In order to overcome fluctuations due to atmospheric effects the laser and optical components are enclosed in separate box structures made from hardboard-foam plastic panels, and
the laser box has ventilation apertures to maintain a steady flow of air.   After initial
warm-up the direction of the beam has been found to fluctuate by only about 1 microradian
over a 10 minute period.   Plastic tubing surrounding exposed regions of laser beam within
the second box also reduces atmospheric effects.   Since very high stability of the components is required, these are designed carefully on kinematic principles, and all are mounted
on a stiff steel honeycomb table supported on air mounts.   A photograph of the system with
an X-ray mirror in place is shown in Fig 3.

Initial measurements were made by manual stepping of the prism along the mirror axis and
nulling the amplified signal from the photodetector by operating a micrometer tangent screw
to tilt the glass plate.   The system has subsequently been automated by 1) Provision of a
stepper-motor drive to the prism carriage, 2) Servo-controlling the null setting of the
optical micrometer;   instead of a manual micrometer, a cam wheel driven by the geared-down
servo motor bears on the tangent arm of the tilt plate, 3) Measurement of the tilt angle by
means of a moire grating-type linear transducer (by Heidenhain) bearing on the arm, 4) Computer control of the complete measurement cycle;   stepping of the prism, reading and storing
the optical micrometer output, calculating the profile, and printing out the data, with subsequent automatic plotting of the curves.   Fig 4 is a block diagram of the control system
based on a PET computer and Ealing micropositioning equipment.

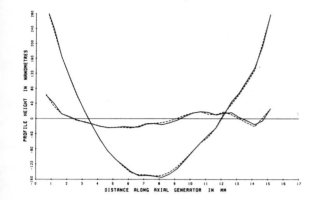

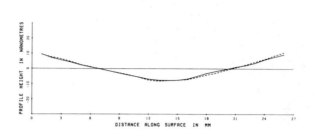

Fig 5. Profile and error curves of ellipsoidal surface recorded in two
opposite directions

Fig 6. Profile measurement across optical
'flat' recorded in two opposite
directions

Operation and performance of the system

The mirror under test is first adjusted in its mount so that the laser beam strikes the
mirror surface at normal incidence midway along the profile to be measured. Without focusing optics, the Gaussian beam cross-section is approximately 0.9 mm in diameter at the surface. The beam is then stepped along the mirror from one end by ½ mm increments, pausing
for 3 seconds to allow vibrations to die down before the servo-controlled optical micrometer
is automatically read. Readings are stored in the computer, which integrates the slope and
normalises the resulting profile height curve by a linear regression procedure. This
removes any variation in initial slope caused by errors in setting up. Since the profilometer was being used to check the perfection of figure of well-defined conicoid surfaces
manufactured by diamond turning, the computer was also programmed to calculate the difference between design and measured profile curves. Fig 5 shows both profile and error curves
for an ellipsoidal mirror. The dotted curves are repeat runs recorded with the mirror
turned end-for-end and re-aligned. The close agreement between curves indicates that slide-
way errors are having no appreciable effects since these would cause asymmetric differences
in profile. Repeatability of measurement is to within ± 2 nanometres, and this is limited
principally by mechanical instabilities introduced through air-borne vibration, and not by
electrical noise.

To check that the instrument was not introducing significant systematic errors, it was
modified to measure optically flat surfaces whose profile could be checked interferometrically (but not to such high accuracy). Since the reflected beam is not drawn out into a
line in this case, it is necessary to magnify it by 10 X before it falls on the detector, to
reduce power density. Greater precautions in alignment are also necessary, for reasons
described in reference (1). Fig 6 shows profiles of a near-flat surface measured by this
means in two opposite directions. Agreement between them is good and the figure of the
surface agrees with that predicted by Fizeau interferometry, which relies upon comparison
with a standard calibrated optical surface.

Application to monitoring of mirror polishing.

The laser autocollimation system has been used over the period of one year to monitor
successive stages in the polishing of X-ray microscope mirrors. The majority of the measurements were made before the system was fully automated, when the time to complete one
measurement run was approximately 12 minutes. With full automation this has been reduced
to 1½ minutes. Fig 7 shows a series of successive profile error curves recording the progress of correcting a mirror by localised lapping.

High resolution profilometry

The system as described averages the slope over a distance of 0.9 mm so that surface ripples
of a wavelength below this will not be detected. It is possible to increase the spatial
resolution at the expense of accuracy, by focusing the beam down to a smaller cross-section
using a positive lens interposed between the beam-splitter and the tracking prism. The
power of the lens must be kept low enough for the focused beam not to change size appreciably
as it is tracked across the mirror. For the X-ray mirror application the Gaussian beam was
focused down to 105 µm diameter, a value that increases by only 14% when defocused at the
extreme ends of the 15 mm long mirror. Since the reflected beam must not return through

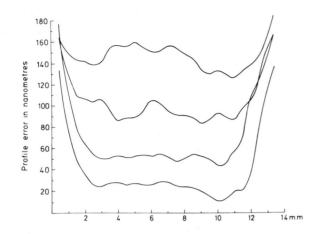

Fig 7.　Profile error curves for X-ray mirror after successive stages of polishing

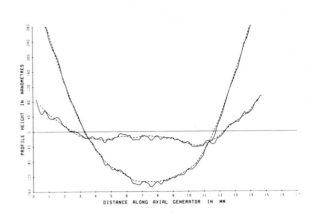

Fig 8.　Profile and error curves for hyperboloidal surface recorded with:
——— 0.105 mm dia. probe,
- - - - 0.9 mm dia. probe

the lens (otherwise it would constitute a 'catseye' reflector insensitive to mirror slope) the upper part of the lens was removed and the test surface slightly tilted to enable the reflected beam to pass clear of it.　Realignment of the detector system was then necessary.

Comparison measurements recorded for the X-ray hyperboloid mirror, at both high and low resolution, are shown in Fig 8.　While both profiles have the same general shape, the high resolution curve shows an undulation of the surface that is smoothed out when the larger probe is employed.　With five times the number of readings (the probe is not overlapped), the high resolution process is slower and more liable to suffer from statistical uncertainties, but it undoubtedly reveals small scale structure while measuring overall shape to comparable accuracy.

<u>Two-axis system for measurement of optical flats</u>

An instrument for laser profilometry in two dimensions is also under development for the precision measurement of optical flats.　Instead of moving a pentagonal prism to scan the surface, two hinged periscope prisms are employed to step the beam in raster fashion over it as shown in Fig 9.　Periscopes have the property that entrant and emergent rays remain parallel to one another independently of the orientation of the periscope.　In the instrument, two fused silica periscope prisms are held in arms mechanically joined by rotation bearings acting like shoulder and elbow joints.　The beam can thus be directed on to any point on a flat surface within a circle of radius equal to twice the periscope length. This arrangement has the advantage that the optical path through the prisms to the test surface remains constant during scanning, so that no change in the illuminated area occurs when tracking.

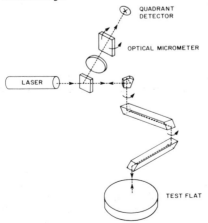

Fig 9.　Schematic of laser autocollimator for measurement of optical flats

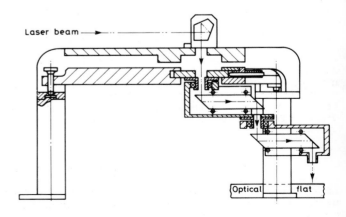

Fig 10.　Sectional drawing of double periscope system for tracking across optical flat

An outline engineering drawing of the scanning system holding the periscope prisms is shown in Fig 10.   It consists of a three-armed aluminium alloy gantry on which is mounted a pentagonal prism for directing the laser beam down the central axis.   The two rotatable arms are held by a subsidiary structure which is mounted within the gantry, with translation and tilt adjustments for alignment.   The optically worked prisms are also adjustable within their holders.   A combined radial and thrust ball race is employed in each joint, in order to minimise rotation of the prisms other than about a vertical axis.   The laser and auto-collimator section of the system, which is mounted on a vibration-isolated table, is not shown.   For measuring the slope variations in two coordinate directions a quadrant photo-cell is employed, together with a two-axis tilt plate to null the signals from the two quad-rant pairs.   Since the range of slope variation is likely to be small, it may be possible to use a direct reading linear position-sensitive detector instead of a null-setting optical micrometer.

Measuring surface slope in two directions over a complete optical flat will require many steps of the laser probe.   The number can be reduced, at the expense of resolution, by enlarging the scanning beam by a beam expander.   It is planned to use a diameter of 3 mm for the present instrument.   Preliminary measurements without beam expansion has checked the mechanical and optical practicability of the system, but further work is necessary to assess its performance critically.

Conclusions

Slope measurement by laser autocollimation has been shown to give very high precision in profile measurement of smooth surfaces.   The uncertainty of measurement on X-ray mirrors has been reduced to ± 2 nm over a 15 mm length, using a computer-controlled automatic system.   Theory indicates that this uncertainty could be further reduced by improvement of mechanical stability and local environment.   A system for flatness measurement over an extended area is also being developed.   The systems of measurement are digital and give direct print-out of data, a distinct advantage over optical methods involving interference fringe assessment.   The sensitivity of measuring the height variation, and the lateral resolution can be varied, and there exists the possibility of measuring rougher surfaces ie. engineering finishes, by reflecting the laser beam at a high angle of incidence.   Although instruments have so far been developed only for the measurement of nearly straight genera-tors, it should also be possible to extend the technique to steeper curved surfaces.

Acknowledgements

The authors are indebted to D W Robinson for help in computer programming and to P R Breakwell in the design of the periscope system.   H G Loe assisted in design of the servo system.

References

1.   El Sayed, A.A. and Hume, K.J.   "Assessment of the accuracy of flatness measurement", The Production Engineer, Vol. 53, pp. 395-401, 1974.
2.   Ennos, A.E. and Virdee, M.S.   "High accuracy profile measurement of quasi-conical mirror surfaces by laser autocollimation", Precision Engineering, Vol. 4, pp. 5-9, 1982.

Coherent optical measurement of surface roughness and its application in the paper industry

P.Richter, E.Lőrincz, I.Péczeli, F. Engard

Technical University, Budapest
1521 Budapest, Budafoki ut 8., Hungary

Abstract

A coherent optical method to measure the surface characteristics of rolling material e.g. paper on line without mechanical contact in industrial enviroument is presented. This method is based on the optical heterodyne detection of light scattered into a single radiation mode by the rough surface. The detection is followed by an appropriate signal processing that makes characterization of the surface smoothness by parameters called "correlation lengths" indicating the average size of uniform surface elements. These parameters show good correlation with other, widely used static surface roughness measurement methods. Furthermore they give the possibility of deeper insight into the surface structure of different types of paper.

Introduction

Rough surfaces may posess highly complicated structure with several scales of roughness present on the same surface. Methods to characterize these surfaces are usually integrating or statistical in their nature and use either mechanical contact or interaction with electromagnetic radiation /or both/.[1,2,3]

Mechanical methods are usually slow and in the case of soft surfaces they necesserily distort the surface under investigation.

The other class of methods - that may be called optical methods - usually requires taking samples /static methods/ and observe the structure under magnification.

Dynamic optical measurement of surface structure in the presence of high ambient light level by taking data from a large area of the surface of rolling material, giving the possibility of various evaluations of the data received are the purpose of the method and experiment described in this paper.

The method that we call 3S method /Statistical Surface Structure/ is based on the measurement of backscattered light from the surface by the optical heterodyne detection technique and on statistical evaluation of the signal of the detector. The actual experiments were carried out on paper surfaces of different types and the results were compared with widely used types of measurements.

Theory

Distinction of rough and smooth surfaces by optical means always depends on the wavelength λ used and there is a continuous transition between rough and smooth surfaces[4]

By Rayleigh the surface can be considered smooth if the distortion $\Delta\phi$ of the phase front of the wave scattered from the surface relative to the incident wavefront is much smaller then 2π and it is rough if $\Delta\phi >> 2\pi$

It can be easily shown that

$$\Delta\phi = 2\pi \frac{h\cos\theta_i}{\lambda/2} \tag{1}$$

where θ_i is the angle of incidence of the wave and h is the variation of surface height

Fig 1. shows schematically how the surface influences the propagation of the incident wave.

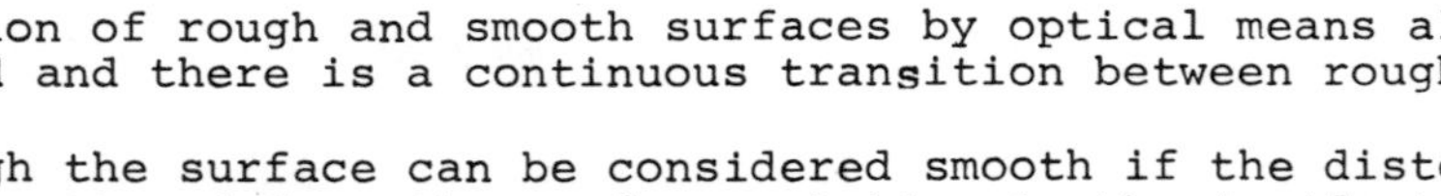

smooth surface	intermediate surface	rough surface
reflection only	reflection and scattering	scattering only

Fig 1. Transition from reflecting to scattering surface

In the following the possibility of gaining information from the surface characteristics in the case of rough and intermediate surfaces will be considered.

It is well known that in the case of rough surfaces the amplitude a of the scattered wave, that can be considered as the sum of components from random elements of the surface, is a random variable with Rayleigh probability distribution

$$p(A)=\frac{A}{\sigma^2}\exp(-\frac{A^2}{2\sigma^2}) \qquad 2\sigma^2=<A^2>$$ (2)

A much wider class of surfaces belongs to the intermediate case where the amplitude A of the scattered wave outside a narrow cone about the direction of specular reflection is also a Rayleigh distributed random variable.

In the general case when A is a random variable and I=R.A is also random variable where R is a wave with constant amplitude and phase. Furthermore any part of the complete scattered wave and I selected by geometrical constrains are random variables with the same distribution as of A. Such geometrical selection is carried out for instance by using the optical heterodyne technique[5] where the quantity I is measured, and scattering into a radiation mode determined by the reference wave R is selected as shown schematically in Fig 2.

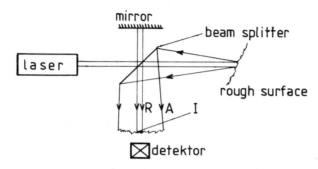

Fig 2. Detection of backscattered light by optical heterodyne technique.

In the following the only assumption made is that the quantity I is a random variable with probability distribution P/I/. P/I/ can be realized by measuring the distribution of I for different surface elements.

$$P_o=\int_{I_o}^{\infty} P(I)dI$$ (3)

denotes the probability of $I>I_o$ in these measurements.

Considering two independent surface elements B and C the probability of $I>I_o$ from both elements is

$$P_o(B,C)=P_o(B)P_o(C)$$ (4)

If a large surface area contains n independent elements then the probability of $I>I_o$ for the whole area is $(P_o)^n$.

The basic idea of the rough surface parameter measurement presented here is to find the size of independent elements and characterize the surfaces by them.

If a surface is illuminated with a light spot that moves along the surface the probability that $I>I_o$ holds for a length X is

$$P_o(X)=(P_o)^{\frac{x}{\Delta x}}$$ (5)

if the surface consists of independent elements with linear dimension ΔX (i.e. $n=\frac{x}{\Delta x}$)

ΔX is called correlation length. For smooth surfaces ΔX is large for rough ones ΔX is small. $P_o(X)$- s are shown in Fig 3.

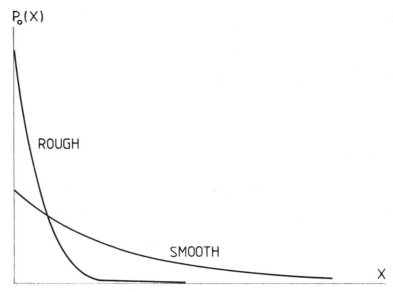

Fig 3. The probability distribution $P_O(X)$

The conditions among which the independency or correlation of the surface elements are considered have still to be clarified. This can be done by introducing the parameter $\Delta X_{min}=$ resolution. If $I>I_O$ fails over a length $X<\Delta X_{min}$ then the element is still considered to be correlated and the beginning of a new independent element occurs only if $I<I_O$ holds for $X>\Delta X_{min}$. Therefore no element size smaller then the resolution can be measured and the correlation lengths measured always depend on the resolution used in the measurement.

The surfaces measured may have highly complex structure that can not be characterized by a single correlation length. In this case

$$P_O(X)=\sum_i Q(P_O)^{\frac{X}{\Delta X_i}} \qquad (6)$$

gives the probability function where Q_i are the weights corresponding to each ΔX_i.

An easily evaluable method for the comparison of surfaces seems to take moments of $P_O/X/$ e.g. the first moment $\bar{X}$

$$\bar{X}=\frac{\int XP_O(X)dX}{\int P_O(X)dX} \qquad (7)$$

that gives the average size of the independent elements.

For the Rayleingh distribution, valid in most practical cases

$$P_O=\alpha\exp(-\frac{I_O^2}{<I^2>})$$

$$\qquad (8)$$

$$P_O(X)=\beta\exp(-\frac{I_O^2}{<I^2>}\frac{X}{\Delta X})$$

where $<I^2>$ is the avarage value of I^2, α and β are normalization factors.

The quantities described above characterize the macroroughness of surfaces that seems to have imprtance in various practical applications.

In the following the technique of measuring $P_O/X/$, Q_i and ΔX_i and $\bar{X}$ will be described and experimental results on paper surface samples will be presented.

Experiment

The experimental set-up is shown in Fig 4.

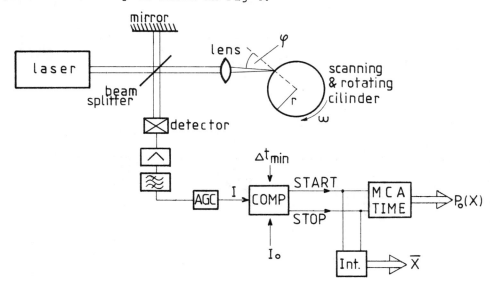

Fig 4. The experimental setup

The light source is a commercial 5 mW He-Ne laser thats d=2 mm diameter beam is split into a direct reference wave R projected directly onto the surface of the detector and another part that is focused onto the moving surface that is investigated. Paper samples are measured in such a way that they are fastened to a rotating cylinder with angular velocity ω that is simultaneously scanning along its rotation axes therefore an area of about 5o cm^2 takes part in the measurements. This set-up is a model for rolling material in technological processes.

The focal length of the lens is f=1,5 cm that gives a

$$D=1,22 \frac{\lambda \cdot f}{d} = 6.1o^{-3} \text{ mm diameter diffraction limited focal spot.}$$

The direction of the focused beam is at an angle to the normal of the surface of the cylinder therefore at the focal point the surface posesses a velocity component $v_{\parallel}$ parallel to the light beam.

$$v_{\parallel} = r\omega\sin\varphi \tag{9}$$

therefore due to the Doppler effect the backscattered light collected by the same lens is shifted in frequency by Δf

$$\Delta f = \frac{2v_{\parallel}}{\lambda} \qquad (\sim 2,5 \text{ MHz}) \tag{1o}$$

The backscattered light is also projected onto the surface of the detector where that part of it - A - that is spatially coherent with the reference beam R produces by interference a time varying light intensity

$$I = A \cdot R\cos 2\pi\Delta ft \tag{11}$$

In this way if the signal of the detector is fed into an electic filter tuned to frequency Δf the electronic signal will be proportional to I at the output /for simplicity it is denoted I in the following/. This set-up - using the optical heterodyne detection principle - automatically carries out the geometrical selection mentioned in the previous section, because only that part of the scattered light interferes with reference beam R that belongs to the same radiation mode.

With special self-aligning arrangements the necessity of interferometric precision alignment can be avoided. The advantages of the heterodyne detection are: its excellent

261

selectivity relative to background light /the component of the background light radiated into the radiation mode selected by the reference beam is negligable/, the good signal to noise ratio because of the high frequeny nature of the signal and the amplification due to measuring I=R.A instead of A^2 /R≫A/ and the possibility it provides for remote measurement /the apparatus - except the lens can be removed several tens of meters from the spot of the measurement/.

Information is contained in the modulation of the Δf frequency signal therefore the bandwidth B of the filter has to be carefully selected in order not to loose information /B≈1 MHz/. The form of the signal is shown in Fig 5.

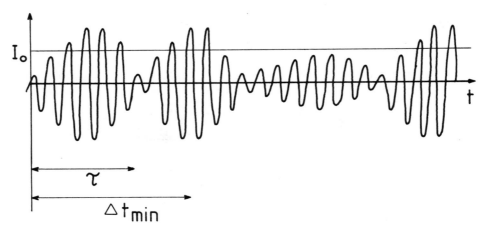

Fig 5. The heterodyne signal from backscattered light.

The principle modulation contains wave packets of approximately equal lengths $\tau \approx D/r\omega$ that is the time needed to move the surface the smallest correlated length inherent in this measurement - the diameter of the focal spot. This is the effect of laser speckle noise: The amplitude of the backscattered component that interferes with R is the sum of wavelets with random phase because of the random nature of the surface element illuminated at the focal spot. This gives the lower limit of the resolution of the measurement, because the size of the smallest element to be considered independent can not be smaller the D /i.e. the resolution can not be better then determined by the resolving power of the lens./.

After careful examination one can observe that the maxima of these wave packets are also modulated - and this longer period modulation contains information on surface structure.

As shown in Fig 4. the output of the filter is fed into an AGC circuit that removes the difference in reflectance between surfaces to be compared and then to a comparator that compares the incoming signal to an externally adjustable level I_o, usually set at $I_o = \langle I^2 \rangle^{1/2}$

It also posesses an option for setting resolution Δt_{min} to pre-determined value. The output of the comparator is a START pulse when I raises above I_o and a STOP pulse when I falls below I_o and stays below for a time period longer then Δt_{min}. $\Delta t_{min} > \tau$ is chosen so the speckle modulation is eliminated . By

$$\Delta X_{min} = r . \omega \Delta t_{min} \tag{12}$$

it determines the spatial resolution of the measurement.

The START and STOP pulses are fed into a multichannel analyzer measuring the frequency of the time periods between the START and STOP pulses i.e. the size distribution of independent surface elements P_o /X/.

The P_o /X/ curves shown in Fig 3. are of actual measurements on two paper samples .

On the basis of eqs /6/ and /8/ the measured curves are decomposed by a fitting programme using a microcomputer and the correlation lengths ΔX_i and their wights Ω_i are received.

Another possibility is to feed the START and STOP pulses directly into an integrating unit that provides on line the first moment $\bar{X}$ of the probality function.

<u>Results</u>

Measurements were carried out on 15 different paper samples. The surfaces of these samples had been tested by static methods - the Bendtsen and the FOGRA-KAM numbers were given. /The Bendtsen method measures air leakage through a knife edge and the paper surface pressed against it, the FOGRA measures optically the percentage of a plain glass surface that gets into perfect contact with the surface of the paper pressed against it./ These two measurements do not give correlating results and the FOGRA method seems to have a better correlation with the printability of the paper.

Table 1. shows these data and the results of our measurements. It is quite clear that there is a good correlation between the results received with our 3S method and the FOGRA method numbers. Fig 6. shows the relationship between the FOGRA KAM numbers and the first moments $\bar{X}$ at two different ΔX_{min} resolutions.

It is very interesting to see that the linear connection occurs at $\Delta X_{min} = 74$ /um.

Table 1.

Sample	Description	Grammage	Bendtsen ml/min	FOGRA KAM % 25kp/cm^2	FOGRA KAM % 5o kp/cm^2	FOGRA KAM % 75 kp/cm^2	3S $\bar{X}$/ um/ $\Delta X_{min}=74$ /um	3S $\bar{X}$/ um/ 133 /um
	Uncoated							
1	Newsprint	47,o	1223	8,4	17,o	23,7	296	583
2	"	44,3	111	12,2	2o,8	27,7	3o8	995
3	"	43,9	1oo	14,6	24,6	31,5	336	1322
4	"	5o,5	214	12,o	2o,2	26,8	3o3	1141
5	"	4o,o	187	12,6	21,5	28,7	324	1194
6	"	58,5	192	14,9	25,1	32,4	334	1878
7	"	5o,4	171	12,6	21,3	27,7	337	1o86
8	Supercalandered paper	59,2	24	22,9	36,1	44,9	659	97o1
	Coated							
9	Gravure paper	7o,o	23	32,8	58,7	71,4	657	5384
1o	"	67,1	15	29,7	55,2	68,8	588	11963
11	"	46,5	29	32,7	56,8	68,8	588	17824
12	Offset paper	5o,3	14	27,o	46,3	57,5	498	6653
13	"	67,4	18	16,o	33,6	46,6	438	34oo
14	"	7o,7	26	17,5	34,7	47,6	448	3674
15	Dull-finished paper	97,4	32	7,7	14,7	2o,2	265	452
16	Offset paper	11o,1	34	18,2	36,o	48,6	474	3914

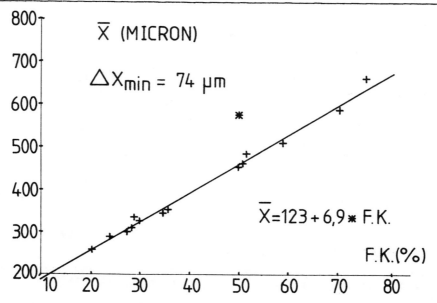

$\bar{X}$ (MICRON)

$\triangle X_{min} = 74 \ \mu m$

$\bar{X} = 123 + 6,9 * F.K.$

F.K.(%)

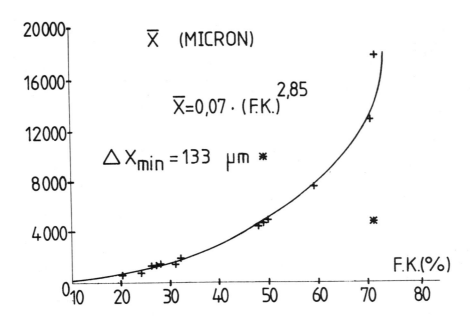

Fig 6. Correlation between the FOGRA-KAM numbers and the first moment $\bar{X}$ of the probability distribution for two values of resolution $/\Delta X_{min}/$

i.e. the entirely different FOGRA method seems to be sensitive to the surface structure on the scale of 75 μm. At this resolution the FOGRA numbers at different pressures all show a linear relationship to $\bar{X}$ /Fig 7./

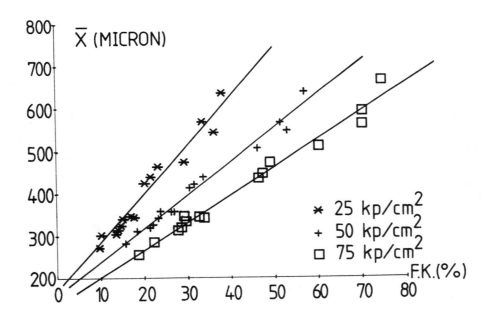

Fig 7. Correlation between the FOGRA-KAM numbers at different pressures and the first moment $\bar{X}$ of the distribution.

On the basis of the results shown in Fig 7. using an appropriate calibration the FOGRA-KAM numbers can be received by a remote non-contact measurement on rolling material.

Using the parameters ΔX_i and Q_i can give a deeper insight into the structure of the surface. Table 2. shows these data for a basis paper, the same sample with one layer of coating and two layers of coating.

Table 2.

Sample	$\bar{X}$ /um	ΔX_1 /um	Q_1	ΔX_2 /um	Q_2
Basis paper	17o	17o	1	-	-
Basis paper + 1 layer	169	153	o,97	7oo	o,o3
Basis paper + 2 layer	228	177	o,94	1o26	o,o6

The effect of coating is the appearance of a longer correlation length, while the structure of the basis paper is gradually disappearing applying more coating.

Conclusion

A measurement technique and apparatus has been described that makes high speed non contact measurement of structure statistics of rough surfaces by observing backscattered light from the surface possible. Using the optical heterodyne detection technique the measurement can be carried out with low power laser in the presence of high ambient light level from a distance that makes the 3S technique applicable in industrial environment as well. The reduced data /$\bar{X}$/ from the information received by the method provides characteric numbers of the surface on line that have been received by static procedures before; in the case of paper surfaces FOGRA-KAM numbers that are in good correlation with the printability of the paper. /It is worth to note that by applying pressure using a transparent cylinder e.g. glass, the surface structure at the compressed area can be examined./ Furthermore a more detailed processing of the data can give more information on the surface namely on the size distribution of correlated areas that may have significance in practice during carrying out technological steps like applying coating to the surface. 3S technique can be therefore used for technology experiments in the laboratory just as well as for monitoring and process controlling on the production line.

Acknowledgements

The authors wish to thank Dr Lars Nordman of Central Laboratory of the Finnish Pulp and Paper Industry and Dr S. Annus of the Research Institute of the Hungarian Paper Industry for helpful discussions and providing samples.

References

1. Chapman, S.M., "The Measurement of Printing Smoothness", Pulp and Paper Magazine of Canada Vol 48, pp 14o-15o, 1947

2. Ginman R., Makkonen T., Nordman L., "Profile Measurements on Printing Papers", XII. IARIGAI Conf., Versailles 1973 pp 1-28

3. Kapoor, S.G., Wu, S.M., Pandit, S.M., "A New Method for Evaluating the Printing Smoothness of Coated Papers", TAPPI 61 /6/, pp 71-74, 1978

4. Ishimaru, A., Wave Propagation and Scattering in Random Media, Academic Press, 1978

5. Kingston, R.H., Detection of Optical and Infrared Radiation, Springer Verlag, 1978

Optical Profilometer for Monitoring Surface Contours of Si Power Devices

H.P. Kleinknecht and H. Meier

Laboratories RCA Ltd., Badenerstrasse 569, CH-8048 Zürich, Switzerland

Abstract

The fabrication of power transistors and thyristors involves the etching of deep grooves (30-100 μm) into the Si wafers. These grooves separate electrically the individual devices from one another, and they are the site where the high collector field meets the surface and where the passivating glass and oxide layers have to be applied. Therefore, the groove depth has to be monitored and controlled in manufacturing to ±5%. Since the dimensions are large compared to the wavelength of light, and since the surfaces can be rough (etch pits) the usual interference techniques can not be used. The light section microscope technique, used up to now, is slow, inaccurate and operator dependent.

We describe an optical profilometer essentially consisting of a high-power microscope equipped with a laser attachment which automatically steers and holds the microscope focussed to the sample surface. The automatic focussing uses the beam of a He-Ne laser, which is coupled into the microscope by beam splitters and goes through the objective lens to form a fine spot (0.5 μm) on the sample surface. From there it is reflected back through the objective and projected into a focus outside the microscope. The position of this focus is a function of the sample surface height, and it is sensed by an arrangement of apertures and photodiodes, which drive a feedback servo-motor for microscope focussing. An electronic depth gauge (LVDT) resting on the shoulder of the microscope measures the vertical excursions of the microscope, which follow the sample contour.

This instrument is fast, accurate (± 1 μm) with a range of well over 100 μm and can be connected directly to a data terminal.

1. Introduction and Objective

There are numerous cases in research, development and manufacturing, where one has to measure and monitor the surface contours of polished, etched or machined parts. Correspondingly, the number of testing methods which have been developed for doing this is very large[1]. Among all of these techniques the instruments using a mechanical stylus are the most direct ones[2]. However, in many instances the possibility of damage to the sample by the stylus, the time requirement for such measurement and/or the lateral resolution limit given by the stylus radius can not be tolerated. In these cases touchless optical profilometers are required.

The objective of the work reported here is the development of an optical profilometer for measuring silicon wafers in power device manufacturing. In particular, the essential dimensions of the surface contour of Si mesa transistor wafers and thyristors (Figure 1) have to be monitored and controlled in the mesa etch operation. This control is critial, because the depth of the moat surrounding the mesas affects the surface shape along the line, where the high field region of the collector junction meets the surface. Therefore, this is the location where the passivating glass-SIPOS-oxide combination has to be effective for obtaining high-voltage capability of the device.

As indicated in Fig. 1, the moat depths range from 40 to 100 μm and have to be measured to ±1 μm, and the bottoms of the moats are rough, covered with etch pits and are not planar at all. Hence, optical interference techniques[1,3] are not applicable. On the other hand, interference accuracies are not required for that particular application.

Up to now the measurement of the moat depths is done in the factory with a light section microscope[4], which, however, is slow, subjective and not very accurate.

2. Approach

Our approach for an optical profilometer for the above application essentially follows the schemes indicated in references 5 to 9. It is outlined in Figure 2. The beam of a He-Ne laser (λ = 0.6328 μm) is brought to a focus at "1" in Fig. 2, is coupled into a high-power optical microscope by means of a beam splitter cube and projected on to the sample surface at "2" by the objective lens L_1. The reflected laser light goes back through L_1 and via a second beam splitter to a focus at "3". The position of focus "3" along the beam is strongly dependent on the surface height of the sample, i.e. on the distance between the sample surface and the objective lens. For small changes in sample height, z,

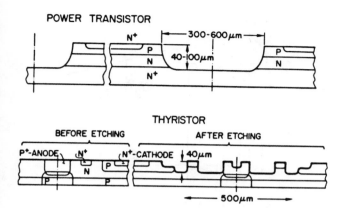

Figure 1.   Cross-section of Silicon Power
            Device Wafers

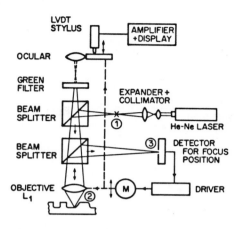

Figure 2.   Principle of the
            Optical Profilometer

and for an objective magnification, M, one gets a change of the position of focus "3":

$$z_3 = M^2 2 z \qquad (1)$$

We use objectives with magnification between M = 64 and M = 25.6, which for a sample height change of z = 1 μm give focus changes, z_3, of 8 mm and 1.3 mm, respectively. The box in Fig. 2 called "Detector for Focus Position", which is an important part of our design, will be described in detail in the next section. It gives a d.c. signal, roughly proportional to z_3, which in a feed-back loop by means of a d.c. motor corrects the vertical microscope position such that the focus "2" is always kept at the sample surface. The vertical micro-scope movement, as it follows the sample surface contour, is picked up by the stylus of an electronic depth gauge, a linearly variable differential transducer (LVDT), which rests on the shoulder of the microscope.

The use of a complete microscope enables the operator to select the desired measurement position on the wafer and to observe the laser spot on the sample during the test. For this purpose we use regular bright-field illumination with white light (not shown in Fig. 2) and observe through a green-filter (Fig. 2) which allows most of the white light to pass while it blocks most of the red laser light, such that the laser spot on the sample is visible but far too weak to do eye damage.

3.   Detection of Focus Position

For this we have chosen a technique which uses two apertures and two sensors, one some distance in front of and one behind the focus to be measured[6,7]. This is shown in Figure 3. One uses a third beam splitter which deflects half of the intensity towards the closer sensor 1, while the other half goes straight towards the more distant sensor 2.   The aper-tures covering the sensors are symmetrically located in longitudinal position around the zero position of the focus. The unfolded beam and the two apertures for S1 and S2 are sketched in Figure 4a. In zero position both sensors receive the same fraction of the laser intensity, and the difference amplifier will give a zero signal. A forward or back-ward deviation of the focus "3" (z_3 in Fig. 4a) will favour the light falling on sensor 2 or sensor 1, giving a positive or negative signal, P_2-P_1, respectively, which can be used to drive the microscope up or down for bringing the focus back to zero.

The above described focus position detector works in principle with circular pin hole apertures at S1 and S2 [6,8]. However, a problem is encountered when the sample surface is tilted, as indicated in Figure 5a. Under these circumstances the reflected beam may partly or totally miss the apertures giving a signal which is too low for measurement. And, this can also cause erroneous readings. Fortunately, the Si power device wafers considered as measurement objects in our case have a surface tilt which is predominantly oriented in a plane perpendicular to the primary direction of the moat or the mesa edge. This is shown in Figure 5b, where the moat runs horizontally and tilt is in the vertical plane. This kind of tilt can be accommodated to a large degree by using vertical slit apertures instead of pin holes. In practice it is always possible to rotate the wafer under test such that the moat to be measured lines up as in Figure 5b.

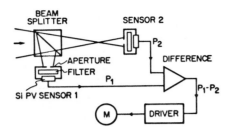

Figure 3. Detector for Focus
           Position

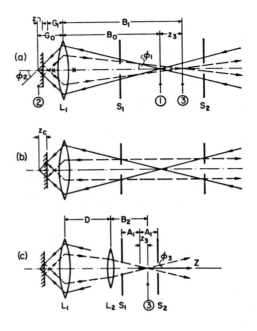

Figure 4. Unfolded Beams of
           the Profilometer

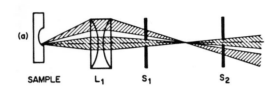

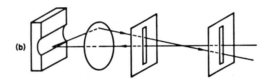

Figure 5. The Tilt Problem:
           Slit Apertures

4. Design Analysis and Optimization of the Optical System

An important problem is the range of the sample surface height, z, within which the system can respond properly. In particular, if z reaches a certain limit, the focus "3" goes to infinity (see Figure 4b), and beyond that the beam becomes divergent leading to $P_2-P_1 < 0$ (for $z > 0$) causing run-away. A calculation shows that for our microscope tubus length of 256 mm and for our highest magnification of 64x one gets a range of 31 μm, which is not satisfactory for our application. As a remedy for this we have added a weak additional convex lens, L_2, with focal length F_2 = 150 mm in front of the third beam splitter (see Fig. 4c and 8). With the insertion of the extra lens L_2 one now gets a range of 440 μm, which is more than enough for our application. In order to compute (and optimize) the difference signal P_2-P_1 as a function of z, we have to calculate the fraction of the laser beam going through the two slit apertures. This was done on a PET computer, and the function $P_2-P_1 = f(z)$ was plotted for various parameters. The optimization of these plots together with some experiments has resulted in the choice of parameters listed in Table I. Figure 6 gives in dashed lines the plots of P_2-P_1 versus z, as calculated for the parameters in Table I.

This function is the response of the focus detector. It is always positive for negative excursions z and negative for positive z, as required for feed-back control. Near z = o the curves are very steep, more so for higher magnification than for lower one. And for very large z of either polarity, the signal becomes low which limits the range in particular for high magnification. But there is no run-away.

The solid lines in Figure 6 are experimental plots corresponding to the theoretical ones. They were obtained by connecting the analog output of the LVDT to the x-input of a recorder and connecting the output of the difference amplifier to the y-input. The servo-loop was interrupted and the fine z-motion of the microscope was operated by hand. The theoretical

Table 1

Optical Tubus Length	B_o	= 256 mm
Magnification (Objective & Tube)	M	= 25.6-64
Focal Length of L_2	F_2	= 150 mm
Distance L_2 to L_1	D	= 135 mm
Distance S_1 to "3" = "3" to S_2	A_1	= 39 mm
Halfwidth of Slits	A	= 0.25 mm
Convergence of Input Beam at "1"	$\emptyset_1$	= 0.013 rad

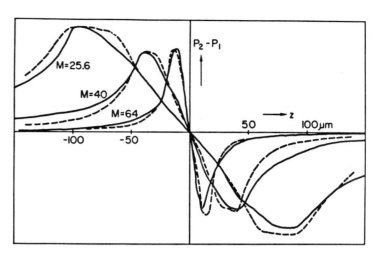

Figure 6. Response Function of the Focus Detector: Experiment (solid lines) and Theory (dotted lines)

curves were fitted to the experimental ones by choosing suitable amplitude factors. We have also inserted a factor (of 1.23) in the z-axis (the same for all three magnifications) to obtain a good fit. The agreement in curve shape and in the behaviour of the three objectives is seen to be good.

5. The Complete System

Figure 7 is a block diagram of the whole test set. The laser is a 2 mW He-Ne (λ = 0.6328 μm). We use a Zeiss Epi microscope with a d.c. motor, M, attached to the fine focus control via a friction clutch.

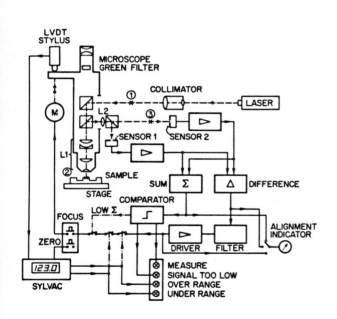

Figure 7. Block Diagram of the Complete System

The sensors 1 and 2 are enclosed in light tight boxes with the slit apertures, and line band-pass filters (Mells Griot 03 FIL 037) are inserted between slit and sensor. In this way no white light from the microscope illumination or room light can reach the sensors, while 75% of the laser light gets through. The sensor assemblies are mounted on little stages with the possibility of adjustment in longitudinal direction and laterally perpendicular to the slit direction. The whole laser light paths from the laser to the microscope and back to the sensors are enclosed in a light tight compartment in order to fulfill the laser safety requirements.

The light sensors are photovoltaic Si cells of 5 mm diameter. Their signals feed into d.c. amplifiers. The difference between these two signals is further amplified by a driver stage which operates the servo-motor, M.

The servo-loop is normally open, and it is closed only when the operator pushes the "Focus" button. This prevents time consuming run-away. Run-away is further avoided by a circuit which interrupts the loop for low signal. This can happen on a steep mesa slope or if the sample surface is too far away. For this purpose, the sum of the two sensor signals is going to a comparator which operates a relay switch interrupting the feedback loop for low signal. The comparator also controls the indicator lamps "Measure" and "Signal too low".

The vertical excursions, z, of the microscope are measured with a commercial LVDT Model SYLVAC D25 consisting of a stylus resting on the shoulder of the microscope and a digital LED display reading to tenths of microns. This display can be set to zero by operating the "zero" push-button, and the LVDT from then on reads the difference with respect to that z-value. The Sylvac digital unit also gives special signals when the reading is above a preset upper limit or below a preset lower limit. These signals are again used to interrupt the servo-loop in order to prevent the microscope excursions to go beyond safe limits and also to trigger the indicator lights "Over Range" and "Under Range".

Figure 8 is a photograph of the complete test set.

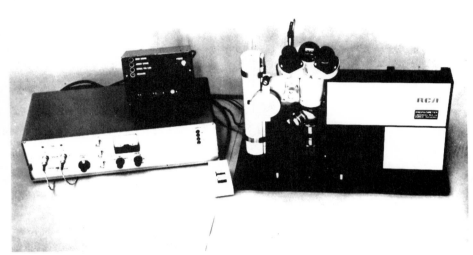

Figure 8.   The Optical Profilometer

6.   Quantitative Tests

Figure 9 shows a recorder trace of a scan across a sample consisting of three gauge blocks (100 μm, 75 μm and 50 μm thick) glued with nail polish to a flat Si wafer. The vertical coordinate of the trace comes from the analog output of the LVDT and the horizontal is the time as the sample stage was driven with a synchronous motor. As can be seen by comparison with the vertical scale taken directly from the LVDT display, the traces check very well with the nominal values of the gauge blocks, if we subtract about 1 μm for the thickness of the glue. The circular points are transferred from a Dektak trace taken of the same sample.

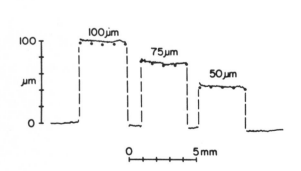

Figure 9.   Scan Across Three Gauge
            Blocks

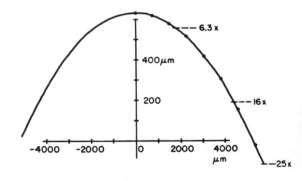

Figure 10.   Scan across a Cylindrical
             Lens

Figure 10 is a trace taken in the same way across the surface of a cylindrical lens which had been sputter coated with aluminum. The trace is an ellips, not a circle, because the x- and y-scales of our plot differ. The points on the ellips are calculated assuming a perfectly circular cylinder with a radius (obtained by fitting) of 22.04 mm.

Other traces of the same kind for three objectives and a Dektak trace are given in Figure 11 and 12. The curves are vertically displaced from one another for clarity. Figure 11 is a trace of a Si wafer into which several steps have been cut by multiple passes with a diamond saw followed by heavy etching. The surfaces are still rather rough. As a consequence we see a lot of noise in the low magnification trace due to the local tilt. The 40x curve is smooth and compares well with the Dektak trace. The oscillations in the deepest step are real, coming from the saw marks. Figure 12 is a set of scans across a moat on a Si wafer of the type shown in Fig. 1.

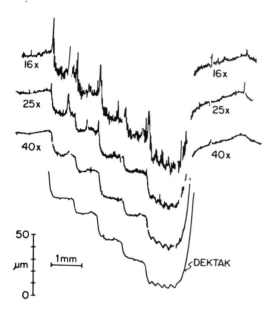

Figure 11. Scan Across Steps on a
Si Sample

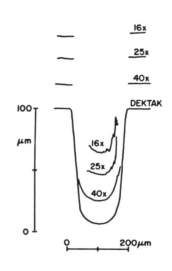

Figure 12. Scan Across a
Moat on a Si
Power Wafer

For measuring power device wafers (Fig. 1) in manufacturing, it is important to be able to measure through photoresist layers, because when the moat depth is tested, the mesa tops are still covered with photoresist, in order to be able to re-etch the wafers if necessary. Fortunately, the laser beam is only weakly reflected from the air-photoresist interface, 3%, while the reflectivity at the photoresist-Si interface is 20%. Hence, it is expected that the laser profilometer will look at the Si surface, "not seeing" the photoresist. This is actually confirmed by visual observation through the ocular at high magnification when the automatic focussing works on a photoresist covered wafer.

Tables II and III list a number of measurements made on two different kinds of Si power device wafers, one of them with photoresist covered mesa tops, along with very schematic sketches of the geometry indicating with arrows the points along the moat structure where the measurements were taken. These sets of measurements were repeated five times at locations near the periphery and in the center of each wafer. Also, the measurements of Table III were repeated after exchanging the 16x by the 40x objective. At each location on the wafer, the first measurement was done on the mesa top on one side, zeroing the display. Then, measurements were done at the bottom of the moat followed by a measurement at the mesa top on the other side. This latter measurement should come back to zero, if the wafer is smooth and the stage is leveled. The data show that it does so to a precision of $\pm1\,\mu$m. The values taken at the moat bottoms vary from one location on the wafer to another, because the moat depths do actually vary by a few μm. In Table II the repetition after switching from 16x to 40x was done as closely as possible at the same location, and the agreement of the two blocks of data, including the moat bottoms, is in the average also close to $\pm1\,\mu$m. This indicates that our profilometer has a precision of $\pm1\,\mu$m.

The precision of $\pm1\,\mu$m derived from the above tests includes errors due to the nonuniformity of the surface and to the limited precision of the sample stage. A direct test of the precision and accuracy inherent in the profilometer itself was made by mounting a planar mirror on a long lever arm whose height was controlled by a piezo-element (Figure 13). The actual height change of the lever arm was measured with a second LVDT, which in a separate experiment had been compared with the built-in LVDT and had been found to agree within

Table II

Photoresist on Mesas

Location of Wafer	Objective	Top	Right	Bottom	Left	Center
Depth in μm	16x	0	0	0	0	0
		-89	-92	-82	-90	-88
		-85	-79	-68	-84	-84
		-93	-95	-87	-93	-90
		0	0	0	+1	0
	40x	0	0	0	0	0
		-89	-92	-82	-88	-87
		-89	-82	-65	-84	-85
		-91	-94	-86	-92	-89
		0	0	+1	0	+1

Table III

No Photoresist, 16x

Position on Wafer		Top	Right	Bottom	Left	Center
Depth in μm	M	0	0	0	0	0
		-32	-34	-36	-34	-36
	M	-1	-3	0	0	-1
		-29	-32	-33	-31	-34
	M	-2	-2	0	0	-2
		-18	-17	-14	-14	-20
	M	0	-1	0	0	0
		-30	-28	-33	-34	-34
	M	0	-1	0	-1	-2
		-34	-28	-31	-33	-35
	M	-1	-2	0	-1	-2

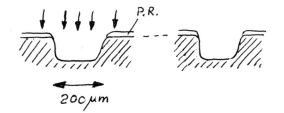

200 μm

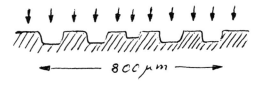

800 μm

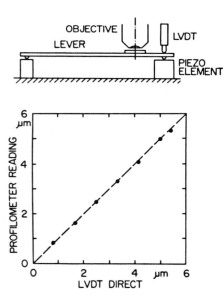

Figure 13. Accuracy Test of the Profilometer

better than ±0.1 μm. In this way the mirror could be elevated in a controlled manner between 0 and 6 μm. Figure 13 gives the readings of the profilometer as a function of the "true" surface height of the mirror, as measured by the second LVDT. The points in Fig. 13 are mostly averages from 6 measurements, 3 times going up and 3 times down. The deviations of these averages are about ±0.1 μm for most points. The standard deviation of the 6 points from their mean is also about ±0.1 μm. From these tests we conclude that the accuracy of the profilometer inherent in the optics and in the feedback loop (without sample movement) is better than ± 0.2 μm.

We also have done a few tests with non-mirror, non-Si samples. For instance we could measure the thickness of paper labels. Another non-Si sample was the measurement of Riston Dry Film Resist on a printed circuit board (about 12 μm). This could be measured easily at steps of this resist on the copper plated front face as well as on the bare rear face of the board. Also the thickness of metal layers on printed circuit boards could be measured.

7. Conclusions

We have developed a laser profilometer for optically reflecting or scattering surface contours in the range of up to 200 μm with an accuracy and a precision of about ±1 μm, which is mostly limited by the surface quality of the samples and by the precision of the sample stage. The accuracy and the precision inherent in the optics and in the feedback was found to be 5 times better. The sampling area can be as small as 0.5 μm.

The instrument described above has been specifically designed for testing and monitoring the depths of the moats surrounding the mesas of Si power transistor and thyristor wafers in the manufacturing process. However, we can see that the same instrument or the same principle can be applied to the measurement of surface profiles of other kinds of samples which are too deep and not smooth enough to be handled by interference techniques.

Acknowledgement

The problem addressed by this work was put to our attention by R.U. Martinelli (Princeton) and Jo White (Mountaintop). We would also like to thank for stimulating discussions with J. Sandercock (Zürich), H. Hook (Princeton) and F. Stensney (Mountaintop).

References

1. J.M. Bennet, Appl. Opt. 15, 2705 (1976) and references therein.
2. J.M. Bennet and J.H. Dancy, Appl. Opt. 20, 1785 (1981) and references therein.
3. G.E. Sommargren, App. Opt. 20, 610 (1981).
4. H.E. Keller, Microscope 21, 59 (1973).
5. F.T. Arecchi, D. Bertani and S. Ciliberto, Opt. Comm. 31, 263 (1979).
6. D.P. Bortfeld, I. Gorog, P.D. Southgate and J.P. Beltz, RCA Engineer 26, 75 (1980).
7. Y. Fainman, E. Lenz and J. Shamir, Appl. Opt. 21, 3200 (1982).
8. Exploratory work for measuring TV gun parts at RCA Princeton (H. Hook) and RCA Lancaster (J. Bleacher), personal communication.

Diameter measurement of optical fibers and wires by laser

Andras Podmaniczky

Department of Atomic Physics, Institute of Physics, Technical University,
H-1521 Budapest, Budafoki ut 8., Hungary

Abstract

A novel method is presented for the measurement of optical fibers with diameters of 10-120 µm. The method is based on the detection of resonances in the intensity of dye laser light backscattered from a fiber in the far-field when tuning the wavelength. It has been experimentally confirmed that not only a thick (~120 µm) communication fiber produces Fabry--Perot resonances, but a thin (~13 µm) doubly-clad imaging fiber as well. The obtained accuracy is better than $\pm$ 0,1 µm.

Also, the basic features of the WIRE GAUGE instrument based on laser beam scanning are shortly outlined.

Introduction

There are two, rather popular methods for diameter measurement of fibers and wires. The first one[1] is based on the analysis of the angular distribution of the forward scattered laser light when a fiber is illuminated with a laser beam perpendicularly to its axis. This method has a $\pm$ 0,25 µm accuracy, but it becomes impractical to apply it to measuring fibers with much larger diameters, because there are too many fringes in the scattering pattern.

In case of the second method[2] the specimen to be measured is scanned with a focused or unfocused laser beam, and there is a negative pulse in the detector output when the scanning beam is intercepted by the fiber or wire. The time duration of this pulse serves as a measure of the diameter. The tipical accuracy of this method is $\pm$ 0,5 µm.

But there are quite a few many measuring problems in the field of fiber manufacturing where higher accuracy is required. These involve ellipticity measurement of thick communication fibers and accurate diameter measurement of thin (~13 µm) imaging fibers used in image transmitting boundles. At the Bell Laboratories Ashkin and his colleagues developed[3] the "near-field resonant backscatter" (NFRBS) method. This method is based on the detection of the Fabry-Perot (F-P) and surface wave (SW) resonances observable in the backscattered light when the wavelength is tuned. As large as $\pm$ 0,01 µm accuracy was demonstrated in relative diameter measurements of a thick communication fiber when detecting the F-P resonances.

We extended the resonant backscatter technique to far-field measurement of thin (~13 µm), doubly-clad imaging fibers (FFRBS method) with better than $\pm$ 0,1 µm accuracy. Also we developed a WIRE GAUGE instrument for the measurement of wires with $\pm$ 0,4 µm accuracy, in which laser scanning technique is applied.

Fabry-Perot resonances of a thin doubly-clad fiber

As our main goal was to measure thin fibers we checked at first, that whether they show F-P resonances at 180° scattering angle. Fig.1 shows that there are four rays which can contribute to 180° scattering. In Fig.2 we show the angular distribution of the intensity of the backscattered laser light in the vicinity of 180° angle at normal incidence for two incident polarization directions. The direction is referred to the axis of the fiber. A simple ray tracing calculation shows that the contribution of ray III is nonnegligible, but for orthogonal polarization its contribution can considerably decreased. Fig.2 gives an experimental confirmation of this. A further experimental confirmation can be reached if the centric part of a thin fiber is regarded to be an almost plane-parallel F-P resonator (see Fig.3) and the length of the resonator is changed by scanning the angle of incidence (SAI method). Introducing the weighted average of the refractive indices for a doubly-clad fiber as

$$\bar{n} = \frac{n_1 t_1 + n_2 t_2 + n_3 r}{t_1 + t_2 + r} \tag{1}$$

a minimum intensity in the backscattered light a $\theta=0$ is obtained when

$$2d \, \bar{n} \, \cos\beta \approx m \, \lambda/2 \quad ; \quad \beta = \sin^{-1}\left(\frac{\sin\alpha}{\bar{n}}\right) \tag{2}$$

Here n_1, n_2 and t_1, t_2 are the refractive indices and thicknesses of the claddings, respectively, r is the core radius of the fiber with refractive index n_3, m is an odd integer, and λ is the wavelength of light. When writing down Eq.2 we took into account the phase changes of ray I and II due to passage through the phase lines and due to reflection from the surface of an optically denser medium, and that $\bar{n}$ is close to n_3, because the thicknesses of the claddings (~0,7 μm) are small compared to the core radius (~5 μm).

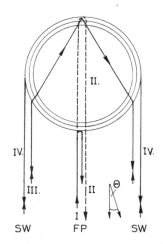

Figure 1. Rays giving 180° backscattering

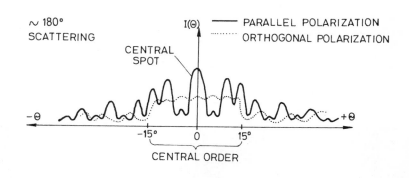

Figure 2. Angular distribution of the intensity of the backscattered light for a 13 μm doubly clad fiber at normal incidence (λ=514,5 nm)

As our fibers are of small diameter, the angles of incidence that give minima are rather large. Therefore we chose to rotate the fiber itself, perpendicularly to its axis, as it is shown in Fig.4. Interference fringes obtained in this scattering geometry are shown in Fig.5 and Fig.6 for a thin and a thick fiber, respectively. In Fig.5 the F-P resonances, along the θ=0 line, (see Fig.4) are much deeper for orthogonal polarization then for parallel polarization. The interference of rays I and II gives the F-P resonances in vertical direction. The interference of some other rays incoming close to rays I and II give the large-scale intensity modulation in horizontal direction (see Fig.5.a). The small-scale intensity modulation is caused by the contribution of ray III and rays close to it, in accordance with a ray tracing calculation, the details of which will be published elsewhere. The fringes obtained for a 120 μm diam fiber are, of course, much more similar to F-P fringes.

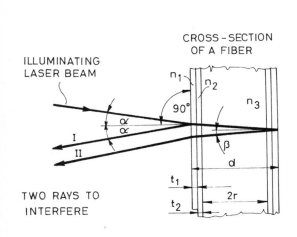

Figure 3. Analogy between a fiber and a Fabry-Perot resonator

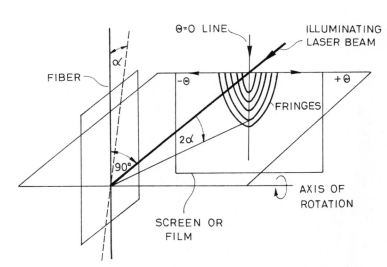

Figure 4. Generation of "F-P" fringes in backscattering by rotating the fiber

Reading out the $2\alpha_1$ and $2\alpha_2$ angles for two adjacent maxima along the θ=0 line from Fig.5, the diameter can be obtained as

$$d = \frac{\lambda}{2\bar{n}} \left(\cos\beta_1 - \cos\beta_2 \right)^{-1}$$

(3)

This equation could serve for diameter measurement, but the accuracy of measuring the α_1 and α_2 angles is enough for only an estimation with 2-3 µm accuracy. For example, Eq.(3) gives 13,5 µm for a 12,2 µm fiber, the nominal value of which is determined by the FFRBS method. Also, the accuracy of the SAI method decreases in case of thick fibers because the resonances are much closer in angle.

We may conclude now, that our thin doubly-clad fibers do show definit F-P resonances in the far-field, and neither the contribution of the surface waves nor the contribution of ray III can smooth out these resonances. This is rather important, because the NFRBS seems to be hardly applicable to measuring such a thin fibers like ours. In case of the NFRBS method a magnified image of the fiber is produced and narrow slits are used for the separation of the different components of the backscattered light. For thin fibers, much larger magnification is needed to perform spatial separation and the apparatus becomes sensitive to fiber movements. Although, the FFRBS method used by us does not make possible to detect the SW resonances thus to provide extreme resolution, but it offers less experimental difficulty and much greater tolerance to fiber movements because of not using an imaging lens.

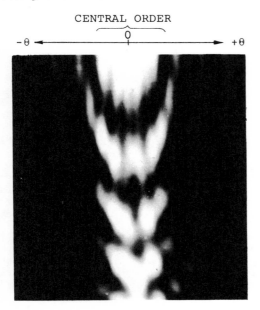

a/ ORTHOGONAL POLARIZATION

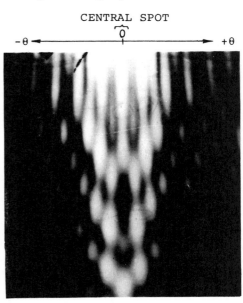

b/ PARALLEL POLARIZATION

Figure 5. Interference fringes in backscattered light for a 12,2 µm doubly clad fiber, obtained by the SAI method at λ=514,5 nm

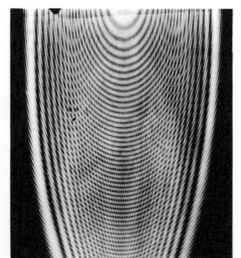

Figure 6. Fringes for a 120 µm communication fiber (parallel polarization)

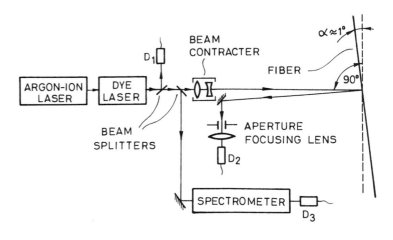

Figure 7. Block diagram of the apparatus used for FFRBS measurement

Diameter measurement by FFRBS method

The block diagram of the measuring apparatus is shown in Fig.7. Detector D_1 monitors the dye laser intensity, detector D_2 measures the intensity of the backscattered light at $\theta=0$ angle, and a spectrometer with a resolution of 0,1 nm serves for wavelength calibration. An incidence angle of $\alpha=1°$ was chosen to avoid the use of a semitransparent mirror when detecting the backscattered light.

A tipical resonance spectrum obtained for a 14 μm doubly-clad fiber is shown in Fig.8. Knowing the average refractive index and the $\delta\lambda$ wavelength separation between two adjacent maxima or minima, one can calculate the diameter. But we found that there is at least one resonance, the wavelength of which is shifted in place by a few tenth of a nanometer. This shift is probably caused by a small but extra phase shift suffered by ray II at a certain wavelength when traversing the claddings. Therefore all the available minima and maxima of the measured spectrum are to be used in the evaluation procedure. This can be done, for example, in the following way.

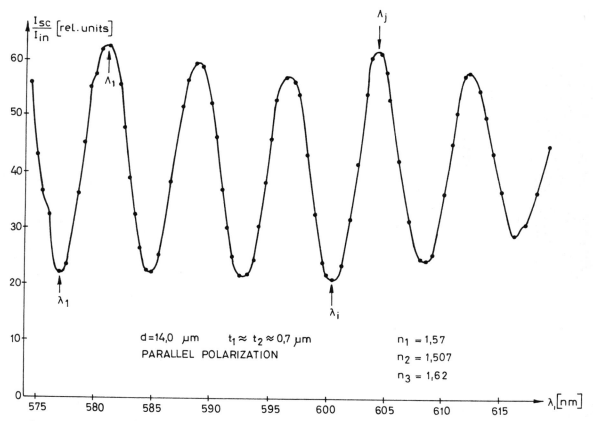

Figure 8. Fabry-Perot resonances of a thin, doubly-clad fiber at 180° scattering angle

Let us denote the wavelengths of the measured minima and maxima by λ_{im} and Λ_{jm}, respectively. λ_{1m} is the smallest among λ_{im} and Λ_{1m} refers to the first maximum following λ_{1m} in the direction of increasing wavelengths. From Eq.2 the diameter is (at $\alpha=1°$)

$$d = m_1 \frac{\lambda_{1m}}{4\bar{n}} \tag{4}$$

where m_1 is the order of interference for the first minimum. We can calculate several values of m_1 from the measured values of λ_{im} and Λ_{jm} according to the relations:

$$m_{1i} = \frac{2(i-1)\lambda_{im}}{\lambda_{im} - \lambda_{1m}} \qquad (i=2,3,\dots) \tag{5}$$

$$m_{1j} = \frac{(2j-1)\Lambda_{jm}}{\Lambda_{jm} - \lambda_{1m}} \qquad (j=1,2,\dots) \tag{6}$$

In Eq.4 we use the averaged value of $m_{1i,j}$:

$$\bar{m}_1 = \frac{1}{N+M} \sum_{i,j}^{N,M} (m_{1i} + m_{1j}) \qquad (7)$$

which may be regarded as an approximately correct value. Using the averaged value of m_1 we can calculate the orders of interference and the wavelengths for the minima and maxima:

$$m_i = \bar{m}_1 - 2(i-1) \qquad (8)$$

$$M_j = \bar{m}_1 - (2j-1) \qquad (9)$$

$$\lambda_{ic} = \frac{\bar{m}_1}{m_i} \lambda_{1m} \qquad (10)$$

$$\Lambda_{jc} = \frac{\bar{m}_1}{M_j} \lambda_{1m} \qquad (11)$$

These equations come from simple resonance conditions. In order to check the fit of the calculated wavelengths to the measured ones we calculate the averaged differences as

$$\overline{\delta\lambda} = \frac{\left| \sum_{i=2}^{N} (\lambda_{im} - \lambda_{ic}) \right|}{N-1} \quad ; \quad \overline{\delta\Lambda} = \frac{\left| \sum_{j=1}^{M} (\Lambda_{jm} - \Lambda_{jc}) \right|}{M} \qquad (12)$$

Measurements on different fibers and the calculations of the averaged differences show that a fit of $\overline{\delta\lambda} < 0,1$ nm and $\overline{\delta\Lambda} < 0,1$ nm can always be obtained. Calculations show that if there were an error of only one in $\bar{m}_1$ ($\Delta\bar{m}_1 = 1$), this would spoil the fit giving three times larger values of $\overline{\delta\lambda}$ and $\overline{\delta\Lambda}$. This shows that Eq.12 defines a rather good measure of fitting. But an error of less than one in $\bar{m}_1$ causes an error of less than 0,09 µm in diameter. Of course, the correct value of the averaged refractive index must be known for absolute measurement.

At last, Fig.9 shows how the fiber diameter changes when the rate of revolution of the reel is changed. The deviation from the calculated curve may probably be attributed to taking the fiber samples at non-steady state.

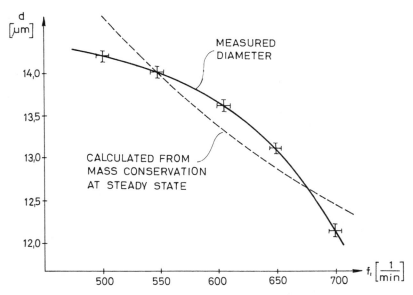

Figure 9. Fiber diameter as a function of reel speed

The WIRE GAUGE system

The operation of this system is based on laser beam scanning. The system has been originally developed for the measurement of tungsten wires used in electric bulbs. In case of tungsten wires the majority of wires is in the 10-70 µm range. The necessity of measuring 10-20 µm wires is the main reason why a focused scanning laser beam is used in the optical subunit, the scheme of which is hown in Fig.10. An acousto-optic modulator is incorporated in the system in connection with detector D_1 to perform intensity stabilization of the laser thus eliminating the effect of aging and warm-up. The focusing lens (1) together with a beam expander forms a 10 μm laser spot in the object plane. Detector D_2 produces an object pulse when the scanning beam is intercepted by the wire. The time duration of this pulse is compared to the duration of the etalon pulse produced by the same laser beam scanning along an etalon. The etalon plate has seven chromium stripes of precalibrated width for each of the measuring ranges. A given stripe is positioned so that the etalon pulse coincides in time with the object pulse thus eliminating the effect of the angular nonlinearity in scanning.

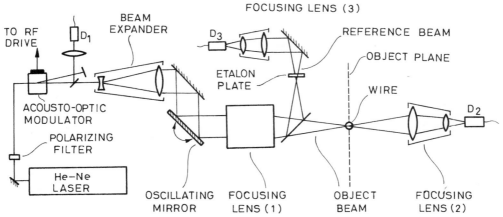

Figure 10. The optical scheme of the WIRE GAUGE system
(By permission of MIKI-Instrumentation)

Table 1. Basic technical data of the WIRE GAUGE system

Diameter ranges	10 - 15 μm	Indicator unit
	15 - 25 μm	Displays-6 digit LED readout in μm of:
	25 - 50 μm	a/ Instant diameter
	50 - 75 μm	b/ Average diameter on 100, 1000 samples
	75 - 100 μm	Preset function: tolerance setting
	100 - 200 μm	Warnings : NO WIRE, NO LASER
	200 - 400 μm	Overstepping tolerances freezes output
Scanning rate	100 scan/sec	Microprocessor electronics
Accuracy	± 0,4 μm	Interface to character printer for docemen-
Repeatability	± 0,3 %	tation
Laser	He-Ne, 5 mW	Alarm: HI, GO, LOW
		Out of focus position of wire
		Out of scan center position of wire

The indicator unit is μP based and it also includes the RF card, the mirror drive card, the pulse processing card, and the power supplies. The basic technical data are summarized in Table 1.

The system has been operating at the TUNGSRAM Ltd. in off-line mode for almost two years.

Conclusions

It has been experimentally confirmed that very thin (~13 μm) doubly-clad image transmitting fibers do show Fabry-Perot resonances when scattering laser light at 180° angle. The SAI method provides only a 2-3 μm accuracy in diameter measurements. The FFRBS method gives an accuracy of better than 0,1 μm and it can be applied with much less experimental difficulty in comparision to the NFRBS method.

Also, some basic features of the WIRE GAUGE system based on laser beam scanning are outlined.

Acknowledgements

The author wishes to thank Attila Markus, Miklos Barabas, Laszlo Csillag and Geza Hegyesi for their help in the experiments, in making samples, and in discussions.

References

1. Smithgall, D.H., Watkins, L.S. and Frazee, R.E., "High-Speed Noncontact Fiber-Diameter Measurement Using Forward Light Scattering", Appl.Opt., Vol.16, pp. 2395-2402. 1977.

2. Statistics Laser Gauge, Beta Instrument Co. Ltd., Bucks, England.

3. Ashkin, A., Dziedzic, J.M., and Stolen, R.H., "Outer diameter measurement of low birefringence optical fibers by a new resonant backscatter technique", Appl.Opt., Vol.20, pp. 2299-2303. 1981.

Advances in non Doppler Laser Velocimetry and selected applications

Helmut Selbach

Polytec GmbH & Co., Siemensstrasse 15, 7517 Waldbronn 1 , W.-Germany

Abstract

A number of publications concerning conventional Laser Doppler Velocimetry and showing their rich range of applications have appeared in the last decade. For various situations this technique turned out to be very difficult or not applicable. Looking for other techniques the Laser Two Focus Method has extended the situations where laser velocimetry can be applied to. Due to recent progress in this technology the range of applications continues to broaden. Some ideas of the progress of the last years is given by presenting the hardware techniques of today's instruments and by presenting selected experiments.

Introduction

Within the last decade laser velocimeters have gained acceptance as a method for non-intrusive investigation of fluid flows. The Doppler-difference anemometry was applied first and became a very popular laser diagnostic tool in the field of fluid- and aerodynamics. Certain areas, for example turbomachinery research, proved to be very difficult as the flare rejection capability of the laser Doppler was not sufficient. These difficulties have lead to the development of the Laser-2-Focus Velocimeter. By replacing the Doppler fringe-probe volume by two highly focused beams many drawbacks could be overcome. Based on an idea of Tanner |1| Schodl |2-4| designed an instrument which can be applied to an almost routine basis to areas which were considered as difficult applications.

The Laser Doppler difference velocimetry

The Laser Doppler difference velocimetry is based on the principle of optical heterodyning. A laser beam is divided by a prism splitter into two parallel beams which are focused by a lens. The measuring volume is formed at the place where the two beams intersect. When particles pass through this probe volume the light waves are scattered ad Doppler shifted. When two different portions of the scattered wave having associated Doppler shifts are mixed at the photodetector a time varying light intensity variation with the difference frequency of the two waves can be sensed. This intensity variation can be analyzed if the beat frequency does not exceed 100 MHz. The electronic systems available today are limited to this frequency resolution.

The fringe model greatly simplifies the analysis of Laser Doppler systems. If a particle is illuminated by two focused equal intensity Gaussian profile laser beams a set of stationary fringe planes are formed in the crossover region as shown in figure 1.

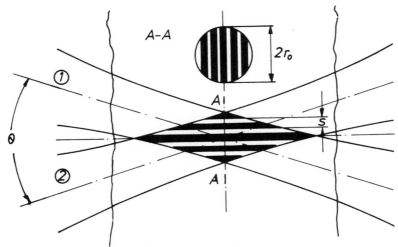

Figure 1. Laser Doppler probe volume

If a particle passes through the fringe planes, it scatters light whose intensity is modulated at the frequency

$$F_d = \frac{2v}{\lambda_0} \cdot \sin\left(\frac{\theta}{2}\right)$$

v is the particle's velocity perpendicular to the fringe planes, λ_0 the wavelengths of the laser light and θ the intersection angle of the two beams.

The geometrical size of the detection volume is the volume enclosed by the surface where the intensity falls to $1/\ell^2$ to its maximum value. The actual probe-volume from where signals are observed is usually smaller than the geometrically defined probe volume as the observable part can be restricted by appropriate optical stops in the receiver optics and a further restriction is introduced by thresholds within the electronic processors which are adjusted to process only signals that exceed a given amplitude. Thus only signals from particles passing through the most intense region of the probe-volume will be processed. Typical effective diameters of the measuring volume are about 150 microns and mostly the lenghts is between 1 and 2 mm.

The Laser 2-Focus principle

The probe-volume of the Laser 2-Focus Velocimeter is formed by two highly focused laser beams as shown in figure 2.

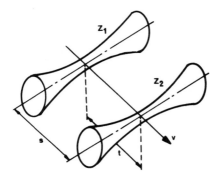

Figure 2.   L-2-F probe volume

Typical diameters of the probe volume which are achieved by focusing a Gaussian profile laser beam are in the order of 10 microns. The two focal volumes are about 250 microns apart. The length of the probe-volume usually has the same value as the intra-spot distance.

The backscattered light from particles passing through each spot is focused onto two photomultiplier tubes. The time of flight t for particles to traverse this 'optical gate' relates to its velocity as the separation s of the two spots is known.

The probability that the same particle will produce a signal in both channels depends on the trajectory of the particle in the L2F probe volume during transit. This probability is rapidly reduced as the direction of the two focal volumes does not match the flow direction. It is therefore essential to rotate the orientation of the two spots.

Problems in laser velocimetry

The performance of a laser velocimeter depends upon its ability to provide a signal that can be processed by the electronic system. Usually flare light, which is always present when measurements have to be performed in narrow channels, reduces the signal contrast. The incident laser light is reflected and scattered from access-windows and bulk surfaces. When it hits the receiver the background noise will be increased. To optimize the efficiency of a laser velocimeter it should be configured for the special application problem. Very often the accessibility to the flow channel does not allow to configure the laser velocimeter at optimum. When the access is restricted one must operate with a small aperture of the sending and receiving optics. On the other hand a large collecting aperture allows more scattered illumination to be collected. A problem especially associated with high speed fluid flows is the flow seeding. According to figure 3 only particles with very small diameters are able to follow a flow practically without inertia. The particle size necessary to get a sufficient signal-to-noise ratio is about 1.2 microns in diameter for a Laser Doppler Velocimeter operating in a back-scatter arrangement |5|.

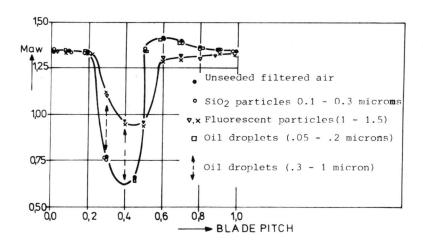

Figure 3. Relative Mach Number against Blade Pitch

The Laser Doppler Velocimeter versus the Laser-2 Focus Velocimeter

The significant differences between the Laser-Doppler and the Laser-2-Focus relate to the geometry of the probe volume.

For the same aperture the L2F offers a greater light intensity per unit area than the LDV. Because of this using the same laser power smaller particles can be detected and used.

Because of the optical principles involved the depth of field of a LDV system is approximately 10 to 20 times greater than that for the equivalent L2F with a similar f-number. Because of this, for a given particle size, the L2F offers a more practical approach to the close to wall measurements.

The optical design of the L2F is such that the expanded output of the final focusing beams are very insensitive to refractive index inhomogeneities and variations of time thereof between the front-lens and the measuring volume.

Because the detection volume is smaller in the L2F system the data rate is inherently lower. This is compensated by the ability of the L2F to respond to a smaller size range of particles. In naturally occuring aerosols small particles are more likely to occur. These smaller particles follow the flow more closely. If a laser velocimeter system is able to respond to the smaller particle size distribution then no artificial seeding is necessary. The same particle is less likely to make a stop- and a start signal as the turbulence increases. The theory requires that for a 2-dimensional flow measuring device all vectors in the x-y plane must be determined iven if there are large z-components. These two restrictions result in the ability to measure turbulent flows that are only below 30 % in turbulence intensity.

Optical set-up of a Laser-2-Focus Velocimeter

To realize the theoretical limits of the L2F system all the optical components must be polished to a very high degree of accuracy. If this is not the case it leads to a degradation of detected signal to noise. The same requirements for a Laser Doppler Velocimeter are only required when the optics-system has a f-numbers greater than 4.

After having made a careful study of various optical configurations the system outlined in figure 4 turned out to give the best signal to noise ratio.

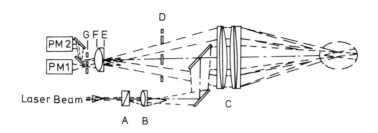

Figure 4.

The incoming laser beam is split by a prism A to form two beams. One of these continues along axially wheras the other diverges. The beam waists at the splitting point are transformed into a plane one focal length forward the lens B. Thus two parallel beam waists are generated. These waists intern are transformed by the lens system C into the measuring volume which may be some distance away.

If the beams diverge at the prism by an angle θ then the spot size r in the measuring volume is determined by the equation

$$r = \frac{f_0 \cdot \lambda}{\pi \cdot R} \cdot \frac{f_{ext}}{f_{int}}$$

where f_0 is the focal lengths of lens B and λ the wavelength of the laser. R is the diameter of the incoming laser beam at $1/e^2$ intensity. f_{ext} is the right hand focal length of the lens system C and f_{int} the left hand focal length s of this lens system.

The beam separation is calculated according to

$$s = 2 \cdot f_0 \cdot tg\left(\frac{\theta}{2}\right) \cdot \frac{f_{ext}}{f_{int}}$$

When particles pass either of the focused spots in the measuring volume light is scattered in all directions. A portion of this light is collected by the lens system C and re-imaged in the plane E. The optical lens system F reimages and enlarges the plane E onto the pinholes G. Two very sensitive high speed PMT's are located behind these pinholes.

The measuring direction is perpendicular to the axis of the two measuring spots. Because the spot size is very small compared to the spot separation only particles travelling through the measuring volume and are too within a small angular range of the measuring direction are detected. In order to obtain data in two dimensions it is necessary to rotate the measuring spots. Thus it is possible to build up a two dimensional flow distribution. It is necessary to rotate the beam splitting prism and the pinholes within the receiving optics together. During rotation it is necessary to insure that the system remains aligned and free from aberration.

Rotation of the measuring volume requires a high degree of mechanical accuracy and is achieved internally by a small stepping motor.

In order to minimize scattered light other than that from the measuring volume stops are used in the detection systems. This maximises the detected signal-to-noise. Any multiple reflections within the optical system are minimized by additional optical stops. This also helps to reduce crosstalk between the two detecting channels.

L2F signal processing

There are two methods to process the pulses from the photodetectors (one photomultiplier for each focus).One is based on a digital correlator and the other on a direct time of flight measurement and multichannel analysis.

With correlation techniques the similarity between the start- and stop-pulses is determined as a function of the time shift between them. At an orientation α of the two foci the pulse stream from the start-photodetector may be represented as the function $F_1(t_1)$ with the arbitrary time coordinate t_1. If the pulse-stream detected with the second photodetector is $F_2(t_2)$, with the time coordinate $t_2 = t_1 + T$ the correlator will compute

$$k(T,\alpha) = \int_0^{T_m} F_1(t_1) \cdot F_2(t_1 + T)\, dt_1$$

where T_m is the time needed for the measurement. The digital correlator which from the theoretical point of view is the best suited instrument for data processing has unfortunately some drawbacks when applied to high speed fluid flows or to periodic flows. In the first case a high speed correlator with a sufficient number of bins is not available. When doing measurements in periodic flows, which are for example present in pumps and turbines, it is necessary to relate the time of flight measurement to the phase of the periodic phenomenon. For this application a multi-station correlator is necessary which, for the time being is also not available.

In most practical fluid flows the transit time from one focal volume to the other is short compared with the mean interval between particles. One can therefore directly measure the time of flight and process this value with a multichannel analyzer. The measured time-of-flight value is interpreted as the address of a cell within the multichannel analyzer's memory. Each new measured time of flight B added to the appropriate memory cell, so that a spectrum of measured time intervals vs. time of flight is obtained.

L2F data reduction

If the three dimensional velocity probability density funcion $p(v,x)$ at a point $\vec{x}$ is known the position dependent mean values $F(x)$ can be derived

1. $$\bar{F}(\vec{x}) = \iiint_{-\infty}^{+\infty} F(\vec{v})\, p(\vec{v}, \vec{x})\, dv_1\, dv_2\, dv_3$$

In case of the L2F Velocimeter, the velocity information along the optical axis is averaged during the measurement. Equation 1 can be transformed to read

2. $$\bar{F}(\vec{x}) = \iint_{-\infty}^{+\infty} F(v_1, v_2) \int_{-\infty}^{+\infty} p(\vec{v}, \vec{x})\, dv_3\ \ dv_1\, dv_2$$

This equation shows that the computation of any two dimensional moment of the velocity vector is only possible if the measured two dimensional probability density function $p^+(v_1, v_2, \vec{x})$ fulfils the condition

3. $$p^+(v_1, v_2, \vec{x}) = \int_{-\infty}^{+\infty} p(\vec{v}, \vec{x})\, dv_3$$

This means in practice that the v_1 and v_2 components of all velocity vectors $\vec{v}$ must be measured, even for large v_3 components.

Using the L2F velocimeter, the components of arbitrary velocity vectors $\vec{v}$ lying in the x_1, x_2 plane can be measured for direction α and magnitude $v_\perp$ as shown in figure 5

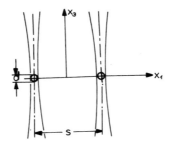

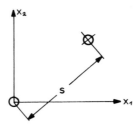

Figure 5. Geometry at the L2F probe volume

As outlined before the L2F does not measure the velocity magnitude $v_\perp$, but the time of flight t.

4. $\qquad\qquad t = \dfrac{s}{v_\perp}$

Transformation of equ. 2 in polar coordinates and using eq. 3 and 4 yields

5. $\qquad \bar{F}(\vec{x}) = \int\limits_{0}^{2\pi}\int\limits_{0}^{\infty} F(\alpha,t)p^{+}(\alpha,t,x) \cdot \dfrac{s^2}{t^3} \ dtd\alpha$

At a fixed position $\vec{x}$ equation 5 indicates the probability for the direction of the flow velocity in the angular range α to $\alpha+d\alpha$, and its magnitude $v_\perp$ in the assigned time window t to t+dt. The data points H_{ij} accumulated in a digital correlator or in a multichannel-analyzer are a measure at the probability of the velocity within the time window $t_j t_o \ t_j+\Delta t$ and angular range α_i to $\alpha_i +\Delta\alpha$. Therefore

6. $\qquad H_{ij} \sim p^{+}(\alpha_i,t_i,\vec{x}) \cdot \dfrac{s^2}{t^3} \ \Delta t \ \Delta\alpha$

To calculate integral functions according to equ. 5 a numerical integration procedure can be performed.

If the number of sample points is limited, this is for example the case when a digital correlator with 64 bins is used to measure high speed low turbulence flow, the mathematical treatment can be done as described in |6|.

When for a flow investigation the variables required are only the components of the mean velocity vector and the turbulence intensities an accumulation scheme as suggested by Schodl |7| can be used. This scheme needs only one tenth of the data and shortens the measuring time.

L2F-Design for 3-dimensional measurements

Previously we have only described a system that is capable of measuring velocity components perpendicular to its optical axis. In order to measure in the third dimension a second measuring volume is required. By using the green and blue lines from an argon-ion laser two measuring volumes can be defined at an angle to one another as shown in figure 6. The two volumes are coincident with one another and converge towards one another from the output lens at an angle of 2γ.

Because of the finite depth of field of the measuring volume particles travelling at an angle between the two measuring volumes ß can still be detected. It is only required that the particles travel in the plane that contains the two volumes A and B.

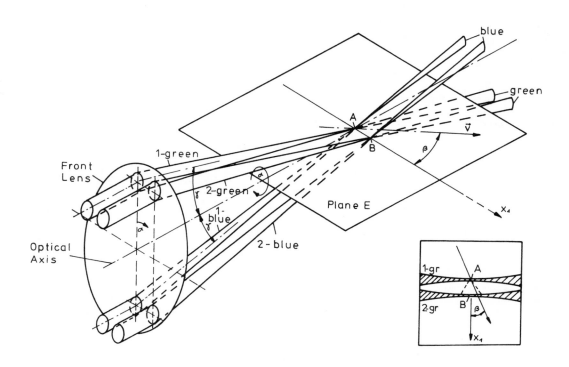

Figure 6   Beam System of the 3D time-of-flight velocimeter

When the velocity vector is either in the plane defined by the two green spots or the two blue spots only time of flight histograms will be obtained in one color. When the measuring volume is rotated around the optical axis in the way previously described for a specific velocity vector a signal will be detected firstly in one color and then in the other color as the rotation continues.

A rotating by an angle α_{bl} causes the intersection of the beams with the previously defined plane to change. The axis defined by drawing a line between the two green and blue measuring volumes cuts through the plane E (Figure 6) at different points during rotation. At one angle α_{bl} the line between the two blue measuring volumes lies in the plane E and at another angle α_{gr} the line between the two green measuring volumes lies in the plane E.

By accumulating time-of-flight histograms as described before the mean flow angle will be different for the green and the blue beams. From this information the angle between the plane E and the true velocity vector $\vec{v}$ can be determined according to

$$\alpha = \frac{\alpha_{gr} - \alpha_{bl}}{2}$$

The measurement also gives you the angle α as defined in figure 6.

$$\beta = \text{arc tg} \frac{\sin\alpha}{\text{tg } \gamma}$$

The magnitude of the velocity vector $\vec{v}$ is determined according to

$$|\vec{v}| = \frac{|\vec{v}_{b1}|}{\cos\alpha \cdot \cos\beta}$$

A three dimensional time of flight velocimeter has been built by Schodl and could be successfully applied to an axial flow compressor.

Applications

To reduce the fuel consumption of diesel engines turbocharger compressors are used. The efficiency of this compressor impacts directly on the specific fuel consumption of the engine. Computer programs for designing the impellers of the centrifugal compressor have been developed but they have to be proven. To do this the L2F is the appropriate instruments. The depth of the passage where measurements have to be made in is in the order of some millimeters and the velocity of the air flow in the order of 200 m/sec. In this situation flare light is present from the compressors casing and the access-window at the same time. Inside the impeller passage measurements as close as 400 microns to the impeller's surface could be made with the L2F. The impeller rotated at 80 000 rpm Non intrusive flow mapping with a time-of-flight velocimeter helped engineers to improve the performance of their turbochargers and led to a saving in fuel costs of about 800 $ per truck per year [8]

The insensitivity to windows could be demonstrated in Munich. The time needed to measure the flow at one point in a blow down channel was 10 seconds. Access to the point of investigation was possible via two plexiglass windows. One was 50 mm thick the other 100 mm. The working distance was 1.6 meters. The channel of the collection lens was only 100 mm. The laser was operated at power of 200 mW.

At high fluid flow speed the L2F always shows a better performance than the Laser Doppler. The velocity in hydraulic machines as water pumps and water turbines is only a few meters per second. The accessibility to these machines however is very limited. One has to operate with small apertures. The smaller intersection angle of the two laser beams of a Laser Doppler Velocimeter, the longer is the measuring volume. It was demonstrated [9] that even at low flow speeds the L2F could outperform the LDV.

Acknowledgements

Part of the development of the L2F electronics was funded by the German Bundesministe-rium für Forschung und Entwicklung. The author would like to thank Dr. R. Schodl from the DFVLR for valuable discussions. He is also grateful to Heinz Lossau from Polytec GmbH for the permission to publish this paper.

References

1. Tanner, L.H., Journal of Scientific Instruments, Vol.4, 1967, pp 725-730

2. Schodl, R., "A Laser Dual Beam Method for Flow Measurement in Turbomachines", ASME Paper No. 74-GT-157 (1974)

3. Schodl, R., "Laser Two Focus Velocimetry for use in Aero Engines", Lecture No. 4 in AGARD LS-90 (1977)

4. Schodl, R., "On the Extension of the Range of Applicability of LDA by Means of the Laser-Dual-Focus (L2F) Technique," Proceedings of the LDA-Symposium Copenhagen,(1975)

5. Straziar, A. and Powell, J. A., "Laser Anemometer Measurements in a Transonic Axial Flow Compressor Rotor", Measurement Methods in Rotating Components of Turbomachinery, ASME, (1980)

6. Ross, M. M., "Transit Laser Anemometry Data Reduction for Flow in Industrial Turbo-machinery," Proceedings of the 3rd International Conference on Photon Correlation Techniques in Fluid Mechanics, (1979)

7. Schodl, R., "A Laser-Two-Focus Velocimeter for Automatic Flow Vector Measurements in the Rotating Components of Turbomachines", Measurement Methods in Rotating Components of Turbomachinery, ASME, (1980)

Application of laser Doppler velocimetry to flame propagation studies

Kyung-Kook CHO and In-Seuck JEUNG[*]

Combustion Laboratory, Department of Aeronautical Engineering, Seoul National University
Shinrim-dong, Kwanak-ku, Seoul 151 Korea

Abstract

Gas flow speed along the combustion chamber longitudinal axis of a suddenly propagating flame into a quiescent propane-air premixed combustible gas charged inside an open combustion chamber, and ignited by the usual capacitor discharge ignitor, under the initial conditions of room temperature and atmospheric pressure, was investigated by using a one component laser Doppler velocimetry (LDV) equipped with a 15 mW He-Ne gas laser and a 40 MHz Bragg cell based frequency shifter and a frequency tracker of maximum dynamic slew rate of 10 MHz/msec.

The measured velocity profile of pre- and post-combustion period are demonstrated with a flame front propagating shapes visualized by high speed schlieren photography. Their results show reasonable agreement if the refraction effect due to a flame front is fully considered. The flame front just approaching and passing through the LDV measuring volume has the role or effect of a suddenly moving concave lens in the optical path of the LDV system.

Introduction

Nearly two decades have passed since Yeh and Cummins[1][†] applied a laser spectrometer system on the measurement of a fully-developed laminar pipe flow of water at first in 1964, of which the concept was the same as the "reference beam mode" optical configuration available at the commercial market nowadays. From the early seventies, Whitelaw and his coworkers have studied the application of the LDV system with the "fringe mode" optical configuration on a laboratory scale stationary burner flame[2] and further on an industrial burner flame.[3] The progress of research works during the last decade has increased the potential to apply LDV systems to many situations in combustion science, such as a motored and fired internal combustion engine problem[4], a real jet engine problem[5], etc. It is pointed out that still many of the LDV system applications in combustion problems have been limited to stationary flame studies rather than transiently propagating flame studies. Also the behavior in the vicinity of a propagating flame to LDV measurements has not been clearly understood, even though several noble attempts were made to elucidate the effect of a flame front, with convoluted phase boundaries, in the case of a turbulent burner flame[6] and a supersonic emerging flame from a model combustion chamber similar to the industrial jet engine[7].

This paper, which was motivated by these considerations, presents measurements of one component gas velocity history of pre- and post-combustion period along the longitudinal axis of the model combustion chamber and brief discussions of the refractive index effect by propagating flame front to the LDV measurement, when made with LDV system in a transiently propagating flame into the quiescent propane-air premixed combustible gas. Also the flame propagating shapes were visualized by high speed schlieren photography.

Instruments and Method

Measurements were made in a 100 x 100 x 395 mm rectangular shaped model combustion chamber. This chamber contained a pair of anti-reflection film coated optical glass windows over the whole chamber front, allowing an unobstructed view through the chamber. A sketch of the model combustion chamber, as seen by the photomultiplier site of the LDV system, is shown in figure.1. Upper two conduits are the premixed gas intake cock and exhaust cock at right-hand and left-hand side, respectively. At the right surface of the chamber, the strain-gauge type pressure transducer was attached to record the static chamber pressure, to see whether it remained at atmospheric pressure throughout the entire combustion process. The ion gap probe was inserted at the center point of the geometric mid-plane of the chamber to detect the flame arrival time. Needle shaped spark electrodes of 1 mm diameter were introduced at the lower side of the chamber, which were energized by a capacitor discharge ignitor. The upper pot at the top of the chamber can be removed after propane-air premixed gas refill before the start of each experiment run to maintain a constant pressure combustion process. The lower pot, at the bottom of the chamber, is the scattering powder seeding pot, which disperses nominally 1 μm diameter magnesium oxide powder inside the chamber after each premixed gas refill. A detailed sketch of it is shown in figure.3. All of the measurements reported in

---

* Present address ; Department of Mechanical Engineering, University of Minnesota, 111 Church Street, S.E., Minneapolis, MN 55455 U.S.A.
† Numbers in superscript designate references at the end of paper.

this paper were made at the positions marked as the open circles shown in figure.1. Measurements are reported for traverses along two perpendicular axes, say, they are marked as measuring position #1, #2, #3 along the vertical axis of the chamber center-line from 30 mm above the spark electrodes to 145 mm above the spark electrodes, where position #3 is spatially the same position as the tip of the ion gap probe. Along the horizontal line, measuring positions #2, #4, #5, #6, #7, #8 are marked with 10 mm distance except the position #8, which is close to the chamber wall.

Figure.2 is a schematic sketch of the LDV system optics arrangement and the chamber. A Kanomax-Japan one component forward-scatter LDV system (27 series, under the licence production of TSI, U.S.A.) was used for all of the measurements. The He-Ne gas laser (Spectra-Physics 124A) was operated at 15 mW in the 632.8 nm wavelength band. A transmitting lens of 241 mm focal length was used; this gives a beam cross angle of 11.58° and a fringe spacing δ of 3.1 μm, which indicates close to unity visibility for a 1 μm diameter scattering particle and a 3.1 μm fringe spacing.[8] 40 MHz Bragg cell based frequency shifter (TSI 9180) was used mainly for the purpose of easy transient tracking of the frequency tracker and also for the purposes of the directional ambiguity removal and the pedestal noise removal. Doppler burst signal, collected by the receiving lens of 243 mm focal length and focused on the photomultiplier (RCA 4526) by focusing lens, was processed by a phase locked loop (PLL) frequency tracker (TSI 1090) whose maximum slew rate is 10 MHz/msec and capture range is 0.5 MHz. Analog output from the frequency tracker provided the instantaneous Doppler shift frequency f_D which is related to the instantaneous flow velocity by the following equation,

$$U = \frac{\delta}{1/f_D} = \frac{\lambda \ f_D}{2n \ \sin\theta} \quad .$$

The frequency tracker ouput data, photomultiplier output data, and ion gap probe output data were recorded simultaneously on a storage oscilloscope (Tektronix 7613/5113), externally triggered by the spark ignition signal.

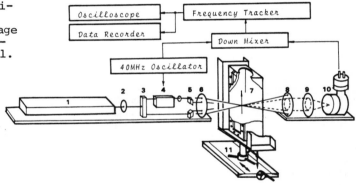

Figure.2. Optics arrangement

1.15mW He-Ne laser head 2.Polarization rotator
3.Beam splitter 4.Bragg cell and compensator
5.Beam reducer 6.Transmitting lens 7.Combustion chamber 8.Receiving lens 9.Focusing lens
10.Photomultiplier tube 11.Traverse unit

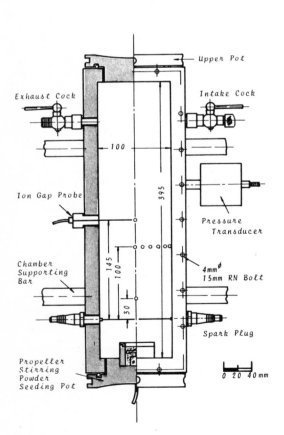

Figure.1. Rectangular combustion chamber

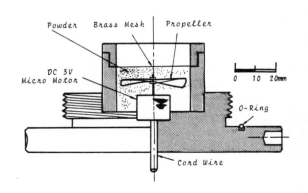

Figure.3. Detailed drawing of propeller stirring powder seeding pot

As propane-air premixed gas was mixed under atmospheric pressure, and was delivered into the chamber by the water displacement method, the propeller stirring powder seeding pot as shown in figure.3 was used to disperse scattering particles in the chamber after each gas refill, until the data rate of the frequency tracker attained 2×10^4 - 5×10^4 samples/sec typically[¶], this corresponds to particle concentrations of approximately 5×10^3 - 10^4 particles/cm³.

High speed schlieren photography was taken to visualize the flame propagating shapes, to measure the flame propagation speed, and to compare the burning speed. Schlieren optics arrangement was typical horizontal Z-type optics alignment and schlieren pictures were taken by using high speed motion camera (Hitachi 16HM) at the photographing speed of 3,000 pictures/sec.[9]

Results and Discussions

A typical LDV measurement result is shown in figure.4, which is composed of 20 runs multiple data plotting of gas velocity records measured after ignition. Measuring position was #2, which is 100 mm above the spark electrodes. This position corresponded to the position where the flame propagation speed was defined as the representative value. From the measurements at position #3, the LDV data, when the flame was just touching the LDV measuring volume, as the flame passed through, was determined as the value that corresponded to the photomultiplier signal decreasing suddenly and the LDV signal dropping steeply as indicated in figure.4. This moment coincided with the peak time of the ion gap probe signal. This sudden decrease in the measured velocity could be caused by the phase change in the incident laser beams. The discussion concerning the apparent pseudo-velocity is referred to here as the result of the gradient effect of refractive index in the vicinity of the flame front. The measured velocity signal before the flame front reached the LDV measuring volume was fairly reproducible, but after the flame front passed the LDV measuring volume it was widely dispersed, resulting in a maximum value of standard deviation of approximately 10 % near the end of combustion process. Also drop-out phenomena occured just after the flame passed the

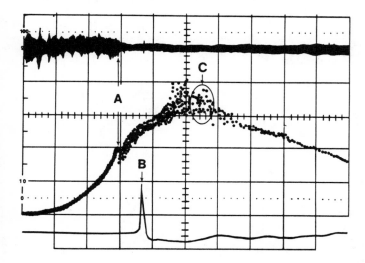

Figure.4. Velocity-time data pairs

Oscillogram is composed of 20 runs multiple data of gas flow speed of transiently propagating flame after ignition. Measuring position ; #2. Propane-air mixture ratio ; 5.7 % volumetric. Horizontal axis ; 10 msec/division, Vertical gas flow speed ; 3.1 m/sec/division.
A ; Flame front arrives at LDV measuring volume.
B ; Flame front arrives at ion gap probe.
C ; indicates end of combustion process.
Upper trace ; photomultiplier signal
Second trace ; LDV output records
Lower trace ; ion gap probe signal

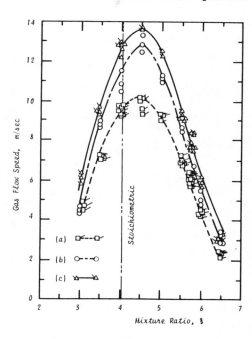

Figure.5. Gas flow speed versus propane-air mixture ratio

(a) Unburnt gas flow speed in front of flame front
(b) Flame propagation speed at the same position as LDV
(c) Burnt gas flow speed at ion gap probe position

___
[¶] Although it was realized that seeding of scattering particles by the streaming flow would have been preferable, this was not done for some reasons, because of the complicated law of combustible gas treatment and the construction fund of gas dilution system, and also because particle concentration is still a reasonable number concentration.[2]

LDV measuring volume, which could be due to the slight but rapid movement of LDV measuring volume and rescattering by condensation of water vapors from the products of combustion[9,10].

Unburnt gas flow speed just in front of the propagating flame, the flame propagation speed measured at the same position as the LDV measuring volume (this position is corresponding to measuring position #2.), and the burnt gas flow speed when the flame front was passing through the ion gap probe are compared in figure.5. General quantitative trends agree reasonably between each measurement, but the difference value between the flame propagation speed and the unburnt gas flow speed in front of the propagating flame, which becomes a burning speed of the specific premixed combustible gas, is strongly biased due to the combined effects of the negative apparent pseudo-velocity, the asymmetry of flame propagating shape, the cellular flame pockets in the rich mixture region, and the other uncertainties, compared with the results from the traditional methods[9,11].

Figure.6 is the set of LDV signal records along the horizontal line of the chamber, which shows the velocity history of each measuring position. It is remarked that the steep dropping of LDV signal through the vicinity of the flame front is clearly shown in each record, but the record of position #8, near the chamber wall, is somewhat smooth and cannot be easily recognized. It may be due to the fact that the gradient of refractive index in the vicinity of the flame front near the chamber wall is steep distribution across the horizontal axis but weak distribution along the vertical axis, which would be imagined from the schlieren photography (refer to figure.8).

The result of the LDV measurements along the horizontal axis is plotted in figure.7. All of the curves are almost coincident from 10 mm apart from the chamber wall, which could be suggesting that the gas flows outside the flame front surface along the chamber wall. The coincidence of the parabolic velocity distribution and the measured burnt gas flow speed at the position of the ion gap probe tip assures the certainty of this measurement.

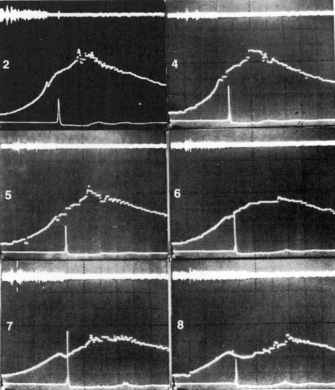

Figure.6. Gas flow speed signal records

Measured along the combustion chamber horizontal axis of 100 mm above the spark electrodes. Measuring positions are #2, 4, 5, 6, 7, 8. Other legends are same as in figure.4.

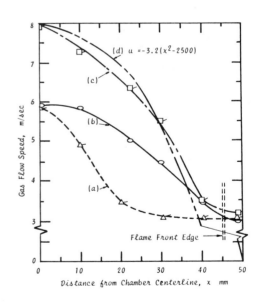

Figure.7. Gas flow distribution

(a) Measured at 29 msec after ignition.
(b) Measured when the flame front arrives at the LDV measuring volume.
(c) Measured when the flame front arrives at the ion gap probe.
(d) Parabolic flow distribution
Other legends are same as in figure.6.

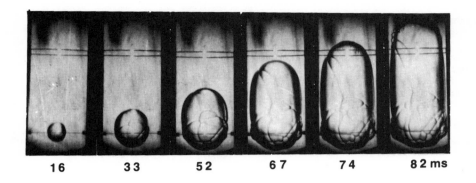

Figure.8. Flame propagation shape

Propane-air mixture ratio ; 6.5 % volumetric
Time is shown after the ignition.

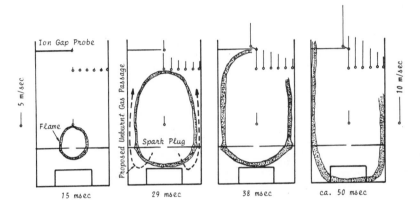

Figure.9. Combustion gas flow distribution and time history
in laboratory coordinates

Propane-air mixture ratio ; 5.7 % volumetric
Time is shown after the ignition.

The high speed schlieren photography shown in figure.8 indicates the gas flow along the trough space between the flame and the chamber wall, and also the development of cellular flame pockets. The schematic gas flow distributions are attached as time goes on in figure.9 concerning the propane-air volumetric mixture ratio of 5.7 %.

The apparent pseudo-velocity referred here is known as one of fundamental uncertainties for the LDV application on the convoluted phase boundaries, and moving surfaces with very steep density or temperature gradient,[6,7,9,10] but still it is not fully understood. Here, by the simple geometrical optics calculation of laser beam rays, it is revealed that the possible apparent pseudo-velocity term is its magnitude of 10 % of measured velocity values in this LDV optics arrangement, of which the value is nearly 3 times higher than the burning speed of propane-air premixed gas listed in references.[9,11]

Concluding Remarks

Among the many examples of LDV application research in combustion problems, this study would be attained as a LDV application to the transiently propagating flame, single-event combustion research for better understanding of combustion and flames, even though the apparent pseudo-velocity occured by the temperature gradient of the flame front is still large enough comparing with the burning speed. However, this approach would become a role of opening the method of the burning speed measurement after the artificial removal of the apparent pseudo-velocity or by different optics system.

What is happening the whole combustion chamber flow field is very difficult to interpret from the measurements at a single point. Hence, two components simultaneous measurement with the coincident measurement of flame propagation speed through the whole flow field would be a hopeful future work. Also, it is necessary to study further on the control of the flame propagation to be suitably stable to apply the LDV system even on the transiently propagating flame in single-event combustion problems.

Nomenclature

f_D ; instantaneous Doppler frequency
n^D ; index of refraction of fluid
U ; instantaneous velocity

δ ; fringe spacing
λ ; wavelength of laser
θ ; half beam cross angle

Acknowledgement

The authors wish to acknowledge that the partial work of this study was performed by the Korea Science and Engineering Foundation fund under the KOSEF grant.

References

1. Yeh,Y. and Cummins,H.Z.;"Localized Fluid Flow Measurement with a He-Ne Laser Spectrometer," Applied Physics Letter, vol.4, no.10, pp.176-178, 1964.

2. Durst,F.,Melling,A.,and Whitelaw,J.H.;"The Application of Optical Anemometry to Measurement in Combustion Systems," Combustion and Flame, vol.18, no.2, pp.197-201, 1972.

3. Baker,R.J.,Bourke,P.J.,and Whitelaw,J.H.;"Application of Laser Anemometry to the Measurement of Flow Properties in Industrial Burner Flames," 14th Symposium (International) on Combustion, The Combustion Institute, pp.699-706, 1972.

4. Rask,R.B.;"Laser Doppler Anemometer Measurements in an Internal Combustion Engine," SAE Technical Paper Series 790094, 1979.

5. Smart,A.F.,and Moore,C.J.;"Aero-Engine Application of Laser Anemometry," AIAA Journal, vol.14, no.3, pp.461-585, 1977.

6. Hong,N.S.,Jones,A.R.,and Weinberg,F.J.;"Doppler Velocimetry within Turbulent Phase Boundaries," Proceedings of the Royal Society of London - A, vol.353, pp.77-85, 1977.

7. Schafer,H.J.,Koch,B.,and Pfeifer,H.J.;"Application of LDV Techniques to High Speed Combustion Flows," International Congress on Instrumentation in Aerospace simulation Facilities, IEEE publication 77 CH1251-8 AES, pp.31-39, 1977.

8. Adrian,R.J.,and Orloff,K.L.;"Laser Anemometer Signals : Visibility Characteristics and Application to Particle Sizing," Applied Optics, vol.16, pp.677-684, 1977.

9. Jeung,I.-S.; A Study on the Transiently Propagating Flame of Propane-Air Premixture, Ph.D. dissertation, Seoul National University, 1982. (written in Korean)

10. Strehlow,R.A.,and Reuss,D.L.;"Effect of a Zero g Environment on Flammability Limits as Determined Using a Standard Flammability Tube Apparatus," NASA CR-3259, 1980.

11. Lewis,B.and von Elbe,G.; Combustion, Flame and Explosions of Gases, 2nd edition, Academic press, New York and London, p.389, 1961.

Some Industrial Applications of ESPI

Ole J. Løkberg

Physics Department, The Norwegian Institute of Technology
N-7034 Trondheim, Norway

and

G.Å. Slettemoen

The Foundation of Scientific and Industrial Research (SINTEF), Div. 34
N-7034 Trondheim, Norway

Abstract

Electronic Speckle Pattern Interferometry - ESPI - may be described as dynamical hologram interferometry based on video recording, processing and presentation. The technique has been used to solve various problems in Norwegian industry and industrially related research. Work has been mainly concentrated on vibration testing and measurement, but recent measurements on crack opening and locations of high strains have also attracted interest.

Introduction

Hologram interferometry has a great potential as a measuring and non-destructive testing tool in industrial research and production. Still, although two decades have passed since its invention, the technique is mainly confined to research laboratories. The few examples of industrial on-site use deal with costly and critical products like aircraft tires and heat exchangers(1). Apart from the usual conservatism and sceptisism against strange, new methods, some of this non-acceptance stems from the slow and cumbersome filmprocessing which prohibits for example on-line inspections. Even the new thermoplastic recording process is too slow for many purposes.

We have for the last decade worked to improve and extend the use of a holographic interferometry system based upon direct videorecording and reconstruction. The system which consequently works in real-time is commonly called electronic speckle pattern interferometry - ESPI - although TV-holography might have been a name with better sales appeal. The ESPI-principle was almost simultanously described by three groups in 1970-71 (2-4), there is also a fourth paper appearing a bit later (5).

In this paper we very briefly describe the basic principles of the ESPI-technique, discuss its strong and weak points and illustrate its potential with some applications from Norwegian industry and industrial related research.

Description and Evalution of ESPI

The main building blocks of our ESPI system is illustrated in the symbolic flow chart on fig. 1.

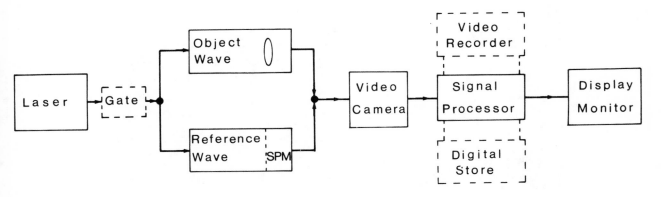

Fig. 1. ESPI-structure

The optical part of ESPI is best described as an image holography set-up where the object- and reference waves interfere in-line by means of a beam-combining device. The recording step is performed by the photoelectric action of the TV-camera while the reconstruction is done by electronic prossessing on the videosignal. The videostore(s) is necessary whenever we want to compare different videoframes. Finally, the prossessed videosignal is converted into a "reconstructed" image by the TV-monitor display. As in hologram interferometry, the image of the testobject will be covered with a fringe pattern representative of the positional change of the object's surface during (or between) the exposure(s) for example, J_0^2-fringepattern for a vibrating object, while a double exposure provides a $\cos^2$-pattern.

For a more complete description of the ESPI-techniques, the interested reader may consult refs. (6-9).

The main advantage and strength of the ESPI-system are due to the videosystem, which produces 25 exposed, developed and reconstructed holograms each second. Combined with the short TV-exposure (40 msec), the high sampling rate makes the ESPI a very stable real-time holographic interferometer. This holds especially true for vibration analysis, but also in deformation studies the recording of a reference state is effected within fraction of a second. The fast presentation rate enables us to implement effectively many interferometric techniques which never left the publishing stages in ordinary holography (e.g. sinussoidal phasemodulation).

Another interesting property of ESPI, which has not really been exploited so far, is that the interferometric information at one stage is available as a videosignal (analogue or digital). For one thing, this simplifies the incorporation of computors into the system. In addition, the videosignal may be relayed or transmitted over any distances to the analyzing station, e.g. from the moon back to earth (if so desired). We may also store individual ESPI-holograms by means of a videorecorder and do the subtraction by a digital store or computor to get the interference between any states of the deformation. Finally, we should note that the electronic reconstruction very effectively removes such problems as emulsion shrinkage, correct repositioning etc.

Other positive properties of ESPI-operation are; display on a large TV-screen, extremely low running expenses and the system may operate in roomlight.

On the negative side, the most notable drawback is the speckled appearance of the image, which is caused by the low resolution of the videosystem. For vibration analysis this does not cause problems as speckle-averaging can be done either on the documenting photos or in near real-time by an image-processor. The resulting quality can be quite impressive as witnessed by the stroboscopic ESPI-recording of a vibrating steelplate (fig.2).

Fig. 2
Stroboscopic ESPI, plate vibrating at f=3280 Hz

In deformation studies, speckle averaging is possible but is rarely performed. However, the fringe-patterns are here cos^2-pattern with constant contrast and sufficient fringes (30-40) can usually be resolved. In additon, one should remember that magnifying and zooming is simple and that one is allways able to pick out more details by direct observation on a dynamical TV-image than by looking on still pictures of the screen image.

Commercial ESPI-systems tend to be somewhat costly, but in this context one should remember that a high experimental and measuring throughput is possible. For vibration analysis, holographers with some insight into videotechniques may put up an ESPI-set up quite cheaply. For applications where a reference frame is needed, digital stores are the only realistic alternative. Especially as they are often designed for doing image processing operations as well. Digital stores have been costly so far, but prices are coming down as competition increases.

Application areas of ESPI

A. Vibration measurements

Our work has so far concentrated on perfecting ESPI for vibration studies as we have allways felt that ESPI would be a superior tool compared to ordinary holographic techniques. The real time presentation at high sampling rate is perfectly suited for observation of repetitive events. ESPI allows us to use effectively sinusoidal phase modulation - SPM. By SPM we get a greatly extended amplitude measuring range in ESPI, from below 10^{-5} of a wavelength using photoelectric lock-in techniques (10) to above 200 wavelengths (11). The vibration phase is also provided, in many cases as contours of constant phasevalues (12). The stability is very good at normal TV-exposures; if we decrease the exposure time further, the stability becomes almost dramatically good (13). In this context, note that by using the so-called speckle reference ESPI (14), very good stability is obtained at normal TV-exposures (15). The same set-up may also be used with lasers having very short coherence length (15) and to directly compare the vibrations (and deformations) of different objects (16). Most vibration measurements are done by time average ESPI with SPM, but non-sinusoidal or unsymmetric vibrations have to be examined by stroboscopic techniques. As already demonstrated by fig. 2, the quality of such recordings can be outstanding.

We have used ESPI for many applications both in biomedicine and industry. From an industrial point of view, the application of an ESPI at a Norwegian gas-turbine manufacturer is probably most interesting. However, as this project has already been fully described elsewhere (17), the interested reader should consult this reference. A more recent example is studies of a modell propeller whose vibration patterns were of interest due to some highly unexpected cracks developing in the real propeller. The modell (ϕ=30 cm) was examined both in air and water. In air the entire propeller was first examined to look for individual variations. Fig. 3a shows one of the resonance pattern of the entire wheel, where one clearly see the coupling between the various blade. One representative blade was thereafter singled out for closer inspection, fig. 3b.

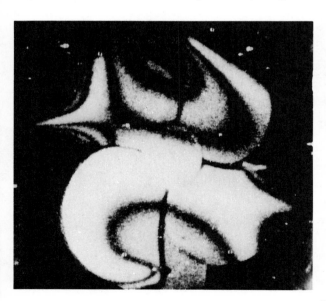 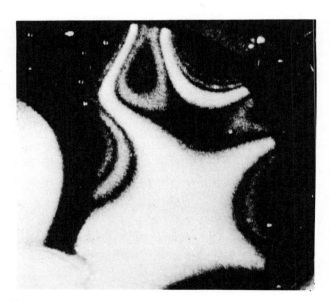

Fig. 3

a) Propeller vibrating at f=1346 Hz   b) Blade vibrating at f=4398 Hz

The vibration patterns were mapped with amplitude - and phase values from 500 to 20 000 Hz; in addition the Q-values of the main resonances were measured. These measurements together with the necessary photographic documentation was done by one man in half a day. The propeller was thereafter submerged in water and the inspection repeated. The strong damping in water gave less clearly defined resonances. In addition, sufficient excitation at higher resonances was difficult to obtain with the excitation source mounted on the center axis. We could not excite the blade directly, as this invariably lead to wrong conclusions, due to "forced" vibrations at the excitation area. (In general we have found it extremely difficult to obtain truely neutral and representative excitations of objects). The investigations in water have therefore been put off until a more powerful shaker is provided.

A more difficult object was presented by very thin (15 μm) and large (40 x 40 cm^2) aluminium foils stretched over a frame to simulate ship plates. The foils were extremely microphonic, picking up e.g. speech very easily. In addition, their near specular surface was difficult to illuminate effectively. To top off the problems, the frame had to be placed on the floor outside the floating table. The fringe patterns were however sufficiently good to give an unexpected result. Some of the previously recorded Chladni patterns indicated contours of constant phase instead of nodal lines.

Other examples of industrial interest would be vibration measurements on underwater transducers, piezoelectric crystals, exhaust pipes, optical fibers and objects heated above + 800°C.

B. Deformation testing and measurements

As already mentioned, to follow slow or two-step changes in the object, we need a video memory for storing the TV-frame representing the reference state of the object. So far the university economy has not allowed us decent videostores, but during the last year we have borrowed two digital stores (W. Vinten Ltd. and a Quantex DS-30). We have found the equipment very useful for many industrial purposes - in fact, we have already been asked to build several ESPI set-ups for use in local laboratories. We do not intend to compete with hologram interferometry in the surveillance of a large area by one shot, the ESPI simply does not have the resolution capability for this. In many practical problems, however, one knows the approximate position of strain consentration (mechanical hot spots), or cyclical loading allows us to move the ESPI around on the object. For such purpose we are constructing a compact ESPI set-up which can be placed on the structure itself. Measurements will be done on areas ranging in size from 2 x 2 to 10 x 10 cm^2 with an accuracy which should be near or equal to the strain gauge. Needless to say, there are some practical problems involved in the implementation of such a compact ESPI set-up. Most of the problems, however, have already been solved.

Concluding Remarks

We have described some industrial applications of hologram interferometry using direct videorecording and processing - ESPI. Although the quality and resolution of fringe patterns of a ESPI - system cannot compete with conventional holography, its speed and ease of handling makes it attractive for many industrial purposes. ESPI may also present the only realistic alternative if the sensitivity of holography is required for on-line inspection. Furthermore, we emphasize that processing and evaluation of video images by now is a well established and sophisticated technique where standard, commercial equipment is available.

Acknowledgment

The authors would like to thank Dr. B.M. Watrasiewicz, W. Vinten LTD. and Dr. K. Gåsvik, SINTEF for instrumental assistance.

References

1. SPIE, vol. 349: Industrial Applications of Holographic Nondestructive Testing (1932).
2. J.N. Butters and J.A. Leendertz, Opt. Laser. Technol. 3, 26 (1971).
3. A. Macovski et al, Appl. Opt. 10, 2722 (1971).
4. O. Schwomma, Austrian pat. no. 2988 30 (1972).
5. U. Köpf, Messtechnik 4, 105 (1972).
6. Proc. Eng. Uses of Coh. Optics. ed. E.R. Robertson (Cambridge Univ. Press), pp. 155-169, (1976).
7. Speckle Metrology ed. R.K. Erf (Acad. Press). pp. 111-157, 234-245, (1978).
8. O.J. Løkberg, Phys. Technol. 11, 15 (1980).
9. R. Jones and C. Wykes, Holographic and Speckle Interferometry (Cambridge Univ. Press) (1983).
10. K. Høgmoen and O.J. Løkberg, Appl. Opt. 16, 1369 (1977).

11. O.J. Løkberg and K. Høgmoen, J. Phys. E. $\underline{9}$, 347 (1976).
12. O.J. Løkberg and K. Høgmoen, Appl. Opt. $\underline{15}$, 2701 (1976).
13. O.J. Løkberg, Appl. Opt.,2377 (1979).
14. O.J. Løkberg and P. Svenke, Opt. and Lasers in Eng., $\underline{2}$, 1, (1931).
15. G.Å. Slettemoen, Appl. Opt. $\underline{19}$, 616 (1980).
16. G.Å. Slettemoen and O.J. Løkberg, Appl. Opt. $\underline{20}$, 3467 (1981).
17. O.J. Løkberg and G.Å. Slettemoen, Appl. Opt. $\underline{20}$, 2630 (1981).

INDUSTRIAL APPLICATIONS OF LASER TECHNOLOGY

Volume 398

Session 6

Laser Measurement Techniques II

Chairmen
Armin Felske
Volkswagen AG, West Germany
P. Schuster
AOL-Dr. Schuster GmbH, Austria

Flame diagnostic using Optoacoustic Laserbeam Deflection and Frequency Modulation Spectroscopy

W. Zapka*, P. Pokrowsky**, E.A. Whittaker, A.C. Tam and G.C. Bjorklund

IBM Research Laboratory
5600 Cottle Road, San Jose, California 95193

Abstract

Two laser-based techniques, Optoacoustic Laserbeam Deflection (OLD) and Frequency Modulation (FM) spectroscopy, were introduced for flame diagnostics. Flame temperature was measured by OLD with 1.5% precision and flame temperature profiles were taken. Ground state and excited state sodium within an aspirated flame were detected by FM-spectroscopy. Present limits of sensitivity are absorptions of $1.5 \cdot 10^{-4}$ and ground state sodium densities of $3.3 \cdot 10^{6}$ cm^{-3}.

Introduction

Research in combustion phenomena has grown considerably over the past decade and various techniques have been applied to study laboratory flames, as well as practical combustion devices (see the review article 1). Physical probes like thermocouples or gas-sampling tubes have nowadays been replaced by laser-based techniques, which are noncontact and nonperturbing by nature. Among the great variety of such laser-based tools there is the major group of spectroscopic techniques that can provide flame temperature and species concentration within the flame. Fluid flow velocity can be measured e.g. by laser Doppler velocimetry. The high complexity of most of these diagnostic tools has stimulated us to introduce two new handy and sensitive laser-based flame diagnostic methods. One of these techniques is based on all-optical acoustic velocity measurement (this technique is referred to as Optoacoustic Laserbeam Deflection (OLD) in the following) and provides flame temperature and fluid flow velocity. Frequency modulation (FM) spectroscopy, the other flame diagnostic tool introduced here [3], has been used to monitor species concentration within the flame.

Flame diagnostic by Optoacoustic Laserbeam Deflection

The optoacoustic generation of an acoustic pulse, e.g. by laser induced air breakdown or by resonant absorption of laser light is a noncontact process and may be performed inside a flame. Such acoustic pulses have been detected by Allen et al. [4] and Tennal et al. [5] These authors, however, used microphone-detectors, which necessarily were located outside of the flame, resulting in the disadvantage that the acoustic signal had to traverse the noisy and fluctuating outer parts of the flame before detection. The OLD-technique described here encompasses laser-induced generation as well as laser-based detection (which can be performed within the flame) of the acoustic signal and is the first totally noncontact optoacoustic measurement technique for flame diagnostic.

Experimental set-up and discussion of results

The experimental set-up for OLD-measurements is shown in Fig. 1. The flame studied was a premixed propane-air flame (supply rates 0.5 l/min of propane and 10 l/min of air). A pulsed laser-beam (1.06 μm, 10 ns, ∼ 200 mJ) from a Quanta-Ray Nd:Yag-laser was focused into the flame by a lens of 25 mm focal length to produce a weak transient plasma spark via gas breakdown, which acted as the source of an acoustic pulse. The propagation of the acoustic pulse was monitored by one or more weakly focused HeNe-laser beams, which were all parallel to the excitation beam, the different displacements from this being finely adjustable. Whenever the acoustic pulse traversed such a probe laser beam it caused a transient deflection of the probe (therefore the term 'Optoacoustic Laserbeam Deflection' (OLD)). Using knife edges to block half of the probe laser beams before detection at the fast photodiodes we observed the transient probe-beam deflections (which provided the time of arrival of the acoustic pulse at the probe-beam position) as characteristic transient variations of the detected probe light intensity (see Fig. 1).

---

* On leave from Max-Planck-Institut für Biophysikalische Chemie, Göttingen, West Germany
Present address: IBM Deutschland GmbH, GMTC, Sindelfingen, West Germany

* * World Trade Visiting Scientist

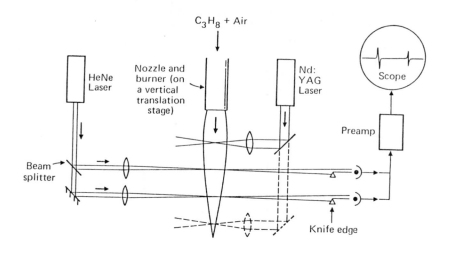

Fig. 1   Experimental set-up for OLD measurements in a flame

A plot of propagation distance x, i.e. the displacement between excitation beam and one probe beam, versus propagation time, i.e. time of arrival, clearly indicated an initial supersonic propagation of the acoustic pulse decaying into linear sound propagation at propagation distances beyond 1 cm (see Fig. 2). We therefore employed a set of two parallel HeNe-probe laser beams, separated by an accurately known distance Δx (in our experiments typically 12 mm), with the closer one being displaced from the excitation beam by 1.5 cm, to measure the acoustic propagation velocities v_d and v_u in a downstream, respectively upstream, direction as

$$v_d = c_o + v_f = \Delta x \: / \: \Delta t_d \qquad\qquad\qquad (1)$$

$$v_u = c_o - v_f = \Delta x \: / \: \Delta t_u, \qquad\qquad\qquad (2)$$

where c_o is the linear sound velocity, v_f the gas flow velocity, both being averaged over the region between the two probe-beams. Δt_d and Δt_u are the differences in arrival times of the acoustic pulse at the two probe-beam positions, which were both downstream, respectively upstream from the excitation laser beam.
Measurement of Δt_d and Δt_u provided both, v_f and c_o. The flame temperature T, spatially averaged over the region between the two probe-beams, was obtained from

$$T = c_o^2 \cdot \overline{M} \: / \: \left\{ R \left[1 + R \: / \: \overline{C}_v \: (T) \right] \right\} , \qquad\qquad\qquad (3)$$

with $\overline{M}$ the average molecular weight of the gases in the flame, R the universal gas constant and $\overline{C}_v(T)$ the temperature dependent averaged specific heat at constant volume. A typical gaseous composition for a premixed propane-air flame like ours is 72% N_2, 16% H_2O, 10% CO_2, 1% O_2 and 1% CO [6,7]. Since $C_v(T)$ is given in literature in analytic form for all of these gases, both the averaged values $\overline{M}$ and $\overline{C}_v(T)$ may be calculated and equation (3) be solved to yield the temperature T.

Typical experimental values, obtained on the axis of the premixed propane air flame at a distance 10 cm from the burner nozzle were v_f = 10.3 + 5 m/s and c_o = 940 + 5 m/s yielding T = 2100 + 30°C. This corresponds to a precision of 1.5% for temperature measurements. As mentioned above the temperature as derived from OLD-measurements is spatially averaged over the region between the two probe-beams. However, the spatial resolution in the two orthogonal directions is high, as described elsewhere [2], and enabled us to perform flame temperature mapping by scanning the assembly of excitation-beam and probe-beams across the flame. Such a temperature profile as measured on our horizontally burning premixed propane-air flame is shown in Fig. 3. The temperature profile agreed in shape with thermocouple measurements. However, the thermocouple readings were smaller by some 15% (probably due to conduction or radiation losses in the thermocouple).

Some advantages, besides the obvious simplicity of both experimental set-up and data reduction, are that the strong noise due to combustion processes, which is in the frequency range below ∼ 100 kHz, can be filtered out without influencing the acoustic signal (> 500 kHz). Furthermore the temperature obtained from the OLD-measurements directly represents the translational temperature, whereas most of the spectroscopic techniques yield only vibrational or rotational temperatures. Moreover, the precision of 1.5% for OLD-temperature measurements is fairly high.

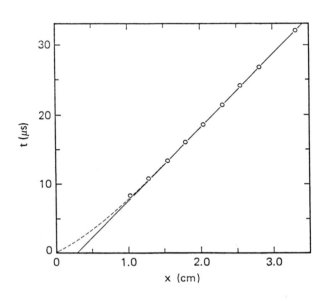

Fig. 2   Time of arrival t of the acoustic pulse (measured from the time when the excitation pulse is fired) as observed for different displacements x of the probe laser beam

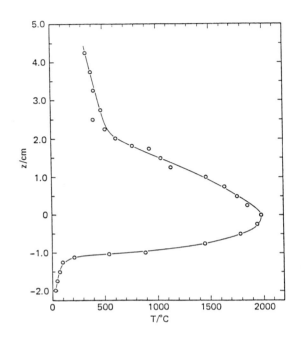

Fig. 3   OLD measured temperature profile of our horizontally burning propane flame. Data were taken on a vertical line crossing the flame axis 9 cm from the burner nozzle

Flame diagnostic by FM-spectroscopy

Direct absorption methods like the traditional Atomic Absorption Spectroscopy (AAS) have the advantage of being easily calibrated and the spectra obtained being easily interpreted. However, the major disadvantage is the limited sensitivity as compared to indirect laser spectroscopic methods such as resonance fluorescence [9], laser induced fluorescence [10], optoacoustic [11] and optogalvanic spectroscopy [12] and coherent Raman spectroscopy [13]. Laser absorption methods provide higher sensitivity than traditional AAS as was shown in the mid-IR using a tunable lead-salt diode laser [14]. Flame diagnostic by laser absorption spectroscopy in the visible wavelength range has been performed by now with low sensitivity on only major flame components (CO [15], OH [16], NH_2 [17]). With the introduction of Frequency Modulation (FM)-spectroscopy high sensitivity detection of absorption and dispersion [18,19] (as well as of gain [20,21]) have become possible. Since FM-spectroscopy may be performed with any single-mode laser (e.g. diode laser, dye laser, frequency doubled dye laser) the accessible wavelength range extends from the UV to the IR. Here we report on the first application of FM-spectroscopy in flame diagnostic [3].

The principle of FM-spectroscopy has been described in detail in ref. 18 and may be briefly reviewed as follows. A single-mode laser beam is frequency modulated (FM) when passing through an electro-optic phase modulator, which is driven at radiofrequency (RF) of typically several ten MHz to several GHz. As result lower and upper frequency sidebands are put on to the transmitted laser beam. Whenever this FM-laser beam is affected in a sample in such a way that the upper, respectively lower, sidebands experience different absorption or gain or that the relative phase of carrier and sidebands is changed, this results in an amplitude modulation of the beam. The amplitude modulation can be detected by a fast photodiode yielding an electrical signal (for small absorptions and phase changes)

$$S(t) \sim I_o \, e^{-2 \, \delta_o} \left\{ 1 + (\delta_{-1} - \delta_{+1}) \, M \cos \omega_{RF} t + (\phi_1 + \phi_{-1} - 2\phi_o) \, M \sin \omega_{RF} t \right\} \qquad (4)$$

with I_o the laser light intensity, M the FM modulation index. δ_n and ϕ_n are the absorptions, respectively phase shifts, with n = -1, 0, +1 indicating the lower sideband, carrier, and upper sideband. Heterodyning the photodiode electrical signal S(t) by using a local oscillator being in-phase with the RF-signal which drives the electro-optic modulator reveals the in-phase part of the signal S(t) (that part proportional to $\cos \omega_{RF} t$) which contains information about the absorption of the spectral feature.

Experimental set-up

Our experimental arrangement for FM-spectroscopic detection of sodium atoms aspirated into an air-acetylene flame is shown in Fig. 4. The single-mode laser beam of typically a few mW power from a Coherent 699-21 dye laser was modulated by a LiTaO$_3$ electro-optic modulator (EOM) before it passed through the flame F, of length L = 11 cm, which was produced by an air-acetylene slot burner (from a Perkin Elmer Model 303 Atomic Absorption Spectrometer). The transmitted laser beam impinged on a Motorola MRD-510 photodiode, the output was amplified by a Q-Bit 538 signal amplifier and fed into a Mini-Circuits ZFM-4 double balanced mixer. A Hewlett-Packard 8690B oscillator both drove the EOM and served as local oscillator for the double balanced mixer. The output of the double balanced mixer was amplified and displayed on an oscilloscope as function of the dye-laser frequency.

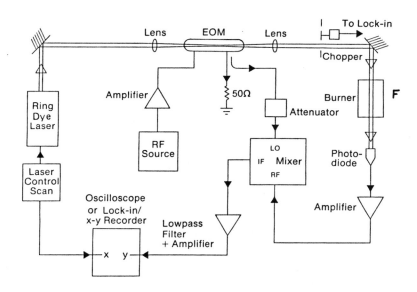

Fig. 4   FM-spectroscopy set-up for detection of sodium in a flame

Experimental results

A typical signal as obtained by keeping the RF-frequency constant at 450 MHz and sweeping the dye laser frequency through the Na 3s-3p D$_1$ resonance at 589.0 nm is shown in Fig. 5 for the case of a 5% absorption of the laser light by the sodium aspirated flame. Since the RF-frequency was small compared to the effective linewidth (FWHM) of 8.3 + 0.3 GHz (see below) of the above transition the observed signal exhibited the derivative shape as expected[18]. Shot noise limited performance was not achieved, but rather both radiofrequency interference and residual amplitude modulation from the electro-optic modulator determined the noise signal, which was measured with no aspirated solution. From the measured data we deduced a signal-to-noise (S/N) ratio of about 250. Higher sensitivity was achieved at higher RF-frequencies, as expected from theory[3]. At 2 GHz a sodium absorption $\alpha L = 0.17$ gave rise to the in-phase FM-signal shown in Fig. 6a, which was detected by lock-in technique (see Fig. 4) because of increased RF interference. The background trace, which is given in Fig. 6b, was observed with the flame not being aspirated. Sodium contamination in the burner caused the small residual resonant signal including the large spikes which were the result of small bursts of sodium. The remaining noise level resulted in a S/N of 1100, providing an ultimate sensitivity of $0.17/1100 = 1.5 \cdot 10^{-4}$ absorption units.

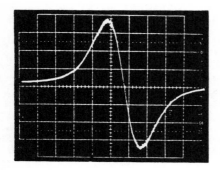

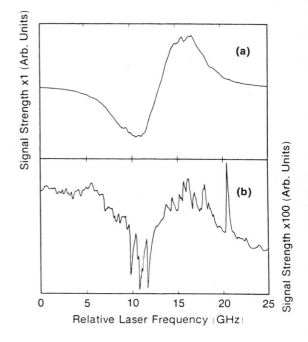

Fig. 5   Typical in-phase FM-signal of the
         NaD_1 resonance in a NaCl aspira-
         ted air-acetylene flame. Corres-
         ponding absorption was 5% and RF
         modulation frequency 450 MHz.

Fig. 6   (a) In-phase FM-signal at 2 GHz
             modulation frequency for an
             aspirated flame with absorp-
             tivity αL = 0.17 at the
             NaD_1 resonance. Laser fre-
             quency scan was 25 GHz.
         (b) Noise signal for Fig. 6(a),
             vertical sensitivity increa-
             sed by a factor of x 100.

For absolute measurement of sodium concentration a calibration of the FM-probe was
performed. The NaCl-concentration of the aspirated solution was varied over a wide range.
Direct measurements of the αL of the flame (using conventional measurement techniques)
were made at three sufficiently high sodium concentrations, and since these data revealed
a linear dependence, the αL of the weaker aspirated solutions were deduced by linear
extrapolation from their measured NaCl-concentrations. A log-log plot of sodium absorptivity
αL, respectively corresponding sodium number density, versus measured in-phase FM-signal
amplitude is presented in Fig. 7. Over the investigated range of about 3 decades we observed
an almost linear dependence (the straight line has slope 1). The deviations from the
straight line at high and low concentration seem to be due to absorption saturation,
respectively to sodium contamination. The open circle at $3.3 \cdot 10^6$ atoms/cm^3 indicates the
detection limit of our present set-up when using 2 GHz modulation frequency and a multi-
pass-arrangement (5 passes through the flame).

In the limit of $\omega_{RF} \ll \Gamma$, Γ the linewidth of the spectral feature under investigation,
and under the assumption of Lorentzian lineshape, the observed derivative shaped in-phase
FM-signal (see e.g. Fig. 5) has extrema located at

$$\Delta_{1,2} = \pm \frac{\Gamma}{\sqrt{3}} + \text{terms of order } \left(\frac{\omega_{RF}}{\Gamma} \right)^2 \qquad (5)$$

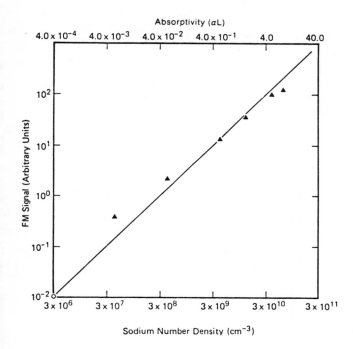

Fig. 7   In-phase FM-signal amplitude
          versus sodium number density
          in the aspirated air-acetylene
          flame. The straight line has
          a slope of one and the open
          circle represents the detec-
          tion limit with the present
          experimental set-up.

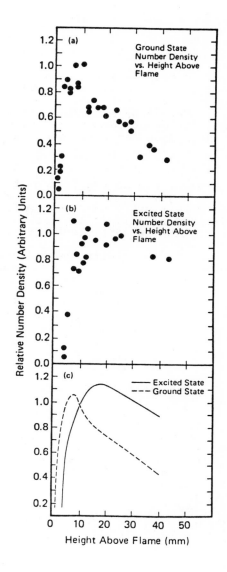

Fig. 8:   Map of sodium distribution in
           a lean air-acetylene flame.
    (a)   Ground state atomic distribu-
           tion with a 10% absorbing
           solution.
    (b)   Excited 4d state atomic distri-
           bution with a highly concentra-
           ted solution aspirated into the
           burner.
    (c)   Smooth curve comparison of 8(a)
           and 8(b). The vertical scales
           for these two curves are not
           the same.

From Fig. 5 we thus deduce an observed linewidth (FWHM) of 8.3 ± 0.3 GHz. Corrected for the Doppler-width of 3.4 GHz (at ~ 1900 K) and 1.77 GHz hyperfine splitting, the net pressure broadened width of 6.4 ± 0.3 GHz agrees fairly well with literature data of 7.5 GHz, derived by extrapolating to atmospheric pressure the pressure broadening coefficient of N_2 on the Na D_1 line [22]. Vice versa linewidth measurements of well defined transitions may be employed for flame temperture measurements using FM-spectroscopy.

FM-spectroscopy was employed for concentration mapping of the flame for both ground state sodium (on the Na 3s-3p, 589.0 nm, resonance at $\alpha L \approx 0.1$) and excited state sodium (on the Na 3p-4d, 568.3 nm, resonance at 1/10 saturated solution of NaCl corresponding to roughly 10^{13}-10^{14} Na atoms/cm^3. The experimental data in Fig. 8 clearly show the existance of a zone of low number density of excited state sodium, stretching up to 5 mm above the burner and indicating a low temperature zone.

Conclusion

We have introduced two methods for flame diagnostics:

a)   The Optoacoustic Laserbeam Deflection (OLD) technique is attractive for its simplicity and relatively high precision of 1.5% for temperature measurements. Spatial resolution is sufficient for flame temperature mapping.

b)   FM-spectroscopy with the present set-up can detect absorptions as low as $1.5 \cdot 10^{-4}$ with S/N of one which is an improvement of an order of magnitude as compared to traditional AAS, and allows for measurement of Na ground state densities of down to $3.3 \cdot 10^6$ atoms/cm^3 in a 5-pass cell at a cell length of 11 cm. Absolute calibration of the FM-signal versus absorption is easily performed and shows a linear dependence. The high spectral resolution of FM-spectroscopy allows for accurate linewidth measurement, which may be used for flame temperature determination. Furthermore transient data with high temporal resolution may be obtained [20] due to the high signal bandwidth.

References

1.   J.H. Bechtel and A.R. Chraplyvy, Proc. IEEE 70, 658 (1982)
2.   W. Zapka, P. Pokrowsky and A.C. Tam,, Optics Letters 7, 477 (1982)
3.   E.A. Whittaker, P. Pokrowsky, W. Zapka, K. Roche, G.C. Bjorklund, J. Quant. Spectrosc. Radiat. Transfer (to be published).
4.   J.E. Allen, W.R. Anderson and D. Crosley, Opt. Letter 1, 118 (1977)
5.   K. Tennal, G.J. Salamo, R. Gupta, Application of Photoacoustic Spectroscopy to Combustion Diagnostics, in Technical Digest, Second International Topical Meeting on Photoacoustic Spectroscopy, Berkeley, 1981 (unpublished).
6.   H.P. Hooymayers and C.Th.J. Alkemade, J. Quant. Spectrosc. Radiat. Transfer 6, 847 (1966).
7.   A.C. Gaydon, H.G. Wolfhard, Flames, Their Structure, Radiation and Temperature, Wiley & Sons, New York (1979).
8.   G.J. Van Wylen, Thermodynamics, Wiley & Sons, New York (1959).
9.   L. Pasternack, A.P. Baronavski and J.R. McDonald, J. Chem. Phys. 69, 4830 (1978).
10.   J.A. Gelbwachs, C.F. Klein and J.E. Wessel, Appl. Phys. Letters 30, 489 (1977)
11.   J. Allen, W.R. Anderson and D. Crosley, Opt. Letter 1, 118 (1977)
12.   P.K. Schenk and J.W. Hastie, Opt. Eng. 20, 522 (1981).
13.   L.A. Rahn, L.A. Zynch and P.L. Mattern, Opt. Commun, 30, 249 (1979)
14.   D.T. Cassidy, J. Reid, Appl. Opt. 21, 1185 (1982).
15.   R.K. Hanson, P.A. Kuntz and C.H. Kruger, Appl. Opt. 16, 2045 (1977).
16.   K.C. Lück and F.J. Müller, JQSRT 17, 403 (1977).
17.   R.M. Green and J.A. Miller, JQSRT 26, 313 (1981).
18.   G.C. Bjorklund, Opt. Lett. 5, 15 (1980); G.C. Bjorklund, W. Lenth, M.D. Levenson and C. Ortiz, SPIE 286, 153 (1981).
19.   W. Zapka, M.D. Levenson, F.M. Schellenberg, A.C. Tam and G.C. Bjorklund, Opt. Lett. 8, 27 (1983).
20.   W. Zapka, P. Pokrowsky, F.M. Schellenberg and G.C. Bjorklund, Opt. Commun. 44, 117 (1982)
21.   M.D. Levenson, W.E. Moerner and D.E. Horne, Opt. Lett. 8, 108 (1983).
22.   R. Kachru, T.W. Mossberg and S.R. Hartmann, Phys. Rev. A 22, 1953 (1980).

Innovative nondestructive techniques for measuring the refractive
index of a lens using laser light

R.S. Kasana and K.-J. Rosenbruch

Physikalisch-Technische Bundesanstalt, Bundesallee 100, 3300-Braunschweig
Federal Republic of Germany

Abstract

Some nondestructive techniques for measuring the refractive index of a lens have been
reported. The salient feature of this innovative approach is to avoid the use of two miscible liquids as commonly used in the liquid immersion method. Wavefront shearing interferometry and the coherent speckling phenomenon have been applied for determining the focal
length of the test lens immersed in different types of liquids. The reported method is
quick to perform and adequate for producing good accuracy as compared with existing
techniques.

Our intentions and the main objectives of our efforts are to point out that the present
approach is economic as well as less time-consuming, especially when the type of glass in
a lens is to be identified. The preparation and the selection of the required liquids for
the miscible liquid immersion technique is a time-consuming process and moreover, it needs
a large number of trials to achieve the desired results. In contrast to this, the proposed
method does not need two miscible liquids.

A mathematical equation has been derived for calculating the lens' index which depends
upon the focal lengths of the lens in different liquids and their refractive indices. It
has also been deduced that the plano-convex lens and a thin lens behave in the same way and
produce identical results. For N liquids there will be $N(N-1)/2$ ways of calculating the
lens' index. Henceforth, the present technique will be termed the nondestructive nonmis-
cible liquid immersion technique.

Introduction

The refractive index of a lens can be calculated by knowing its power and the strength
parameters. For such purposes, instruments such as spherometers and vertometers can be
used, but this is seldom the case because of their limited accuracy. Optical techniques are
therefore employed for higher accuracy and better results. The liquid immersion method is
generally used for measuring the refractive index of a simple lens.[1] The collimation technique is also adequate for quickly recognising the type of the glass in a lens just by
knowing the lens' index.[2] But owing to its recursive nature, some trials and errors are introduced. This is also limited by the inaccuracies involved in the measurements of the
various parameters such as the equivalent power of the collimator, the strength parameters
of the lens, the displacement of the object from the front focal plane of the collimator,
etc. Apart from the factors discussed here, these techniques require two miscible liquids
or compounds. Sometimes it becomes a problem to manage such liquids in sufficient quantities.

In view of these limitations, the authors have put forward some simple and quick methods
for determining the lens' index. The techniques discussed are based upon shearing interferometry and the coherent speckling phenomenon which are nondestructive in nature and give
adequate accuracy. The focal length is affected by the presence of the glass cell with flat
sides and the thickness of the liquid column inside the glass cell. The change in focal
length due to these factors has been taken into accout. If the walls of the cell are parallel to each other, its focussing effect is reduced.

Experimental arrangement

The optical components and the devices are mounted and adjusted as shown in fig.1. The
light from the laser source is filtered out and is collimated by the spatial filtering
unit. The collimation is tested by the plane parallel glass plate (Murty shearing interferometer).[3] The test lens inside the glass cell immersed in the liquid converges the light
to a point on the plane mirror. It also collimates the light reflected from the mirror.
The position of the mirror decides the exact focal plane.

In place of the Murty shearing interferometer, the Ronchi grating or a glass diffuser
plate can also be used. The grating or the diffuser itself locates the exact focal plane.
Hence, the plane mirror is replaced either by the grating or by the diffuser.

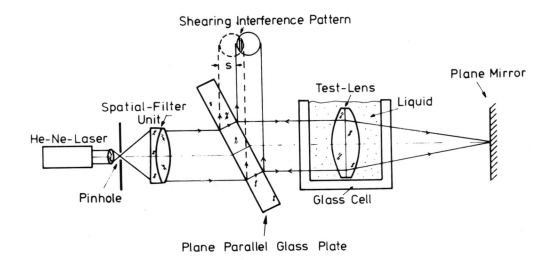

Figure 1. "Optical system for observing the defocussing and measuring the focal length."

<u>Theory</u>

The focal length of the test lens immersed in a liquid inside the glass cell is the main parameter to be discussed and measured. In determining the lens' index, its exact focal length should first be measured accurately with the different types of liquids in the glass cell. The theoretical formulation can then be developed as follows:

The general expression for the focal length of a lens in the presence of the liquid around the lens can be written as:

$$(1/f) = (n - n_L)(C_1 - C_2) + (n - n_L)^2 \cdot tC_1C_2/n \tag{1}$$

where n and n_L are the refractive indices of the test lens and the liquid, respectively.
C_1 & C_2 are the curvatures of the test lens
t and f are the thickness and the focal length of the test lens.

If we use the two liquids separately, the focal lengths of the lens would be

$$(1/f_1) = (n - n_1)(C_1 - C_2) + (n - n_1)^2 \cdot tC_1C_2/n \tag{2}$$

and

$$(1/f_2) = (n - n_2)(C_1 - C_2) + (n - n_2)^2 \cdot tC_1C_2/n \tag{3}$$

Dividing equation (2) by (3) and by means of further transformations, we have

$$n^2 \left[(1 + K)(f_1 - f_2) \right] - n(1 + 2K)(n_1f_1 - n_2f_2) + K(n_1^2 f_1 - n_2^2 f_2) = 0 \tag{4}$$

where $K = tC_1C_2/(C_1 - C_2)$ \hfill (5)

The roots of the quadratic equation (4) are as follows:

$$n = \frac{(n_2f_2 - n_1f_1)}{(f_2 - f_1)} \left[1 + \frac{-1 \pm \sqrt{1 + \alpha}}{2(1 + K)} \right] \tag{6}$$

where $\alpha = 4(n_1 - n_2)^2 (K + K^2) f_1 f_2 / (n_1 f_1 - n_2 f_2)^2$

or $\qquad = K(1 + K) f_1 f_2 \left[2(n_2 - n_1) / (n_2 f_2 - n_1 f_1) \right]^2$ $\qquad\qquad$ (7)

Thus equation (6) is the general equation for finding the lens' index. It can now be modified for particular cases as follows:

I. For a thin lens

I. For a thin lens

We have t = O, which implies that the value of K as well as of α becomes zero. Hence,

$$n = (n_2 f_2 - n_1 f_1) / (f_2 - f_1) \qquad\qquad (8)$$

II. For a plano-convex lens

We have C_1 or C_2 = O, which makes both the factors K and α equal to zero. Therefore,

$$n = (n_2 f_2 - n_1 f_1) / (f_2 - f_1) \qquad\qquad (9)$$

Thus, equations (8) and (9) show that a thin lens and a plano-convex lens have the same mathematical equation for the value of their refractive index. The lens' index relation can be written in a more general way as

$$n = \frac{n_j f_j - n_i f_i}{f_j - f_i} \qquad\qquad (10)$$

where i and j indicate the different liquids.

To implement equation (10) for determining the lens' index, the focal length of the test lens with i^{th} and j^{th} liquids should be measured. Accuracy in the measurements depends upon how exactly the value of the focal length has been determined. Some liquids of known indices are selected and the focal length of the test lens with each liquid is determined independently. The possible theoretical error Δn in estimating the value of the lens' index n can also be calculated as:

$$\frac{\Delta n}{n} = \frac{n_j \Delta f_j + n_i \Delta f_i}{n_j f_j - n_i f_i} + \frac{\Delta f_j + \Delta f_i}{f_j - f_i} \qquad\qquad (11)$$

As is obvious from this equation (11), the possible error can be reduced to a great extent by choosing liquids of higher refractive indices. When the refractive indices of the liquid and of the lens match each other, the effective focal length of the lens has become infinity. In other words, the combination of the lens and the liquid will behave as a plane parallel plate and this plate does not focus the incident parallel light wavefront. For testing the parallelism of the wavefronts and observing the defocussing involved, the Murty shearing interferometer has been employed.

Murty shearing interferometer

This is a simple plane parallel plate of glass as shown in fig.1, where it acts as a double shearing interferometer for the plane wavefronts coming-out from the spatial filter unit and from the test lens side. Hence, the parallelism of the wavefront emerging from the spatial filter unit is checked on the side facing the incident light. The light converging to a point by the test lens incides on a plane mirror and this point acts as a point source. The reflected spherical wavefront again passes through the test lens which acts as an autocollimator and collimates the light. The parallelism of this wavefront emerging from the lens is also checked by the same shearing interferometer.[3] The defocussing involved is observed in the form of fringes in the shearing interference pattern.

The parallelism of the wavefront depends upon the position of the point source on the plane mirror. The amount of defocussing can be calculated by using the theory of wave-abberations.[4]

$$\Delta f = \frac{1}{2} \cdot \lambda f^2 / s \cdot \Delta X, \tag{12}$$

where λ - wavelength of the laser light, f - focal length of the test lens,
s - shearing factor, ΔX - distance between two fringes occurring in the overlapping region of the laterally shearing wavefronts. The factor (1/2) accounts for the double passing of the light.

If there are N fringes in the overlapping region D between these two shearing wave-fronts, the equation (12) can be written as:

$$\Delta f = \frac{1}{2} \cdot \lambda f^2 / s \cdot (D/N)$$

$$= \frac{1}{2} \cdot (\lambda/s)(N/D) \cdot f^2 \tag{13}$$

But the possible error in focussing can also be calculated by differentiating the equation (1) for a thin lens as:

$$\Delta f = f^2 \cdot (C_1 - C_2) \cdot \Delta n \tag{14}$$

Comparing equations (13) and (14), we have

$$\Delta n = \frac{1}{2} \cdot (\lambda/s)(N/D) r_1 r_2 / (r_2 - r_1) \tag{15}$$

for a plano-convex lens

$$\Delta n = \frac{1}{2} \cdot (\lambda/s)(N/D) r \tag{16}$$

where r - radius of curvature of the curved surface.

There should not therefore be any error involved either in focussing or in the refractive index value of the lens if zero fringe exists in the overlapping region between the two shearing wavefronts. However, it is too difficult to observe a fringe-free space. Hence, a new criterion has been adopted for eliminating the focussing error involved during the investigations.

The shearing factor s can also be calculated by simple geometric optics and comes out as:

$$s = t' \cdot \sin 2w \left/ \sqrt{n'^2 - \sin^2 w} \right. \tag{17}$$

where n' - refractive index of the glass plate, t' - thickness of the glass plate,
w - angle between the parallel beam incident on the glass plate from the lens' side and its normal to the plate surface.

For the fringes with a good contrast, the angular size of the point source should be of the order of (λ/s). The angular size approaches the diffraction limit if s is quite a large fraction of the diameter of the wavefront.

Experimental observations

The observed defocussing defect is shown in the form of photographs in fig.2. The number of fringes corresponds to the amount of defocussing involved during the investigations.

The glass cell in which the lens is inserted also affects the focal length of the lens. If the sides of the cell deviate from the optical flatness and parallelism, some undesired fringe pattern is obtained which indicates the focussing action of the cell. This error in focal length is troublesome and arises due to the focussing nature of the cell, therefore to eliminate this error, a good quality cell should be used. In that case, there should not be any fringe pattern when the plane wavefront incides on the glass cell.

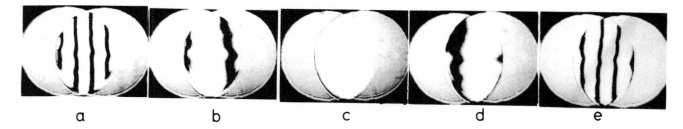

Figure 2. Photographs of the different values of the defocussing; (a) $\Delta f = -1.50$ mm,
(b) $\Delta f = -0.50$ mm, (c) $\Delta f \cong 0.00$ mm, (d) $\Delta f = +0.50$ mm, (e) $\Delta f = +1.50$ mm,

The experiment for a plano-convex lens has been performed. The lens has a focal length in air equal to 308.33 mm, radius of curvature $r_1 = 161.003$ mm and a thickness t = 4.22 mm. Its calculated refractive index is 1.5222. The diameter of the lens is equal to 25 mm. The experiment has been performed by using the laser light, hence, the refractive indices of all the liquids used in the experiments have been determined by a Pulfrich refractometer with the help of laser light. The values of the focal length and the refractive index of the lens with the different liquids have been tabulated in table 1 below:

Table 1. (A)

No.	Liquids/Medium	Refractive index of the medium	Measured focal length in mm
1	Air	1.0000	308.33
2	Water	1.3317	845.24
3	Nonan	1.4041	1364.01
4	Terpentine	1.4775	2958.16

(B)

No.	Possible set(i,j) of two media	Lens' index $n = (n_j f_j - n_i f_i)/(f_j - f_i)$	Possible error in estimation
1	Air-Water	1.52219	0.00010
2	Air-Nonan	1.52212	0.00005
3	Air-Terpentine	1.52218	0.00002
4	Water-Nonan	1.52206	0.00011
5	Water-Terpentine	1.52217	0.00002
6	Nonan-Terpentine	1.52221	0.00003

Special care should be taken in determining the exact value of the focal length. In practice, the distance measured is not the correct focal length of the test lens, but the distance between the focal plane and the back surface of the glass cell. The actual focal length which has been discussed in the text is given by

$$F = s'_p + t_G/n_G + t_L/n_L + s'_f \qquad \ldots \tag{18}$$

where s'_p – axial distance between the principal point and the back vertex of the lens.
t_G – thickness of the back of the cell.
t_L – axial thickness of the liquid column between the lens and the back of the cell.
s'_f – distance between the back surface of the cell and the focal plane.
n_G & n_L – refractive indices of the glass cell and the liquid, respectively.

The general expression for the factor s'_p has been calculated and is given by

$$s'_p = (t/n)(1 - r_2/r_1) \tag{19}$$

For a plano-convex lens we have

$$s'_p = t/n, \tag{20}$$

and for a biconvex lens of equal surface curvatures

$$s'_p = t/2n \tag{21}$$

The parameters of equation (18) are clearly indicated in fig.3

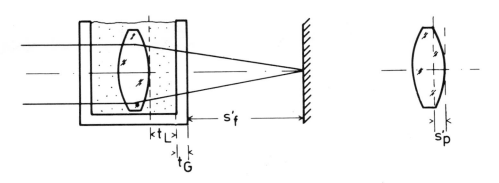

Figure 3. Description of the actual measurable parameters.

By turning or rotating the lens along its vertical axis, the principal point is coincided with the lens' optical center which has been decided by observing the movement of the focal point. The principal point overlaps the lens' center if the focal point remains stationary, irrespective of the turning of the lens. Under these conditions the distance between the principal point and the back vertex of the lens has been measured as having the value of s_p'. Its value is independent of the refractive index of the medium around the lens and so it is measured without pouring the liquid inside the cell.

Obviously, the focal length of the lens-liquid combination inside the glass cell varies with the refractive index of the liquid. With the increase in the liquid index, the effective focal length of the system is also increased. According to the mathematical formulation (13), the inaccuracy or the error in the focussing (i.e. Δf) given by a fraction of a fringe (say, 1/10th of a fringe) increases by f^2 and reaches several millimeters in the case of terpentine. The exact value of s_p' is therefore important only for the low index media viz. air, water etc.. For liquids of nearly the same refractive index as that of the lens, the value of s_p' does not play any significant role. In such a case, the parameter s_f' is sufficient to produce the desired result.

The refractive index is affected by the variation in the temperature by a factor of $2-6.10^{-4}$ per degree. Hence, temperature control within the limits of ± 0.1 °C is recommended for performing the experiments.

Ronchi grating as a wavefront shearing interferometer

Apart from the Murty shearing interferometer, the Ronchi grating test can also be used as a shearing interferometer. The spherical wavefront is diffracted by the Ronchi grating into various identical diverging wavefronts which interfere and produce the fringes. These fringes can be observed on a screen placed at any arbitrary distance from the grating. This is essentially the lateral wavefront shearing interference phenomenon.[5,6]

Briers has interpreted this type of shearing in terms of Ronchi test formulation.[7] The fringe spacing depends upon the grating position along the optical axis and the defocussing can be expressed as:

$$\Delta f = R.g/\Delta X \tag{22}$$

where R - distance from the focus to the sheared wavefront, g - grating element and
ΔX - fringe spacing

If the spatial frequency of the grating is ν and N fringes appear in the overlapping region of the linear distance D, the defocussing parameter can be written as:

$$\Delta f = R.N/\nu D \tag{23}$$

To deduce a relation for the focal length, we may compare equations (13) without factor (1/2) and (23) as follows:

$$f^2 = R.s/\lambda.\nu \tag{24}$$

Thus we conclude that the defocussing or the focal length of the test lens can be easily determined by the Ronchi test formulation. In this case two spherical wavefronts are laterally displaced for generating the sheared interference. The exact position of the focus is a point where the lens converges the light and either this point or its plane can be located by using the Ronchi grating.

Coherent speckling spots

A simple technique of coherent speckling can also be applied for locating the exact focal plane of the test lens. The exact focus can be traced out by observing the broadening in the sizes of the speckling patterns or the spots. It has been noticed that the size of the speckling spots is maximum when the glass diffuser is located exactly in the focal plane, while this decreases if the diffuser plate moves away from the focal plane in either the positive or the negative direction. The screen for observing the speckling spots is fixed at any arbitrary distance. The size of the speckles also depends upon the diffuser grains and hence only one diffuser plate should be used for a particular set of investigations.

Results and discussions

In the present case the plane parallel plate of glass of the thickness $t' = 8.99$ mm and the refractive index $n' = 1.4571$ has been introduced at a shearing angle $w = 45°$. The wavelength of He-Ne laser light is 6328 nm. A plano-convex lens has been used as a test lens whose parameters are: focal length $f = 308.33$ mm, thickness $t = 4.22$ mm, radius of curvature $r = 161.003$ mm, diameter = 25 mm and the refractive index $n = 1.5222$.

The theoretical prediction that the error Δn in the estimation of the lens' index decreases if the factor $(f_j - f_i)$ increases, has been experimentally verified. Hence, liquids of higher indices are recommended.

The refractive index also depends upon the wavelength of the light used. To eliminate the possibility of error due to the wavelength dependence, the refractive indices of all the liquids are measured with the He-Ne laser light. There is therefore no chance of any error, as the experiment is also performed with the same light.

It is obvious from the theoretical formulation that the uncertainty Δf is directly proportional to the square of the focal length. Hence, for the higher values of the focal length f, the uncertainty in locating the focal plane becomes several millimeters. This has also been experimentally verified.

The effect of the lens' aberrations and the relative aperture have been also investigated. It has been found that the spherical aberrations are the functions of the focal length. The longitudinal spherical aberration is directly proportional to the f-number. Hence, for a particular value of the lens' aperture, the amount of the longitudinal spherical aberration varies with the focal length. Thus with liquids of a higher refractive index, the longitudinal spherical aberration increases because of the large focal lengths.

It has been also noticed that the wavefront aberrations are in inverse relation to the focal length. This is reduced to a great extent with liquids of higher refractive indices.

Acknowledgements

The authors wish to thank Mr. H. Reuschel for his technical assistance in carrying out the experimental investigations. Dr. Kasana is on leave from the Indore University of India under the Indo-German Academic Exchange Programme and he acknowledges the financial support of the Deutscher Akademischer Austauschdienst.

References

1. Bergmann-Schaefer, "Lehrbuch der Experimental Physik Band III Optik." (Walter de Gruyter Berlin, New York 1978) p. 416
2. Smith, S., Appl. Opt. 21 755 (1982)
3. Murty, M.V.R.K., Appl. Opt. 3 531 (1964)
4. Welford, W.T., "Aberrations of the Symmetrical Optical System." (Academic Press, Inc. London Ltd. 1974) p. 74
5. Lenouvel, L. and Lenouvel F., Rev. d'Opt. 17 350 (1938)
6. Bates, W.J., Proc. Phys. Soc. (London) 59 940 (1947)
7. Briers, J.D., Opt. Laser Techn. 11 189 (1979)

Laser-based particle sizer

David Waterman and Barry Jennings

Electro-Optics Group, Physics Department, Brunel University,
Uxbridge, Middlesex, U.K.

Abstract

An apparatus for rapid particle size analysis has been developed, utilising the phenomenon of electric birefringence. A principle use is in analysing polydisperse colloid suspensions, where for a monomodal distribution, a complete size distribution plot is generated. This method can produce data in terms of an equivalent spherical diameter, but a unique feature is that if the particle shape is known then the results obtained give data on the true major dimension of the particles in the medium.

Electric birefringence

The majority of particles in the micron and sub micron size range are optically or geometrically anisotropic. When in a dilute suspension, the particles are randomly oriented, due to Brownian forces. The medium as a whole is then isotropic. However if orientational order is imposed by application of an electric field the medium becomes doubly refracting[1]. This phenomenon is known as electric birefringence or the Kerr (1875) effect. Thus the suspension may be considered to behave as a uniaxial crystal, with the optic axis parallel to the electric field direction. If linearly polarised light impinges on the medium at 45° azimuth to the direction of the electric field, the light may be considered to be transformed into two components, one polarized parallel and one polarized perpendicular to the electric field. On traversing a length l in the birefringent medium, a phase difference between the two orthogonal components is established. This is called the optical retardation δ and is given by

$$\delta = \frac{2\pi l \Delta n}{\lambda}$$

(1)

where the birefringence $\Delta n = n_{\parallel} - n_{\perp}$, $n_{\parallel}$ and $n_{\perp}$ are the principal refractive indices parallel and perpendicular to the electric field and λ is the wavelength of the light in vacuo. The resultant light emerging from the system is elliptically polarised. This is transmitted through a quarter-wave plate with its fast axis at 45° azimuth to the electric field and the elliptical polarization is restored to linearly polarized light. The angular deviation of the polarization direction from the initial 45° is a measure of both the magnitude and sign of the birefringence. This is detected as a variation in light intensity by transmission through a further polariser, offset by a small angle α from the crossed position. If the electric field E is applied as a short duration square pulse, the change in light intensity corresponding to the birefringence of the medium is in the form of a transient signal. A representation of this signal is shown in fig.1.

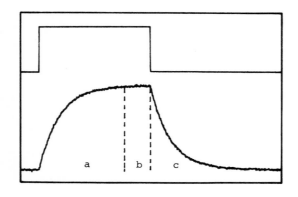

Region a. The electric field interacts with any particle dipole moments causing orientation.

Region b. Equilibrium orientation where the Brownian disorienting forces are balanced by the electrical orienting force.

Region c. Field free relaxation.

Figure 1. Field pulse and transient optical signal.

In a viscous fluid, the decay or field free relaxation of the birefringence is characteristic of the particle shape and major dimension. It is the analysis of this decay rate that provides the means of obtaining a size distribution for the constituent particles in the medium.

Principles of particle size analysis

For a monodisperse suspension, the relaxation is represented by a single exponential decay process, directly dependent on the rotary diffusion coefficient (D) of the particles. Typically sols are polydisperse and the decay is a multiexponential process representing all sizes of particles oriented by the electric field. This decay process is described by the equation.[2]

$$\Delta n_t = \sum (\Delta n_o)_i \exp(-6D_i t) \qquad (2)$$

where subscript i denotes the parameters for a specific size species and Δn_t is the birefringence at any time t after termination of the electric field pulse, when t = 0 and $\Delta n_t = \Delta n_o$. In theory the decay curve contains all the information required for a complete description of the total size species contributing to the decay. However this ideal case is not realisable in practice. The most accurate experimental data in the decay are those close to t = 0. Therefore the technique is to obtain the initial slope from a plot of the logarithm of the normalised birefringence of the decay curve against time. This slope (S) is equal to $-6\bar{D}$, where $\bar{D}$ is an experimental average rotary diffusion coefficient, which depends on the nature of the orientation mechanism of the particles and the specific experimental conditions. The total size species that are aligned by the orienting force and hence contribute to the decay process are determined by the pulse duration and amplitude(E). Generally the larger particles possess a greater charge separation, and will be subjected to a greater torque in the electric field and hence may be oriented in low amplitude fields. The smaller particles require a larger E for the same degree of orientation. Thus in poly-disperse sols the total size species contributing to the relaxation process may be determined by variation of the electric field amplitude.

In the majority of aqueous colloidal sols, induced dipolar mechanisms may be assumed to predominate. However permanent and induced dipole moments have different dependencies on particle dimensions[1,3]. If both are present they may be isolated by using bursts of sinus-oidal waves of such frequency that the permanent dipolar mechanism cannot contribute to the orienting torque. The size distribution for the majority of colloidal suspension may be adequately matched by a log-normal[4] curve, which has two characteristic parameters m and σ. These parameters are obtained from two values of $\bar{D}$ relating to two defined experimental conditions, which, given pulse durations sufficient for equilibrium orientation are, (i) a single low amplitude field E_o and (ii) a high amplitude field E_∞. Values for E_o and E_∞ are determined as follows. With increasing E the magnitude of the equilibrium birefringence Δn_o initially increases in direct proportion to E^2. It then deviates from this quadratic dependence until the value of Δn_o reaches a maximum; the orientation having reached statistically complete alignment of all the particles in the bulk suspension.

The method, in essence, is to capture two transient birefringence signals for low and high field conditions, to calculate the initial slopes from the decay curves and hence obtain two values of D. From these m and σ may be evaluated and a log-normal distribution of particle sizes plotted.[5,6,7]

The particle size analyser

An outline of the particle sizing apparatus is detailed in fig.2.

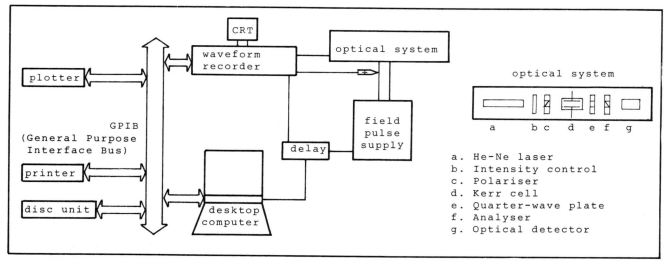

Figure 2. Block diagram of complete system with optical system in detail.

Optical assembly

Optical components.

The optics are contained in a rectangular matt black, anodised aluminium box. Each separate component is attached to a perspex mount which slides into accurately milled slots along the box sides. From equation 1. it can be seen that the magnitude of the optical retardation signal is inversely proportional to the wavelength of the light. However the ideal light source should be compact, stable, noise free and produce a high intensity, well collimated output. The optimum source is provided by a laser. A 2mW linearly polarised He-Ne laser is used in this case, the output of which passes through a rotatable polariser to enable adjustment of the initial intensity before transmission through the rest of the optics. The polariser and analyser are high precision calcite prism polarisers of Glan-Thompson design, manufactured from Schlieren free material, with 90% visible transmission and an extinction ratio of 1×10^{-6}. These are held in divided heads, adjustable to 1.7×10^{-3} radians. A similar holder contains the 633 nm wavelength quartz quarter-wave plate.

Kerr cell. The amplitude of the birefringence signal is directly related to the optical path length in the sample and to the particulate concentration, although the theory strictly applies to non-interacting particles and hence dilute systems. For optimum signal and transmission through the sample, a path length of 50 mm is chosen. The Kerr cell is manufactured to rigid specifications from optical quality glass, and is of open rectangular construction with internal dimensions 50 x 15 x 18mm. The end windows are fused, and doubly annealed to reduce strain birefringence to a minimum. The electrodes are of 50 x 15mm stainless steel; 3mm thick, placed with a separation of 2mm. These are held vertically along the cell length in an accurately machined, glass reinforced Teflon block. Inlet and outlet tubes are fed through the top of the electrode assembly, so that the sample under test may be introduced to the cell by syringe or may be continuously pumped through the cell as in an on-line system. The total sample volume is 2.5 ml.

Optical detector. After final passage through the analyser the light impinges onto the photodetector. This is a photomultiplier tube with 11 dynodes in a venetian blind structure and of antimony caesium emitting surfaces. The current output at the photomultiplier anode is converted to a voltage by either of two methods. Switch-selectable load resistors may be used in the following manner. The low field transient signal has a relatively long relax-ation time but low amplitude and therefore may have a large signal to noise (S/N) ratio. Increasing the value of the load resistor increases the voltage and decreases the detector circuit bandwidth hence limiting the high frequency noise present in the signal. However care must be taken to ensure that the time response of the detector circuit does not affect the measured relaxation time of the optical signal, and that the linear response is limited to small values of anode current. These last two limitations may be overcome by using an operational amplifier as a current to voltage converter. This increases the detection circuit bandwidth, hence passing the noise present in the signal. A computerised signal averaging technique may be used to increase the S/N ratio. This method also has the advantage of providing a simple means of 'backing off' the initial d.c. level by applying a variable voltage to the second input of the operational amplifier.

Data recording and analysis

Data recording.

The electrical signal from the detector is captured by a waveform recorder, which has maximum sensitivity of 100 mV full scale and in the case of small signals, a preamplifier is used. The waveform recorder is a Datalab DL 1080 transient recorder, which has dual channels and a sampling rate variable from 20 MHz to 50 Hz. This is a digital instrument so that the signals at the inputs are sampled at equal time intervals and the analogue signal is converted to a digital value and stored as an 8 bit binary number representing the amplitude of the signal at discrete moments in time. The information held in store is constantly output via a digital to analogue converter to a CRT monitor, where traces of the recorded waveform may be inspected visually or photographed. It may also be transferred over the general purpose interface bus, (GPIB), to a digital plotter for hard copy or via the desktop computer to the floppy disc unit for permanent storage. The computer acts as the GPIB controller and data analyser. After mathematical analysis the computed results may again be transferred on the GPIB, to the printer for copy of tabulated figures or rapid printout of the size distribution plot, the disc unit for storage, the plotter for accurate drawing of graphical data or via the DL 1080 for display on the monitor. The computer may also be used to initiate the field pulse by sending a trigger to the delay unit, which then triggers the transient recorder and after a short delay the field pulse supply. The electric field is applied to the electrodes in the cell and particle orientation occurs resulting in optical birefringence in the sample and a corresponding electrical signal from the detector to channel 1 of the DL 1080. The field pulse is monitored on channel 2 after attenuation by a 1000:1 high voltage probe.

Data analysis. The digital information in the two channels is passed over the GPIB to the computer. An edge detection machine code routine processes the values from the field pulse to give time locations for the rising and falling edges and so provide fiducial points for calculating the pre-field light intensity and the beginning of the decay process. Computer software has been written to determine the logarithm of the normalised birefringence values along the decay curve and calculate the value of $\bar{D}$ from the initial slope for that particular field amplitude. Fields corresponding to E_o and E_∞ are applied.

The resulting transient signals (fig.3) are transferred to the computer and the respective values of $\bar{D}_o$ and $\bar{D}_\infty$ determined. The operator may then key-in the particle shape from a menu displayed on the VDU. This may be equivalent spherical diameter (e.s.d), disc, oblate ellipsoid, prolate ellipsoid or rod and in the case of ellipsoid or rod an estimate for the axial ratio or diameter respectively must also be inserted. Values of the median m and a shape parameter σ for a two parameter size distribution are then calculated. The majority of polydisperse colloidal systems exhibit a log-normal size dependence and the values of m and σ are used to generate a full log-normal size distribution plot.

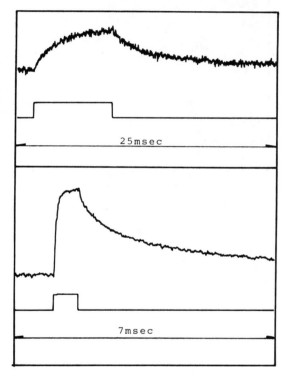

Waveform 1. Transient, after 'signal averaging', for the low field E_o region exhibiting the longer relaxation time attributable to larger particles.

Applied d.c. pulse.

Waveform 2. Transient signal for the high field E_∞ region. All particles are aligned and the fast initial relaxation is indicative of the smaller particles present.

Applied d.c. pulse

Figure 3. Typical recorded waveforms.

Sizing data

A suspension of the clay mineral halloysite was prepared in aqueous dispersion at approximately 0.01% solids concentration, using an ultrasonic homogeniser. The cell was filled with 2.5ml of the sample and a further aliquot prepared for examination by electron microcopy. Electric birefringence transients were obtained for low and high field conditions. Short bursts of sinusoidal waves at a frequency of 1kHz were used to inhibit any effect from possible permanent dipolar mechanisms. The pulse amplitude and duration for E_o was 100V for 3ms and for E_∞ was 1.2kV for 600μs respectively.

Computer analysis of the decay rates produced values of $24s^{-1}$ for $\bar{D}_o$ and $151s^{-1}$ for $\bar{D}_\infty$. At this stage had the particle shape been unknown the computer analysis system would have produced an equivalent spherical diameter. However the particle morphology was known and a more accurate value of particle size was obtained. A prolate ellipsoid of axial ratio 6 was used and the computed values of m and σ were calculated as shown in table 1. The electric birefringence values for m and σ were used to generate a cumulative size curve.

Particle lengths from electron micrographs were measured, and a size histogram produced. From this values of m, σ, and the mean size were calculated as detailed in table 1.

A comparison of the curves and histogram is shown in fig.4. The agreement is very satisfactory.

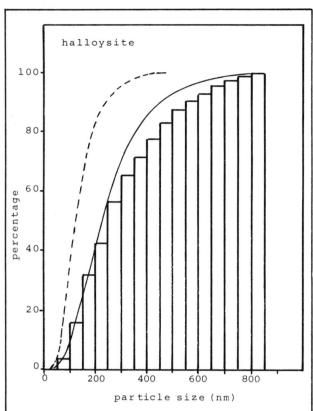

Figure 4. Comparison of size histogram data from electron micrographs for the halloysite dispersion, with size distribution curves from electric birefringence data, using theories for spheroidal (broken) and ellipsoidal (full) particles.

Table 1. Data for clay mineral halloysite.

$$\bar{D}_o = 24 \ s^{-1} \quad \bar{D}_\infty = 151 \ s^{-1}$$

	m (nm)	σ	mean (nm)
sphere	129	0.6	150
prolate ellipsoid	234	0.6	272
histogram	228	0.8	314

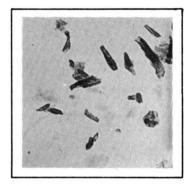

Figure 5. Electron micrograph of halloysite particles, showing ellipsoidal shape.

Conclusion

It is evident from this test sample that the common use of e.s.d. as used in the majority of commercial particle sizers, is extremely inaccurate for highly assymetric geometries. If the shape is known then electric birefringence theory and experiment allows an accurate estimate of size distribution to be rapidly obtained. Unlike traditional light scattering techniques, this method is impartial to small amounts of contamination, and should thus provide an online sizing procedure, suitable for commercial situations.

Tests have been carried out on a number of samples including biopolymers, minerals, pigments, synthetic polymers, liquid crystals, bacteria and biological macromolecules. The full size distribution curves are plotted in under 60 seconds from the first measurement.

References

1. Fredericq, E. and Houssier, C., Electric Dichroism and Electric Birefringence. Clarendon Press, Oxford 1973.
2. Benoit, H., Contribution a l'etude de l'effect Kerr présénté par les solutions dilutées de macromolécules rigides. Ann. Phys., Paris, 6, pp.561-609. 1951.
3. Mandel, M., The electric polarization of solutions of rod-like, charged macromolecules. Mol.Phys., 4, pp.489-496. 1961.
4. Hastings, N. and Peacock, J., Statistical Distributions, Butterworths, London. 1975.

5. Morris, V.J., Foweraker, A.R. and Jennings, B.R., Particle size distributions from transient electric birefringence data: I. Polydisperse rods by two-parameter distribution functions. Adv. Mol. Relaxn. and Int. Proc., 12, pp.65-83. 1978.

6. Morris, V.J., Foweraker, A.R. and Jennings, B.R., Particle size distributions from transient electric birefringence data: II. Polydisperse discs by two-parameter distribution functions. Adv. Mol. Relaxn. and Int. Proc. 12, pp.201-210. 1978.

7. Morris, V.J., Foweraker, A.R. and Jennings, B.R., Particle size distributions from transient electric birefringence data: III. Polydisperse ellipsoids by two-parameter distribution functions. Adv. Mol. Relaxn. and Int. Proc. 12, pp.211-220. 1978.

Analysing the dynamics of fast-moving objects using a pulsed laser diode

Raimo Ahola, Risto Myllylä

University of Oulu, Department of Electrical Engineering, SF-90570 Oulu 57, Finland

Abstract

A new method is suggested for analysing the dynamics of fast-moving objects. The basic differences, compared with the optical rangefinders and radar equipment available, are the continuous measurement of the distance and the possibility of measuring the alteration in distance and of measuring the velocity and acceleration of the object continuously and simultaneously. The above mentioned characteristics are attained using the well-known time of flight method with an exceptionally high pulse repetition rate and with a new type of an analog time to amplitude converter (TAC). The velocity and acceleration voltages are formed by derivation from the continuous distance voltage in respect of time. Preliminary results with a simple prototype have proven the method to be highly usable in many applications, e.g. in sports training and printing technology.

Introduction

Nowadays there are available a host of different kinds of electronic or optoelectronic devices capable of measuring distances quickly and accurately [1,2]. Many people are also familiar with, or have at least some kind of 'experience' of devices which are used to measure the velocity of an object, e.g. a traffic radar used by policemen. But there are very few commercially available devices which are able to measure both the distance to an object and the velocity of that object simultaneously and continuously.

The three most commonly used optical methods used to measure a distance, are interferometric, phase comparison and time-of-flight method. All of these methods have some unique features, e.g. the time of flight method is the only one which always gives an unambiguous distance result by a single measurement.

The velocity measurement of an object is usually based on the detection of the Doppler frequency shift of the signal backscattered from the moving object. The Doppler technique can be used not only with optical signals but also with microwaves and ultrasonic waves. In practical applications the Doppler technique has some drawbacks, such as the difficulty in measuring a low velocity, difficulty in finding out the direction of the velocity, difficulty in measuring the distance, and in some applications, the transverse (or partly transverse) velocity of the object causes errors. Most of the above mentioned difficulties can be eliminated, but it makes the device more complicated and more expensive.

When analysing the dynamics of fast-moving objects, two main groups can be distinguished. One of them is made up of applications where the motion of the object is directed back and forth and where the amplitude of the motion is thus quite small. As examples of these kind of applications one could mention the analysing of the fluttering of a paper band in printing technology and analysing the shaking of some machines. In some applications, the movement of the object is directed mainly in one direction, and the interesting characteristic of the object may be e.g. how the velocity changes when the object goes away or comes closer. A very good example of the latter group of applications is analysing the velocity of a sprinter during his run in respect to time or distance.

There are already devices available which can be used to measure and analyse small back and forth motions of an object. They are based on e.g. the measurement of the point of a light spot on the surface of the object, the measurement of the shape of a light strip on an object, change in the intensity of a light signal when the distance is changed, or the measurement of the Doppler frequency shift of laser light backscattered from the object [3,4].

If the object is moving mainly in one direction only, the available methods of measurement are also different. For example, in analysing the velocity of a sprinter during his run, there have been attempts to use a microwave radar, an ultrasonic radar, a host of photoelectric switches placed across the running track and video cameras, but none of them has so far given satisfactory results.

In this paper we suggest a new method of analysing the dynamics of fast-moving objects*. The method is based on the time of flight of a light pulse and a very high measurement rate. According to the results attained by a simple experimental device, the method can be used in

*Applications for Finnish patents, Nos. 821835 and 824240.

analysing the dynamics of the objects no matter if the object is moving back and forth or in one direction only.

The measurement of the distance, the alteration in distance, the velocity and the acceleration

The principle of the method

The block diagram and the operation principle of the experimental device is shown in fig. 1. The time of flight of the light pulse is measured by a new analog time to amplitude

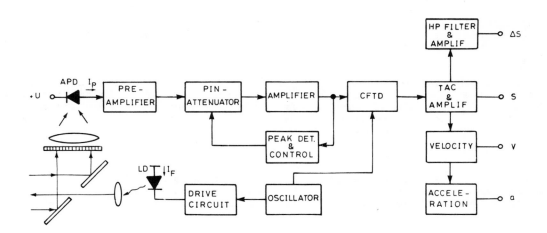

Figure 1. A schematig diagram of the experimental device.

converter (TAC) whose operation is based on a very stable pulse repetition rate. That is why the drive electronics of the laser are triggered by a crystal controlled oscillator. The timing information related to the emitted light pulse (START) is taken from the oscillator directly and not for example from the emitted light pulse. That causes an extra delay which gradually changes with temperature and age, but it is hardly a problem in measuring the dynamics of fast-moving objects. If absolute accuracy of the distance measurement is necessary, the receiver can for example be duplicated so that there are two identical receivers, one for the emitted and one for the reflected light pulse. The light pulse emitted by the laser is collimated into a proper beam divergence and a portion of the backscattered light is detected by the receiver.

Using a so called constant fraction timing discriminator (CFTD), the timing information of the START and STOP pulses are directed to the time to amplitude converter (TAC). With the TAC and proper amplification and filtering, the time of flight of the light pulses are converted to a continuous analog voltage (signal s in fig. 1). The voltage has a known proportion to the distance of the object and it can be analysed using 'standard' equipment. Using a high pass filter the dc component of the voltage can be removed. The ac component of the voltage describes the back and forth motion of the object and the voltage can easily be amplified to a proper level for standard equipment (signal Δs in fig. 1).

The velocity and acceleration voltages (v and a in fig. 1) are formed by derivation from the distance voltage in respect of time and filtered by low pass filters. These voltages have a known proportion to the velocity and acceleration of the object, respectively, and can thus also be easily analysed using standard equipment.

The transmitter

According to the method suggested in this paper, the primary information on the dynamics of the object is the continuous distance information. To be able to measure the distance with satisfactory resolution, the averaging of primary distance results is necessary. Fast moving objects on the other hand call for a fast measurement, and that is why the pulse repetition rate of the transmitter must be very high and it thereby allows averaging, too.

The best choice for a light source in respect of modulation would be the so called double heterostructure CW laser diode, but the peak output power (about 1 - 20 mW) of the laser would hardly be high enough. Single heterostructure pulse laser diodes have a high peak output power (typically more than 10 W), but they have quite a low maximum duty factor (0.02 - 0.2 %) which limits the maximum prr to the kHz-range. A compromise between the former laser types is a DH-type pulse laser whose peak output power is about 0.1 - 1 W and the maximum duty factor about 1 - 10 %.

A DH-type pulse laser which has a nominal peak output power of 500 mW with 2.5 A forward current and whose maximum duty factor is 5 % has been used as the light source for the experimental prototype. The drive circuit of the laser consists of VMOS and bipolar transistors, and it is capable of driving short (FWHM $\simeq$ 5 ns) current pulses with a very high prr (well beyond 1 MHz). Because of the short current pulses, the laser can be modulated up to a pulse repetition rate of 10 MHz without exceeding the maximum duty factor. When the laser is driven with very short current pulses, the peak output power becomes lower, e.g. the laser used had the peak output power of about 35 % of its maximum when the width of the current pulse was 5 ns. (With a 3 ns pulse width the output power is practically zero.)

Because the transmitter is capable of operating with a very high prr the distance range selected may in some cases limit the maximum prr, if only one light pulse at a time is allowed to be in the way. For example, if the distance range needed is 150 m, the maximum prr is about 1 MHz.

The light pulses emitted by the laser are gathered into a proper beam divergence with a simple optics. If the target is passive, the beam divergence may be as small as possible. If an active target (e.g. a corner cube) is used, the divergence of the emitted beam (and of course the field of view of the receiver) must be chosen and the device aimed so that the target is all the time within the beam.

The receiver

The receiver part of the device is very similar to those of 'ordinary' range finders operating on the time of flight method. High prr does not cause any special demands. The wide field of view needed in some applications is realized simply using a large area APD as a detector.

The pre-amplifier of the receiver is a so called transimpedance amplifier, constructed using discrete bipolar transistors and having a bandwidth of 150 MHz and an input noise current density of about 3 pA/$\sqrt{Hz}$. To make the amplitude variation of the received pulses smaller, the receiver is equipped with an automatic gain control (AGC). The AGC-operation is accomplished using a PIN-diode attenuator, an RF amplifier, a peak detector and control electronics as depicted in fig. 1. The effect on timing of the amplitude variation, which goes over the dynamics of the AGC, and the very fast variations of the amplitude are eliminated using a constant fraction timing discriminator (CFTD). The dynamics of the AGC are more than 30 dB.

The CFTD

A simple constant threshold timing discriminator is not satisfactory because of the limited dynamics of the AGC and the finite risetime of the received pulses. (The risetime of the pulses are typically at least 3 ns, which corresponds to a distance of about 45 cm). One solution to the problem is to use a so called constant fraction timing discriminator (CFTD), which makes the timing decision on the leading edge of the pulse always at the same fractional point (e.g. at a 20 % point) with regard to the amplitude of the pulse. With proper arrangement, the discriminator is insensitive to both the amplitude and rise time variations. If the rise time of the received pulse is, for example, 3 ns, the walk caused by the CFTD can be within 1 cm when the amplitude variation is about 1:20 [5]. The constant timing pulses are directed to the time to amplitude converter (TAC).

The TAC

The method suggested for analysing the dynamics of fast-moving objects demands a very good resolution and accuracy for the measurement of distance and especially for the measurement of the alteration in distance. That is why it is quite obvious that only an analog method for measuring the time of flight of a light pulse gives a good enough resolution. On the other hand, 'smooth nonlinearities' of the distance measurement do not make the accuracy of the measurement of the velocity and acceleration very much worse. (The nonlinearity is generally regarded as the main drawback of the analog methods used in measuring short time intervals).

A very common analog method to measure short time intervals is based on charging a capacitor with a constant current during the time interval to be measured. If the current and the capacitance are constant and the current switches ideal, the voltage change accross the capacitor is directly proportional to the time interval.

A new kind of time to amplitude converter which effectively makes use of the very high prr of the system has been constructed for the experimental device. The principle of the

TAC is depicted in fig. 2 and in fig. 3 one possibility for realizing the principle is shown.

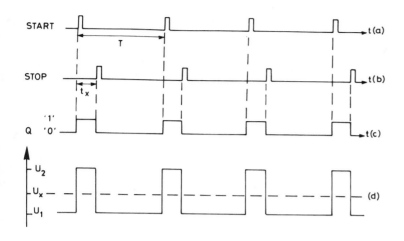

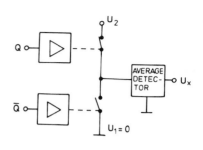

Figure 2. The principle of the time to amplitude converter (TAC) based on a precise time interval (T) between the successive light pulses. t_x is the time interval to be measured and U_x the corresponding voltage.

Figure 3. One possibility for realising the principle shown in fig. 2. When Q or $\bar{Q}$ is '1' the corresponding switch is on and when '0' the switch is off.

(A TAC constructed according to fig. 3 was used in the experiments). According to fig. 2 the light pulses are triggered precisely at constant (crystal controlled) time intervals T. After a time t_x, the backscattered light pulse is detected at the receiver. With an ECL RS-flip-flop for example the START and STOP pulses are combined to a signal shown in fig. 2c. With a proper circuit, the logical levels in fig. 2c can be converted to the signal in fig. 2d.

As shown in fig. 2, the time of flight of a light pulse, t_x, is converted to the duty factor of the square wave in fig. 2d. The average value of the signal, U_x, is directly proportional to the time interval t_x. Using the symbols in fig. 2, U_x can be written in a form

$$U_x = (U_2 - U_1)\frac{t_x}{T} + U_1 \tag{1}$$

If the TAC is constructed as in fig. 3, where the reference voltage $U_1 = 0$, the formula (1) gets the form

$$U_x = \frac{U_2}{T} t_x \tag{2}$$

In the simpliest case, the average detector in fig. 3 is accomplished using a RC low pass filter. When the time constant RC is selected to be much longer (e.g. 100...1000 times longer) than the time interval between successive pulses T, the nonlinearity caused by the averaging circuit is small and 'smooth'. If, for example, T = 1.4 µs (the distance range about 200 m) and the averaging time constant RC = 440 µs (the figures are those used in the experimental device), the maximum theoretical deviation from a straight line in the distance scale is about 3.6 cm. By increasing the prr (1/T) or the time constant (RC) the nonlinearity can be made smaller.

The results obtained in testing the whole device, (which are presented later in this paper), show that the TAC used is very efficient. Special advantages which could be mentioned are the relatively simple construction of the TAC, and the possibility of defining the 'scaling' factor (e.g. U_2/T in formula 2) easily and precisely.

The continuous measurement of the distance and the alteration in distance

As can be seen in fig. 1, and was mentioned in the introduction, the absolute accuracy in the distance measurement has not been the main goal in the experimental device. Very often the nature of analysing the dynamics of fast moving objects is such that the information about the absolute distance between the device and the object is immaterial.

Based on equation (2) and the known amplification, the relationship between the distance and the corresponding voltage has in the experimental device been adjusted to 1 mV $\triangleq$ 3 cm, when the range is about 200 m (T = 1,4 μs). The upper frequency limit (3 dB) of the distance voltage is about 360 Hz (RC = 440 μs). The continuous distance voltage can thus be analysed by 'standard' equipment (xy/t recorders, tape recorders, spectrum analysers, oscilloscopes etc.).

If the object is moving back and forth with a small amplitude, and the movement is analysed relatively far away from the object, the dc component of the distance voltage can cause some problems. In such cases, the dc component can be easily removed with a high pass filter, and the weak ac signal can be amplified to a proper level for 'standard' equipment used in analysing the motion.

The measurement of the velocity and acceleration

The velocity and acceleration voltages are formed by derivation from the distance voltage in respect of time and filtered by low pass filters with an upper frequency of about 20 Hz. The derivation causes very high demands on the resolution, the differential linearity and the integral linearity in 'short range' of the distance measurement. The derivation itself is a relatively simply method of forming the velocity voltage with a known relation to the velocity of the object.

The formation of the acceleration voltage by twice derivating from the distance voltage in respect of time emphasizes even more the 'differential' errors in distance voltage than derivating once. Obviously the result is also worse.

Some examples of the results

In all measurements represented in following, the prr of the experimental device has been about 0.7 MHz (the distance range about 200 m) and the upper frequency limit of distance measurement about 360 Hz. As was already mentioned earlier, in the distance measurement the resolution, the differential linearity and the integral linearity in 'short range' are very critical characteristics. In fig. 4, there is an example of the nonlinearity and resolution of the distance measurement within a range of 1 m measured at the distance of about 60 m. (A corner cube with an aperture of about 38 mm was used as object in the measurement). On the ordinate in fig. 4, the deviations of the measured distance results are shown compared with the results obtained by linear curve fitting. The standard deviation (σ) between the results in fig. 4 is about 1.2 mm. (The results are not essentially dependent on the distance at least up to 60 m.)

The results in fig. 4 show that the TAC used in the device is very efficient. The total standard deviation of 1.2 mm in a range of 1 m corresponds to an error of about 8 ps in time, and that includes all the errors present in the measurement.

In fig. 5, the result obtained directly by a yt-recorder is shown, when the object was a corner cube fastened to the end of a 'mathematical pendulum' moving back and forth.

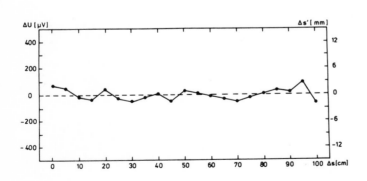

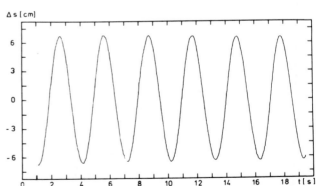

Figure 4. The deviation of the measured distance results compared with the results obtained by linear curve fitting. The object (a corner cube) was at a distance of about 60 m from the device and it was moved at 5 cm intervals over a range of 1 m.

Figure 5. The amplitude of the motion of a 'mathematical pendulum' measured at a distance of about 40 m from the pendulum (A corner cube was fastened at the end of the pendulum).

The distance between the object and the device was about 40 m. The ordinate in fig. 5 is the voltage Δs in fig. 1. The lower frequency limit (3 dB) of the voltage is about 0.007 Hz and the upper frequency limit (with one dominant pole) about 7 Hz (roughly 100 times greater than the dominant frequency of the pendulum). The plot in fig. 5 shows that also very small motions can be analyzed quite far away from the object.

In figures 6 and 7 (correspondingly to fig. 5) the velocity and the acceleration of the 'mathematical pendulum' are shown. (Figures 5, 6 and 7 are not measured precisely at the same time and hence the amplitude of the pendulum was not the same in different measurements). Based on fig. 6, it can be roughly estimated that the resolution of the velocity measurement is about 0.03 m/s peak-to-peak (with an upper frequency limit of about 20 Hz). In fig. 7, the resolution is roughly about 0.3 m/s^2 peak-to-peak (with an upper frequency limit of about 20 Hz).

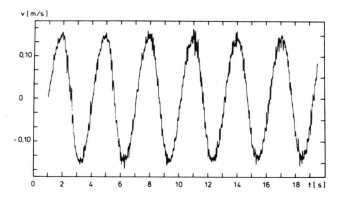

Figure 6. The velocity of the motion of a 'mathematical pendulum' measured at a distance of about 40 m from the pendulum.

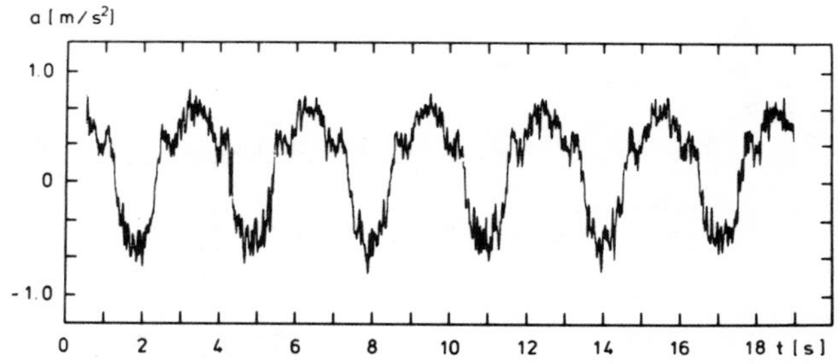

Figure 7. The acceleration of the motion of a 'mathematical pendulum' measured at a distance of about 40 m from the pendulum.

Examples of potential practical applications are shown in figures 8 and 9. The plot of figure 8 is obtained by an xy-recorder when the object was a corner cube fastened to the back of a running man. The device was placed behind the runner at a distance of about 4 m and the velocity and distance voltage were connected to the inputs of the xy-recorder. Because of the wide beam divergence of the transmitter and the wide field of view of the receiver, the corner cube remained inside the beam all the time during the running process.

The plot in fig. 9 represents the transverse motion (fluttering) of the paper band in a printing machine. The measurement was accomplished by recording with a tape recorder the ac distance voltage obtained from the device and by playing back the recordered signal with a lower speed via a yt-recorder. For such applications as in fig. 9, a distance range of about 200 m is of course far too large. If the system is optimized to measure small back and forth motion of relatively small distances (e.g. 0 - 2 m) the resolution is obviously much better.

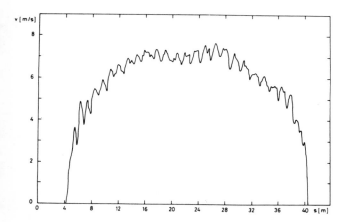

Figure 8. The velocity profile of a man when he was running a short distance with a corner cube fastened to his back.

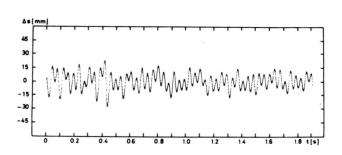

Figure 9. The fluttering of the paper band in a printing machine. The signal was first recordered with a tape recorder and then played back at a lower speed via a low pass filter and a yt-recorder. (The upper frequency limit was about 25 Hz).

Conclusions

Despite the seemingly 'discrete' nature of the distance measurement by using the time of flight method, with appropriate modifications it is very well suited also for analysing the dynamics of fast moving objects. Using a very high pulse repetition rate, averaging, and an efficient time to amplitude converter the information of the dynamics of the object can be extracted from the continuous distance information. By taking into account the special features in different applications, the performance of the device can be optimized.

Acknowledgements

The authors wish to thank 'The Foundation of Finnish Inventions' for financial support in the construction and testing of the experimental device and in patenting. One of the authors (R.A.) wishes also to thank the 'Koneen Säätiö' foundation for a personal grant for research work.

References

1. Laurila, Simo H., Electronic Surveying & Navigation, John Wiley & Sons Inc., USA 1976.
2. Zetsche, Hans, Elektronische Entfernungsmessung (EDM), Konrad Wittwer KG, BRD 1979.
3. Chijiwa, M., Morii, T., Measurements of Vibrations in Audio Components Using a Laser Vibrometer, Disa Information, no. 27, January 1982.
4. Nilsson, N-Å, Bennerhult, O., Vibration measurement on Moving Objects with aid of Laser-Doppler-Vibrometry (LDV), Nordisk Akustisk Sällskap NAS-82 proceedings.
5. Myllylä, R., Kostamovaara, J., A Positron lifetime spectrometer with improved linearity and resolution, Positron Annihilation edited by P.G. Coleman, S.C. Sharma, L.M. Diana, pp. 889-891, North-Holland 1982.

Theory and structure of a novel laser interferometric lens-centring system

Dong Tai-huo,   Yu Zhong-ru,   Yuan Jun-yi

Optical Instruments Dept., Zhejiang University
Hangzhou, Zhejiang, People's Republic of China

Abstract

This paper is described a high precise lens-centring apparatus for use with the altitude variation interferometric theory. It also analysed some basic principles for the centring accuracy. The possibility with higher cement-centring accuracy was proved in theory and obtained in practice. And the centring problem for high accuracy in manufacture and assembly of medium and large aperture was solved well.

Introduction

This paper is concerned with the problems of the theory and apparatus of lens-centring. Since 1970s, with the rapidly developing science and technology especially in the space technology and microelectronic industry, set a still higher demand on the performance of optical instruments. At each of these optical systems, as is well known to optical engineers there is a common characteristic that their resolution have been approaching the diffraction limiting even achieved, for example, the plano-field apochromatic lens which used in high magnification microscope for research, high contrast and high resolution multi-spectrum photogrammetric lens, specific wide-angle photographic lens used in LSI fabrication of sub-micron photo-mask lens and projection-print lens etc.

For the purposes to research and make such high precision lenses, it needs to overcome a series of technical difficults. One of them is the severely centring technique of lens during cementing and assembling their elements.

The techniques which are used in manufacturing usually are observing and measuring the centring images of inspected sphere by reflection or refraction. But the accuracy of classic optical centring method usually only about 0.01—0.02mm, it cannot suit the needs of the centring accuracy of above described high precision lenses.

So that, it is an improtant subject in optical engineering for precision correcting or measuring of lens eccentricity with accuracy of 1—2µm on magnitude.

Recently, the photo-electronic and laser interferometric methods are used in optical centring field to instead the old optical method. Welford had assembled a 140° telephotographic lens for the National Argome Lab. by Tropel interferometric system, and Carnell had modified and improved this interferometric system to measure the surface altitued variation and assembled a bubble-chamber 110° telecentric photographic lens with this centring system whose unsymmetry deviated from theoretical axis was less than ±1µm on any observing direction and was pleased in the results. In Bell Lab., they used a computer to correct decentring optical elements separately by semiautomatical process, the obtained accuracy was ±1µm (1 arc sec.).

This paper describes the experimental result of a new laser interferometric lens-centring system according to the surface altitude variation interferometric theory.

Some basic principles for the design

In designing the new laser interferometric lens-centring apparatus we evolved certain princeples that we could use it to improve the precision and sensitivity for measuring and correcting as follows.

1. The height-difference between two surface positions which caused by the decentre of the inspected sphere regarded as a information of eccentricity, and introduced it into the laser interferometric system to measure.

In terms of the new definition of lens-eccentricity published by the Western Germany (DIN) in 1976, the eccentricity of single surface is presented by the angle between the normal line of the top point of the optical centring and locating axis, i.e. the inclined angle of the surface ω .

The purpose of the most inspecting and correcting items is to measure this inclined angle of the surface. There are three measuring methods. From Fig.1 we can see:

A. to measure the distance between the both boundary of the reflective images of the inspected sphere centre with the classic optical methods, i.e. value $2c = (\overline{oo}_1 + \overline{oo}_2)$,

B. to measure the angle variation of the reflective light from point A at the height D/2 with the laser interferometric system,

C. to measure the surface altitude variation Δh caused by the eccentricity of inspected surface near the point A at the height D/2 by laser interferometric system.

Obviously, the accuracy of the third method is better than the others.

For the inspected surface we suppose that

$$\alpha = Sin^{-1}(D/2R) \tag{1}$$

where: α —— the centring angle
$\quad\quad\quad$ D —— the testing circle diameter where the laser beam incidenting
$\quad\quad\quad$ R —— the curvature radius of the testing lens,
let $c = \overline{oo}_1$ then

$$c = \omega \cdot R \tag{2}$$
$$\Delta h = 2c \cdot Sin\alpha \tag{3}$$
substitute form (1) and (2) into (3), then
$$\Delta h = \omega \cdot D \tag{4}$$

The displacement value of the reflective light exit pupil $\overline{F_1 F_2}$ is

$$\Delta L = 2c \cdot Cos\alpha \tag{5}$$

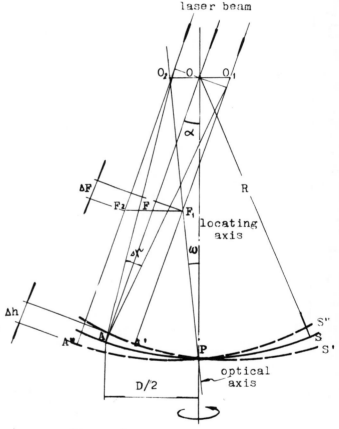

Since the displacement we mean the variation of reflective angle which caused by the eccentricity as angular aberration

$$\Delta\gamma = \frac{2c \cdot Cos\alpha}{R} = 2\omega \cdot Cos\alpha \tag{6}$$

So that the inspecting and measuring of the eccentricity is as follows.

$$C = \frac{\Delta\gamma \cdot R}{2Cos\alpha} \tag{7}$$

$$\omega = \frac{\Delta\gamma}{2Cos\alpha} \tag{8}$$

and the axial displacement of reflective light exit pupil ΔF is

$$\Delta F = \Delta h = \omega \cdot D \tag{9}$$

In fact, the aberration increment of reflective light in the optical path near the point A is 2(Δh), and in the interferometric system when the interferometric fringe changes from the maximum (light fringe) to the minimum (dark fringe) the aberrational increment will be coincide with the following equation,

$$2\Delta h = \pm N\frac{\lambda}{2} \tag{10}$$

and the eccentricity should be

$$\omega = \frac{N\lambda}{4D} \cdot \rho'' \tag{11}$$

$$C = \frac{N\lambda}{4D} \cdot R \tag{12}$$

Figure 1. Principle of decentre inspection for eccentric surface

2. The principle of the test-at-edge

From equation (6) $\Delta\gamma = 2\omega \cdot Cos\alpha$, we see the angular aberration is directly proportional to the cosine of the centring angle and the sensitivity of the eccentric testing is reduced with the testing aperture   increase, it is very detrimental. However, in equation (4), the surface altitude variation is directly proportional to the testing aperture, thus using the test-at-edge method one can get better sensitivity and accuracy.

We define that the aberrational increment which emerged during the central interference fringe variates from maximum (light fringe) to minimim (dark fringe) in the interference concentric circles pattern is called eccentric testing sensitivity (N=1) S_ω or S_c

$$S_\omega = \frac{\lambda}{4D} \cdot \rho''$$ (13)

$$S_c = \frac{\lambda}{4D} \cdot R$$ (14)

Thus, according to the image quality, the optical designer can pre-calculate the theoretical allowable eccentricity to define the counted total number (N) of central fringes changing from light to dark and vice versa, in other words the counted total number is the sum of light-dark cycle add dark-light cycle. Each cycle is 1 bit in number counting.

$$N_\omega = \frac{4\omega D}{\lambda}$$ (15)

$$N_c = \frac{4CD}{\lambda R}$$ (16)

Through experiments, we have proved that for medium aperture lenses of the radius of curvature within a range of -500mm to +5oomm during the testing, the eccentric testing sensitivity will be less than 2μm (about 1 arc sec.). For photo-mask lens and projection printing lens this accuracy is quite satisfactory. For example, we calculated the eccentric testing sensitivity of two lenses. The results are shown in table 1, 2 and 3.

Table 1. Centring accuracy of a photolithographic lens

surface	radius of curvature (mm)	testing diameter (mm)	eccentricity		number of fringes N	centring sensitivity S_ω/S_c
			C (μm)	ω (sec.)		
1	41.49	35	2	9.6"	10	1"/0.5μm
2	-38.4	35	2	10.4"	11	1"/0.5μm
3	110	35	2	3.6"	4	1"/0.5μm

Table 2. Centring accuracy of a large operture photographic lens

surface	radius of curvature (mm)	testing diameter (mm)	eccentricity		number of fringes N	centring sensitivity S_ω/S_c
			C (μm)	ω (sec.)		
1	90.8	157	2	4.5"	10	0.5"/0.5μm
2	1310	52	13	2"	1.5	1.3"/8μm
3	-2400	65	26	2.2"	2	1.1"/13μm

Table 3. Accuracy of cement centring

cement lens	radius of curvature (mm)	centring angle	number of fringes N	eccentricity	
				ω (sec.)	C (μm)
upper	47	15°	2	2.5"±0.5"	0.6±0.1
	-45	-18°	2	2.5"±0.5"	0.6±0.1
lower	-45	-15.5°	2	2.5"±0.5"	0.6±0.1
	-360	6.9°	1	1.25"±0.5"	2.3±0.1

3. In order for easy to observe and measure, and keep the operator's eyes in comfortable condition and convenient to use the photo-automatic measurement, the principle of the interferometric fringe number counting replaces the classic methods which are observing and controlling the rocking magnitude of the sphere centre image.

4. In theoretically, the laser has enough coherent length, but for the different curvature radius R, the position of the exit pupil in the inspected path of the interferometric system will be changed largely with the variation of the curvature radius R, generally as for the lens R =-500mm to +500mm. Therefore, in principle, the adjusting possibility of the exit pupil must be provided to get approximate equi-width interferometric fringe for various R in the viewing screen. In our system the diameter of the central interferometric circle is about 10mm. Fig.2 shown the real-time fringe contours of each measuring surface.

upper surface

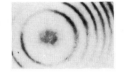
cemented surface

lower surface

Fig.2   Interference pattern of cemented lens.

5. The laser is a higher monochromaticity and well coherence source, so that we have used a polarized laser interferometric system, because it avoids the parasitic interference which caused by reflection of the nontesting surface effectively. Therefore, this apparatus is suitable to the concentricity adjusting of cemented doublet and triplet even a whole optical system.

The optical system of the polarized laser interferometric apparatus

Fig. 3 is the scheme of an interferometric lens-centring system, and Fig.4 was its analogic experimental mounting and was used when the authors began their project work. The polarized light from the He-Ne laser 1 was splitted into two beams P and S by the polarization spectroscope 3; the components P and S were projected into the reference path and inspected path respectively. The component S passed through the adjustable lens 4, $\lambda/4$ wave plate 5, and deflected by the centring angle operator 6, finally arrived in the surface of centerening lens 7, and then returned along the original path to the polarization spectroscope 3; from there S passed through the polarization plate 12, projection lens 13, mirror 14 and finally projected on the viewing screen 15. The P component came from 3 was passed through the $\lambda/4$ wave plate 9 and adjustable lens 10, then returned along the original path by the combined together with S in inspected path to form the interference fringes as a concentric circles pattern in the viewing screen 15.

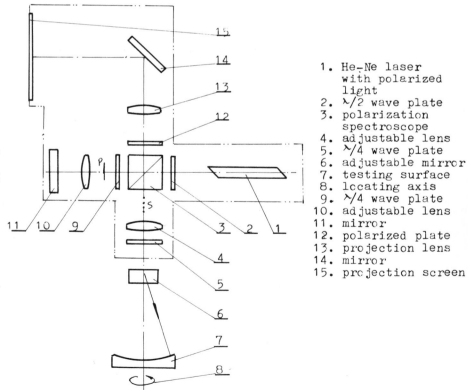

1. He-Ne laser with polarized light
2. $\lambda/2$ wave plate
3. polarization spectroscope
4. adjustable lens
5. $\lambda/4$ wave plate
6. adjustable mirror
7. testing surface
8. locating axis
9. $\lambda/4$ wave plate
10. adjustable lens
11. mirror
12. polarized plate
13. projection lens
14. mirror
15. projection screen

Figure 3. Principle of interference system.

In the optical system, each of the λ/4 wave plate is positioned in 45° inclined from the linear polarization directions of the P and S components respectively. Thus the polarization directions of both returning beams were rotated a π/2, and hence they passed the polarization spectroscope 3 again. In the meantime the polarization spectroscope 3 also eliminated the returned beams which were directed to the source of the He-Ne laser, thus made the laser works steadily. It also eliminated the parasitic interference which was caused by the reflective diffused light of the unworking surface. The contrast of interference fringes will be much improved.

There are two adjustable lenses 4 and 10 were inserted into their paths for adjusting the relative position between both exit pupils. In other hand, the projection lens 13 had properly adjusted to change the relative distance between exit pupil and interference field to obtain an interference pattern with an approximately constant fringe size on the screen.

Figure 4   The analogic experimental mounting

Some problems in operating

1. Due to the reduced difference between the two refractive indexes of elements, in the classic optical centring method, the contrast of the spherical centre image on the cementing surface is very low, thus caused the difficulties of visible observing and operating even cannot operate well. In the present system, the contrast of interference fringe of cementing surface depends upon the difference of the refractive indexes between the cement-layer and both lenses. The refractive index n_c of various balsams is from 1.4 to 1.7. In our system, we had calculated the theoretical contrast curves of interference fringes at the cement layer, for the most common balsams that can display their visible interference frings at the cemented surface, their contrast is well. (see Fig.2 and 5)

2. The advantage of the laser interferometric lens-centring technique by height-difference method must be ensured by a precise lens-centring mechanism (Fig. 6) which consists of a high precise rotating axle assembly and an adjusting mechanism with which successively adjusting the lens surface to a concentric position corresponding to the geometric axis. It is quite evident that the lens in the holding device needs precise

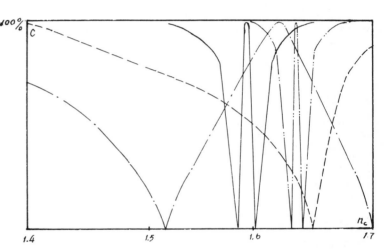

Figure 5 Contrast of interferometric fringe as functions of refractive index of balsam n_c and difference of the refractive index

———·——·——   △n=0.17741
————————   △n=0.08300
——————   △n=0.07640
··——··——··—   △n=0.00330

adjusting movement with five degrees of freedom; the rotation around the locating axis two displacements at the plane of perpendicular to the locating axis, two swings with a given centre. When operator is rotated by hands (also the centerening surface) we'll observe that the whole interference fringes rotated around the field center, and its period equal to the rotary period of the geometric axis. Utilize this phenomenon we can operate for rough centring. After then, adjusting the fringe to a non-fluctuation condition, that means the precise centring has been achieved and the lenses assembly was free from eccentricity.

3. Obviously, the accuracy of the eccentric inspection is the core of this subject. Except the above-mentioned factors, there are some factors as follow which could have an effect on accuracy of eccentric inspection. They are the accuracy of axle assembly, the distinguished exactitude during the interferometric fringes from light to dark or vice versa, the partial manufacture error of the optical surface and influence of the preceding surface eccentricity

on the consequent surface etc.

For this reason, we use a set of high precise semi-kinematic axle assembly where the radial pulsation and axial pulsation is less than 0.5μm and suppose that one can read out reliably once of fringe variation. According to these conditions, we had make some experiment with our apparatus and get a satisfactory result.

We selected a lens whose surface R=-45mm, D=16mm and made a precision coordinated movement on a plane which was perpendicular to the geometric axis of the lens-centring operator and read out the corresponding eccentricity (in fluctuation of fringe). According to calculation, the theoretical centring sensitivity of this lens surface is S_c=0.46μm, and in practical measurement the average sensitivity is S_c= 0.45μm, hence the experimental extreme error is 3σ=±0.1μm, and this result is very perfect.

4. To work normally, the enviroment of work place should be shock isolated, the room temperature variation should be as slow as possible, and the room illumination condition also should be dimming than normal.

Figure 6   The lens-centring oparator

Summarized reference

1. Jena Review 1978.3.
2. Hofmann Ch., Jena Jb 1960. .341
3. New Developments in Optical and their Application in Industry papers to the SIRA conference 1976. 15-1
4. DIN 3140 1976.6
5. JOERA 1968.5. No.14
   JOERA 1973.10. No.5
   JOERA 1975.12. No.2
6. Hopkins H.H, Tiziani H.J, Brit.J. Appl. Phys. 1966.17.33
7. Jaunet G., Marioge J.P., Farfall F., Mullot M. and Bonine B., Optique 1975.6.353
8. Jaunet G. and Marioge J.P., Optics and Laser Technology 1978.10.148
9. J.O.S.A. 1948.38.343
10. Jaunet G., Marioge J.P., Farfall F., Mullot M. and Bonine B., Journal of Optics (Paris) 1978.9.31
11. Topical Meeting on the Use of Optics in Microeletronics TB 2-1
12. Hormed.F., Optical Production Technology, Adam Hilger, London 1972.195
13. Carnell K.H., Optica Acta 1974.21.615
14. Optica Acta 1976. No.5 341-346
15. Optic 1962.19.14
    Optic 1959.16. 652-659
16. Wang GuoQiang, Technical Bulletin of Northwest Optical Factory (in Chinese) 1977. 2-3. 1-44
17. Dong Taihuo, Xu Xiaoda, Paper (in Chinese) presented in jointed 1982 meeting of OSC and CIS.
18. Shan Yuekang, Master degree dissertation (in Chinese), 1981, Optical Instrument Dept. Zhejiang University, China.

INDUSTRIAL APPLICATIONS OF LASER TECHNOLOGY

Volume 398

Session 7

Laser Material Processing I

Chairmen
Lorand J. Wargo
WEC Corporation, USA
László Erdélyi
AOL-Dr. Schuster GmbH, Austria

Applied electronics and mechanics in industrial lasers

Lorand J. Wargo

President WEC* CORPORATION, 4399 Gollihar Rd.
Corpus Christi, Texas 78411 USA

Abstract

Opto-mechanical and electronic criteria are analyzed in relation to the design of super-stable resonators and laser heads used in industrial environment. Examples of faulty design due to engineering shortcuts are demonstrated and results of such defects are presented. Influence of industrial environment on applications of lasers is studied and proper choice of support equipment such as power supplies and coolers is recommended. Comparison of most widely used power conditioning circuits and cooling methods is presented.

Introduction

It is difficult to envision how a human being would look like today if in primarial times light would not have assisted in the evolution of nature. Later in the history of humankind when the light fantastic was created and an other milestone was reached in harnessing the light energy, few of us had the exceptional opportunity to be at the right place at the right time and take part from the very beginning of this creative history of the light fantastic. I am proud to say that Dr. Theodore Maiman, one of the principal pioneers in lasers was my friend and tutor. I had the opportunity to work at Hughes Aircraft Company, when the first laser was born, and successfully demonstrated principles of Light Amplification by Stimulated Emission of Radiation. Since then this powerful light energy was applied to a wide variety of good uses and I wish to explore one of these uses, which is related to the industrial field of applications.

Applications engineer, the magic matchmaker

Who is capable to determine whether it is simpler to match or find a laser to do a certain job or to have a job to be done and find a laser to do it? It is definitely a job for a special engineering talent. The one who knows lasers and is intimately connected with them, at the same time has a good background in material working and chemistry. Because for a laser to do a job it must be matched to the best possible level. First of all materials working by laser involves two important factors. First, the cooperation of the material we want to work, namely its absorption characteristics at the selected wavelength because energy has to be absorbed by the material to be worked in order to produce changes in the state of the material, in its composition, volume, shape or any other characteristic. It is quite certain that there is never a perfect match. As we know, that whenever we wish to drill a hole or cut a plate or weld two pieces together, the junction of the material, or the interface has to absorb energy. And if this energy is to be supplied by the laser, not only the absorption characteristics of the materials involved at the interface must be very closely matched with the wavelength, but also the type of the pulse profile and rate should be selected carefully. Of course the second factor is the precise selection of the amount of the energy, that leads to successful operation. There are many second order characteristics which influence successful material working and we will deal with them later. It is the applications engineers duty to analyze the properties of the materials and the characteristics of the laser, and decide whether such favorable match can be expected, and after all the analysis, he will have to go down to the laboratory and make an entire series of tests to confirm the feasibility of the task. It is almost impossible to do measurements in restricted time frame that would determine all the applicable characteristics of compatibility.

Way-way back at the dawn of laser technology the matchmaking was very easy, no matter what we tried to machine it had a perfect match, since there was only one laser in existence namely the good old workhorse, the ruby laser. And we worked with the ruby laser practically in every possible application. Some of them turned out to be good, some of them turned out to be duds, but certainly we have welded, we have drilled and we had cut with the beam of the ruby laser. First we started with less sophisticated materials, such as the basic metals, then the hard to work metals and we wound up drilling diamonds with the ruby beam. We have tried an enormous number of materials which of course did not cooperate and we were not able to achieve any success with them. We certainly had excellent results with some sophisticated materials used for turbine blades for aircraft engines, with diamond dyes, stainless steel materials and we have learned in a very short period of time that for different approaches of material working the selection of pulse width and energy profile are very important.

Later with the advent of YAG, CO_2 and a variety of different lasing materials and different types of lasers, the choice was more difficult, because now we had a variety of wavelengths and we were able to match more and more materials working processes. Certainly today the ruby laser is more of a rarity and is being used only for very special applications. The YAG predominates the solid state field and the CO_2 the gas laser field. As a matter of fact, several companies are trying to approach the material working with combination of two beams and even trying to form a sequence of programmed pulses and CW beams from a CO_2 laser to achieve very complex machining tasks.

With the advent of Q switching the laser applications engineer had a great hope that now with the millions of watts of peak power the materials working would become relatively simple. Unfortunately it did not turn out to be the case. An extremely short laser pulse such as a Q switched pulse evaporates the surface of the material and forms a plasma barrier over this surface and therefore it is only suitable for applications where the surface of a material or a coating has to be removed. Such applications are found primarily in semiconductor industry and also in the photographic industry in preparation of photo masks. Such a photomask has a quartz or glass base, called the pill and then chromium is deposited on the surface. This thin film forms a photomask. For mask repair, this chromium can be very easily removed from the substrate by the application of a very short Q switched laser pulse. We have built several of such machines by using Q switched YAG or harmonically upconverted and Q switched YAG into the green. The same methods can be used for fusing thermocouples, cutting lines on semiconductor wafers and surface hardening of materials. Short nanosecond pulses are also used with remarkable success in micro machining of the upper layer of a material to be machined simultaneously with more conventional long pulse method.

The initiation of the surface activity by Q switched pulse will prepare the noncooperative surface for greater absorption of the lower peak and higher energy pulse. In such matter, materials which are very little absorbing at the fundamental laser wavelength become absorbent. This method of Q switching or a Q spoiling is successfully used with continuous wave laser and not only with pulsed lasers. In continuous wave laser the Q switching method increases the peak power of each consecutive pulse and material removal is more prompt than with the source laser without the Q switched pulsing.

Applied sciences in industrial lasers

In the simplest approximation, imagining a laser system as a black box, receiving a certain amount of energy in any one of the conventional forms such as electrical energy or kinetic energy and out of the black box we receive heat, sound, mechanical vibrations and laser light. We have specifically set up this order because of the amount of heat is the greatest output of a laser system. If we do realize that a laser system works with 2 to 4% efficiency in case of solid state laser, or maybe 10 to 12% efficiency in a case of a gas laser, we must realize that all the rest of the energy must be still removed from the laser system, and all that energy will be distributed primarily into heat, accustical shock, and mechanical vibrations. Since we have established the characteristics of the input output of the black box we can analyze which one of the scientific disciplines will govern certain parts of this black box. First of all at this point we wish to exclude from our analysis the optical part. We will look at mechanical and electrical or electronic parameters. We will determine how much of the output energy will go into the variety of byproducts of the laser action. Let us examine first the mechanical structure inside the black box.

The mechanical structure of the power supply and cooler of the laser system will be treated under a separate paragraph. Lets start looking at the mechanical characteristics of the laser head itself. Since the laser head will have to contain a resonator it comes to our mind immediately that if we do not want any instabilities in the resonator, we better separate it from the mechanical enclosure. The simplest and most efficient way of separating the resonator from the housing of the laser head and from the machinery it is attached to, is a good stable kinematic suspension which is relatively simple and it is quite stable. This kinematic suspension provides good isolation of the structural members which are the interface members of the outside housing of the laser head attached to the machine and the resonator which has to be as stable as possible if uniform results are expected from the laser. Such kinematic suspensions can be designed with front end or rear end compensation. With the proper design of orthogonal slots and screws on the bottom of the resonator plate, having the freedom of moving independently from the housing which is also provided with the similar orthogonal slots, a stable and semi-rigid connection between the resonator and the housing can be achieved. The purpose of this suspension is to let the housing of the laser head follow the distortion of the machinery it is mounted on, without translating this distortion into the structure of the resonator. (Fig.1) Since we know that with temperature changes during the daytime or nighttime operations, the machinery the laser head is mounted on will expand and contract. This motion is directly translated to the resonator and distortion of the optical path would result, and instability would set in. For this reason, isolation of the optical resonator from the base mounting plate and the machinery it is mounted on is one of the most important factors in achieving longterm stability. (Fig.2,3,4)

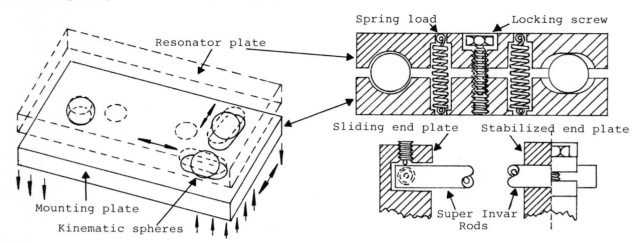

Figure 1. Kinematic plates

Figure 2. Super Invar suspension

The mechanical suspension of the optical elements within the resonator itself is the second most important consideration. In mechanical choices of suspension of the resonator itself, in case of a simple Fabry-Perrot or similar resonator utilizing a mirror, an output window and a pumping cavity, the relative position and stability of the two optical elements determining the resonator length are of great importance. These suspensions will have to be motionless during the operation of the laser and the optical boresight should not suffer any changes of position if the resonator is exposed to outside mechanical or thermal influences. One of the highest demands in the design of industrial lasers is mechanical stability. Since we know that industrial lasers do not operate in air conditioned laboratories with filtered air, thermal stabilization, and mounted on a large granite slab, these lasers have to work out in the shop mounted on a monstrous machine which is shaking and vibrating and changing temperature without any pattern and yet the laser has to perform a job of the level of optical precision. Therefore it is important to achieve great mechanical rigidity and stability within the resonator itself. We have tested several optical suspensions to be used in resonators and so far after many experiments and tests have found that the majority optical mounts presently used in industrial lasers are inadequate if subject to changes due to vibrations of mechanical stress or temperature. We have had in the laboratory a laser which was mechanically so unstable that if someone lifted one end of the laser head from the table, it stopped lasing. Can you imagine such a laser performing in the environment of industrial applications? Very few manufacturers take the luxury of utilizing the best in the mechanics for the design of the optical resonator. Unfortunately this trend cannot be blamed all on the manufacturers because the customer who wants less expensive devices forces the manufacturers of different brands of lasers into a very stiff competition. Such competition is very healthy as long as it does not cut across the lines of sound engineering practices and we all know from our previous experiences that the same way as the automobile today is primarily designed by fashion designers the lasers are designed by the marketing people.

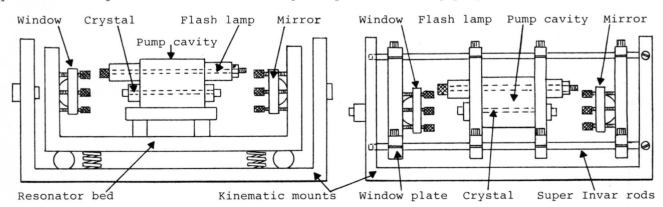

Figure 3. Kinematic Laser Head

Figure 4. Super Invar Laser Head

Cost consciousness is a great goal but it is certainly not a license to come up with cheap and unreliable devices. We have spent a lot of time and effort in designing most stable laser structures. For the past ten years we have field tested one particular design which proves to be very promising and seems to be good enough to consider it state of art concept. The evolution of this super stable mechanical optical mount started with setting up the most desirable characteristics an optical mount could have. First of all our desire was to maintain the boresight or the beam in the optical center and on the surface of the optical element. The second goal was to maintain the center of the optical element be it a mirror or a window, or a Pockells Cell or a diagonal, without translation or displacement of the beam, because once the beam is translated by the introduced motion, the stability of the resonator will suffer from externally applied vibrations, shocks, or thermal differential expansion. An other guideline in the development of this super stable optical mount was the fact, that once the optical element is properly aligned in position, it should remain there. Such stabilization can be achieved by different locking methods. With these guidelines in mind we started searching for a geometry that favors the conditions, and we found that there is one very simple easy to produce mechanical device which does have a point which will never move, no matter how the element is rotated, namely a sphere. Therefore we have choosen a sphere as basis for the design of the mount. We have placed an optical element in the middle of the sphere. From the optical surface of this element we were reflecting the laser beam. The center of this optical surface is also the center of rotation of the sphere. Placing such a sphere in a corresponding matching socket, we have obtained a ball socket combination. Applying a second half of this hemispherical socket we are able to rotate the sphere, between the two hemispherical sockets, as long as they are apart. The original design incorporated a sherical optical mount with a double hemispherical socket combination with the active optical surface positioned in the center of rotation of the sphere. Subsequent improvements have brought simplifications of the attachments to manipulate the sphere within the sockets and a combination of locking devices which would not perturb the optical elements position, once it was properly aligned. It become evident that the same effect as a complete sphere in two hemispherical sockets can be accomplished by using only a half of the sphere and only one hemispherical socket. (Figures 5,6)

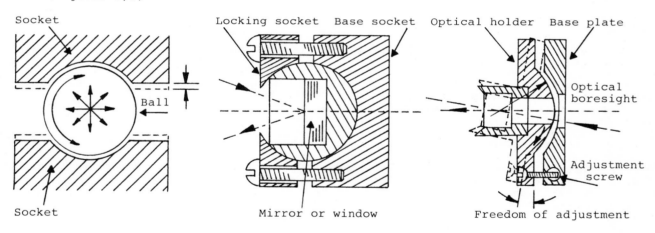

Figure 5. Typical ball joint   Figure 6. Optical ball joint   Figure 7. Ball socket mount

This simplification also improved the production, assembly and end use of the optical mount. The ultimate design incorporates a mounting plate provided with a conical socket on one or both sides and in these cones, a ball segment is able to move freely. The ball segment is in contact with the mounting plate in a tangential circle and is held in intimate contact with the cone through pressure excerted by the X-Y adjustment screws and a coil spring around the locking screw. (Figure 7)

During the X-Y adjustments the spherical surface is free to move in X-Y planes in the conical socket. Once alignment is obtained, the locking screw compresses the spring and locks the sphere into the socket, preventing any further changes in alignment. It is feasible to mount two optical holders on two opposite sides of the same mounting plate. The segment of the sphere is made in a fashion that its flat reference plane intercepts the center of the sphere and therefore any optical surface in contact with this flat will indeed have no nutation or translation of the optical axis during adjustment. To obtain similar action, the adjustment and the locking screws are formed into spherical sockets. By this method one can produce large structures from light and inexpensive aluminum, and the spherical inserts, screws and sockets are made of super Invar. Since no dissimilar metals are used to support the optical members, no differential expansion is introduced into the well balanced alignment system. These are the reasons behind the unusually exceptional thermal and mechanical stability of this resonator system. (Figures 8,9,10)

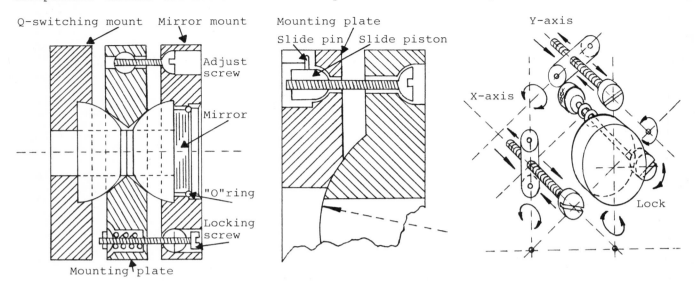

Figure 8. Compound dual mount   Figure 9. Adjustment screw   Figure 10. X-Y-lock optics

Opto-mechanical recommendations

Our summarized recommendations for the choice of successful industrial laser head are the following:
1. Rugged head enclosure with gasketed interfaces, to prevent impurities from entering the inside of the laser head.
2. Kinematically suspended resonator either with or without super Invar suspension.
3. Easy access to the optical alignment screws and a positive pressure locking screw for the optical mounts.
4. Lamp change without opening of the pumping cavity.
5. No need for realignment after lamp change.
6. All optical elements with telescoping or bellows type light guides to prevent accidental exposure and contamination of optical surfaces when the laser head is open.
7. All access covers in the laser head interlocked with the energy dump circuits of the power supply.
8. Mechanical or electromechanical beam stop in the resonator, with normally spring closed position.
9. X-ray shields in high voltage (20 KV +) gas (CO_2) laser heads to protect bystanders.
10. All electrical contacts covered with high voltage insulation, top and bottom side of the head and Laser-On warning light on the outside of the head.

Electrical and electronic consideration

Industrial laser systems, just like any other laser systems require power input into the pumping cavity to sustain operation. This power input can be as unusual as sunshine in sun-pumped $Nd+3$ YAG lasers or kinetically stored energy in a large flywheel, driving a high voltage generator, charging pulseforming networks or as simple as a flashlight battery. In portable laser systems predominantly a battery pack, in rechargeable or replaceable form is used. Transportable lasers systems normally use vehicle storage battery of 12 or 24 VDC. Airborne laser systems are supplied by on board polyphase generators and sea going systems utilize a selection of either on board DC or 50-60 or 400 Hz AC. Industrial applications depend primarily on industrial power networks and therefore the primary power for small systems is single phase 115 or 230 VAC 50 or 60 Hz. For medium size systems, delta connected 208 or 440 VAC, three phase supply is common. For larger systems 480 VAC polyphase 50 or 60 Hz AC power network is used. None of the mentioned power input is suitable to drive directly a solid state or gas laser. Except for the Tungsten iodine, halogen cycle lamp pumped, CW. $Nd+3$ YAG lasers, limited to about 3 watt output maximum. All other laser systems require power conditioning equipment, commonly called power supply. The purpose of power supply is to match the electrical characteristics of the feeding network to the load network, be it a gas discharge tube of a CO_2 laser head, or a flash lamp or arc lamp of a solid state laser pumping cavity. It is easy to see that the low voltage supply network with its constant voltage characteristics will not match the variable impedance and high voltage requirements of a 30-50 KVDC, CO_2 plasma tube. Due to the variety of possible applications we should group the typical industrial systems first into two major divisions. Such as pulsed solid state and CW, gas laser systems. All systems use additional auxiliary electrical or electronic control and timing circuits, but these circuits are quite similar for all systems and therefore we can treat them as one group. Such auxiliary circuits represent the starting triggering or igniting circuits and simmer or keep-alive supply circuits. An other circuit type is the safety interlock circuit that controls the turn off procedures of the power conditioning equipment. Timing circuits provide clock and feedback informations for the proper pulse discharge and recharge in pulsed systems and sequencing in continious wave lasers.

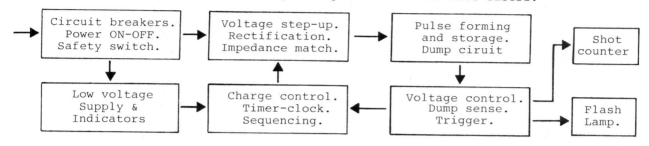

Figure 11. Typical pulsed laser power supply block diagram.

Pulsed industrial laser charging circuits

In pulsed industrial lasers, the primary sources of the pumping optical radiation are one or more flash tubes. These flash tubes are gas filled discharge tubes, utilizing Xenon or Krypton fill. In order to operate these flash lamps we have to provide a certain amount of energy, which is stored in the pulse forming network capacitors, and also a higher voltage trigger pulse to initiate the discharge. The trigger pulse ionizes the gas column which than becomes conductive and will discharge the pulse forming network through the flash lamp. An other way of discharging a pulse forming network through gas discharge tubes is simmering, or also known as keep-alive method. The simmering method utilizes an additional power supply to supply low current high voltage continuous electric field through a ballast resistor across the gas column of the flash lamp. When the flash lamp breaks down due to this high E field applied across the gas column, a second high voltage supply provides the necessary lower voltage, higher current "simmering" or "keep alive" discharge at a very low energy rate. (Figure 12.)

Parallel or cavity trigger   Series injection trigger   Keep-alive or simmer turn-on

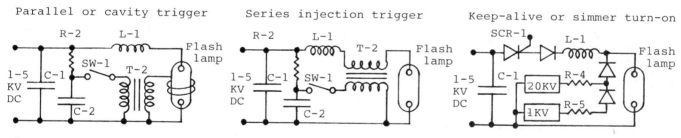

Figure 12.   Trigger and turn-on methods

Energy stored in the storage capacitors in the pulse forming networks will have to be replenished every time the laser is operated. At low repetition rates this is a minor task but if higher repetition rates are required the circuits providing recharging of the pulse forming network become more and more complex. A large variety of charging circuits is known to the state of the art. The first charging circuit used in lasers was a resistance charged fixed voltage charging circuit. This circuit has a very low efficiency, not higher then 50% because of its operational principles. Such a circuit is shown in figure 13, and it consists of a step-up, fixed voltage ratio transformer, preceded by a power switch, indicator light and circuit breaker and it is connected to the primary input power. The transformer T-1 is a step-up transformer with its primary matching the requirements of the network it is connected to. The secondary of this transformer is required to have the necessary voltage step-up characteristics in order to provide the voltage required to charge up capacitor C-1. Usual requirements for flash lamps in industrial lasers are approximately 1000 to 5000 volts across one or several flash lamps connected in series. Therefore the secondary of this transformer will have to have the necessary voltage applied to the rectifier bridge which through a series charging resistor will charge capacitor C-1. We can see that in the initial phase of charging the entire voltage will be dropped across the charging resistor, since the pulse forming capacitor is totally discharged and its impedance is zero. Therefore the charging resistor will have to conduct the current which will be limited only by its resistance and will have to withstand the entire source voltage supplied by the rectifier bridge. At the end of the charging cycle, after many time constants, the voltage across the pulse forming network capacitor will be equal to the peak voltage

341

supplied by the transformer secondary and rectified by the full wave bridge. During this charging time R-1 will dissipate the same amount of energy as is deposited in the pulse forming networks capacitor. Therefore theoretically such a charging circuit can not be more efficient then 50%. Additional losses within the circuit such as the foreward drop losses in the rectifier bridge, capacitive losses and magnetic losses and also resistive losses in the transformer will make this 50% theoretical efficiency much worse. Normally in a well designed charging circuit of this type, the efficiency runs about 35 to 40%. Simple power supply circuits provide triggering directly from the pulse forming capacitor through a drop resistor charging up a trigger capacitor. Upon demand of closing a switch, this capacitor is discharged through the primary of a trigger transformer. This transformer generates a high voltage pulse which ionizes the gas column in the flash lamp and discharges pulse forming capacitor through a pulse forming inductor. It is evident that the length of the pulse and the amount of the energy is governed by three major factors. The capacitance of the pulse forming network capacitor, the inductance of the pulse forming network inductance, and the dynamic characteristics of the flash lamp. This type of power supply circuit is mainly used for low pulse rate and medium power output lasers because of the lack of efficiency and the length of the charging cycle governed by the resistance of the charging resistor and the capacitance of the pulse forming capacitor.

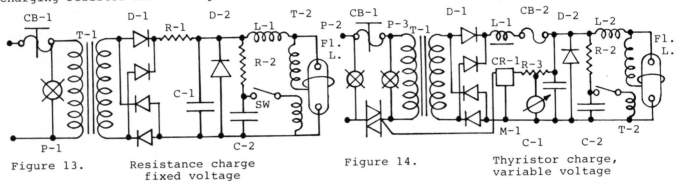

Figure 13.   Resistance charge
fixed voltage

Figure 14.   Thyristor charge,
variable voltage

Figure 14 shows a more complex power supply which is capable of higher pulse rates and variable output pulse energy. This is a thyristor charging variable voltage power supply circuit. This charging circuit, similarly to the previously described one, has a step up high voltage transformer and a full wave rectifier bridge. However, the primary of the transformer is turned on and off by a voltage comparator and a programming circuit which senses the high voltage across the pulse forming capacitor. It turns on and off in variable firing angle mode the thyristors in the primary of this circuit. An other variation is utilizing high voltage thyristors or SCR-s in the secondary circuit of the power transformer, which are turned on and off in firing angle controlled mode by the comparator and control circuits. The rest of the circuit is essentially the same as the previous one as far as the triggering method and the triggering mode of the flash lamp. The advantages of this circuit are that with a firing angle controlled input primary or firing angle control output from the secondary of the transformer, one can achieve a predetermined or preset voltage requirement across the pulse forming networks capacitor or capacitors. The operation of this circuit is quite straight forward in the initial charging cycle. The pulse forming capacitor is discharged and therefore its impedance is virtually zero. Turning on the SCR at the proper firing angle on the primary side or on the secondary side of the transformer will provide a very late initial so called soft start firing pulse for the SCR-s or thyristors providing only a minute amount of energy to be passed through the transformer; thus virtually matching the low impedance of the pulse forming network capacitor into the power network. During the charging cycle, the firing angle is advanced gradually to the point where at the end of the cycle it is totally cut off after it has achieved a peak somewhere in the middle of charging cycle. Upon demand, the flash lamp or flash lamps are fired, the capacitor discharges through the pulse forming inductor and the cycle starts again. The voltage sensing comparator senses zero voltage on the pulse forming capacitor and it starts the cycle from the beginning with a very delayed firing angle going through the cycle again, until total cutoff. Since these power supply circuits, utilizing thyristor or SCR charge, have the advantage of rapid recycling and a much higher efficiency due to the fact that losses are primarily straight-forward losses of the forward conduction of the control and rectifying elements and the loss factor of the transformer. We know that all these losses are relatively small. Typical loss factor of about 20% yields about 80% efficiency for a well designed power supply of this type. The thyristor or SCR type charging power supply has the advantages of being able to operate at fairly high pulse rates and at fairly large output energies per pulse. It certainly has the disadvatages of introducing transient conditions into the power supply network and generate considerable electromagnetic noise radiation. An other type of charging power supply is the variable transformer input supply with servo motor drive, figure 15. In this type of power supply the input network is connected to an isolation type variable secondary transformer known under names as

Variac, Powerstat etc. The principle of operation of this type of power supply is soft turn-on and constant reflected impedance characteristics to the network. The charging current remains unchanged and the transformer reflects the impedance level of the load into the impedance of the network supplying the power. This type of variable transformer input provides an ideal match of the impedances because the timing and drive circuit for the motor provide constant current supply during charging and constant initial zero transfer by the virtue of the driving the movable contact on secondary down to zero point at the start of every charge cycle. When demand signal arrives to the control circuit because charging is required, the motor drives the sliding contact up away from the zero point and increases the output voltage of the transformer by engaging more and more turns on the secondary. This is followed by a proportional increase of the voltage on the PFN capacitor. When the voltage control circuit compares the preset or demand voltage with the error signal voltage supplied by the PFN network and finds the two voltages identical it drives the motor down to zero point, engaging a safety switch. This guarantees, that at the next charging cycle the recycling will start with zero voltage on the secondary. This type of power supply is very reliable, it is quite simple and requires minimum of electronics circuitry to control the charging cycle. It also represents almost perfect match of the impedance represented by the pulse forming capacitor into the outside network supplying the power. At any point during the charging cycle it drives the sliding contact of the secondary variable transformer just a minimum of error signal above the capacitors voltage and it guarantees this way an almost perfect constant current charge. The triggering circuits and timing circuits of this power supply are identical to the previously described power supplies. Although we can use series injection triggering transformer instead of parallel triggering transformer to ignite the lamp. This type of power supply is not suitable for high reprate due to the fact that it takes 2 to 4 seconds for completion of the variable transformer cycle.

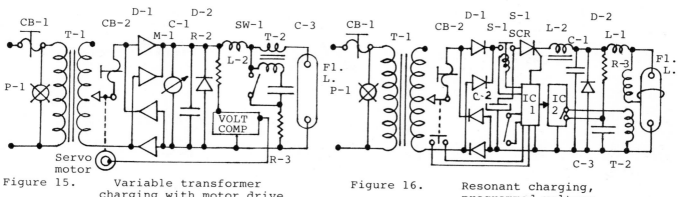

Figure 15.   Variable transformer   Figure 16.   Resonant charging,
              charging with motor drive   programmed voltage

The next circuit (Figure 16) represents a resonant charging, programmed voltage circuit. These circuits are highly favored in high pulse rate industrial lasers. The advantages represented by a resonant charging, programmed voltage power supply are many. First, it does not inject into the input line any appreciable noise or transients during the charging cycle. Second, this power supply compared to all other power supplies has the highest possible efficiency, well over 85%. This is quite significant due to the fact that in industrial applications we are normally talking about fairly large energy outputs from the laser head, lets say about half a kilowatt, and the efficiency of the laser being approximately 1 to 3%, about 25KV to 30KV input power is required. Each percent of efficiency means 300 Watts. A 10% loss would be 3KV, 20% 6KV, which is six times as much as the laser output. Third, the charging circuit is the same for single or polyphase input. Fourth, in polyphase operation, the regulation of output power is better then $\pm$ 1%. The input stage of this supply is a variable transformer circuit, similar to one in figure 15. It can manually set without the aid of the motor and once it is set to the desired voltage it does not require any resetting of the control of the variable transformer. Presetting the secondary voltage on the transformer T-1, provides charging current to the rectifier bridge D-1 and storage capacitor C-2, which in turn supplies the resonant charging pulse forming network C-1, L-1. This resonant charging circuit is operated by a high voltage SCR which fired charges up the pulse forming capacitor C-1, through the resonant charging inductor L-2, to twice the voltage across the storage capacitor C-2. This of course backbiases the SCR and drives into hold-off. The resonance of the inductance and capacitance is set to be approx. 80% of the total cycling time at the highest repetition rate produced by the laser power supply. The actual resonant charging occures by turning on the gate drive IC-1 of the SCR which is connected to the storage capacitor C-2, the SCR turning on conducts a current through itself and through the resonant charging inductor L-2 until the voltage across the pulse forming capacitor C-1, is equal to the voltage on the storage capacitor C-1. But during this charging cycle a magnetic flux builds up a magnetic flux field in the resonant charging inductor. After the equal voltage condition it will continue collapsing. This collapsing magnetic field generates an electromotive force in the winding of the inductor L-2 which continues driving its generated current into the pulse forming capacitor C-1, charging it up to twice the voltage of the storage capacitor. The resonant charging inductor stores $\frac{1}{4}$ of the total energy deposited into the pulse forming capacitor and the resonant time constant of the inductor and the capacitor will determine the fastest pulse rate produced by this power supply. Upon completion of the charging cycle the SCR control circuit provides a starting pulse for the delay generator IC-2. This delay generator produces a trigger pulse and ionizes the gas in the flash lamp and conducts away the charge from the pulse forming network. When the voltage on the PFN capacitor drops to zero, the sensing circuit of the recharging drive circuit starts the cycle again, firing the SCR voltage, doubling the voltage on the pulse forming capacitor, fires the trigger circuit again, and keeps on going at the rate set by the timing clock. This type of power supply represents the highest economy in design of high pulserate high power industrial laser power supplies. Charging and discharging curves for all described power supplies are displayed on figure 17.

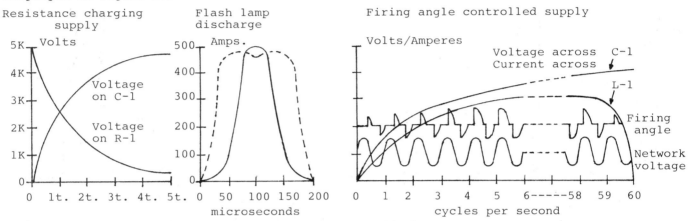

Figure 17.   Power supply charge and discharge curves

Industrial lasers often use highly powerful power supply circuits based on chopper conditioning of the input power. Such power supplies often have transformer-less input, single or polyphase rectified full wave, charging up two storage capacitors. The storage capacitors are connected to two SCR-s which alternately turn on and off the current from the storage capacitor pair to the primary of a power transformer producing quasi sine wave alternate current to the primary. The secondary of this step-up transformer is usually connected to a full wave bridge which in turn supplies the charge current to the pulse forming network capacitor. Advantages of such supplies are; first, it does not require power transformer connected to the input power network. Second, since the chopper can utilize frequencies which are much higher then the standard line frequencies the magnetics involved are relatively small due to the high chopping frequency. Thirdly, the high chopping frequencies require smaller filtering capacitors and the ripple voltage on the pulse forming network is also smaller. The charging cycle starts with sensing of zero voltage on PFN capacitor by the comparator circuit, which compares it with preset, desired voltage and sensing the need for charging turns on the drive to the gates of the SCR-s. The SCR-s provide push-pull chopping and generate primary current in the step-up transformer. The rectified current of the secondary charges up the pulse forming network capacitor to the point of the preset value and the voltage comparator turns off the chopping SCR-s. Firing the lamp via trigger circuit, the voltage on the pulse forming capacitor drops to zero, and

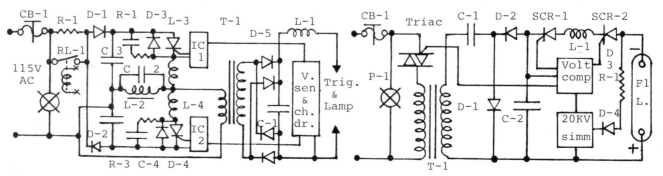

Figure 18.   SCR-chopper power supply   Figure 19. Shortcircuit proof power supply

the cycle starts all over again. These high frequency chopper power supplies are extremely efficient and relatively small compared to the low frequency units. Chopper power supplies can operate in class A, B or C depending on restrains of the application. The circuit of such chopper power supply is shown on figure 18. An other power supply often used by the industry for its simplicity and short circuit proof load characteristics is the reactively controlled constant current charging supply with resonant current limiting, shown on figure 19. The operation of such a power supply is described on figure 20. The main advantage of this type of power supply in industrial use, is that in case of lamp hang-up or short circuit in the laser head, the power supply is protected and so is the load.

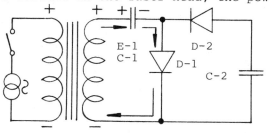

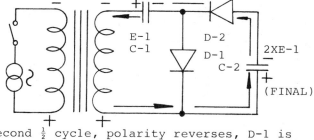

Switch closes, C-1 charges up to E-1 (PEAK) through D-1. D-2 is reverse biased and does not conduct during this $\frac{1}{2}$ cycle. Energy stored in C-1 is $\frac{1}{2}C(E-1)^2$

Second $\frac{1}{2}$ cycle, polarity reverses, D-1 is back biased and does not conduct. E-1 across C-1 adds in series with E-1 on transformer and charges through D-2 the C-2

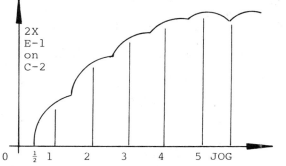

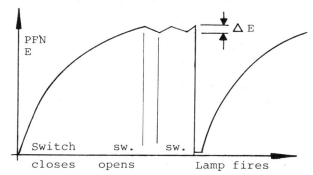

Since the transfer of energy is limited by C-1, the ratio of C-2/C-1 will determine the number of cycles required to charge C-2 to 2 x E-1

If one choses L of the transformer to resonate with C-1 at f (60Hz) in case of short circuit across C-2 will cause resonance, and no current will circulate.

Figure 20.   Short-circuit proof, resonance limited charging power supply

CW and gas laser power supplies

The power supplies used with CW or gas lasers are basically the same as the pulsed power supplies, because the power conditioning and impedance match requirements are similar. The difference in the CW or gas laser power supplies is due to absence of pulse forming network, since they are continuously operated and not pulsed. This difference requires constant current source for supply of CW, arc-lamp or plasma tube of the gas laser. Block diagram of such a CW or gas laser is shown on figure 21. Gas lasers or CW lasers require a current source rather than a voltage source. Such current source is shown on figure 22.

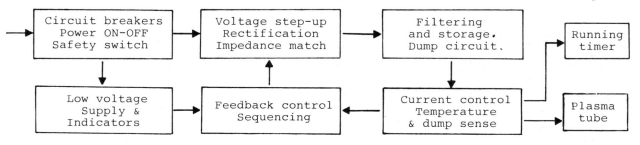

Figure 21.   Typical CW or Gas laser power supply block diagram

a single phase power input network is connected through a circuit breaker to the primary of a transformer controlled by Triac (bydirectional thyristor). The transformer secondary is connected to a full wave rectifier and through a current limiting choke or inductor and a second circuit breaker provides a current through resistors R-1 which is a current sensing shunt. From resistor R-1 through a blocking diode, the current source is connected to the CW lamp or the plasma tube. The 20KV high voltage supply provides the initial ionization through blocking diode D-2 and resistor R-3. Filter capacitor C-1 provides a ripple free DC current supply to the lamp. IC-1 regulates the amount of firing angle control through the Triac and maintains a constant current through the lamp. Figure 23 shows the basic variation of the same gas laser supply with a three phase input stage. The three phase input stage connected via step-up transformer T-1 and controlled by three SCR-s and three pass diodes, charges up storage capacitor C-1 and supplies the lamp or plasma current for the laser. The three SCR-diode pairs SCR-1-2-3 and D-1-2-3 are in the three legs of the three phase transformer secondary and are controlled by a three phase firing angle control circuit. When even higher current is demanded, because of change in the impedance of the laser tube or laser arc lamp, the voltage across shunt resistor R-1 changes the constant current control network advances or retards the firing angle of the SCR-s. This three phase input circuit also.has an auto start 20KV igniting supply protected with a diode and a ballast resistor.

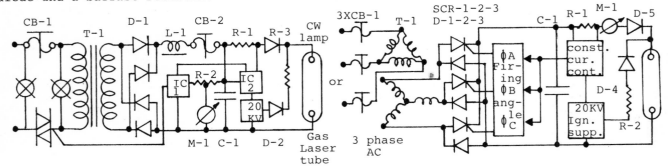

Figure 22. CW or Gas laser power supply   Figure 23. CW or Gas laser three phase
with auto start   power supply with auto start

Cooling considerations in industrial lasers

In the initial chapter of this presentation we have indicated the lack of efficiency in solid state or gas industrial lasers. The initial statement indicates that due to this lack of efficiency, large amounts of thermal energy have to be removed from the laser system. The predominant excess thermal energy will accumulate in the laser head and a considerably lesser amount in the power supply itself. One of the favored cooling methods in industrial lasers is liquid cooling. The reason for preference of liquid cooling is the superior heat transfer characteristics of the liquid. Specifically in industrial lasers we will find predominantly the use of cooling water. In industrial laser heads water cooling is applied to the crystal, the flash lamps and the housing of the cavity. Practical coolers utilizing liquid to air, liquid to liquid, or liquid to chiller cooling cycles have a primary loop of closed cycle deionized, demineralized, deoxidized water to cool the elements in the laser head. This primary cycle cooling is then recirculated by pumps through heat exchangers which can be either liquid to air or liquid to liquid. In the primary loop of the coolant we will find several stages of filtering used to remove ions and minerals from the liquid circulating in the primary loop. Several sensors are immersed into the primary loop such as temperature probes, and flow instrumentation. The instrumentation is necessary to prevent overheating and damage to the laser head.

Conclusions

The rapid evolution in applied optomechanical and electronic technology is in continuous fluent state. New solutions are found daily to a variety of complex problems. Introduction of new optical materials and solid state electronic devices opens new avenues of the solutions of old problems. Quest for increase of efficiency and new fields of applications expand daily the repertoire of the state of the art and let us hope that the challenge will never end.

Reference

This paper contains condensed excerpts from the authors upcoming book titled: "Applied laser electronics and mechanics with your microcomputer." The book is a design guide for technical persons how to make full use of microcomputer while designing hardware and circuits, it gives sample programs, and program writing examples and ideas.

The Fabrication of Fine Lens Arrays by Laser Beam Writing

Michael T. Gale and Karl Knop

Laboratories RCA Ltd., Badenerstrasse 569, CH-8048 Zurich, Switzerland

Abstract

Lenticular arrays with near diffraction-limited performance have been generated by laser beam writing. A photoresist film is exposed in a line raster mode by scanning under an intensity-modulated focused laser beam; subsequent development of the resist film produces the desired lenticular relief profile. The writing of exposure patterns by raster scanning is analyzed, and it is shown that very precise positioning of each raster line is required to avoid unwanted modulation terms. A low cost xy positioning table with better than 20 nm dynamic positioning accuracy was constructed to satisfy this requirement, and has been used to generate high quality fine lenticular arrays with lenslets of widths down to 20 μm and relief depths up to about 4 μm.

Introduction

Fine arrays of cylindrical microlenses (lenticular arrays) have applications in a number of optical techniques and devices. The use of lenticular array relay optics in a single chip, solid state color television camera is described in Ref. 1; the required lenticular array must have near diffraction limited performance, with typical lenslets of width in the order of tens of microns and focal length of about 0.25 mm. Further applications include moiré pattern generation[2], lenslet array processors[3], optical diffusers and the general field of optical filtering.

Classical micro-milling or ruling techniques for producing lenticular arrays are not suitable for producing fine arrays with lens widths in the order of tens of microns. They are also inflexible - a change in lens width or depth (focal length) involves retooling. The laser beam writing approach overcomes these limitations. A photoresist coated substrate is raster scanned under a focused laser beam modulated to write an exposure pattern which after development produces the required relief. The raster scan and beam modulation are computer controlled, enabling the exposure parameters to be changed by simply varying the program inputs. The scanning mechanism must be capable of very high dynamic positioning accuracy, since the generation of optically smooth surfaces using a typical focused spot diameter of 1 μm involves positioning tolerances in the order of 10 nm. A low-cost translation table which satisfies this requirement has been constructed using a modified commercial stage coupled to a fast, high-resolution, electromagnetic-driven second stage with interferometer control.

The laser beam writing technique is most suited to the fabrication of fine lens arrays (and other surface relief structures) with up to about 4 μm depth and typical lens widths in excess of 20 μm. The developed photoresist relief structure can be converted to a metal master relief by electroplating techniques, enabling large number of low-cost replicas to be produced by embossing or casting.

Exposure by Raster Scanning

In the raster scan mode the total area to be exposed is split up into lines or columns which are scanned one after the other at constant speed with a small spot of light. The basic parameters describing the scanning motion are (see Fig. 1):

x, y: coordinates
s : line spacing in x-direction
v : scanning speed in y-direction

If we assume that the laser beam intensity remains constant over one line, which is the case when exposing lenticular arrays, the total exposure is a function of x only and can be written as

$$E(x) = \frac{1}{v} \sum_k I_k \, g(x-ks) \qquad (1)$$

where I_k = laser intensity at line k
$g(x)$ = normalised intensity distribution

For an unrestricted focused laser beam $g(x)$ can be well approximated by a Gaussian function

$$g(x) = \frac{2}{\sqrt{\pi} \ D} \exp\left(-4 \ (\frac{x}{D})^2\right) \tag{2}$$

where D represents the focus diameter at 1/e intensity points.

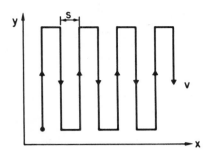

Figure 1.   Raster Scan Exposure

If the ratio of the focus diameter to the raster line spacing is chosen properly, the exposed pattern E(x) is a smooth function, approximating the desired ideal pattern $E_O(x)$ for

$$I_k = \ vs \ E_O(ks) \tag{3}$$

Clearly, functions $E_O(x)$ which change very rapidly cannot be reproduced accurately. The raster scan exposure as described by formulas (1) to (3) acts as a low pass filter on the original function $E_O(x)$ and can be best described by a modulation transfer function (MTF). The MTF, which gives the ratio of E/E_O for a sinusoidal input with spatial frequency f is the Fourier transform of g(x), given by

$$MTF(f) = \exp\left(-\frac{\pi^2}{4} \ (Df)^2\right) \tag{4}$$

This means that a spatial frequency f = 1/D, which corresponds to a period equal to the focus diameter is reproduced with a reduction in amplitude by a factor of 0.085. For half that frequency f = 1/2D, (period 2D), the MTF is still 0.54. 10% or less loss requires $f < 0.207 \ D^{-1}$ (or periods > 5D). In addition to the loss of amplitude, the sampling of the original function as described by equation (3) may have effects (such as moirés and beats) which must be studied for any particular case.

For the exposure of lenticular array patterns, high resolution is only needed to achieve a sharp edge defined at boundary of two adjacent lenslets. Clearly, this can best be achieved with a focus which is as small as possible . On the other hand, as will be discussed in more detail below, a small focus is not desirable for a smooth, slowly varying exposure as required for the rest of the exposure that forms the shape of the lens surface. Any additional modulation in the exposure leads to unwanted straylight. Typically, a surface smoothness of a few percent is required for good performance.

To achieve a smooth exposure, the raster line spacing s and the focus diameter D must be chosen in an appropriate ratio. If D is much smaller than s, the individual raster lines will show up in the exposure pattern. Fig. 2 (solid lines) illustrates the situation for D = $\sqrt{2}$s. The developed photoresist contains a periodic surface relief which leads to unwanted diffraction effects. The modulation amplitude can be calculated using equations (1) and (2) for a flat exposure with I_k = const.

$$E_{max} = C \cdot \sum_{k=-\infty}^{\infty} \exp\left(-4 \ (\frac{ks}{D})^2\right) \tag{5}$$

$$E_{min} = C \cdot \sum_{k=-\infty}^{\infty} \exp\left(-4 \ (\frac{ks+s/2}{D})^2\right) \tag{6}$$

The ratio of the modulation amplitude E_s to the average exposure E_m is then given by

$$E_s/E_m = (E_{max} - E_{min})/ (E_{max} + E_{min}) \tag{7}$$

The numerical evaluation of equations (3) to (7) is shown in Fig. 3, solid curve. For most practical applications, E_s/E_m should be smaller than one percent, which leads to the condition

$$D > D_{0.01} = 1.44s \tag{8}$$

Since E_s/E_m is a very strong function of D/s, any violation of condition (8) may have severe consequences.

Another important consideration is the tolerance on the x positioning. As illustrated in Fig. 2 (dashed curves), an error in the position of one of the Gaussian exposures

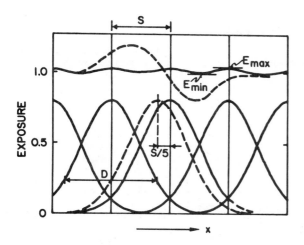

Figure 2. Constant Exposure for D/s = $\sqrt{2}$

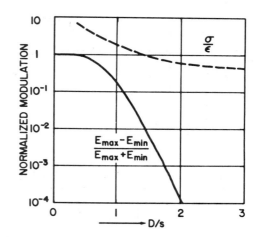

Figure 3. Normalized Modulation Amplitude as Function of D/s

causes an additional unwanted modulation. At a given y-value we may assume that the x-positions of the individual exposures vary randomly with an rms variation ε . We can then compute the rms deviation σ of the exposure from its nominal value normalized to unity for various values of the ratio D/s. The result is also plotted in Fig. 3 (dashed curve). As expected the effect of random positional errors decreases with increasing focus diameter. For a "reasonable" focus size of D = 1.44s, however, we observe a rms variation of exposure which is just equal to the relative rms variation in the x-position dx/s. This means, that a 1% accuracy in the exposure requires an accuracy of the positioning which is better than s/100. For practical values of s which typically are around 1 µm, this leads to very stringent conditions for the x-positioning, which must be accurate to tens of nm. A further increase of D/s does not help much and also would lead to long writing times or reduce resolution.

It should be noted that for exposing binary images, such as required for instance for generating chrome masks, positioning is far less critical. Here, the tolerances are simply given by the required accuracy of the edges and lines in the pattern. Compared to this, the exposure of well defined relief structures in photoresist is one or two orders of magnitude more critical.

Recording System

The recording system is shown schematically in Fig. 4. A HeCd laser (λ = 442 nm) is suitable for exposing most high-resolution, positive photoresists. The laser beam passes through an acousto-optic modulator and is focused by a microscope objective onto the surface of a resist-coated glass substrate. Focused spot diameters of down to about 1 µm are readily achieved by suitable choice of the objective and beam diameter. The resist-coated substrate is mounted on a motorized xy positioning table interfaced through a controller unit to a small computer (Commodore PET) which generates the commands required to move the substrate in a line raster scan movement. The controller executes these commands in sequence and also relays data to the modulator to set the laser beam intensity to the required value at the beginning of each line scan. The required relief profile and the exposure/development parameters such as the non-linear resist development characteristic and the writing scan speed are programmed into the computer, enabling each scan line to be written with the exact exposure required to produce the desired relief profile after resist development.

The most suitable photoresists for the fabrication of fine relief structures are high-resolution positive resists such as the Shipley AZ or Waycoat HPR series commonly used in semiconductor IC lithography and as holographic recording media. Uniform, striation-free resist films with thicknesses up to about 4 µm can conveniently be produced by spin coating onto glass substrates. For the generation of 3-dimensional relief profiles, an exact knowledge of the resist development characteristic (developed relief vs. exposure) is necessary to compute the exposure pattern required to produce a given relief pattern. This is a function of numerous system parameters (resist film preparation, exposure writing speed and scan line spacing, development procedure etc.) and is best determined for a given writing system by a series of calibration runs. A typical development characteristic is shown in Fig. 5 - for computation, a good approximation to this characteristic is given by a function of the form:

$$E = a\, t^{0.25} + b\, t^{1.25} \tag{9}$$

where E: exposure (Energy/area)
       t: depth
       a,b: experimental constants for a given wavelength, resist and development
               procedure.

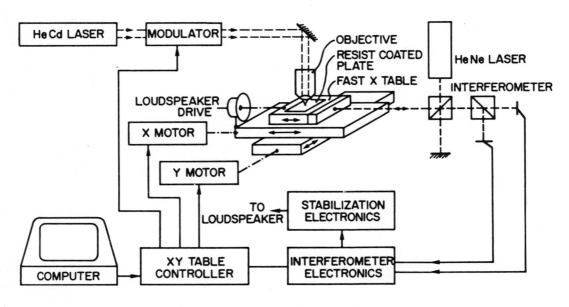

Figure 4.   Recording System

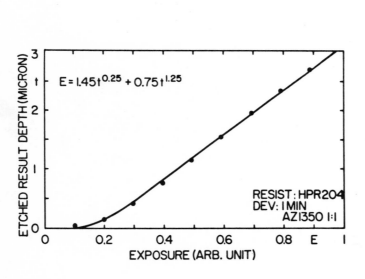

Figure 5.   Typical Resist Development
            Characteristic

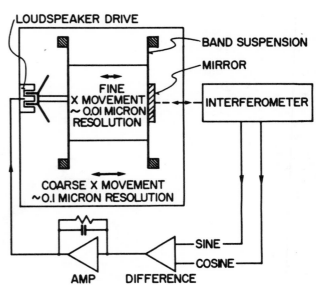

Figure 6.   Fast x-Translation Stage

As shown in the previous section, the generation of an optically smooth relief profile imposes extremely severe positioning tolerances on the mechanical scanning system. Typical parameter values for writing lenticular arrays

spot diameter $D \sim 1 \, \mu m$
line spacing $s = 0.63 \, \mu m \; (\sim D/1.6)$
x positioning tolerance (rms) $\sim s/100 \sim 6$ nm.

We describe here a low-cost solution to this problem using a commercial xy positioning table modified by the addition of a fast, high precision x-movement. The commercial unit (Anorad, NY) is a conventional 6"x6" travel motorized xy translation table with each stage running in a precision track bearing and driven by a lead screw coupled to the dc motor. The stage positions are sensed by integral optical encoders (0.5 μm resolution) and relayed to the control electronics where they are compared to the programmed positions to obtain the motor drive signals. The x stage control has been modified by the addition of an external interferometer (see Fig. 4) which can be switched into the control loop to replace the optical encoder signals. A static positioning accuracy of better than $\lambda/8 = 0.08 \, \mu m$ ($\lambda_{HeNe} = 633$ nm) is then obtained. However, when the y stage is moving, positioning errors in the x-axis arise due to factors such as bearing noise, vibrations etc. Although the x-motor drive feed-back loop acts to correct these errors, the response time is limited by the inertia of the motor and table. Measurements show dynamic x-positioning errors in excess of $\pm 0.2 \, \mu m$ during a line y scan at speeds of about 4 mm/sec.

This performance is considerably improved by the addition of a second x translation stage with limited travel, but fast enough to compensate the positioning errors (Fig. 6). This fast stage is mounted on top of the xy table, secured on 4 spring metal arms allowing a very smooth and rotation-free movement in the x-direction. It is driven by a small loudspeaker coil movement and has a maximum travel of about $\pm 10 \, \mu m$. The laser interferometer mirror is attached directly to this stage and the correction signal is derived as the difference of the cosine and sine signals from the interferometer electronics. The allowed stable positions of the stage are then given by

$$\cos(4 \pi x / \lambda) = \sin(4 \pi x / \lambda) \qquad\qquad (10)$$

Depending on the sign of the overall gain in the feedback loop, two sets of stable x-positions are obtained (Fig. 7).

$$
\begin{aligned}
x_k &= (k + 1/4) \lambda/2 \quad \text{for positive gain} \\
x_j &= (j - 1/4) \lambda/2 \quad \text{for negative gain}
\end{aligned}
\qquad (11)
$$

The spacing between discrete x values, $\lambda/2 = 312$ nm, is comparable to the smallest focus size of the writing laser and sufficient for most applications (arbitrary x-value positioning can be obtained with modified electronics). The active stabilization using proportional and integral feedback, is switched on at the beginning of each scan line at a moment when the positional error is zero to ensure locking in to the correct x_k or x_j value, and removed during movement from one scan line to the next.

The improved performance of the commercial xy table equipped with this fast x-stage is summarized in Table 1. The dynamic positioning error magnitude is dependent upon the (y) writing speed and is shown for a number of values. At a writing speed of 4 mm/sec, as suitable for lenticular array fabrication, an improvement from 80 nm to 10 nm (stabilized) is obtained. These errors are predominantly at relatively high frequencies (~ 200 Hz) - the improvement in the low-frequency positioning accuracy is an order of magnitude better. The quoted errors are the rms values ε; amplitude excursions of up to $\pm 3\varepsilon$ may be observed in practice.

Table 1. Dynamic Positioning Errors (rms)

y writing speed	ε without stabilization	ε with stabilization
4 mm/sec	80 nm	10 nm
2	55	7
1	40	7
0.5	30	7

The effect of dynamic positioning errors can be further reduced using a 2-pass writing technique. In the first pass the smooth part of the exposure pattern (Fig. 8a) is exposed using a coarse focus ($D \sim 5 \, \mu m$) obtained by defocusing the spot. (Care must be taken to

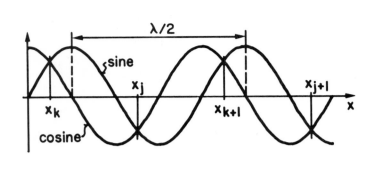

Figure 7. Stable x Positions x_j and x_k

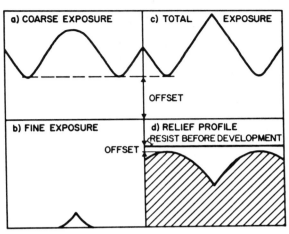

Figure 8. 2-pass Exposure Technique

adjust the optics, such that the focused and defocused spots are concentric). The second pass uses the fine focus ($D \sim 1$ μm) to record the remaining exposure (Fig. 8b). The sum of the two exposures approximates the ideal exposure pattern (Fig. 8c) which corresponds to the desired surface profile (Fig. 8d). The coarse focus exposure, with considerable overlap of the scan lines, further reduces the effect of dynamic positioning errors as well as of other error sources, such as variations in the y line scan velocity. The result is a lens with a considerably smoother surface than that obtained with an equivalent 1 pass exposure.

A typical lenticular array about 15x15 mm^2 in size requires about 15 hours recording time. This is acceptable since large numbers of high-quality copies can be generated from a single original. After development of the exposed resist pattern, a thin gold layer is sputtered onto the relief surface and a nickel electroform is generated. This metal master relief can then be used to produce copies by hot-embossing into thermoplastic sheet (e.g. PVC) or by casting techniques.

An Experimental Result

Lenticular arrays have been used for imaging periodic stripe patterns in color video cameras with a single image sensor (Ref. 1). Color stripes are required for encoding the chroma information. In addition, the lenticular array serves as a one-dimensional optical diffuser destroying spatial frequencies in the picture which are close to the chroma carrier frequencies. The main problem with the lenticular array needed in this application is straylight, which produces crosstalk between the different chroma channels. Any roughness of the surface or deviation from the ideal cylindrical shape will produce straylight. A lower limit for the amount of light which is not focused into the ideal line is determined by diffraction. This part may be kept at a reasonable value ($<$ 10%) by proper choice of the lens parameters. The additional part of straylight due to the limitations in fabrication should then not exceed 15% for a satisfactory performance of the array in the camera.

For the camera which we have built, using a CCD image sensor (pixel size 30x30 μm^2) and vertical color stripe encoding, a lenticular array with the following dimensions is used in a 1:1 imaging configuration:

ℓ = 45 μm: width of one lenslet
f = 0.26 mm: focal length in air
t = $\ell^2/8f(n-1)$ relief depth (n=1.52, refractive index)
  = 1.87 μm

An area of 15x15 mm^2 of lenticular array was recorded in a 3 μm thick layer of HPR 204 photoresist. At a recording speed of 10 mm/sec with raster spacing s = 0.633 μm this takes about 15 hours. Focus diameter for the first pass was about 1 μm and 5 μm in the second. After a 1 min development the surface relief shown in Fig. 9 was obtained. From this curve we may estimate the deviation from the ideal parabolic shape to be within $\pm$50 nm. (The difference between the ideal and observed relief depth of 1.65 μm is mainly due to the Dektak profilometer measurement using a mechanical stylus with a finite radius of curvature.)

A more direct evaluation of the lenticular array is obtained from an imaging experiment. A high contrast stripe pattern with a line to opening ratio of 2:1 is imaged using the array. The image is magnified through a 10x objective onto a linear image sensor (Reticon). Figure 10 shows the observed intensity pattern (solid lines) together with an ideal (not diffraction limited) curve (dashed line). The light not falling into the rectangular ideal distribution may be called straylight. It contributes to crosstalk and amounts to 17.6%. About half of that amount is due to diffraction and the rest to the fabrication process. The performance of the array is fully adequate for use in a single chip CCD color camera. It should be noted, that without active stabilization of the x-positioning crosstalk figures below 40% could not be reached and no useful arrays were produced.

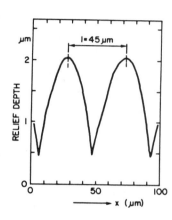

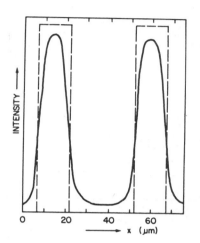

Figure 9. Measured Surface Relief of Lenticular Array

Figure 10. Image of Stripe Pattern - Measured (solid line) and Ideal (dashed line)

Acknowledgements

The assistance of J. Sandercock, H. Schütz and H. Neeracher in constructing and operating the recording equipment is gratefully acknowledged.

References

1. Knop, K., Color TV Camera using a Single Solid-State Image Sensor, Proc. SPIE, Vol. 236, p. 441 (1980 European Conference on Optical Systems and Applications, Utrecht).
2. Livnat, A., and Kafri, O., "Moiré Pattern of a Linear Grid with a Lenticular Grating", Opt. Lett. Vol. 7, p. 253, 1982
3. Glaser, I., "Lenslet Array Processor", Appl. Optics, Vol. 21, p. 1271, 1982.

Laser cutting of metallic and nonmetallic materials with medium powered (1.2 kW) CW lasers

Erich H. Berloffa, J. Witzmann

VOEST-ALPINE AG, Finished Products Division, 4010 Linz, Austria

Abstract

Beside the principle setup of cutting equipment the "Riefen" structure itself as well as the specific cut-pattern obtained for metallics are discussed. A practically useful formula for nonmetallics is given. Possible instant lens destruction is cited and an explanation tried to be given.

Introduction

Material cutting with medium and high-powered CO_2-lasers is now well established in the forefield of industrial application.[1] There exist many drawbacks in producing fine cuts and a detailed understanding of the phenomena is still missing.

Cutting setup and cutting geometry

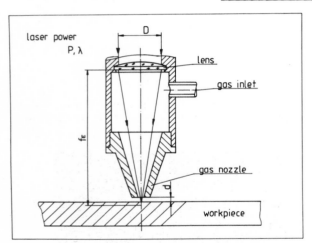

Figure 1.   Schematic cutting setup.

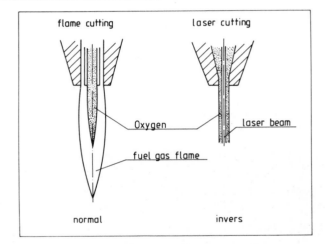

Figure 2.   Heating geometry.

Optics

The energy concentration commonly is accomplished by a simple planoconvex lens. (See Figure 1). ZnSe with the appropriate coating for CO_2-laser radiation has proved itself very well as lens bulk material. In practice an effective focal length of 5 inches is widely used for reasons of best matched parameters' beam waist[2] and focus depth as well.[3]

But where an enhanced cutting speed is required, for instance with thin sheet metal, more often a 2.5 inch lens is used, although, due to the lower focal depth, an optimum nozzle - workpiece distance, d, is much more critical to keep.

Nevertheless, for plastic cutting sometimes a 10 inch focal length is very useful because of the wider focal depth.

With our laboratory work the position of the focusing plane was established close to the workpiece surface, but still underneath it (appr. 1.5 mm maximum).

For evaluating the laser intensity in the beam waist area, a Gauss distribution was assumed. Using the lens F-number[4] $F = f_E/D$ the intensity can be derived[5] as

$$J = \frac{P}{4.68 \ F^2 \ \lambda^2} \tag{1}$$

where P denotes the incident laser power and λ the wave length. With P = 1200 W, F = 6.7

and $\lambda = 10.6 \cdot 10^{-4}$ cm equation (1) yields $J = 5.08 \cdot 10^{6}$ Wcm^{-2}.

Gas nozzle

Whereas the gas nozzle design in the flame-cutting area has undergone a long ranging history [6] and by now has reached a certain maturity this isn't the case in the laser cutting field. The additional condition that the envelope of the radiation field with its specific geometry has to penetrate the nozzle orifice prevents an assimilation to flame-cutting nozzles. Furthermore, the oxygen, when interacting with the specimen under cutting, not only provides the necessary chemical energy, but fulfills a cooling function, too. D. Schuöcker recently worked out a special theory treating this more in detail.[7] Perhaps the heating geometry (see Figure 2) stresses best the differences between flame-cutting and laser-cutting. Whereas with flame-cutting the fuel gas forms the outer shell and oxygen is fed in central parts, the relationship with the laser-cutting process reveals inverse. Laser energy forms the beam nucleus and the chemical energy is guided peripherally. We, therefore, may speak of an inverse heating geometry.

Metal-cut pattern description

High intensity and an extremely small focal area result in a very fast temperature rise on the sheet surface and temperature gradient respectively. With the previously cited F-number a steel surface is heated up to 1000° C in $1\,\mu$sec [8] and the temperature gradient reveals an amount of 3300° C/mm as can be derived from the narrow heat effected zone (0.3 - 0.4 mm). This puts a laser cut ahead by a unique set of advantages as there are:

- o extreme small heat effected zone as mentioned above, normally smaller than 0.4 mm
- o narrow kerf width, with no other cutting technology available
- o excellent edge sharpness
- o kerf surface perpendicular to sheet plane

"Riefen" structure

As with the flame cutting process, the kerf surface of a laser cut reveals a microstructure, the so-called "Riefen"-pattern (German notation). A typical pattern produced with our laboratory machine is shown in Figure 3 and Figure 4.

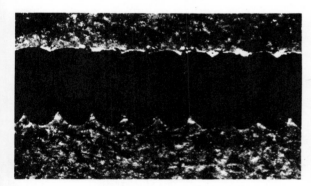

Figure 3.   "Riefen" structure - sheet surface sight

Figure 4.   "Riefen" structure - kerf surface sight

The peculiarities in the above Figures are: excellent periodic structure, not only restricted to the imaged section, a remarkably clear, but asymmetric contour which outlines the kerf and a strong hint according to the fine structure within one single "Riefe" (see Figure 4) that there might have taken place a spiraling process. We investigated this strange periodicity more in depth. By correlating the X-Y table speed to the spacing of homologous points the following proportionality could be found:

$$s = 0.167 \cdot v \qquad\qquad (2)$$

s ... periodicity (mm)
v ... table velocity (m/min)

This law validates with high accuracy that one was able to determine the actual cutting speed when the machining process was over. For the measurement a common stereomicroscope with an eye-piece micrometer was applied. A further consideration reveals that the proportional factor in (2) corresponds to a time constant which with the appropriate scaling equals to 10 m sec. This value in turn is related to the doubled line frequency 100 Hz. Careful mechanical investigations of the whole experimental setup didn't give any reasonable correlation, so it was inferred the 100 Hz were most likely induced by the laser power supply itself modulating the spatial distribution of the optical power. No fast response power monitor was available for a direct check.

G.C. Lim and W.W. Steen [9] in the meantime have reported of strong 300 Hz power fluctuations (up to 40 % of average power) due to the driving power supply of the plasma discharge. We may emphasize that this "Riefen" fine structure is anisotrope, too, i.e. X-direction and Y-direction pattern differ from each other. From a general view "Riefen" patterns are rather specific to the equipment in use, despite being very helpful for getting a deeper understanding of the infocus physics.

Cut pattern classification

For categorizing a laser cut one can use the guidelines worked out for flame-cutting [10, 11] with high confidence. The experienced practitioner judges a cut mainly by visual inspection. Uniform "Riefen" fine structure throughout the X-Y table (laser) path in conjunction with slow or missing slag formation on the lower cutting edge (see Figure 6) and small heat effected zone are the main criteria.

Figure 5 shows an acceptable cut quality. The sample material is 7.5 mm mild steel and the cutting speed was adjusted to 1 m/min.

Figure 6 reveals extensive dross formation on the lower sheet page. Additionally there can be seen a deviation from the straight course of "Riefen" fine structure which indicates an unbalanced relationship between sheet traversing speed, laser power and gas pressure either.

Figure 5.   7.5 mm cut mild steel

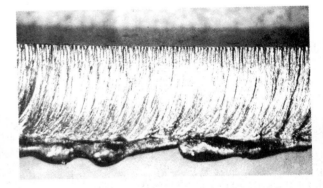

Figure 6.   Extensive dross formation

Specific cut behaviour

The general rule that the cutting speed is directly proportional to laser power and inversely proportional to the kerf cross section [12, 13] still holds with metal cutting, but the requirement of minimum dross formation forces to take oxygen pressure into account, too. In our laboratory we made the experience that it is easier to prevent dross occurence in the medium sheet thickness range (2 - 6 mm) than it is, let's say, with 1 mm or 10 mm sheets in mild steel. Higher oxygen pressure (2 - 2.5 bar) on the thin sheet side and lower pressure (0.5 - 0.8 bar) for bulky sheets, i.e. 8 - 12 mm, yielded best results in cutting low carbon steels.

When the relation kerf width to sheet thickness extends a ratio of 1 : 10, just to give a rough figure, then the steady and undisturbed removal of molten material may cause a severe problem. To investigate this a little bit more in detail a 10 mm mild steel cut (50 x 50 mm) was picked up carefully from the X-Y table and the solidified metal spheres which ball-bearing like were left in the kerf were examined under a microscope. Figure 7 shows the statistical evaluation. Remarkable is the small amount of spheres with diameters between 0.7 and 0.8 mm (kerf width 0.8 mm - widest spacing within the "Riefen" fine structure) indicating a good cut quality.

From our investigation we may infer that the model developed by D. Schuöcker [7] still holds, but the formation of spheres from the erosion front is an essential part in producing an undisturbed laser cut. Confirmation of this fact was yielded by the observation of high carbon steel processing where the cutting of bulky sheets (up to 16 mm) causes minor problems. A distribution shifted more to the left side - as shown in Figure 7 - might be responsible. An experiment wasn't carried out yet.

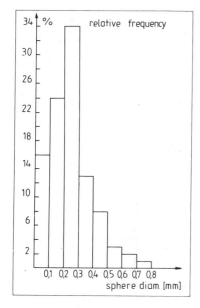

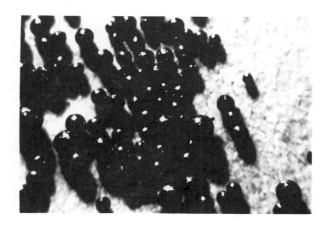

Figure 7.   Statistics of molten spheres   Figure 7a.   Molten spheres

Still another cutting behaviour is experienced with stainless steel, for example with the material X8CrNi18/9 (4301). In Figure 8 a cut of a 6 mm thick specimen is shown.

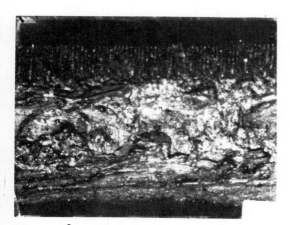

Figure 8.   Stainless steel 6 mm
            oxygen pressure 2 bar

Figure 9.   Stainless steel 6 mm
            oxygen pressure 5.5 bar

The adjusted parameters were: laser power 1.2 kW, cutting speed 1 m/min, oxygen pressure 2 bar. There is obviously to be seen that the typical "Riefen" fine structure is developed solely one fourth of the total sheet thickness, then the erosion process becomes caotic, which results in a broader kerf width and an enhanced heat effected zone on the lower cutting edge. In mild steel cutting the adjusted oxygen pressure would suffice well, but not in this subject case. By rising the O_2-pressure to 5.5 bar, cut-pattern, heat-effected zone and kerf uniformity as well could be improved considerably (see Figure 9).

<u>Cutting nonmetallics</u>

To put it in front a general judge is hardly to be met. To specify some criteria, however, one has to keep in mind:

1) visible material damage either by discoloration or mechanical flaws
2) liberation of toxic exhaust gases
3) poor cutting geometry
4) thermally induced stresses.

For many reasons PMMA (plexiglass) is widely used in the laser material processing. It meets nearly all four above-cited items. It is a simple, but nevertheless efficient indicator for testing laser beam characteristics, because it depolymerizes to an amount of more than 90 %. The oxygen contained in the ester group provides a nearly total combustion of the pyrolitic products.[14] PMMA is a favorite in art design, too.

Wood, especially plywood, is also an excellent candidate for artwork design as to be seen in Figure 10 and 11. No violation of items 1 - 4 was observed.

Figure 10.   Intarsia artwork

Figure 11. Reduction of the writing characters by means of CNC scaling

Although polycarbonate visually looks like PMMA its behaviour in laser cutting is quite different. The material is hardly to depolymerize within the laser focal region. Extensive smoke formation accompanying discoloration and poor cutting geometry are the results. The gaseous pyrolytic products of polycarbonate (lexan, macrolon) are likely phenol, aldehydes, ketones and carbon oxides.[14]

With polyacetale, glass reinforced, good cutting results were found, but exhaust gases smelt of formaldehyde stingingly. The cut pattern resembles much to that yielded from ceramic specimens.

Polyalkanes such as polyethylene (HPPE) and polypropylene mainly violate item 3. With the thermal decomposition long chained fractions as well as cyclic and dehydrogenated compounds are formed and contribute to smoke. Main gaseous final products are carbon oxydes and water.[14]

Excellent cutting results were obtained by machining glass reinforced polyester material. It meets pretty well items 1 - 4 of the previously cited criteria. As primary combustion product styrene can occur. This material is difficult to be machined by mechanical methods.

Ceramic materials are extremely sensitive to induced stresses. The performance in laser cutting this material by CO_2-lasers mainly depends on the material composition itself. We cut 5 mm ceramic flags and got an excellent cutting geometry, but the induced thermal stresses still remain a severe problem which might be solvable by thermal tempering processes.

All nonmetallics were cut with inert gases, mostly nitrogen. The authors don't know whether up to now some investigations have been made concerning chemical fractions or radicals produced with laser cutting plastics by inert gases.

For classifying the cutting behaviour of nonmetallics we used the formula [15] for melt cutting, but for reasons of practical simplicity only the measurable incident laser power was taken into account. The unknown absorptivity was enclosed to the parameter L_c (applied laser cutting energy per volume). By measuring the cutting speed v in m/min, the kerf cross section Q in mm^2 and the laser incident power P in kW the following relation was used to define L_c in kJ per cm^3:

$$L_c = \frac{60P}{Q\ v}$$
(3)

where the factor 60 was selected because of dimensional reasons. In table 1 some values are listed. The laser power or cutting speed was adjusted that way that just as much cutting energy was applied as necessary to separate the sheet sample under investigation. All cut experiments in nonmetallics were carried out by using a 5 inch lens or a F-number of 6.7. Differences found in literature might have their origin in another F-number.

Table 1. Cutting energy L_c in kJ/cm^3

Material	Density g/cm^3	L_c kJ/cm^3	Comments, Sources
pinewood	0.5	0.45	calculated from ref. 16
oak	0.85	4.6	calculated from ref. 16
plywood	0.6	3.24	calculated from ref. 16
asbestos	2.5	50-70	calculated from ref. 16
ceramic	1.9-2.8	57-84	calculated from ref. 16
glass reinforced polyester 10 mm	1.45	8.9	VOEST-ALPINE
plywood	0.6	2.5	VOEST-ALPINE
polyethylene PAS PE10 10 mm	0.94	15.5	VOEST-ALPINE
polyurethane PAS PU	1.2	1.96	VOEST-ALPINE, low-chained
polyurethane foam	0.03	0.15	VOEST-ALPINE
polyacetale PAS-LG	1.6	11.1	VOEST-ALPINE, glass reinforced
polyamid PAS 80X	1.12	5.45	VOEST-ALPINE, mixed with ethylene
plexiglass	1.2	2.4	calculated from ref. 16
plexiglass	1.2	3.8	VOEST-ALPINE
polycarbonate (lexan)	1.2	6.8	VOEST-ALPINE
hard rubber	1.4	17.7	VOEST-ALPINE

Requirements to laser cutting machines

Modern laser cutting machines are equipped with numerical controls (NC) which control all processing functions from a center station. For economic reasons it is highly recommendable to outfit the laser machine with a turret punch press, too. This equipment permits a more cost saving manufacture of simple cutting geometries like small holes or squares. If the so-called cutting center is highly automated by robots or houses special devices, an eight- to twentyfold enhancement of productivity [17] has been estimated.

The high position velocities which can be achieved, require a rigid, but nevertheless utmost precise X-Y table construction. If a turret punch module is integrated in the laser machine, vibrations have to be isolated carefully to prevent undesired interference. Naturally a reliable, industrial-proof laser source should provide both long and short time constant optical power. Optical components assembled in unsealed housings must be protected from mechanical impact and induced stresses either. In future advanced cutting heads are to be expected.

Appendix: lens damage

Anybody who dealt with high-power laser cutting might have experienced instant lens destruction. This was investigated a little bit more in depth and the following found:

From the focal area of laser beam - metal interaction there may be repelled tiny molten spheres which are able to penetrate the nozzle orifice and impinge the oxygen downstream lens surface (see Figure 1 and 12) with high kinetic energy. Nevertheless no catastrophic lens damage was observed. But when the same particle adheres, somehow, to the upper lens surface (bending mirror site) instant lens fracture happens most likely.

A possible explanation might be:

In case of a downstream lens surface pollution by a spherical particle the laser energy flux is back-scattered into the bulk material divergently and thus creates no critical interferences. But in case where a droplet lies on the upper lens surface, the plane unity vector of which is directed towards the incident energy flux, Babinet's theorem has to be considered. Local parts of laser energy are focused inside the lens bulk material and thus heavy thermal stress is induced resulting in a lens fraction.

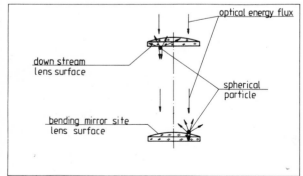

Figure 12. Lens destruction

Conclusions

The inverse heating geometry and laser specific energy concentration are the most essential factors for oxygen assisted metal cutting. The formation of a proper distribution of spheres from the erosion front was found to be highly important for the cutting performance of steels. For the cutting behaviour of nonmetallics four criteria were listed. They should be checked before judgement about the laser machinability is made. Precautions have to be taken that particles don't pollute that lens surface which is faced to the incident energy flux.

Acknowledgements

The authors thank D. Schuöcker of the Technical University, Vienna, and M. Klepp of Chemie AG, Linz, for the helpful discussions. Many thanks to all colleagues in the team for their encouragement and helpful assistance. The research was performed by the VOEST-ALPINE Finished Products Division and partially funded by the "Forschungsförderungsfond der Gewerblichen Wirtschaft".

References

1. Walker R.W., CO_2-lasers: A machining success story, Photonics Spectra, Sept. 1982, p. 65 - 70.

2. Stanley L., Ream, Present and future industrial acceptance of high power CO_2-lasers, laser focus, Dec. 1982, p. 43 - 47.

3. Weber, Herziger, Laser-Grundlagen und Anwendungen, Physikverlag 1972.

4. Melles Griot, Optics Guide 1978, p. 9, Melles Griot Arnhem, Netherland.

5. Duley W.W., p. 110 - 112, Academic Press (1976), New York, San Francisco, London.

6. Westerlund S., High speed cutting with curtain nozzle, colloquium on thermal cutting and flame processes, Sept. 7, 1982, Ljubljana, Yugoslavia.

7. Schuöcker D., Laser cutting of bulk steel (40 mm) due to guided flow of radiation and reactive gas in the workpiece, 4th International Symposium on Gas Flow and Chemical Lasers, Stresa, Sept. 13 - 17, 1982, Italy.

8. Duley W.W., p. 151, Academic Press (1976), New York, San Francisco, London.

9. Steen W.M., Lim G.C., Measurement of the temporal and spatial power distribution of a high power CO_2-laser beam, Optics and Laser Technology 82, p. 149 - 153.

10. German Engineering Standard, DIN 2310.

11. Hermann F.D., Schweißtechnische Praxis, Deutscher Verlag für Schweißtechnik (DVS), Düsseldorf 1979.

12. Duley W.W., p. 266, Academic Press (1976), New York, San Francisco, London.

13. Tikhomirov A.V., Kudrayavtsev E.P., Gas-laser cutting of materials, colloquium on thermal cutting and flame processes, Sept. 7, 1982, Ljubljana, Yugoslavia.

14. Troitzsch, Brandverhalten von Kunststoffen, Carl Hanser Verlag, 1982, p. 27.

15. Sepold G., Teske K., Rothe R., Investigation on laser melt cutting of metals, colloquium on thermal cutting and flame processes, Sept. 7, 1982, Ljubljana, Yugoslavia.

16. Duley W.W., p. 266, Academic Press (1976), New York, San Francisco, London.

17. La Rocca Aldo V., Der Laser als Werkzeug in der industriellen Fertigung; Spectrum der Wissenschaft, Mai 1982, p. 60 - 72.

On the influence of surface condition on air plasma formation near metals irradiated by microsecond TEA-CO$_2$ laser pulses

I. Ursu, Ileana Apostol, Doina Bărbulescu, V. Lupei, I. N. Mihăilescu, A. Popa
[†]A. M. Prokhorov, N. I. Chapliev, V. I. Konov

Central Institute of Physics, Bucharest, Romania
[†]Institute of General Physics, Moscow, Ac. Sci. U.S.S.R.

Abstract

The paper reports a systematic microscopic analysis of metallic surface defects which are in a position to determine the breakdown threshold of a gas in close vicinity of a target surface. Metallic and nonmetallic defects with a vaporization threshold lower than that of an ideal flat metal surface were evidenced which are able to cause the lowering of the gas breakdown threshold in the neighbourhood of the metallic target.

Introduction

Plasma initiation in a gas, in the neighbourhood of a solid target, under the action of powerful, pulsed laser radiation takes place as a result of target vaporization and development of avalanche ionization in vapours[1]. However, there is a significant difference between the phenomena underlying this process in the cases of nonmetallic and metallic targets, respectively.

Thus, as the typical nonmetallic targets (graphite, glass, etc.) are strongly absorbing laser radiation and/or have essentially low heat conductivities, the threshold level of the incident intensity causing the target vaporization, I_{vap}, is always slightly under or in the neighbourhood of the values corresponding to the breakdown of gases, I_{br}. For instance, in the case of standard (microsecond duration) pulses generated by TEA-CO$_2$ laser sources, one gets $I_{vap} \lesssim I_{br} \approx (10^6 \div 10^7)$ W/cm^2. So, when incident laser intensity is getting close to I_{br}, the intense vaporization of nonmetallic target surface is typically taking place, supplying the initial centres for the avalanche ionization process development. Surface defects are therfore of little importance for breakdown development in this case.

On the other hand, for metallic targets, by $I = I_{br}$, as resulting from computations where typical metallic thermal constants are used, not only vaporization is impossible but even the melting of the target material. That is why the vaporization of a surface layer of defects and impurities, at intensity levels lower than threshold values corresponding to ideal metals, was proposed in[1,2] for the interpretation of experimental results obtained from the study of air breakdown in the neighbourhood of metallic targets.

The purpose of our study was the investigation of actual metallic targets surface with the view of evidencing the defects and impurities responsible for the above mentioned lowering of the vaporization threshold.

Experimental

The samples of polycrystalline aluminium, copper and titanium were analysed with a Stereoscan 180 scanning electron microscope having two accesories for elemental analysis: an energy dispersion spectrometer and a wavelength dispersion spectrometer, respectively. Both spectrometers analyse the X rays resulting from the interaction of the microscope electron-beam and sample, the elements contained into the sample being identified as a function of the X-ray quanta energy. The microscope images were obtained by contrast with secondary electrons.

The working parameters were: beam acceleration voltage and current of 20 kV and 100 μA, respectively. The corresponding beam penetration depth into the target is $(5 \div 10)$ μm, while the detection sensitivity, i.e. the ratio of the minimum number of impurity atoms to the total number of atoms inside the analysed volume of material is about 0.2%.

With the view of characterizing the quality of laser irradiated surfaces, the targets were subjected to the action of radiation emitted by a high repetition rate (f $\lesssim$ 100 Hz) TEA-CO$_2$ laser source, generating pulses of 0.3 J and 5 μs total duration.

Results

Previous to laser irradiation

It was shown the samples contain only Al, Cu and Ti respectively, except for oxygen found in the case of Al samples. No C or N were identified, although the afore mentioned back scattered electron method was sensitive enough for this purpose.

The following types of surface defects were shown to exist, their presence being held responsible for the vaporization threshold lowering:

(i) Dielectric centers, appearing lighter on the picture (Fig. 1). The dimensions of the nonmetallic centers were in the range of a few tens of μm and their density was about 10^6 cm^{-2}. These centers can be seen in more detail in Figs. 2 a,b.

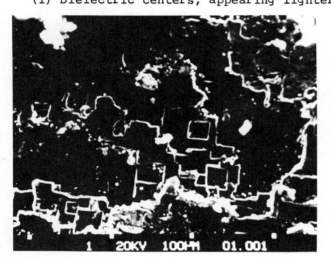

Fig. 1.   General view of a typical aluminium target.

The microanalysis of these regions showed strong lines of metal and its oxide, while the lines of C and N were still missing. So, we inferred that the observed regions contain aluminium oxides, a rather unexpected conclusion since it was known that a compact layer of oxide at the surface of aluminium does not exceed 100 Å in normal conditions[3].

It is worth noticing that centers dimensions and concentration decrease by cleaning the target surface with alcohol. Control experiments have shown that after cleaning operations the breakdown threshold rises significantly.

(ii) In[4] an assumption was made concerning the existence of defects of the target material itself: a thin layer, thermally insulated from substrate of metal flakes, grooves, pores, etc.

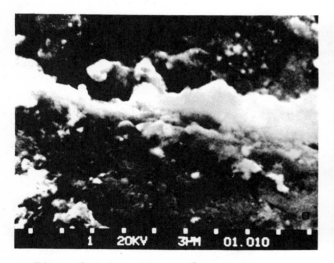
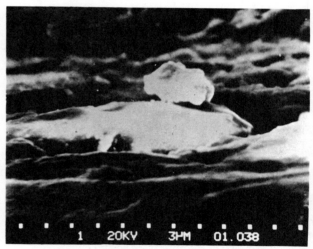

Figs. 2 a,b.   Views of the aluminium target surface at a larger magnification..

Nevertheless, there was no precise statement as to how (in what form) they are organized and what is their surface density.

Our study revealed the existence of such defects. A typical image of such a defect is shown in Fig. 3a and with a larger magnification in Fig. 3 b.

It should be noted that its thickness l is of the order of several μm. The corresponding laser intensity threshold for achieving the vaporization of thin layers of surface defects, thermally insulated from metallic support, having a thickness $l \gg \alpha^{-1}$ (α is the coefficient for light absorption in a metal) can be estimated from

$$I_{vap} \simeq \frac{c\rho T_{vap} l}{\alpha t_1}$$

Here c, ρ are the specific heat and the metal density, respectively while t_1 is the absorp-

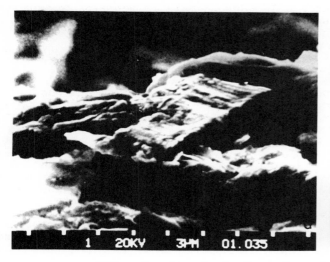
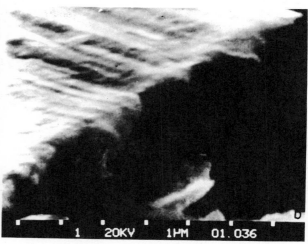

Figs. 3 a,b. Typical surface defects on an aluminium target.

tion time. The computation gives in the case of Al (c = 0.9 J/g.grd, ρ = 2.7 g/cm^3, T_{vap} = 2600°K) for $t_1 \gtrsim 10^{-7}$ s (FWHM of the first peak in the laser pulse, during which the plasma is to be formed or not[5]) and l = 3·10^{-5} cm, I_{vap} = 2·10^7 W/cm^2. I_{vap} thus becomes of the order of magnitude of I_{br} as predicted in the introduction of this paper.

We note that flakes concentration was not too large but actually (1÷2) orders of magnitude smaller than the concentration of nonmetallic centers. Generally speaking, real surfaces always contain a large quantity of different metallic defects. For illustration Fig. 4 shows a rather strange metallic defect in the form of a cylinder with a diameter of approximately 5 μm, isolated from the bulk of the metal.

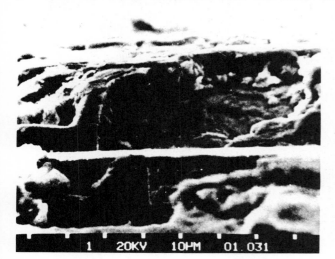

Fig. 4. Metallic cylinder appearing as a defect on an aluminium target surface

At last, we notice that another significant contribution to reducing the threshold could be due to the multiple reflections of radiation inside surface "pseudocavities", as can be seen in Figs. 3 a,b.

Hence, despite[4], the flakes are unlikely to be solely responsible for the decrease of the plasma initiation threshold in the neighbourhood of metallic targets. Indeed, as our study has proven, there is always a large number of defects and impurities covering real metallic surfaces, whose presence gets materialized in the decreasing of the vaporization threshold and thus in the afore mentioned lowering of the gas breakdown threshold in front of metallic targets.

After laser irradiation

We shall further present our results concerning the changes of the target surface induced by laser irradiation and their influence on the plasma production threshold. These modifications become of a special importance when focusing a train of laser pulses onto the metallic target surface. Thus, the action of the first pulses on the metallic surface can alter its structure significantly and consequently the values of I_{br} and I_{vap} for the next pulses.

Such a situation can occur in a chemically active gas medium, e.g. air, when chemical compounds (first of all oxide films) are to be formed, as a result of target heating by the train of laser pulses. It was thus previously observed[6] that, when copper targets were irradiated in air by a high repetition rate pulsed TEA-CO$_2$ laser (f $\lesssim$ 100 Hz) at beam intensity levels lower than the breakdown induction threshold for the single-pulse regime, after a certain irradiation time, when the target was heated up to a certain temperature, T_0, plasma started to appear in every laser pulse. And the higher the T_0 value the lower were the levels of incident laser beam intensity necessary to cause gas breakdown.

We investigated then the surface of the copper samples, irradiated in air by pulsed CO_2 laser radiation, in conditions very close to those afore mentioned (target dimension, mean laser intensity, heating time etc.).

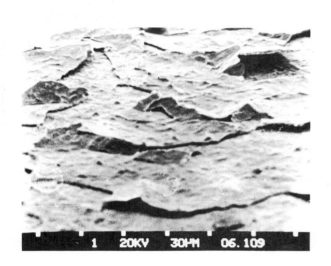

Fig. 5. Image of the surface of a copper target irradiated with a high repetition rate pulsed CO_2 laser up to $T_o \gtrsim 1000°K$

Fig. 5 gives a picture of a typical copper target surface heated under the action of powerful laser radiation. We can observe that: (i) The surface may be considered flat within an accuracy of the order of μm, which is smaller than the incident laser wavelength ($\lambda \simeq 10.6\ \mu m$). This experimental fact provides the possibility to obtain fine interference patterns in the evolutions of target absorptivity, A,as functions of target temperature $A(T_o)$ and oxide thickness $A(x)$, respectively, which were actually reported in[7]; (ii) The oxide layer has low adhesion to the metallic support and in many regions of the irradiation spot it is even thermally insulated from the metal substrate; (iii) cracks are induced into the oxide.

It should be noted that the observed formation of cracks is the first direct evidence that this phenomenon does take place in the case of metal targets heating by laser radiation. In our opinion the cracks formation accounts for the increased oxidation reaction rate in the laser heating case in comparison with isothermic conditions of oxidation as defined in[3].

At last, the decrease of I_{br} with T_o could be ascribed to oxide formation on the metal surface during target heating. Thus, as the oxidation reaction rate is proportional to $\exp(-T_D/T_o)$ (here T_D is the reaction activation temperature), the oxide layer thickness is faster increasing at higher temperatures. The oxide films are then more easily vaporized by pulsed CO_2 laser radiation because:

(i) The oxides can be considered thermally insulated from the metallic substrate for pulse durations $\tau \ll x^2/\chi$. For Cu_2O, the oxide which mainly appears on CO_2 laser irradiated copper targets[8], this condition is fulfilled for $x \approx 1\ \mu m$ (or even smaller x) and $\tau \approx 10^{-7}$ s, the very duration of the first spike of the incident laser pulse. This conclusion is fully sustained by our microscopic analysis that evidenced, as it has already been pointed out, a rather poor thermal contact between the oxide layers and the metallic substrate.

(ii) If the interference phenomena are taken into account, it can be shown that until $x \lesssim \lambda/4n$, where n is the oxide refractive index,which for Cu_2O is about (1.5÷2.5), the surface temperature on the oxide - air border is increasing with x, etc.

Finally, we emphasize that a complex analysis of the structure and composition of the target surfaces gives way to a better understanding of such laser induced phenomena as breakdown plasma and surface thermochemical reactions.

References

1. Barchukov, A. I., Bunkin, F. V., Konov, V. I., and Lubin, A. A., "Investigation of Low-Threshold Breakdown of Gases near Solid Targets Induced by CO_2 Laser Radiation", Sov. Phys. JETP, vol. 39, pp. 469-481, 1974.
2. Smith, D.C., "Gas Breakdown Initiated by Laser Radiation Interaction with Aerosols and Solid Surfaces", J. Appl. Phys., vol. 48, pp. 2217-2225, 1977
3. Kubasshewski, O. and Hopkins, B. E., Oxidation of Metals and Alloys, Butterworth (London), 1962.
4. Walters, C. T., Barnes, R. H., and Beverly III R. E., "Initiation of Laser-Supported -Detonation (LSD) Waves", J. Appl. Phys., vol. 49, pp. 2937-2949, 1978.
5. Ursu, I., Apostol, I., Bărbulescu, D., Mihăilescu, I. N., Prokhorov, A. M., Ageev, V. P., Gorbunov, A. A., and Konov, V. I., "The Vaporization of a Metallic Target by a Microsecond Pulsed TE-CO_2 Laser Radiation", Optics Communications, Vol. 39, pp. 180-185, 1981.
6. Arzuov, M. I., Karasev, M. E., Konov, V. I., Kostin, V. V., Metev, S. M., Silenok, A. S., and Chapliev, N. I. "Investigation of the Absorption Coefficient of Metal Targets Irradiated by Pulse - Periodic CO_2 Laser Radiation in Air", Sov. J. Quant. Electron., vol. 8, pp. 892-897, 1978

7. Arzuov, M. I., Barchukov, A. I., Bunkin, F. V., Kirichenko, N. A., Konov, V. I., and Luk'yanchuk, B. S., "Influence of Interference Effects in Oxide Films on the Kinetics of Laser Heating of Metals", <u>Sov. J. Quant. Electron.</u>, vol. 9, pp. 281-284, 1979

8. Ursu, I., Apostol, I., Mihăilescu, I. N., Nanu, L., Prokhorov, A. M., Konov, V. I., and Tokarev, V. N., "Laser Oxidation of Cu and Ti", <u>Proc. Int. Conf. and School on Lasers and Applications LAICS'82, Bucharest, 1982</u>, Contributed papers, pp. 399-401; to be published in <u>Applied Physics Letters</u>.

YAG LASER MACHINING CENTER

Hans Peter Schwob

c/o LASAG AG, Steffisburgstrasse 1, CH-3600 Thun, Switzerland

Abstract

The different methods of laser drilling of cooling holes in jet engine parts are presented. The successful use of this technology in production is shown by the example of a realized machining center.

1. Introduction

One of the dominant requirements in the development of aero-engines is better fuel economy, which is mainly achieved by still higher combustion temperatures. Two key factors in this trend are the use of heat resistance Fe, Ni and Co based alloys and the introduction of air film cooling between the hot reaction gases and the engine parts. This cool air film is created by the flow of cool air through a dense array of cooling holes at each heat exposed part of the engine. The drilling of these cooling holes is the main task of the described YAG-laser machining center. See fig. 3 and 5 for some parts to be drilled.

Some of the mostly used materials are known under the trade names of Hastelloy, Inconel and Nimonic. In general, all materials require similar laser processing parameters, but can differ quite distinctly in details, e.g. in the surface quality of drilled holes. The thickness ranges from 0,8 mm to several mm, hole diameters from 0,3 to more than 1 mm, the inclination of the hole with respect to the material surface from 90° down to 30° (in some cases 15°) and the number of holes can exceed 100'000 in complicated assemblies (combustion chamber wall).

2. Drilling with lasers

Four modes of laser drilling can be distinguished, which will be described in sequence below. The first two methods use a fixed beam with respect to the hole drilled. In the other two methods, the beam is moved. For each method, typical hole and laser processing parameters are given, but no attempt is made in this paper to explain the physical background.

2.1 Single Pulse Drilling

Single pulse drilling is accomplished with a single pulse of a fundamental mode laser system. The efficiency, the quality and the repeatability of the laser processing can be optimised by strict control of the spatial (high quality mode structure) and the temporal beam distribution (pulse shaping and/or pulse modulation). The resulting high quality fine holes are in general not asked in the type of parts discussed here. The laser too, is normally different from the ones used for the other modes of drilling. In order to show the possibilities of laser drilling, some typical results are given anyway:

Hole diameter range	20 - 250 μm
Ratio depth to diameter	2:1 to 6:1
Max. depth	1.5 mm
Hole tolerance	10 μm approx.
Surface quality	15 μm approx.
Laser pulse energy	0.1 - 10 J
Laser pulse length	0.1 - 0.4 ms
Laser intensity	20 - 50 MW/cm^2
Laser repetition rate	up to 20 Hz

2.2 Percussion Mode Drilling

Percussion mode drilling requires a low order mode laser system which in general consists of an oscillator-amplifier configuration, guaranteeing a highly stable quality of the beam structure. The axis of the laser beam is positioned in the centre of the hole axis and the holes are drilled by a series of laser pulses. The energy and the number of pulses depends on the depth and the diameter of the holes.

General Data

Hole diameter range	0.1 - 1.0 mm
Ratio depth to diameter	2:1 to 15:1
Max. depth	6 mm
Inclination of hole	15 - 90°
Laser pulse energy	2 - 30 J
Laser pulse length	0.2 - 1.0 ms
Laser peak power	10 - 30 kW
Laser repetition rate	5 - 50 Hz

Example 1 : 2mm thick material

Holes perpendicular to surface
Laser repetition rate: 30 Hz

Hole Diameter	Number of pulses per hole	Number of holes per second	Diameter tolerance
0.1 - 0.2 mm	3	10	15%
0.2 - 0.4 mm	3	10	10%
0.4 - 0.5 mm	5	6	10%
0.5 - 0.6 mm	6	5	5%
0.6 - 0.8 mm	7	4	5%

Similar Data are valid for holes in 1.2 mm thick material under an angle of 30° to the surface.

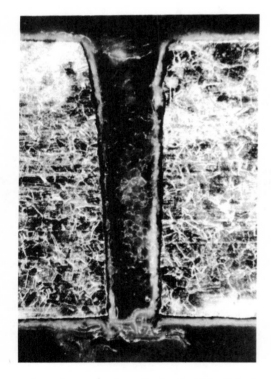

Figure 1. Percussion Mode Drilled hole

Diameter:	0.3 mm
Depth:	2 mm
Production:	10 Holes/s

Figure 2. Percussion Mode Drilled holes

Diameter:	0.5 mm
Depth:	1.2 mm
Inclination:	30°
Production:	5 Holes/s

<u>Example 2 : Deep hole drilling</u>

Hole Diameter	Depth	Number of pulses per hole	Time per hole
0.6	3.5	7	1.0 s
0.6	6.0	12	1.8 s

Figure 1 and 2 show percussion mode drilled holes. Remarkable is in particular its high production rate.

2.3 <u>Trepanning Mode Drilling</u>

Trepanning mode drilling uses the same laser system as in the percussion mode (low order mode laser), but the laser focus point rotates along a circle on the material one to three times per hole. Thus, larger diameter holes with good quality are possible, although due to the larger number of pulses required, the drilling time increases. Rotation of the beam is done by the focusing optics, which is rotated or moved on a circle by a miniature xy-table.

<u>General Data</u>

Hole diameter range	0.4 - 5 mm
Max. depth	10 mm
Hole tolerance	± 20 μm
Inclination of hole	15 - 90°
Laser pulse energy	2 - 5 J
Laser pulse length	0.2 - 0.5 ms
Laser pulse peak power	approx. 10 kW
Laser pulse rate	40 - 100 Hz

<u>Typical Performance Data</u> (different materials)

Depth (Inclination)		Hole Diameter		Number of Pulses	Time per Hole
0.8 mm)		0.7 mm		30	0.5 s
0.8 mm)		2.0 mm		50	2.0 s
1.02 mm)		1.3 mm		45	1.6 s`
1.02 mm)		1.7 mm		50	1.8 s
1.8 mm)	(90°)	0.56 mm		54	0.9 s
3.0 mm)		0.7 mm		50	1.2 s
3.0 mm)		1.0 mm		70	1.6 s
3.0 mm)		1.2 mm		80	2.0 s
3.0 mm)		1.4 mm		80	2.4 s
3.0 mm	(30°)	1.2 mm		180	5.0 s

Examples of trepanned holes are shown in Figures 3 and 4. To mention are the parallelity of the walls, the smooth surface, the sharp edges at the hole entrance and exit and the possible small hole distance, shown in Figure 4.

2.4 <u>Cutting</u>

The same laser system as for percussion and trepanning mode drilling can also be used for cutting. The optic is normally fixed and the workpiece is moving under the control of the CNC but depending on the workpiece, a moving optic may be advantageous.

Hole diameter range	2 mm and more
Max. depth	6 mm
Cutting speed a) 1 mm thick material	700 mm/min
b) 2 mm thick material	200 mm/min
Laser pulse energy	2 - 10 J
Laser pulse length	0.2 . 0.5 ms
Laser pulse peak power	approx. 10 kW
Laser pulse rate	20 - 100 Hz

Figure 3. Ring with trepanned holes

        Hole diameter: 2.0 mm
        Depth: 0.8 mm
        Drilling time: 2 s

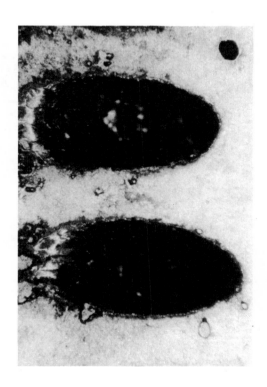

Figure 4. Trepanned Holes

Section (top) and entrance hole (right)

Hole diameter: 0.7 mm
Material thickness: 2.4 mm
Inclination: 30°
Drilling time: 1.5 s

2.5 Quality of Holes

The quality of holes is mostly influenced by the material itself, but depends on the correct processing parameters also. Following is the list of the most important quality criterions.

1) Hole dimensions: Diameter, diameter tolerance, parallelity
2) Surface in Hole: Thickness and structure of Recast Layer, Oxidation
3) Micro-cracks   : Mostly confined to Recast Layer or none at all
4) Deposits of splashes and bars.

3. Machining center for combustion liners and blades

3.1 Description of parts and drilling requirements

The machining center described here, is designed for prototype and small series production of two types of parts, extremely different in size and shape.

The first type are cylindrical parts, mainly combustion liners, as the one which can be seen mounted in Figure 5 and 6. The dimensions are the following:

Diameter : 200 - 1200 mm
Length   : up to 600 mm

The second type of parts are blades, with dimensions not exceeding 200 mm in either axis. There are two additional positioning axis needed for these parts, therefore the mechanical set-up differs slightly for the two types of parts.

The required holes are primarely drilled in the percussion mode as to section 2.2. An optional rotating optic allows the trepanning mode drilling as per specifications in chapter 2.3.

3.2 Mechanical motion system

Figure 7 shows a schematic of the equipment with the positioning axis. The linear axis X, Y and the rotational axis B serve for the positioning of the cylindrical pieces. For the blades, the table B is rotated around the axis E by 90° manually and the axis Z and A are added.

The lasersystem can be manually moved along the axis P and W. The laser beam itself can be positioned by the axis C, U and D. C adjusts the inclination of the hole with respect to the material surface, and D deflects the beam for "drilling on the fly" (see section 3.4). Axis U adjusts the focus distance. Details of each axis are given in the following table.

Axis	Travel	Accuracy	Max. speed
X	630 mm	0.03 mm	3.6 m /min
Y	1250 mm	0.05 mm	3.6 m /min
Z	200 mm	0.05 mm	3.6 m /min
P	200 mm	-	manual
W	400 mm	-	manual
U	50 mm	0.1 mm	75 mm/s
A	360°	0.03°	3.2 rpm
B	not limited	0.01°	3.2 rpm
C	+ 50° (CCW) − 75° (CW)	0.01°	3.2 rpm
D	6°	0.004°	10 ms (positioning time)
E	90°	0.001°	manual

Figure 5. Machining center with combustion liner

        A Combustion liner
        B Laser system

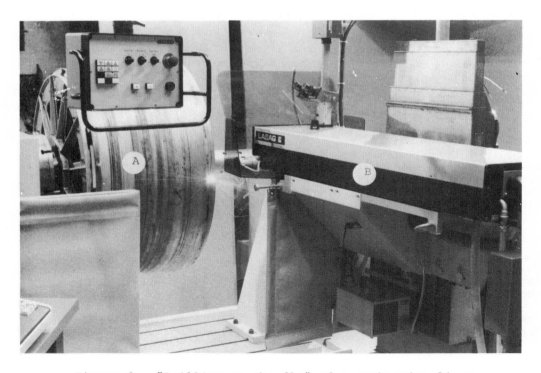

Figure 6. "Drilling on the fly" of a combustion liner

        A Combustion liner
        B Laser system

Set-up for Blades   Set-up for cylindrical parts

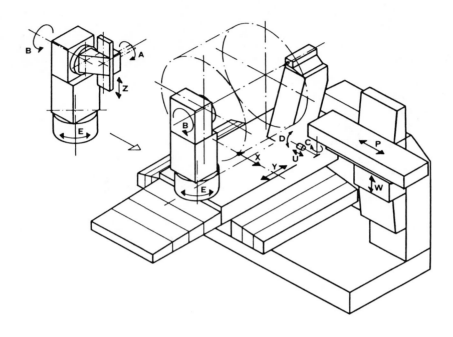

Figure 7. Schematic of equipment with positioning axis.

3.3 Laser system

There are three main requirements for the laser for good drilling results, namely

1) High peak power (10 - 20 kW, adjustable)
2) Low order mode operation
3) CNC controlled laser parameters (which are different for each type of hole)

The used laser system as shown in Figure 6, consists of two Nd:YAG laser heads in a compact oscillator - amplifier configuration, optimized for the above requirements. The data of the system are the following:

General Data

Type of laser	pulsed Nd:YAG
Beam divergence	2 mrad, half angle
Pulse length	0.2 - 10 ms, continuously adjustable
Pulse repetition rate	0.1 - 100 Hz

Maximum pulse energy and corresponding parameters

Pulse length	0.2	0.5	1.0	3.0	10 ms
Max. pulse energy	5	12.5	25	30	40 J
Corresponding pulse rate	40	16	8	6	4 Hz
Max. peak power	25	25	25	10	10 kW

The values for pulse lengths exceeding 1 ms show the systems versatility, which could be used for drilling and welding tasks.

3.4 Beam delivery system

The beam delivery system contains the three CNC controlled positioning axes C, U and D (see section 3.2 and figure 7) and the focusing optic. The focusing optic is protected by a moving film against hot metal droplets. Where advantageous, a gas nozzle can be attached to the focusing lens to assist the drilling operation by gas jet. This effect may be enhanced by using oxygen. For trepanning mode drilling a rotating optic replaces the fixed focusing lens.

For aiming a He-Ne-laser is included in the system. An observation system allows to inspect the hole quality during set-up or production.

3.5 Drilling on the fly

Percussion mode drilling requires several laser pulses onto the same spot. A "step and repeat" procedure would, due to the large workpieces involved, require a long positioning time. Then drilling rates of several holes per second would not be possible. Therefore a special arrangement has been developed and patented, namely the "drilling of the fly". The laserbeam follows the continuous motion of the workpiece, delivering multiple pulses into one hole, and then jumpes back to the next hole position. For that purpose, a laser beam scanner is incorporated.

3.6 Computer numeric control (CNC)

Besides fulfilling the standard tasks required in a machining center, the CNC has to comply with two special requirements:

1. High laser repetition rate (30 Hz standard)
2. "Drilling of the fly" requires

    a) an exact synchronisation of all movements and the laser, and
    b) in spite of the complexity of the problem simple programming instructions.

To reach this goal, the CNC incorporates especially designed hard- and software. Thus the laser is fully integrated into the machining center, the laser parameters and total drilling procedures being programmable.

4. Summary

Lasers have clearly demonstrated their ability of drilling holes in production with the required high quality and rates. Numerical laser control allows to drill holes with different parameters without any tool change. The technical versality and quality on one hand and the low running cost on the other hand, make the laser a very economic tool for a demanding and heavy load production.

MPACT RF EXCITED CO$_2$ LASER DESIGN AND RECENT APPLICATIONS

_.uRS : M. OUHAYOUN and A. ROBERT
S.A.T.

J. MELCER - Paris V University

J.P. PELLAUX - CABLOPTIC company (Switzerland)
now University of Geneva

Abstract

A CO$_2$ waveguide laser in a monolithic RF excited structure has been built at S.A.T.

Besides its ruggedness and simplified design this new structure warrants :

- A longer lifetime for the sealed off laser without electrodes in the plasma.

The laser head is fed through a light, coaxial cable and looks like a big pencil (length = 200 mm, diameter = 25 mm, weight = 200g)

This uncooled low power version emits 1 to 4 W in a 30 s run.

Two civilian applications have been investigated :

- <u>Medical application</u> : The odontological CO$_2$ laser (in collaboration with PR J. MELCER from Paris V University).

This light laser head is especially suitable for the dental surgeon working with the laser in hand :

- On the decayed areas the laser beam sterilizes and hardens the dentinary surface.

- On the mucous membrane the laser beam burns outgrowths.

- <u>Industrial application</u> : Splicing of optical fibers (in collaboration with the CABLOPTIC company).

Due to the high absorption of silica around 10 um, CO$_2$ laser light turns to be a very promising, non-contact fusing agent for optical fibers : CO$_2$ laser splicing technique allows improved control of localization as well as concentration of energy on the area to be fused. This method limits the ionic (OH-) pollution thus improving mechanical resistance.

Introduction

CO$_2$ waveguide laser technology opens a broad range of applications for compact, low power devices : Their output power per unit length can be 2-4 times greater than with a classical (free wave) laser : For a required output power, the sealed-off laser head can therefore be easily fed into small dimensions units or even held as a pencil.

Typical uses of such lasers include :

- military applications.
- micromachining, cutting, engraving of dielectrics.
- (micro) surgery.

We will only outline here civilian applications of CO$_2$ waveguide lasers and first describe the laser technology we used.

Composite A type waveguide laser

First works on RF excited waveguide lasers at S.A.T. referred to composite type, transversely excited structure Al - Al$_2$O$_3$ as proposed by LAAKMAN (1) and Hill (2) (see fig.1)

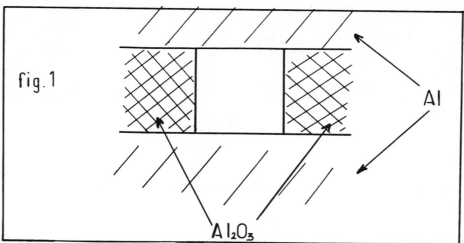

fig.1

Al

Al₂O₃

Here the waveguide channel has a square cross section (2,5 x 2,5mm). It is limited by two metallic and two dielectric walls, and acts both for electrical excitation of the lasing mixture and optical waveguiding within the cavity.

This (optically polished) four piece-structure is fed into an hollow, metallic cylinder (200mm long, 25mm in diameter) used as gas chamber as well as optical bench with mirror mounts at each end.

Ultralow leakage (He) sealing techniques, involving Yag-laser welding of titanium parts are currently used.

Impedance matching of the discharge to the 50 Ω output of the RF network is achieved with a passive circuit.

The uncooled laser head was usually excited at a 10% duty cycle to allow for thermal dissipation.

It emits up to 4W optical power with a 8-10% efficiency on a quasi-gaussian mode (EH_{11} within the waveguide).

.But in a long run the expansion coefficients of different materials may give rise to a slight bending of the waveguide (R$\sim$100m) with lower losses for higher order transverse modes.(This should always be avoided to ensure a high and reproducible concentration of energy in thermal applications).

.Besides, the aluminium electrodes in contact with the gas all along the waveguide can be oxidized via electrical degradation of CO_2 molecules :

$$CO_2 + e^- \longrightarrow CO + O + e^-$$

This limits the recombination reaction and therefore the lifetime of the sealed-off tube.

Monolithic B type waveguide laser

These operating limitations can be overcome in a monolithic waveguide channel made by extrusion or drilling in a dielectric block before proper machining (3) . Fig.2 shows the final form of the waveguide (cross section).

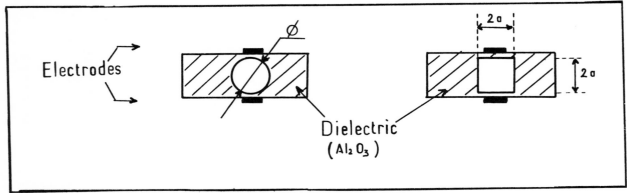

It can be either circular or square shaped :

$$1,5mm \leqslant \emptyset \text{ (or 2a)} \leqslant 2,5mm$$

Linear electrodes are then evaporated onto the substrate and plasma formation in the waveguide occurs via capacitive RF coupling through the dielectric films.

This single-piece excitation structure displays several interesting features :

. Channels'walls roughness and straightness give as low propagation losses as with the composite A structure

. Waveguide's manufacturing time and cost are reduced.

. Without anisotropic distortion on heat dissipation, laser emission can be optimized on a reliable gaussian mode.

. Better cooling of the dielectric walls (with no hot electrode in the lasing mixture) leads to a higher saturation parameter.

. The discharge can be split in several segments for optimum excitation.

. Last, drawing out the electrodes from the plasma region enhances the potential life-time of the sealed-off laser tube as LAAKMAN (1) already suggested.

Medical application : dental surgery

CO_2 lasers are today widely used in surgery : on the human body, absorption of 10 um radiaton by the (highly) hydrated tissues is followed by water vaporization and chemical degradation of cells.

On the skin for example, focused CO_2 laser irradiation leads to a cutting, bleedingless effect.

Commercially available surgical CO_2 lasers devices involve mirrors in an articulated arm to guide the laser to the tissue, until a reliable 10 um optical fibers appear.

The compact waveguide CO_2 laser we present   is an interesting alternative for low power applications such as microsurgery : The surgeon holds now the laser head in hand.

One of us (Pr. J. MELCER) first examined the odontological field (4) for several uses :

- Dental decay treatment

CO_2 laser irradiation turns to be a promising complementary therapy of the decayed area :

Once the decayed cavity has been cleared, the laser beam is focused ($\sim$ 2kW/cm^2) on the dentinal surface. The laser beam sterilizes and vitrifies it on a very small depth (50 um) and secures the living tooth against a later decay while preserving the vital pulp.

It has been proved that the vitrified dentine shows an increased hardness (on Vickers scale) and resistance to acid attack.

Infectious teeth'roots are similarly treated to sterilize and heal the dentinal tissue.

- Gingival surgery

After local anaesthesia, the laser beam burns hypertrophies on the mucous membrane : this aseptic surgery avoids cellular spreading.

We now build watercooled laser heads of similar dimensions, emitting in the 5-10W range for a broader range of applications.

Industrial application : optical fiber fusion

As silica highly absorbs 10 um radiation, we have considered the potential use of low power CO_2 lasers on optical fibers' splicing and lensing operations.

- Optical fiber splicing (in collaboration with the CABLOPTIC Company)

CO_2 laser optical fiber splicing has been already demonstrated (5) and offers several advantages with respect to arc (or flame) fusion techniques :

- The density of optical energy on the fibers ends can be controlled.

- Limited ionic (OH^-) pollution occurs during the splicing time, which improves mechanical resistance and aging properties.

- It is the only fusion technique for fiber bundles splicing.

Our compact CO_2 waveguide laser is adapted to small equipments for working sites.

- Optical fiber lensing [x]

For our own uses, we made reproducible lensing operations on fiber ends through CO_2 laser irradiation : the melting fiber end takes a half-spherical shape (because of surface tension effects) which allows better optical coupling from a laser diode.

To summarize, we have built a compact RF excited CO_2 waveguide laser at S.A.T. Its monolithic desing ensures :

- a stable gaussian mode of emission.
- a longer lifetime for the sealed-off tube.

The light pencil-like laser, which emits up to 4W in a 30s. run, has been used in medical and industrial field.

[x] In collaboration with A. COULON and C. BRODIN (S.A.T.)

(1) K.D. LAAKMAN, P. LAAKMAN SPIE vol.247, 74 (1980)

(2) R.A. HILL et al. SPIE vol.227, 12 (1980)

(3) M. OUHAYOUN and A. ROBERT Brevet 82 12 147 (1982)

(4) J. MELCER et al.   (World Congress on Medical Physics and Biomedical Engineering
                       (Hamburg (1982)

(5) K. KINOSHITA and M. KOBAYASHI Appl. Opt. 18, 19, 3256 (1979)

Fabrication of high resolution metrology target by step and repeat method

Mircea Dusa

Department of Mask Making and Photolithography,
ICCE Bucureşti; Bucharest;Erou Iancu Nicolae 32 B str.
ROMANIA 72996

Abstract

Based on the photolithography process generally used to generate high resolution masks for semiconductor I.C.S, we found a very useful industrial application of laser technology.First, we have generated high resolution metrology targets which are used in industrial measurement laser interferometers as difraction gratings. Second, we have generated these targets using step and repeat machine, with He-Ne laser interferometer controlled stage, as a pattern generator, due to suitable computer programming.Actually, high resolution metrology target, means two chromium plates, one of which is called the" rule" the other one the "vernier". In Fig.1 we have the configuration of the rule and the vernier. The rule has a succesion of 3 um lines generated as a difraction grating on a 4 x 4 inch chromium blank. The vernier has several exposed fields(areas) having 3 - 15 um lines, fields placed on very precise position on the chromium blank surface. High degree of uniformity, tight CD tolerances, low defect density required by the targets, cneates specialised problems during processing.
Details of the processing, together with experimental results will be presented. Before we start to enter into process details, we have to point out that the dimensional requirements of the reticle target, are quite similar or perhaps more strict than LSI master masks.
These requirements presented in Fig.2.

Introduction

This study describes the results obtained using Universal One Barrel step-and-repeat machine, called UER, as a PG when making metrological targets. The quality of the chromium master masks is strongly influenced by the value and stability of temperature and humidity within the working area, and by the variation of the exposure light intensity. In Fig.3., our UER machine.

Because of the importance temperature and humidity have on the final masks, specially in our case when a complet chromium blank (the rule or the vernier) is exposed in 6 - 8 hours, we have controlled permanently the class 1000 clean room area for temperature ($\pm$ 1°C) and humidity (5 %).Also the necessary handlings required by alignment, alignment control are always made under class loo laminar flow bench

The variations in ambient conditions, working material, or in equipment running have been grouped as follws :
1 - optical system performances
2 - programming characteristics
3 - photolithography process performances
4 - measurement performance

Each of these cathegories will be discussed later

Optical system performances

Optical and mechanical system performances of UER projection exposure machine, are presented in Fig.4. We have studied specially, working resolution , depth of focus, exposure light uniformity, and mechanical precision (accuracy/ reproducibility) in order to a better understanding and a real qualification of the exposure machine.

- lens resolving power could be checked with manufacturer resolution test target having 0,5 - 10 um dark clear features on clear dark background. For photolithographic purposes the term " resolution" is better qualified by the" Modulation Transfer Function " of the optical system. Considering the contrast in the image beeing at least 4 (a requirement of most resists), corresponding a modulation of 0,6 and working on MTF curve of the UVWM 1:10/0,35 objective, we have obtained 500 lines/ mm spatial frequency, which means that a pattern width of 1 um with good contrast, could be exposed with UER machine. This actually means that real N.A. is 0,28 instead of o,35 claimed by manufacturer. Same, o,28 value for N.A. has been proved more correct in the exposure- development simulation carried on during this study,

- depth of focus is a dual figure for UER machine.
Considering 0,35 NA or a 0,28 NA resulted from MTF curve we get a Rayleigh depth of focus between $\pm$ 1,3 um and $\pm$ 2,7 um. Based on these figures we have selected for exposure, chromium blanks having maximum 5 um flatness tolerances.
It should be noted that however, UER machine has an automatic focus tracking, with focus accuracy of 1/4 of the wavelength, so the defocus it is not a limiting factor in obtaining acceptable CD tolerances.

- light uniformity is the crucial parameter, because of the existing 5 % variation. To reach necessary uniformity, adjustments have been made to the automatic control of light quantity system, which compensates for different fluctuation of the light flux emitted by 500 W, high pressure mercury lamp.

The mechanical precision of the UER stepper is strongly influenced by temperature fluctuation within the room. The temperature within the room was kept at maximum t_{room} = 2^{0}C, which is normal requirement for the temperature control box of the machine to keep at the photomask level a t_{UER}= $0,1^{0}$C.
This is sufficient to have on the exposed chromium blank a positioning accuracy close to stepper instrumental accuracy, even when working with soda lime or white crown substrates.

Programming characteristics

As we have stated before, the programm developed on the computer and transfered afterwards as an input to the UER, permitted the use of the stepper as a pattern generator, in order to get necessary configuration and precision on the metrological target. The programming data are introduced in the machine via a tape reader on a punched paper tape in ASCI I format and are generated in accordance with available machine programms.
Flow-chart of the program is presented in Fig.5. The most important feature of the program is that the exposure can be interrupted in certain places and continued after a new origin for x/y stage is assumed.
Thus we can get the positioning instrumental reproducibility of the UER of $\pm$ 0,1 um.
For the rule, the exposed area of 90 mm lenght is divided in ten parts, each of them starting with a new origin.
For the vernier, the program has considered the critical parameter(field position accuracy of $\pm$ o,25 um) and has assmumed for each field a new origin.
Exposure programms have been made with minimum step size of 6 um. This gives us best results, even if the total working time is quite long, 6 + 8 hours for each plate.

Photolithography process performances

Process performances will be discussed in connection with : incorning material inspection, materials specification, exposure and development conditions.
Process flow- chart is presented in Fig.6.
Working material is HOYA AR blanks, 4 x 4 in. Plates are coated with positive resist AZ 1350, having 0,63 um $\pm$ 2% thickness.

Prior to exposure we made flatness inspection and selection.Flatness is measured with Sakura Interferometer Flatness gauge which uses He-Ne laser as light source for forming interference patterns between a flat prism surface and the photoplate being measured. The plate is held at 85° to horizontal in order to minimize plate sagging. Even if the normal position of the plate during exposure is different from that used in the interferometer we considered the flatness information as good one and selected for exposure only plates with 5 um flatness overall.

Fluorescence inspection was carried with Leitz Dialux 20 microscope, having special illuminator, 75 watt xenon lamp and two filters called " cubes", namely Hand N. These filters alloows only wavelengths over 510 nm to reach the resist film and to make spin impurities or striation to fluorescee. Because of the incident light (510 nm) the resist film has no damage during fluorescence inspection.Exposure is made at about 80 mJ/cm^2, a normal energy of UER stepper.

In conjuction with the extremely high resolution patterns, we found it necessary to apply a permanent filtration/ recirculation of the developer in a system similar to Millipore Milichem. Filtration level is 0,22 um. Plates development is made at 22°C with 1 :1 dilution for the developer. Temperature for the developer is strictly controled by the thermostatic tank of the chemical bench. Chrome etch and resist stripping are made according to Hoya standard recomandation.
Cleaning procedure is checked by counting defects like bridges, interruptions, resist or chromium spots. We have always obtained less 0,5 def/cm^2 on every plate.

For bettee understanding of the exposure - development process especially for 3 um lines, and also for better control on process variation, an analytical simulation of the lithographic process has been made. The computer programme, is based on the exposure-development mathematical models established by D.H.Dill and co-workers. Programme analyses resist profile configuration in a region of $\pm$ 1 um.Programme resolution is 0,01 um in y and x axis(resist thickness variation is seen on y and line width on x axis).

For the calculation the following data have been used: λ = 436 nm, N.A= 0,28; line width 3 um, Nr= 1,68, A= 0,495 um^{-1};B = 0,086 um^{-1}, C= 0,01 mj/Cm^2, n_{cr}=2,40 and K_{cr}= 2,79.

The resist profile during resist development, from 15 to 15 sec, during complet 1 min development, is presented in Fig.7.
From the simulated resist profiles for complet development, it can be seen that 5% variation in exposure energy produce 0,17 um. variation in line width on bright chrome, and actually no variation on line width on low reflective chrome blanks. It has to be noted that measurements on CD, made after rule and vernier generation, proved same differences in standard deviation and mean values for featune sizes, when 3 um pattern was generated on low reflective/ bright chrome plates.

Measurements performances

Control of the rule and vernier was made for CD tolerances, stepping errors and field position accuracy. Feature size standard deviation is presented in Fig.8.
It can be easily seen,that working on low reflective chromium blanks, gave us better results, actually better than the dimensional requirement of the target.Measurements have been made by most used method and equipment: a filar eyepiece with 0,01 um

resolution. Measurements were carried following NBS recomandation for accurate line width on photomasks:
- koehler illumation
- green filter assuming 546 nm transmitting light
- calibration with high precision Leitz scale
- o,6 N.A. for condenser and O,9 NA for objective(partial coherence
- measurements only on the optical axis
- each measured value being mean value of 10 measurements

Our filar eyepiece has an illuminated cross- hain, which was adjusted to a low illimination level, to minimize edge detection problem.
For stepping errors and field position, measurement were carried on Leitz Mask Comparator.
Measurement precision has been checked through repeatable measurements, among days and within a day, using always two operators.Results presented in Fig.9

Conclusion

An application of industrial laser technology has been made from two points of view:
- generation of metrological targets(similar with difraction gratings) which are to be used for laser interferometry for industrial and metrology measuring instruments;
- usage of a laser controlled machine(/8, He-Ne laser controlled stage of the UER step and repeat projection machine) to generate, through suitable machine programming, the high resolution, high precision target;

The developed method, based on the in- house photolitographic process, proved to be suitable to produce fine line structures with good precision, accuracy and repro- ducibility. Using a computer programme to simulate the exposure, development process, can help for a better control of the process variations and help for obtaining tighter tolerances is manufacturing metrological targets.

Acknowledgments

Generation of the metrology target is a tea, wark. Author like to express his thanks to. M. Iorgulescu for development of simulation programme and his work on UER adjustments and to A. Burcea who supervised UER machine during long time exposures of the target.

REFERENCES

1. J,Jerke, A. Hartman, D. Nyyssomen " Accurate line- width measurements at NBS"- Kodak Microelectronics Seminar Proceedings, 1976

2. H.Rottmann, J.Fiero "Characterisation of an Internal Image Size Standard" Kodak Microelectronics Seminar Proceedings, 1978

3. E.Hryhorenko " A positive approach to resist process characterisation for line- width control"Kodak Microelectronics Seminar Proceedings, 1979

4. W.G. Oldham, A.R. Neureuther " Projection lithography with high numerical aperture optics" - SST may 1981

5. UMVW optical lens system developed for photolithograpnic purposes, Jena Review, 1978/3

6. The UER Universal Sigle Barrel Repeater in industrial use in Japan, Jena Review 1980/2

7. M.Iorgulescu, G.Crăciunescu " Programme for positive resist exposure and deve- lopmrnt simulation " - Paper presented at CAS 82 conference Timis, oct.1982.

FIG.1. Rule/Vernier Configuration

DIMENSIONAL REQUIREMENTS OF THE RETICLE TARGET
Rule & Vernier

Parameter	Value
Plate size	4×4×0,06 in
Max. length	90 mm
Min. feature size	3 μm
Max. feature size	9 μm
Field position tolerances	± 0,25 μm
Admisible stepping error	± 0,25 μm
C.D. tolerances	± 0,15 μm
Defect density	0,5def/cm²

FIG.2. Reticle Requirements

FIG.3. Projection Step & Repeat Exposure Machine

PROCESS MACHINE TECHNICAL SPECIFICATION

Parameter	Value
Reduction scale	1:10
Field dia	8,5 mm
N.A.	0,35/0,28
Working resolution	1,1 μm
Lighting system	λ =436 nm ±5%
Positioning accuracy	⩽0,3μm(2σ)
Reproductibility	± 0,3μm(3σ)
Working speed	4 mm/sec

FIG.4. Process Machine Specification

PROGRAM FLOW-CHART

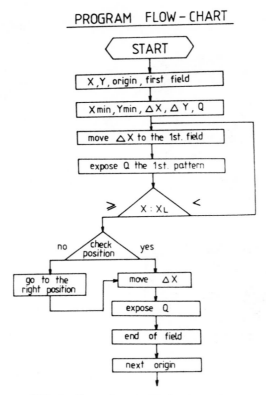

FIG.5. Computer Reticle Generation

PROCESS FLOW-CHART

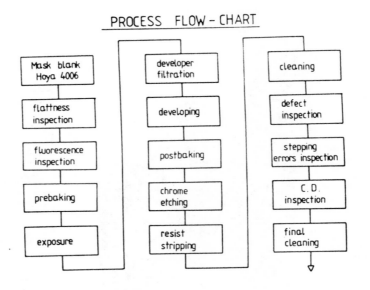

FIG.6. Reticle Manufacturing Process

SIMULATED RESIST PROFILES

A. reflective chrome blanks, 3 μm lines

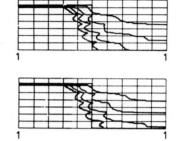

B. low reflective chrome blanks, 3 μm lines

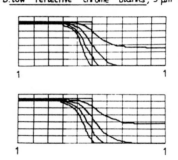

FIG.7. Simulation of the Exposure

FEATURE SIZE STANDARD DEVIATION

a. rule

$X_o = 3 \mu m$
$\overline{X} = 3,066 \mu m$
$\sigma = 0,146 \mu m$

Feature size (μm)

b. vernier

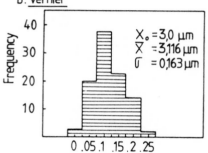

$X_o = 3,0 \mu m$
$\overline{X} = 3,116 \mu m$
$\sigma = 0,163 \mu m$

Feature size (μm)

FIG.8. Dimensional SD

STEPPING ERRORS STANDARD DEVIATION

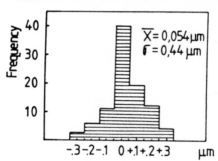

$\overline{X} = 0,054 \mu m$
$\sigma = 0,44 \mu m$

FIELD POSITION STANDARD DEVIATION

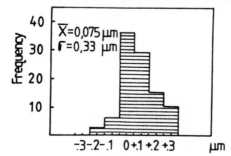

$\overline{X} = 0,075 \mu m$
$\sigma = 0,33 \mu m$

FIG.9. Registration SD

INDUSTRIAL APPLICATIONS OF LASER TECHNOLOGY

Volume 398

Session 8

Laser Material Processing II

Chairmen
Dieter Schuöcker
University of Technology, Vienna, Austria
Norbert Kroó
Central Research Institute for Physics, Hungary

Reactive gas assisted laser cutting - physical mechanism and technical limitations

Dieter Schuöcker

University of Technology, Institut für Nachrichtentechnik, Vienna (Austria)
A-1040 Wien, Gusshausstrasse 25

Abstract

From movies showing the laser cutting process in acrylic glass, it points out that the erosion takes place at a nearly vertical plane at the momentary end of the cut. That plane is covered by a thin molten layer, that is heated by absorbed laser radiation and by reaction. The removal of material from that layer is carried out by evaporation and by ejection of molten material due to the friction between the melt and the reactive gas flow. A computer simulation of that model yields a more detailed understanding of laser cutting and agrees well with experimental investigations.

Introduction

Several models have been developed to describe the mechanism of reactive gas assisted laser cutting.[1] Most of these models ignore two facts: First, they neglect the effects of reactive gas flow, as heating of the workpiece by reaction, ejection of molten material by friction with the gas flow and convective cooling. Second they assume that the area where the laser radiation is absorbed is situated in a plane in parallel to the surface of the workpiece, although that area is situated on a vertical plane at the momentary end of the cut as shown by movies from laser cutting in acrylic glass prepared by ARATA.[2] In order to improve understanding of mechanism and limits of reactive gas assisted laser cutting, a more appropriate model is proposed, that considers the various effects of the reactive gas flow as well as the real geometry of the cut. The analysis of the improved model is presented in the following.

Physical mechanism and theoretical analysis

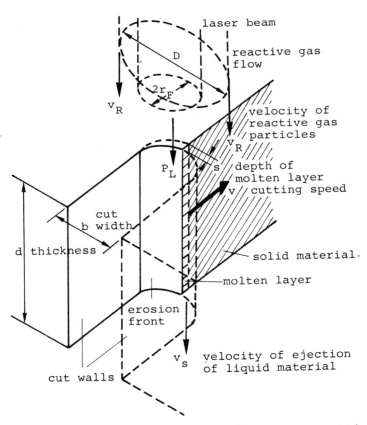

Fig.1. Geometry of reactive gas assisted laser cutting.

The model described here (Fig.1) is based on the fact, that erosion and material removal take place not at a horizontal plane in parallel to the surface of the workpiece, but at the vertical plane at the momentary end of the cut.[2] Laser radiation as well as reactive gas are not absorbed in a plane in parallel to the surface of the workpiece, but are both guided by the walls of the cut and absorbed at the vertical end of the cut.[3] It is assumed, that the erosion front is covered by a thin molten layer, that is heated by the absorbed laser radiation and by reaction between impinging gas particles and the molten metal and loses heat by vaporization, by ejection of liquid material at the lower surface of the workpiece, by heat conduction and by melting of solid material due to the movement of the end of the cut in the direction of the cutting speed.

The cutting process is described by the following quantities (Fig.1): cut width b, thickness of the molten layer s, temperature of the melt T, density of reactive gas particles in the melt n_R, density of the pure metal particles in the melt n_A, velocity of the ejection of the liquid material at the lower surface of the workpiece v_s, and cutting speed v. The cut width b is determined in a first approximation by the diameter of the laser beam at the surface at the workpiece. To determine the other six quantities, six equations are available, namely the balance of the reactive gas particles in the molten layer, the balance equation of the pure metal particles in the melt, the heat conduction equation, the energy balance, the balance of the momentum in a direction perpendicular to the surface of the workpiece and finally a mass balance.

The balance equation for the density of reactive gas particles in the melt includes gain by reactive gas particles impinging on the molten layer (q_R per unit area) and losses by reaction with the rate k_R, by ejection of liquid material at the lower surface with velocity v_s and by evaporation described by the temperature dependent quantity γ_R. Gain by dissoziation can be neglected. The following balance equation is obtained:

$$bdq_R = sbdk_R n_R n_A + sbn_R v_s + bd\gamma_R n_R \qquad (1)$$

The balance of the pure metal particles in the melt includes gain due to melting of solid material during the proceeding of the molten region in the direction of cutting and losses by reaction with rate k_R, by ejection of liquid material at the lower surface of the cut with velocity v_s and by evaporation described by the temperature dependent quantity γ_A (n_0 atomic density of the solid material). From that balance the following equation is obtained:

$$bdvn_o = sbdk_R n_R n_A + sbn_A v_s + bd\gamma_A n_A \qquad (2)$$

The momentum balance of the molten layer in vertical direction includes the gain by friction with the gas flow and the loss by the ejection of melt at the lower surface of the workpiece. That balance yields an expression for the velocity of the ejected molten material (η_R dynamic viscosity of the reactive gas, v_R velocity of the reactive gas, δ_s density of the melt):

$$v_s = \sqrt{\frac{\eta_R}{\delta_s} \frac{d}{sb} v_R} \qquad (3)$$

The solution of the heat conduction equation for a moving point source with energy influx dE/dt and velocity v yields (ρ density of workpiece, c specific heat, K thermal conductivity, $\varkappa = K/\rho c$ thermal diffusity, z distance to the point source in the direction of cutting, K_o Besselfunction of the second kind and zero order)[4]

$$T(z) = \frac{1}{2\pi Kd} \cdot \frac{dE}{dt} \cdot \exp\left(\frac{vz}{2\varkappa}\right) \cdot K_o\left(\frac{vz}{2\varkappa}\right) \qquad (4)$$

Approximately, it can be assumed that at z = b/2 (at the erosion front) a rotational symmetry exists. In that case, the energy lost from the erosion front (temperature T) into the workpiece by heat conduction is found to be

$$\left(\frac{dE}{dt}\right)_K = 2\pi dKT \frac{\exp(-vb/4\varkappa)}{K_o(vb/4\varkappa)} \qquad (5)$$

The depth of the molten layer can be determined from Eq.4 (T_s melting point, $K_1 = -K_o'$):

$$s = \frac{2\pi}{v}\left(1 - \frac{T_s}{T}\right) \frac{K_o(vb/4\varkappa)}{K_1(vb/4\varkappa)} \qquad (6)$$

The molten layer gains energy by absorption of laser radiation and by reaction and loses energy by heat conduction as given by Eq.5, by evaporation from the erosion front, by melting solid material and by ejection of the melt at the lower surface of the workpiece. A numerical estimation shows, that the last two terms can be neglected compared to the first and second term. The following energy balance can be found under use of Eq.5 (ε_V evaporation energy per atom, m_s atomic mass, k Boltzmann constant, A,B,C evaporation constants, ε_R reaction energy per molecule, α_L waveguide attenuation of the cut):

$$P_L(1-e^{-\alpha_L d}) + \varepsilon_R sbdk_R n_R n_A = 2\pi dkT \frac{\exp(-vb/4\varkappa)}{K_o(vb/4\varkappa)} + \varepsilon_v db\left(\frac{133,3}{\sqrt{2\pi kTm_s}} 10^B T^C 10^{A/T}\right) \tag{7}$$

The mass balance of the molten layer includes mass entering the molten region due to the cutting movement and mass lost by evaporation and by ejection in liquid state $[(dE/dt)_v$ energy loss at the erosion front due to evaporation as giben in Eq.7, n_s average density of particles in the melt]:

$$bdn_o v = \frac{1}{\varepsilon_v}\left(\frac{dE}{dt}\right)_v + sbn_s v_s \tag{8}$$

With the approximation $n_s \sim n_o$ and under use of Eqs.3,6, and 7 the following equation results:

$$\frac{bv}{4\varkappa} = \sqrt{\frac{\eta_R}{2\delta_s}}\frac{b}{4\varkappa}\sqrt{\frac{v_R}{d}}\sqrt{1 - \frac{T_s}{T}}\frac{1}{\sqrt{vb/4\varkappa}}\frac{K_o(vb/4\varkappa)}{K_1(vb/4\varkappa)} + \frac{1}{n_o}\left(\frac{133,3}{\sqrt{2\pi kTm_s}} 10^B T^C 10^{A/T}\right) \tag{9}$$

The temperature of the erosion front and the cutting velocity can now be determined by the simultaneous solution of Eqs.1,2,7, and 9. As described elsewhere[5], the two equations for the density of the reactive gas particles and the pure metal particles can be splitted off and treated separately, yielding the maximum energy gain due to reaction between cutting gas and metal that can be obtained under optimized conditions. The maximum energy gain of the molten layer due to reaction is determined by that amount of energy, that is liberated, if all pure metal particles that enter the melt due to the cutting movement are burned. The maximum reactive energy gain writes thus (n_o atomic density of the solid material):

$$\left(\frac{dE}{dt}\right)_{max} = \varepsilon_R n_o vdb \tag{10}$$

The further calculation can be carried out simply under the assumption that the maximum reactive energy gain is obtained by an appropriate adjusting of cutting gas. As shown elsewhere[5] from the four remaining equations two unknown quantities can be eliminated thus yielding two final equations that contain the two unknown quantities temperature of the molten layer and cutting speed. It should be mentioned here that simpler models of laser cutting result in one single equation for the cutting speed. However it is no disadvantage to calculate the temperature of the molten layer because it will be shown later that that quantity determines wether the cut is good or not and which maximum thickness can be cutted.

Theoretical predictions and technical limitations

The numerical evaluation of the equations derived above has been carried out for oxygen assisted cutting of steel.

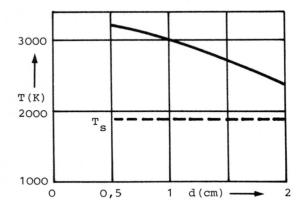

Fig.2. Temperature of the molten layer.

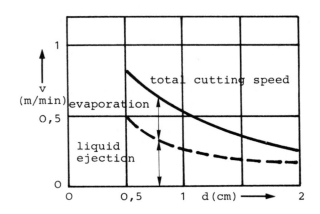

Fig.3. Contributions of evaporation and ejection of liquid material to cutting speed.

A very important result of the evaluation, the temperature at the surface of the molten layer is obtained as shown by Fig.2. It points out that the temperature reaches nearly the boiling point, if the workpiece has a thickness of 0.5 cm and that the temperature falls continously for higher thicknesses reaching the melting point at a thickness between 3 and 4 cm. Because the melting point is an absolute lower limit for the temperature of the molten layer it can be argued that an upper limit for the thickness of the workpiece is imposed by the decrease of the temperature obtained here. That theoretical limit should be 3 - 4 cm for a laser power of P = 1,2 kW as assumed here and for optimized cutting gas conditions. That value agrees well with experimental investigation carried out in the laser application laboratory of Voest-Alpine company, Linz (Austria), where a maximum thickness of 3.8 cm has been obtained for steel.[6]

Fig.3 shows as a most important result of the calculation the contributions of evaporation and of ejection of liquid to material removal and the resulting cutting speed. It points out that for moderate thicknesses evaporation is much more important than at higher values of thickness. That result corresponds to the higher temperature obtained for lower thicknesses. If it is assumed that the well known roughness of laser cuts is related to a discontinous ejection of liquid material, the decrease of the cut quality with increasing thickness can be explained in terms of a change from dominant evaporation to dominant liquid ejection with rising thickness.

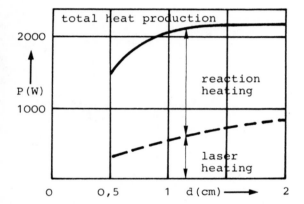

Fig.4. Contributions of radiative and reactive heating to total heat production.

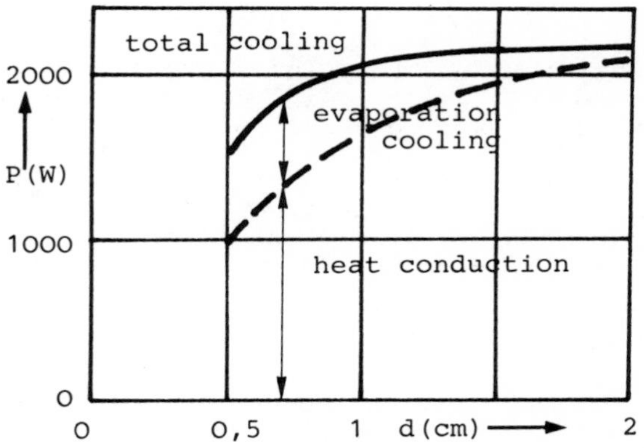

Fig.5. Contributions of heat conduction and evaporation cooling to total cooling.

Fig.4 shows the calculated contributions of radiative and reactive heating to total heat production. It points out that reactive heating prevails strongly radiative heating if optimized cutting gas conditions are assumed. That result agrees with the experimentally found strong influence of reactive gas pressure on cutting performance.[6]

Finally, Fig.5 shows the contributions of heat conduction and evaporation cooling to total cooling. From that diagram, it points out that heat conduction plays a most important role. Therefore the bad cutting performance of highly conducting metals, as copper or aluminum, is not only caused by their high reflectivity but also by there high heat conductivity.

Conclusions

The improved model presented here provides an appropriate description of reactive gas assisted laser cutting. The outcomings of the analysis yield a deeper understanding of the physical mechanism and technical limitations of laser cutting and show some ways to optimize cutting performance.

Acknowledgement

That work has been sponsored by Voest-Alpine AG, Finished Parts Division, Linz, Austria.

The author is indepted to J.Sellner, Linz, Austria, for steady support of that work.

References

1. Duley, W.W., and Gonsalves, J.N., Opt.Laser Technol. 6, p. 78, 1974.
Bakowsky, L., Herziger, G., and Peschko, W., DVS Berichte, Vol. 63, p. 1.
Eneff, T., and Ruge, J., DVS Berichte, Vol. 63, p. 163.

2. Movies prepared by Prof.Arata, shown by Prof.Sona at the occasion of the VDI meeting on laser materials processing, Stuttgart, April 1982.

3. Olsen, F.O., DVS Berichte, Vol. 63, p. 197.
Opower, H., Laser und Optoelektronik 2, p. 47, 1982.

4. Duley, W.W., CO_2-Lasers, Effects and Applications, Academic Press, N.York 1976, p. 259.

5. Schuöcker, D., DPG-meeting, Regensburg 1983.

6. Berloffa, E.H., and Witzmann, J., SPIE 1983 Int. Techn. Conf. Europe, Laser Material Processing I.

Heat treatment of gears in oil pumping units reductor

I.Guţu, I.N.Mihăilescu, N.Comaniciu, V.Drăgănescu

Central Institute of Physics, Bucharest

N.Denghel, A.Mehlmann

"23 August works", Central Laboratory, B-dul Muncii 256
73429 Bucharest, Romania

Abstract

This paper covers laser thermal treatment application to large gears (e.g. in the oil pumping unit reductor). Gear irradiation methods and thermal treatment features are pre - sented. Hardness in the range of $(700 \div 800)$ $HV_{0.1}$ and thickness up to 0.7 mm were obtained. We believe the laser treatment of such gears to provide technical and economic benefits over other methods.

Introduction

Experimental results are reported concerning the treatment of gears in the oil pumping units reductor. The gears are of cylindrical type with Sykers-type V teeth and diameters between $(100 \div 1500)$mm. As they have to withstand high variable stress and high specific con- tact pressures (of about 60 Kgf/mm^2) a pronounced wearing process and the "pitting" pheno- mena both occur. It has been impossible to overcome such difficulties by conventional heat treatment, because of the excessive cost and technical complications involved. In our opi- nion, by its low operation cost and high productivity, laser provides a simple solution to this problem.

Experimental device and methods for gear irradiation

In Figure 1, the gear thermal treatment device is shown. It essentially consists of three parts : the CO_2 laser, the device for laser deflection, focusing and scanning; and the gear translation and ro- tation device. A gas trans - port CO_2 laser of cylindri - cal geometry,[1] was employed that could supply a power of 2KW. The laser beam diameter was 20 mm at 0.5 m distance from the outlet mirror, and the radiation intensity had an even cross section dis - tribution. The laser was equipped with a KCl beam splitter and a 300 W radio - meter for continuous power measurement during gear irra- diation. The device for laser beam deflection, focusing and scanning is shown in Figure 2. Mirror M_2 enables the laser beam to be focused into a 2 mm diameter spot within a distance of about 7 cm. Cam (c) has got a particular pro-

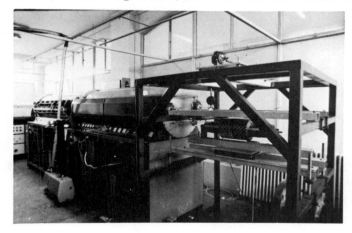

Figure 1. General view of the CO_2 laser device for gear thermal treatment

file which is designed to ensure an even velocity of the beam throughout the scanning am - plitude ; its rotation is 4000 rot/min. The hardened strip width is modified - within the range of $(4 \div 14)$mm ¯ either by modifying length (1) or using cams of various excentri - cities.

The irradiation process of thermally treated gears is shown in Figure 3. Two different methods were used [2] :

a) The individual hardening of each tooth (see Figure 3a). The scanning amplitude of the focused laser equals the tooth flank between $(1 \div 2.5)$ cm/s. Owing to its V-toothed de- sign, the gear also undergoes another rotation, which is synchronized with the translation velocity, so that the tooth surface should keep up under the same irradiation conditions.

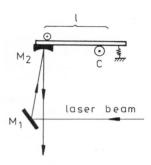

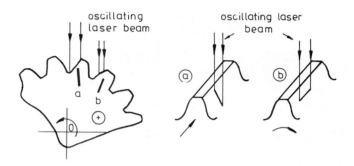

Figure 2. Outline of the laser beam deflexion, focusing and scanning device; M_1-mirror of $R = \infty$; M_2-mirror of $R = 1$ m; c - 1 mm excentricaly cam.

Figure 3. Methods for the irradiation of thermally treated gears: a) hardening of each individual tooth ; b) hardening of one gear portion at each complete rotation.

The method was applied to hardening pinions (gears of diameters up to 220 mm and weights under 100 Kg).

b) The hardening of one gear portion (equal to the scanning amplitude) at each complete rotation (See Figure 3b). The slight overlapping of the hardened strips allows the whole gear to be hardened. The method was used for hardening large weight and diameter gears. The flaw of the method lies in the hardness discontinuities that develop in the strips over - lapping area.

In both irradiation methods, the surface of the tooth was placed in the focus , while the gear spindle was at a certain well - established distance from the laser beam, so that the incidence angle be as small as possible, under conditions of complete tooth surface irradiation.

Some of the gears hardened by the above methods are shown in Figure 4.

Thermal treatment characteristics

Gears used in our experiment were made of the following steel types :

34MoCrNi 15 including C$(0.30 \div 0.38)$%, Mn $(0.40 \div 0.70)$% , Cr $(1.40 \div 1.70)$% , Ni $(1.40 \div 1.70)$%, Mo $(0.15 \div 0.30)$%, S_{max} 0.035%, P_{max} 0.035%.

OLC45 including C$(0.42 \div 0.50)$%, Mn$(0.50 \div 0.80)$%, S_{max} 0.045%, P_{max} 0.040%.

OLC60 including C$(0.57 \div 0.65)$%, Mn$(0.50 \div 0.80)$%, S_{max} 0.045%, P_{max} 0.040%.

Steel denotations are given in accordance with Romanian terminology.

Figure 4. Surface laser-hardened gears

Layer thickness, generally ranged between $(0.4 \div 0.5)$ mm, while hardnesses were of $(700 \div 800)HV_{0.1}$. Of prime importance for gear surface hardening is the macroscopic appearance of the layer. A cross section of a laser hardened gear tooth is shown in Figure 5.

One can notice the evenness of the tooth flank layer. Layer thickness is higher by the top as a result of higher heating of the edge. Whenever such effect is undesirable it may be removed by either modulating laser beam intensity or proper coating.

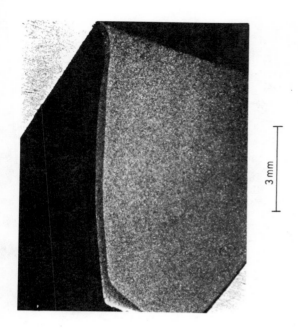

Figure 5. Specific shape of the hardened layer at the gear tooth surface
steel 34MoCrNil5

From the metallographic point of view, the hardened layer shows a very fine martensite structure
throughout its depth (See Figure 6), in all of the
employed steel brands.

For the steel types (OLC45; OLC60; 34MoCrNil5)
which gears are made of, preliminary tests were
performed on the heat treatment regime and hardness
behaviour in the overlapping passes area. The coating material had to be chosen in the first place.
Figure 7 shows that, at similar laser powers,
higher depths are obtained with carbon black coating.

Whichever coating is chosen hardness remains the
same (See Figure 8). In our experiment carbon black
was selected for gear coating

The desired heat treatment depth is achieved by
varying either laser power (See Figure 9) or speed.
At similar laser powers, higher depths were found
to result in alloy steel as compared to carbon
steel.

Hardness in the material depth is shown in Figure 10. It was also found to rise with carbon percentage.

Hardness behaviour in the overlapping passes
area is of particular relevance for the employed
heat treatment technique - Figures 11 and 12 show
the hardness in that area at the material's surface
and in depth.

Figure 6. Microscopic view of laser-hardened layers : a) steel
34MoCrNil5 ; b) steel OLC45.

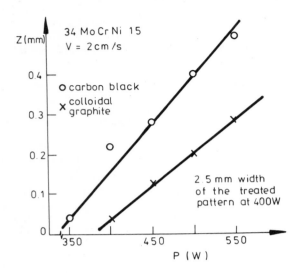

Figure 7. Case hardened depth for various coatings. Circular laser spot.

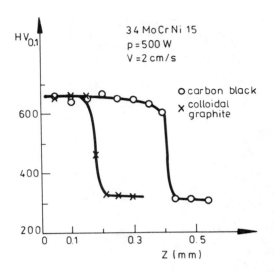

Figure 8. Hardness in the hardened layer's section for various coating materials.

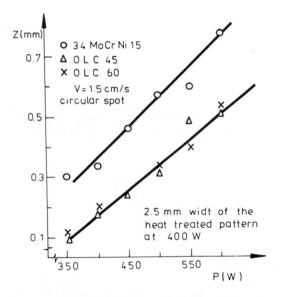

Figure 9. Case hardened depth versus laser power

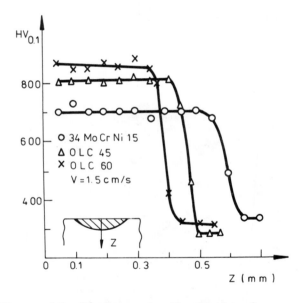

Figure 10. Hardness in the hardened layer's cross section for various steel types. Carbon black coating.

In the former case, it is observed that one hardened strip reduces the hardness of the previous one on about 1.2 mm. Whichever the steel (whether alloy or carbon) or passes overlapping degree, an equal-sized low hardness area is obtained[3]. Such hardness decrease in a strip may be ascribed to circular spot irradiation.

A dramatic hardness decrease is found in depth of the material, under the last hardened strip. The tempering zone extends on about 0.3 mm. As a conclusion, high hardness discontinuities develop in the overlapping area as well as a considerable decrease in the hardened layer's depth.[3] Practically, when gears are hardened using an oscillating beam, this undesirable effect is lessened by obturating the margins of the scanning figure.

For pinion hardening - by the individual processing of each tooth - a scanning amplitude of 9 mm was used on the tooth surface; a speed of 1.2 cm/s and a laser power of 1200W were also used. With a pinion of 2x31 teeth and 8 cm tooth length, the actual hardening time for both sides of the tooth is 9 minutes. With a gear of 620 mm diameter and 8 cm tooth length, hardening was performed on both flancks - by gear rotation and progressive shifts equal to scanning amplitude - in 35 minutes. At a power of 1600 W and 12 mm scan - ning amplitude, a complete gear rotation takes 2.5 minutes. In view of the potential auto- mation of the laser thermal treatment process, the actual gear hardening time may range very close to the above values.

In order to increase efficiency, high power lasers are recommended.

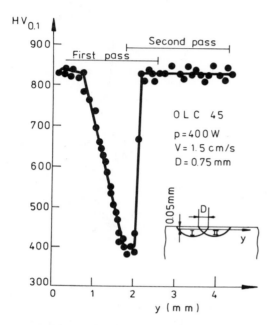

Figure 11. Hardness at sample surface for 0.75 mm overlapping strips.

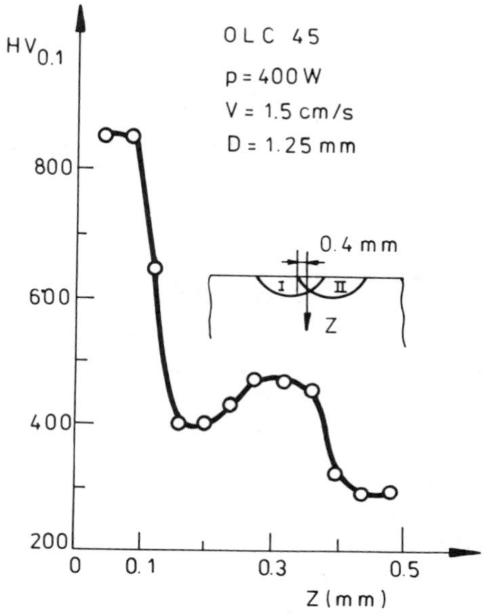

Figure 12. Hardness plot accross over- lapping hardened strips.

Conclusions

Laser thermal treatment application to relatively large weight, large-sized gears may be regarded as highly efficient from the technical and economic viewpoint.

References

1. Guţu, I., Comaniciu, N., Drăgănescu, V., "High Power Gas Transport CO2 laser", Proc. Int. Conf. and School on Lasers and Applications LAICS'82, Bucharest, 1982, Contri- buted papers, pp.175-176.
2. Denghel, N., Mehlmann, A., Guţu, I., Comaniciu, N., Drăgănescu, V., Mihăilescu, I.N., Timuş, C., "Studies concerning the superficial thermal treatment of metallic mate- rials by CO2 laser irradiation", Metalurgia, vol. XXXV, pp.274-278, 1982 (in Romanian).
3. Guţu, I., Mihăilescu, I.N., Comaniciu, N., Drăgănescu, V., Mehlmann, A., Denghel, N., "Research on the laser thermal treatment of large surfaces", Metalurgia, Vol. XXXV, 1983, (in the press) (in Romanian).

Continuous wave laser oxidation of copper

I. Ursu, L. C. Nistor, V. S. Teodorescu, I. N. Mihăilescu, I. Apostol, L. Nanu
[†]A. M. Prokhorov, N. I. Chapliev, V. I. Konov, V. N. Tokarev, V. G. Ralchenko

Central Institute of Physics, Bucharest, Romania
[†]Institute of General Physics, Moscow, Ac. Sci. USSR

Abstract

Oxidation of copper targets under the action of c.w CO_2 laser radiation was studied by transmission electron microscopy (TEM), scanning electron microscopy (SEM) and selected area diffraction (SAD), in conjunction with absorptivity determinations of the metallic samples before and after laser irradiation. A particular attention was paid to characterizing the different stages of the oxidation process.

Introduction

Laser oxidation of metals has become in the latest years an interesting investigation field of laser induced surface phenomena[1]. The understanding of oxide layer growth during metallic targets heating by laser beam is of great importance not only for laser technology development, but also as it gives a new approach to the investigation of surface thermochemical reactions.

Practically, in most papers the main attention was paid to target temperature and to the variation of the metal-oxide system absorptivity with increasing temperature (for instance[2]). The processing of these data has shown[2,3] that the rate of oxidation under the action of c.w. CO_2 laser radiation is about two orders of magnitude higher than in nonlaser experiments. However, this conclusion needs further verification as throughout all stages of its growth up to a thickness of several μm, the oxide layer was assumed to consist of Cu_2O only, while no investigation of the oxide film nature was actually performed.

This paper reports systematic transmission electron microscopy (TEM), scanning electron microscopy (SEM) and selected area diffraction (SAD) investigations of the oxide layer induced by c.w. CO_2 laser irradiation of copper targets, for the purpose of studying its composition and structure. The results are compared with absorptivity data determined before and after laser irradiation. A particular attention was payed to characterize the various stages of the oxidation process.

Experimental

Different types of copper targets with a thickness of (0,3÷1) mm were used: (i) commercial polycrystalline copper disks whose surface was cleaned with acid (HNO_3) and alcohol; (ii) polycrystalline copper sheets, mechanically polished with diamond paste.

Oxidation was performed under the action of a c.w. CO_2 laser heating source developing powers P ≲ 30 W. The targets, whose surface was almost entirely illuminated, were thermally insulated and mechanically supported by chromel-alumel thermocouples welded on their back side. Such a geometry permits to heat the samples to different temperatures up to the melting point and to estimate in the same time the target absorptivity (thermal coupling coefficient) before (A_o) and after (A_1) laser irradiation within an error of ≈ 10% (for experimental details see e.g.[2]). We point out that the A value is also representing an additional checking-up of the oxide film thickness.

Electron microscopy was performed in both transmission and scanning modes using an electron microscope JEM-200CX equipped with a scanning attachment. The accelerating voltage was 200 kV.

For transmission studies two techniques were used for sample preparation: (i) Usual extraction replicas of the oxide layer from the copper disks; (ii) Jet polishing in a suitable acid mixture of the oxidated copper sheets. The oxide layer was masked with a sticky tape which is not attacked by the acid and can be stripped out from the sample without perturbing the oxide layer. The thinning was performed from the metal side.

Electron diffraction on selected areas permitted to determine the type and structure of the oxide on the sample. Finally, the goniometer stage of the microscope was used, when necessary, to tilt the specimen in two opposite directions.

Results

c.w. laser oxidation of copper targets was performed at different temperatures, T, and times, t, to induce different stages of oxidation.

The data concerning the first set of samples (disks of commercial polycristalline copper), namely irradiation conditions, target absorptivity before and after irradiation and oxide layer composition and structure are given in Table I, while typical TEM and SAD images of oxide layers are given in Fig. 1

Table I

No. of sample	P[W] (± 3W)	t[s]	T[°C] ± 5°C	A_0 [%]	A_1 [%]	Type of oxide	Oxide crystallite dimensions [μm]
1	27	79	370	1.1	2.55	Cu_2O and ČuO	(0.02÷0.05)
2	32	112	485	1.25	4.6	Cu_2O and ČuO	≈0.05
3	31	105	505	0.75	9.95	Cu_2O and ČuO	0.05÷0.1
4	27,5	204	800	1.3	26	Cu_2O and ČuO	0.1÷0.2

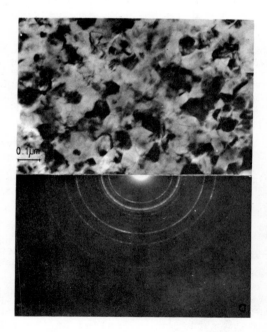 

Figure 1 a,b. TEM and SAD images of oxide layers of samples N° 1 and 4 from Table I.

As one can notice from Table I and Fig. 1, a significant magnification of crystallite dimensions in the oxide film appear as the irradiation temperature increases, which is consistent with an advance of the oxidation process. Though a direct measurement was impossible in this case, but from TEM investigation we can consider that oxide layer thickness increases with increasing the irradiation temperature. Actually, disregardful of the pre-irradiation values which differ from sample to sample (as a function of its surface quality), the sample absorptivity values after irradiation significantly increase with the surface temperature.

Finally, we want to point out that, as revealed by SAD patterns, both Cu_2O and CuO are present in the oxide layer in polycrystalline forms.

The second set of samples, consisting of copper sheets whose surfaces had been carefully polished with diamond paste, were irradiated up to temperatures of (220 ± 5)°C. As a very interesting feature, we stress that the absorptivity determinations showed practically no variation before and after laser irradiation (typically $A_0 \approx A_1 \approx (1.25 \div 1.3)$%), which indicates, in our opinion, an initial stage of the oxidation process. Two scanning electron micrographs (SEM) registered in this case were reproduced in Fig. 2.

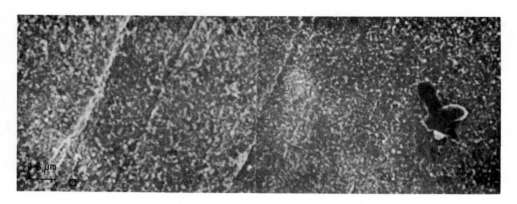

Fig. 2 a,b.   SEM of a sample irradiated to 215°C for 660 s; $A_o = A_1 \simeq 1.3\%$

An oxide layer is evidenced, so thin that scratches caused by the polishing compound are still visible (Fig. 2a). The oxide crystallites are quite small with average dimensions in the 0.05 μm range. In Fig. 2b, a zone was identified where the oxide layer is slightly removed. One can observe the layer edges and estimate its thickness, which appears to be of the same order of magnitude as the crystallite dimensions, namely 0.05 μm.

Another sample from the same group was thinned after irradiation by jet polishing to make it suitable for direct TEM studies. Fig. 3 shows a typical electron micrograph together with the SAD image of a thinned zone.

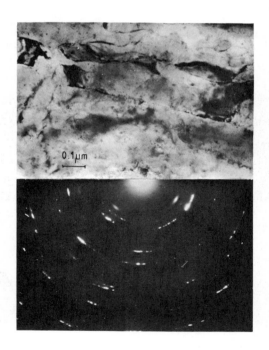

Fig. 3.   TEM and SAD images of a thin oxide layer supported by the copper substrate.
(T = 220°C, t = 1980 s)

As it resulted from the SAD pattern of this region and other zones of the sample, the only oxide type present is Cu_2O.

Fig. 4 gives a TEM image of the Cu_2O oxide layer in which one can observe also the presence of misfit dislocations. Such dislocations appear when a thin lattice layer has to accommodate to the substrate lattice of different lattice constants (for Cu_2O $a_o = 3.6150$ $\overset{o}{A}$ while for Cu $a_o = 4.2696$ $\overset{o}{A}$).

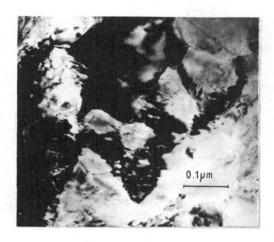

Fig. 4.   TEM of a thin oxide layer revealing the presence of misfit dislocations

    As the copper substrate was polycrystalline no crystallographic relations concerning the growth of the oxide layer on it could be established. But as it appears from the SAD pattern in Fig. 3, the dimensions of the copper grains in the substrate are enough big ($\approx$ 1 μm diameter) to permit an accommodation of the oxide growth in a thin layer.

    Some SEM studies were also performed on targets involving an advanced oxidation process. The target temperature T was near the melting point of the copper substrate. The oxide layer was in this case very thick, as shown by visual inspection, and had a very poor adherence to the support. The oxide layer grows in sheets (one sheet over another) that peel rather easily. Also the target absorptivity increases significantly ($A_o = 2.3\%$ and $A_1 = 58\%$).

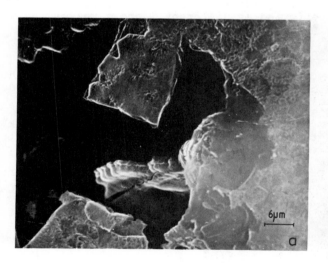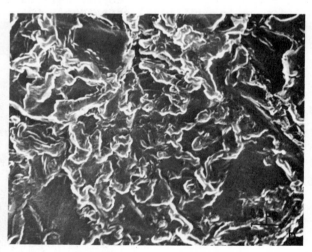

Fig. 5 a,b. The growing in sheets and cracks of the oxide layer (SEM) on the surface of a
sample irradiated up to 1100°C

    Cracks are also present in this case. The enlarged image of one end of such a crack revealed the edge of a growing sheet (Fig. 5a). One can estimate its thickness to be approximately 3 μm. We consider that the average thickness of the oxide layer ought to be much larger, but, unfortunately, no TEM and SAD studies were possible on these samples (the layer was too thick to be studied by direct transmission and had a very poor mechanical strength so no extraction replicas could be prepared.

    The general aspect of the oxide layer (Fig. 5b), very irregular, with flakes and grooves is completely different from that evidenced by the initial stage of oxidation process (as revealed in Figs. 2 a,b).

Discussions and conclusions

Thus, as it was specified before, in the conditions of a moderate target heating $T \leq 300^{\circ}C$ and irradiation time $t \lesssim 1800$ s, a rather uniform oxide layer of thickness $x \simeq 0.05 \mu m$, consisting of Cu_2O only, is growing onto the copper sample surface.

These results are in good concordance with the experimental evidence of the target absorptivity remaining unchanged as an effect of laser irradiation. Thus, if one takes into account the interference phenomena in metal-oxide layer systems, the expression for post irradiation absorptivity A_1 can be written as[2,3]

$$A_1 = A_o + \frac{4\pi(n^2-1)}{\lambda^2} x^2, \tag{1}$$

where A_o is the initial absorptivity of the unoxized metal surface and n is the oxide refractive index. For Cu_2O $n \approx (1.1 \div 2.5)$ and introducing $x = 0.05$ μm, as determined experimentally, we obtain $\frac{A_1 - A_o}{A_o} \leq 4.6 \times 10^{-3}$, which is practically negligible. This result confirms the experimental evidence of no change in target absorptivity for the first stages of oxidation process. We further note that these data shows the oxidation of metal mirrors in the case of $x \leq 0.1$ μm should change their reflectivity, with important consequences for copper mirrors utilization.

As the oxidation process advances the composition of the oxide layer is modified, and both Cu_2O and CuO appear (see Table I). The increase of target temperature during the laser irradiation causes an increase of its surface absorptivity, which is connected also with the oxide film thickness increase, as determined from TEM studies. Also a magnification of oxide crystallite dimensions is consistent with a promotion of the oxidation process.

In a very advanced oxidation stage, near the melting point of copper substrate, the oxide layer is no more uniform and it grows in sheets, one over the other. The average dimension of such a sheet is approximately 3 μm, so that the oxide layer thickness should be very big. A significant increase in the target absorptivity was also observed ($A_o = 2.3\%$, $A_1 = 52\%$).

Further understanding of copper oxidation kinetics under the action of c.w. CO_2 laser radiation needs additional experiments to be performed at different T and t values, as well as comparison with classical oxidation taking place in similar conditions.

References

1. Ursu, I., Apostol, I., Mihăilescu, I. N., Nistor, L. C., Teodorescu, V. S., Prokhorov, A. M., Chapliev, N. I., Konov, V. I., Ralchenko, V. G., and Tokarev, V. N., "A Study of the Compounds which are Induced on the Metallic Target Surface under the Action of a Pulsed Laser Plasmatron", Appl. Phys., Vol. A29, pp. 209-212, 1982.
2. Arzuov, M. I., Karasev, M. E., Konov, V. I., Kostin, V. V., Metev, S. M., Silenok, A. S., and Chapliev, N. I., "Investigation of the Absorption Coefficient of Metal Targets Irradiated by Pulse-Periodic CO_2 Laser Radiation in Air", Sov. J. Quantum Electron, Vol. 8, pp. 892-897, 1978.
3. Kubasshewski, O., and Hopkins, B. E., Oxidation of Metals and Alloys, Butterworths 1962.

The ignition of logging slash from safe distances using a CO_2 laser

M.D.Waterworth

Department of Physics, University of Tasmania, Hobart, Tasmania, Australia 7001

E.R.Rolley

State Forestry Commission, Tasmania, Australia

Abstract

Until recently, tree logging slash, which must be virtually completely removed by burning to allow for regeneration growth, has been ingited by conventional techniques employing man-held drip torches, electrically-controlled igniters, or 'incendiary' matches dropped from helicopters. Whilst satisfactory in many respects, these methods suffer with regard to cost, accessibility and, in particular, safety. Lives have been endangered and indeed lost during several decades of regeneration burning.

The instrument described in this paper permits the safe ignition of logging slash at distances varying between about 100 metres and 1.5 kilometres. It employs a CO_2 laser, beam expander, and Cassegrain telescope to produce a hot spot of the appropriate size and energy density at the required distance. Focussing is achieved by adjustment of the beam-expanding optics. The instrument is fully steerable.

Introduction

Tasmania is one of the more densely forested areas of land on earth. Each year, up to 12,000 hectares of previously logged forest must be regenerated, in order to maintain the supply of timber and to return the forest to its original splendour. In addition, up to 4,000 hectares of native and exotic plantations are established each year by State and private forest owners. The Tasmanian forestry industry is worth about $400 million per annum. Figure 1 shows a typical native forest.

Figure 1.   A typical Tasmanian native forest.

Regeneration is the process by which the forest species harvested from a logging area (coupe) are replaced. To regenerate most forest types, or to establish new plantations, requires the use of hot fires set in fuels left on the ground after logging. The fire removes most of the fuel and prepares a suitable seed-bed on which the new forest can be established. Figure 2 shows a typical fuel area left after logging.

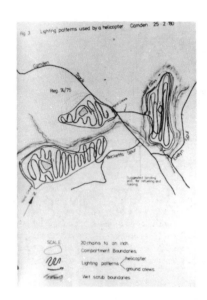

Figure 2.   Fuel remaining after logging.   Figure 3.   Burning plan for a coupe.

     The regeneration burning is carried out in individual coupe areas of size 20-200
hectares.   Figure 3 shows the plan of burning used in a particular coupe.   An intense
fire is lit and produces strong convective activity which is used to control its spread.
The firing techniques used aim to mimic nature's wild fires, which originally produced much
of the existing forest, under controlled conditions.   Figure 4 shows one such fire after
the convective core has been established.

Figure 4.   An established regeneration burn.

     Many different lighting techniques have been developed.   These include hand-held drip
torches, electrical incendiary devices, aerial incendiary devices, and gun-operated incend-
iaries.   All of these techniques have safety disadvantages, there having been a number of
disasters and near disasters in recent years.   In addition, some of the techniques involve
significant costs.

The concept of a safe and economical lighting device

     A safe, economical, and remotely located ignition device would have many advantages over
existing techniques.   Due to such factors as land topology and accessibility, a remote
device could probably not completely replace existing techniques, but would certainly
remove the need for central ignition with drip torches or helicopters.   A survey of coupes
in Tasmania has revealed that, on average, approximately 75% of coupe areas could be
ignited by a remote device employed at several vantage points for each coupe.

     A laser device employing suitable optics, which is mobile and capable of igniting fuels
at distances up to about 1.5 km from roadside vantage points, would satisfy the above
criteria.   Accurate determination of the position of the focussed laser beam is needed,
the spatial resolution required being of the order of 1 metre.   This ensures that no

lighting occurs beyond the coupe extremeties which are normally bounded by narrow fire-breaks or virgin forest. Such a laser device would have the following advantages:

(a) *improved safety in operation;*

(b) *low cost per start point ignited, particularly when compared with aerial techniques;*

(c) *improved control of burning operation;*

(d) *reduced light-up time for a given area compared with ground techniques, assuming virtually instantaneous ignition with the laser system;*

(e) *mobility which allows many burns at different locations to be lit under suitable weather conditions.*

In planning the burning of a coupe, the weather conditions have to be specified precisely. General atmospheric conditions, temperature, humidity, wind speed must fall within certain limits for burning to proceed. Furthermore, the moisture content of the fuel to be burned must fall within certain limits (usually 15%-25%). As will be seen later in this paper, a CO_2 laser system is the most appropriate, in that its operation, transmission, and energy losses are optimised for the environmental conditions for which burning is prescribed.

The ignition of cellulosic materials by radiation

The ignition of cellulosic materials by radiation was studied by the Fire Research Station, Boreham Wood, England, between 1950 and 1962. The published results have been obtained using a low-intensity radiation infrared source. No results are available for the high-power densities delivered by a CW laser beam.

The following points are well established for the ignition of cellulosic materials:

(a) *The dependence of ignition upon the irradiated area is linear and changes only slightly for areas between 1 and 10 sq.cm.; for large areas the intensity required drops rapidly towards a minimum value.*

(b) *For higher incident intensities, the ignition times vary less with the irradiated area for 0.5 sq.cm. < area < 25 sq.cm.*

(c) *The minimum radiation intensity required for ignition increases linearly with moisture content.*

(d) *The minimum radiation intensity required to ignite wood spontaneously is approximately constant for all species. Typical values for a block of wood 2.5cm × 2.5cm × 1cm are*
   18 W/sq.cm. for 2 seconds - Cedar
   19 W/sq.cm. for 4 seconds - Oak

Of importance in regeneration burning in Tasmania is the ignition of eucalypt leaves. Information, plus some experimentation, has indicated that 16 W/sq.cm. is sufficient to obtain instantaneous ignition of these leaves.

The choice of a laser system

At the time of the initial design there were only two types of lasers available capable of delivering several hundred watts of CW power. These were the well-established CO_2 laser operating at 10.6 μ and the YAG modular solid state laser operating at 1.06 μ. The output of a typical high-power YAG laser is multimode with only about 10% in the TEMoo mode. This leads to an increase in beam divergence and the minimum spot size to which the beam can be focussed. As laser power increases, these quantities increase and hence a corresponding increase in target irradiance is not achieved. CO_2 lasers provide the only avenue for obtaining high target irradiances at distances of 1 km or so, because their mode structure is simple and more than 2/3 of their maximum rated power output is in the fundamental mode.

For propagation distances of approximately 1 km and laser powers of, say, 200 watts, the limiting factor which determines the minimum obtainable spot size is diffraction. Thermal blooming can be neglected. At a wavelength of 10.6 μ, it is necessary to project the beam with a relatively large reverse Cassegrain mirror system in order to obtain a small diffraction limited spot at focus. For a diffraction limited spot, the beam's full angle divergence is

$$\theta = \frac{1.22\lambda}{R}$$

The power density at focus is, assuming zero losses in the mirror system,

$$P = \frac{4R^2 P_{out}}{\pi d^2 (1.22\lambda)^2}$$

where P_{out} is the laser output power.

Note that this treatment is valid only for single mode outputs. The table below gives some examples of the diffraction-based theoretical limit for target intensities which can be obtained from commercially available CO_2 and YAG lasers.

Table 1. Theoretical Values of Diffraction Based Parameters for Commercial Lasers

Laser Type	Model No.	Power	d (focal dist.)	R* (in cm)	θ	S (cm)	P (w/cm²)
+YAG Holobeam	260T	20W TEMoo	1 km	7.5	1.724×10^{-5}	1.724	8.57
+CO_2 Korad	KG40	150W TEMoo	1 km	7.5	1.724×10^{-4}	17.24	0.643
CO_2		150W TEMoo	1 km	50	2.586×10^{-5}	2.586	28.56
YAG		20W TEMoo	1 km	20	6.47×10^{-6}	0.647	60.83
CO_2 e.g.Korad	KG45	300W TEMoo	200 m	20	6.133×10^{-5}	1.227	254.67
CO_2			500 m	20	6.133×10^{-5}	3.067	40.6
CO_2			1 km	20	6.133×10^{-5}	6.133	10.15
YAG		200W TEMoo	1 km	12.5	1.056×10^{-5}	1.056	22.85
CO_2 Coherent Radiation		500W TEMoo	1000 m	12.5	1.056×10^{-4}	10.56	5.7
CO_2			500 m	12.5	1.056×10^{-4}	5.28	22.8

*R is the radius of the beam as it leaves the projecting optics, whether this be a mirror or lens system
+   The wavelengths for CO_2 and YAG lasers are 10.6 μ and 1.06 μ respectively

Taking the contents of Table 1 into account, and in addition considering the mechanical, economic and efficiency factors, it was decided that the preferred solution was the acquisition of a CO_2 laser with an output power of not less than 200 watts. Furthermore, a reverse Cassegrain telescope projection system was envisaged.

In arriving at this decision, investigations into the use of CO_2 lasers with the Cassegrain projection system were extended beyond those covered by Table 1. A range of laser power/mirror size combinations were considered viz.a viz. target power density and cost. Table 2 summarises some of the results obtained, and indicates why a 200W laser was selected.

Table 2. Target Power Densities as a Function of Laser Power, Focal Distance and Mirror Diameter

Laser Power (W, TEMoo)	Focal Distance metres	Mirror Diameter cm (inches)	Target Power density W/cm²
300	1000	40 (16)	9.14
300	1000	50 (20)	14.28
300	1000	60 (24)	20.56
300	750	40 (16)	16.24
300	750	50 (20)	25.38
300	750	60 (24)	36.54
300	500	40 (16)	36.54
300	500	50 (20)	57.1
300	500	60 (24)	82.22
150	1000	40 (16)	4.57
150	1000	50 (20)	7.14
150	1000	60 (24)	10.28
150	750	40 (16)	8.12
150	750	50 (20)	12.69
150	750	60 (24)	18.27
150	500	40 (16)	18.27
150	500	50 (20)	28.55
150	500	60 (24)	41.11

Optical and mechanical considerations

The laser acquired for the ignition system is a ruggedised, water cooled CO_2 laser and has the following specifications:

Output power	up to 200 watts TEMoo (CW)
Wavelength	10.6 micron
Mode structure	> 67% TEMoo (CW)
Beam diameter	8 mm (½ power points)
Beam divergence	1.7 milliradians (full angle between ½ power points)

The practical requirements of the system are

(a)   to produce ε focussed spot of radiation over distances 100m < d < 1 km;

(b)   to be as efficient as possible, optically over this range of d, i.e. minimum spill-
over of the beam at all optical surfaces;

(c)   to be both fully steerable and manoeuverable in the field;

(d)   to be 100% safe for the operating crew under all conditions.

With regard to safety, the 10.6 μ radiation leaving the laser remains entirely enclosed
until it leaves the projecting mirror of the reverse Cassegrain telescope.   Here the energy densities are such that no damage can be caused to human tissue.   The viewing optics to
be described below ensure that there is no possible damage to life, human or animal, as a
result of the firing of the laser.

To achieve a focal range of 100 m - 1000 m, some adjustable optics must precede the projection mirror of the Cassegrain telescope.   One may either change the separation between
the two mirrors, which involves considerable mechanical effort, or alternatively procede
the fixed mirror system with some adjustable auxiliary optics.   The latter alternative was
selected because of its ease of operation, both optically and mechanically.

Reference to Figure 5 will indicate schematically the optical layout of the ignition
system.

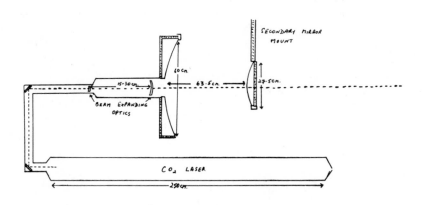

Figure 5.   Schematic layout of laser ignition system

The emergent laser beam is first directed into the main optical system by means of two
45° mirrors.   (At first these were simply uncooled aluminium mirrors, but it was later
found preferable to use uncooled gold-coated mirrors.   A future version of the device may
well eliminate these mirrors completely!)   The beam then passes through two zinc selenide
lenses, the second of these being fixed whilst the first's position is adjustable.   This
provides the focussing adjustment required.   Changing the separation between the two

lenses alters the divergence of the beam emerging from the second lens. This, as well as changing the final focal setting, also changes the illumination of the two mirrors with a resultant alteration in overall efficiency of the system. Table 3 gives some typical figures for focal distance and efficiency as a function of lens separation.

Table 3. *Focal Distance and Efficiency as a Function of Lens Separation*

Lens Separation(cm)	Focal Distance(m)	Energy Loss(%)
16.92	894	0
17.24	805	0
17.56	732	0
17.87	673	1
18.19	622	2
19.46	483	7
20.73	396	11
22.00	340	15
24.54	266	22
27.08	223	28

The focal setting is achieved using an accurately calibrated knurled knob and vernier. To determine where the ignition system is pointing, and hence where ignition will occur, viewing optics are essential and must also be precisely calibrated and, eventually, tied into the focal setting of the main system. Initially, the viewing optics consist of a 7.5 cm f/10 refractor strapped to the main telescope tube. The primary erect image (whose position is related to the object position) is viewed with an eyepiece. At present this position is manually converted to the required setting of the main telescope focussing optics.

Assembly of the complete system

The complete laser ignition device comprises the following components:

1   *A D1310 International 2 tonne bruck with a specially stabilised flat tray as the transit vehicle.*

2   *A modified Bofors MkIII gun mount for carrying the CO_2 laser and all the optics. The mount is fully adjustable in altitude and azimuth, thus permitting a scanning ignition mode of the coupe from a suitable vantage point(s).*

3   *A Radiation Research Model 502, 200 watt, CO_2 laser system, including power supplies, water cooling system, gas supply, etc.*

4   *An 8 kVA petrol generator.*

The components (2)-(4) are mounted on the stabilised tray of the truck. Figure 6 shows schematically the assembly of these components.

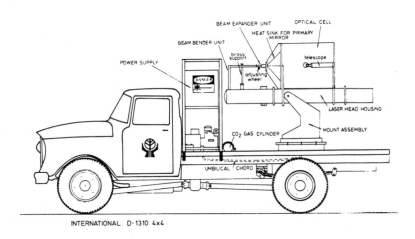

Figure 6. Schematic of assembly on truck

The laser system, which has been especially 'ruggedised' for this purpose, consists of a power console, a portable control unit, and the laser head itself, as mentioned in (3) above. Within the high voltage power console are the gas flow control system, the vacuum pump and vacuum control system, and a number of interlocks and safety circuits. The laser head connects to the power console by a single 3 m umbilical cable. The laser head assembly houses the laser discharge tube 4 m in length and arranged in a fold; the tube and optics are mounted in a stabilised resonator structure.

Testing the device in the field

The laser ignition device is shown fully assembled in Figure 7. In November 1980 the first field tests were carried out. Figure 8 shows, for example, the instantaneous ignition of a canvas sheet at a distance of 500 metres. It has been demonstrated that ignition can be achieved over the range of distances required.

Figure 7. Fully assembled laser ignition system ready for field test

Figure 8. Laser ignition of canvas sheet at 500 metres

During 1981 and 1982 developments have concentrated on the optimisation of the Mark I prototype and on preparing for further field proving trials. Areas under scrutiny include

(a) use of a more reliable and slightly higher powered laser. The sizes of CO_2 lasers available in 1983 are considerably less than the present unit, and it is expected that the beam bending mirrors can be eliminated;

(b) motor control of the altitude and azimuth motions of the Bofors gun mount from a hand-held paddle, together with a push button control for laser firing;

(c) an automatic linkage between the viewing optics and the main telescope optics, ideally using the main telescope in the dual role of a viewing and a projecting system. Alternatively a micro-processor control between the two telescopes is envisaged to allow for terrain-induced variations in focus.

Further field tests are to be carried out during March and April 1983, weather permitting, and some results will be available at the Conference.

Acknowledgements

Thanks are due to Mr.J.L.Hickman of the Tasmanian State Forestry Commission for initially suggesting this project, and to the Commission for its financial support of the project. In addition, thanks go to Mr.A.Ingles and Mr.M.Whitney of the State Forestry Commission for the very significant assistance provided by them on this project.

References

"The Choice of a Laser System for the Ignition of Forest Slash" - P.O.Gourlay (Private Communication

"Some Aspects of the Self Heating and Ignition of Solid Cellulosic Materials" - P.H.Thomas and P.C.Bowes: British Journal of Applied Physics Vol.12, p.222, 1961

"The Ignition of Wood by Radiation" - D.I.Lawson and D.L.Simms: British Journal of Applied Physics Vol.3, p.288, 1952

"Laser Damage in Optical Materials at 1.06 µ" - R.M.Wood, R.J.Taylor and R.I.Rouse: Optics and Laser Technology, June 1975

"Log-Amplitude Mean for Laser-Beam Propagation in the Atmosphere" - F.G.Gebhardt and S.A. Collins Jr. : J.O.S.A. Vol.59,9, pp.1139-1248, 1969

"Turbulence Effects on Thermal Blooming" - F.G.Gebhardt, D.C.Smith, R.G.Bauser and R.S. Röhde: Applied Optics Vol.12, pp.1794-1805, 1973

Nd lasers for processing

T. T. Basiev, B. I. Denker, A. A. Maliutin

Lebedev Physical Institute
Leninsky Prosp., Moscow, USSR

I. Czigány, Z. Gy. Horváth, I. Kertész

Central Research Institute for Physics
H-1525 Budapest, P.O.Box 49, Hungary

Abstract

Annealing of preheated Si-samples by Nd:YAG laser is described. A Nd:YAG laser device with and without acousto-optical Q-switching for trimming, scribing, marking, drilling and medical purposes was used. The development of a portable LiNdLa-phosphate glass laser resulting in 6 % efficiency, Q-switched by LiF with F_2^--color centers is discussed together with some of its applications.

Introduction

Many scientific works have given details of fundamental and advanced quantum electronics but only a few of them have dealt with laser development or applications.

We report on two topics: one connected with different Nd:YAG laser applications, the other with the development and possible applications of pulsed lasers using the new LiNdLa phosphate glass.[1]

Nd:YAG

The Nd:YAG laser development provoked by industry has resulted in a family of devices ranging from 10 to 130 W multimode and 3-13 W TEM_{OO} power with different switching, optical, positioning and frequency converting supplementary units.

Our first industrial laser system was designed for tubular resistor spiralling, the second for thick and thin film trimming with a motorized XY table or XY mirror positioning. Closed circuit television magnification (15-100) x is provided for safe viewing. The power was decreased to the required 0.1-10 W level and in repetetively Q-switched mode 10-100 μm wide cuts at 5-100 cm/s speed have been used. For semiconductor wafer scribing, unstable resonator geometry was used to ensure high TEM_{OO} power. The high level of brightness has enabled pinhole drilling of some μm diameter to be realized: for beam expanders, gas-flow detectors, etc. The beam positioning (by General Scanning mirrors) allows regular, repetetive (up to 10 Kc) drilling and enables filters of 10^5-10^6 holes/cm^2 density to be produced. These filters are able to decrease the uncertainty in accurate microfiltration procedures.

Today's laser application laboratories are very much concerned with laser annealing activities since this technology is extremely promising.

To decrease the resistivity of polysilicon layers used in integrated circuits for interconnections, laser annealing has proved to be a useful tool. The sheet resistance decreases from the initial 100 Okm/◻ to 5-7 Okm/◻, but it should be noted that this low value sheet resistance state is metastable and after high temperature annealing it starts to increase. Laser annealing has been applied in the manufacture of ion-implanted self-aligned p-channel MOS transistors (MOST) in polysilicon layers deposited on SiO_2 films (LAPOS). As a result of laser annealing the average grain-size of the polycrystallites and the mobility of the carriers increases. In our experiments hole mobility values as high as 50 Vsec cm^{-2} were obtained for MOST in laser annealed polysilicon (see Fig. 1).[2]

The average power of the repetetively Q-switched Nd:YAG laser was about 2 W and the beam distribution (cross section) rather smooth - without "hot spots" (but not TEM_{OO}). The heating makes the annealing easier not because less energy is needed to achieve melting but due to the highly increased extinction.

High power laser pulses were used to produce stable damage on the rear of silicon wafers for gettering the unwanted mobile atoms which would otherwise increase the leakage current of the integrated circuits. Our results have shown that laser-gettering is at least as affective as other conventional gettering techniques of lower productivity. Annealing

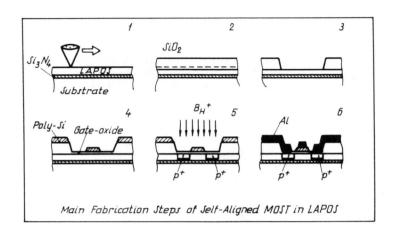

Main Fabrication Steps of Self-Aligned MOST in LAPOS

Figure 1.

investigations with mixed fundamental and second harmonics of Nd-laser offering the possibility to combine or select the solid-phase and liquid-phase epitaxy descriptions are in progress.

The same Nd:YAG laser head with an electro-mechanical shutter and quartz fiber lightguide have been used for gastro-endosopic and other medical purposes.

LiNdLa-glass

The advantage of higher Nd-concentration in laser rods is clear, but the quenching of Nd^{3+} luminescence limits the doping. In the $LiNd_xLa_{1-x}$ phosphate glass (patented by the Lebedev Institute), the weak quenching helps to achieve high (10^{21} Nd/cm^3) concentration without drastic decrease of the upper laser level lifetime.[1] These concentrated neodymium laser glasses have been utilized to develop various low-threshold, high efficiency lasers with air- and water-cooling with and without Q-switching. The aircooled free-running system is pumped in a close, asymmetric reflector[3] by a simple photo-flashlamp.[4] The efficiency achieved at 20 joule pumping is about 6 % using a Ø 6x68 mm rod in a 105 mm long resonator with 85 % output mirror.

The passively - by LiF crystal with F_2^- color centers - Q switched LiNdLa laser delivers 3-50 mj output energy at 3-10 j pumping from Ø 3x50 mm and Ø 2x60 mm rods[5] in 70-1500 mm resonators with mirrors of different curvature and 45-65 % reflectivity. The divergence is nearly diffraction limited without a diaphragm. The output energy, threshold and pulse length can be optimized for the given task by varying the rod surface finish, Nd-concentration, initial Q-switch transmission, output mirror-reflection, -curvature and resonator length. The watercooled (closed circuit) laser device based on the LiNdLa phosphate glass active material can be repetetively pumped up to some 10 cycles. With our economic auto cut off power supply[6] not only higher efficiency but lower heat-loading can be obtained as well. The output energies in a single shot were 400 mj, 15 mj and 3-4 mj in the free running, Q-switched and mode-locked regimes respectively (for mode-locking Kodak and Soviet dyes were used). These portable mini lasers are suitable for many applications from rangefinding and analytics to microprocessing and medical or cosmetic treatment. If LiNdLa-lasers or their second harmonics are used to pump LiF lasers with F_2, F_2^+ and F_2^- color centers one can get tunable output from 0.65-1.25 μm[7] (with KF/F_2^+ and $NaCl-OH^-/F_2^+$ crystals up to 1.77 μm).

Acknowledgements

The authors wish to thank Academician A.M. Prokhorov of the Lebedev Institute, Moscow, and Professor N. Kroó of the Central Research Institute for Physics, Budapest, for helpful discussions and for their active support of the cooperation between the two institutes.

References

1. Denker, B. I., Osiko, V. V., Pashinin, P. P., Prokhorov, A.M., "Concentrated Neodymium Laser Glasses", (Sov.) Quantr. Electr., Vol. 8, pp. 469-483, 1981.(in Russian)

2. Révész, P., Gyimesi, J., Gyulai, J., Czigány, I., Kertész, I., "Process for Semiconductor Device Fabrication", Hungarian Patent 1980; Optika-80 Conference, Budapest

3. Czigány, I., Kertész, I., Denker, B. I., Il'ichov, N. N., Maliutin, A. A., "Reflector for Pulsed Solid State Lasers", Hungarian Patent 33905-541, 1980.

4. Kertész, I., Denker, B. I., Czigány, I., Il'ichov, N. N., Maliutin, A. A., "LiNdLa-Glass Mini-Laser", Laser-Physics Symp. Proc. Helsinki, pp. 82-88, 1981.

5. Basiev, T. T., Denker, B. I., Il'ichov, N. N., Maliutin, A. A., Mirov, S. B., Osiko, V. V., Pashinin, P. P., "A Passively Q-Switched Laser Utilizing Concentrated LiNdLa Phosphate Glass", (Sov.) Quantr. Electr., Vol. 9, pp. 1536-1541, 1982. (in Russian)

6. Cséry, H., "Process and Device for Manually and Automatically Regulated Economic Optical Pumping of Pulsed Lasers", Hungarian Patent H 01 3/03 J-478/82

7. Basiev, T. T., Voron'ko, Yu. K., Mirov, S. B., Osiko, V. V., Prokhorov, A. M., Soskin, M. S., Taranenko, V. B., "Efficient Tunable Lasers Utilizing $LiF:F_2^-$ Crystals", (Sov.) Quantr. Electr., Vol. 8, pp. 1741-1743, 1982. (in Russian)

CW CO$_2$ laser annealing

L. Nanu, E. Cojocaru, I.N. Mihăilescu, L.C. Nistor and V. Teodorescu

Central Institute of Physics, Bucharest, Romania

Abstract

The paper reports on a numerical analysis of the thermal phenomena during cw CO$_2$ laser annealing of implanted silicon. At the surface of the sample the nonlinear heat equation was solved together with diffusion equation for free-carriers density. The results are in a quite good agreement with the reported experimental results.

Introduction

Ion implantation is a well-known method used to dope semiconductors in a controlled and reproducible manner for p-n junctions formation in electronic devices. The implanted ions produce considerable lattice damage, the amount of such damage being strongly dependent on the ion energy and dose. Usually, after implantation, the near-surface region of the sample has an amorphous structure and the dopant is not electrically active because it does not occupy substitutional sites in the crystal lattice. Hence an annealing process is required to restore the crystalline structure and to activate the dopant.

Laser annealing, accomplished by illuminating the implanted sample with laser radiation, is being actively investigated as an alternative to furnace annealing. The physical phenomena involved in amorphous-crystalline transition during pulsed laser annealing are not yet well understood[1,2]. But it is clear now that the continuous-wave laser annealing involves a solid-phase epitaxy with a negligible dopant diffusion. This recrystallization mechanism is well-known from conventional annealing studies. However, the time for regrowth is in the milliseconds or seconds range with laser annealing, as compared to minutes or hours with furnace annealing. Most investigations on laser annealing have been carried out with pulsed or cw lasers with wavelengths of 0.48 μm to 1.06 μm. Within this range the incident photons have energies above the energy band gap of silicon and then the laser light is heavily absorbed in a thin surface layer. In the last few years there were some investigations of laser annealing using a cw CO$_2$ laser and the result was a complete solid phase regrowth of the amorphous layer[3,4].

In this paper we discuss on the annealing of the implanted silicon under the action of c.w.CO$_2$ laser radiation, taking into account the free-carriers diffusion equation.

Physics of laser heating

The wavelength of 10.6 μm corresponds to an energy of the laser photon of about 0.117 ev, lower than the bandgap energy of silicon (1.1 ev). Hence, the absorption of CO$_2$ laser light occurs as a result of free-carriers excitation. This was experimentally illustrated by Miyao et al.[3]. They found that the irradiation time required for full electrical activation under cw CO$_2$ radiation depends on the dopant concentration and that the annealing efficiency increases under bias light (Xe lamp) irradiation. In fact, when a free electron absorbs a quantum of optical incident energy, it is raised to a higher energy state in the conduction band. The excited electrons collide with thermal vibrations and with the other electrons and give up their energy. In time of the order of 10^{-12} s, the absorbed energy is passed onto the lattice. Hence we may consider the optical energy to be instantaneously converted to local heat that diffuses by thermal conduction. Therefore the classical thermodynamic formalism may be used to describe the cw laser heating. For a correct estimate of the temperature distribution in a silicon sample during cw CO$_2$ laser irradiation, the evolution in time of the free-carriers density has to be considered. For a stationary continuous Gaussian circular source, we have to solve the nonlinear equations of diffusion in the cylindrical geometry

$$c\rho \partial T(r,z,t)/\partial t = \nabla[k(n,T)\nabla T(r,z,t)] + Q(n,T)$$

$$\partial n(r,z,t)/\partial t = \nabla[D(n,T)\nabla n(r,z,t)] + R(n,T)$$

(1)

In these equations c is the specific heat capacity, ρ, the density, $k(n,T)$, the thermal conductivity, $D(n,T)$, the ambipolar diffusion coefficient (electrons and holes are assumed to diffuse together) and the source terms are

$$Q(n,T) = \alpha_{FC}(1-R)I_o\exp(-r^2/a^2)\exp(-\alpha_{FC}z) + E_f\gamma(T)n^3$$

(2)

$$R(n,T) = g(n,T) - \gamma(T)n^3$$

with R - the reflectivity, 2a - the laser beam diameter, I_o - the maximum light power density and g(n,T) - the thermal generation rate of the carriers. The free-carriers absorption coefficient is calculated according to Staflin et al.[5] for acoustic phonons scattering, so $\alpha_{FC} = \sigma(T)$ n. $\gamma(T)$ n^3 stands for the carriers recombination rate which is dominated by Auger recombination since it increases as n^3 and the Auger coefficient increases with temperature as[6]

$$\gamma(T) = A \coth(255/T)$$

An accurate estimate of the temperature must not ignore the heat losses from the surface. Therefore the boundary conditions for equations (1) are

$$-k\partial T(r,z,t)/\partial v = \phi$$
$$D\partial n(r,z,t)/\partial v = s\ n \tag{3}$$

where $\partial/\partial v$ denotes differentiation in the direction of the outward normal to the surface, ϕ is the flux across the surface and s is the surface recombination velocity. Initially the sample has a constant temperature T_o and a thermal equilibrium carrier density, n_o. The diffusion equations (1) are nonlinear so numerical methods must be used to solve them.

One can only assess the surface values of the temperature during laser irradiation if one assumes that the laser beam diameter is large as compared with the sample thickness. Hence we neglect the lateral diffusion and for the surface of the sample the equations are

$$\partial n/\partial t = g(n,T) - \gamma(T)\ n^3$$
$$c\rho\partial T/\partial t = \sigma(T)\ n\ I_o(1-R) + E_f\ \gamma(T)\ n^3 \tag{4}$$

At the beginning of the laser pulse there are low excitation levels that the temperature changes very slowly and the recombination term can be neglected in (4). As temperature rises the number of free-carriers will increase by thermal generation. Hence the recombination rate grows and the temperature raises very fast. If we neglect the recombination the equations (4) are

$$\partial n/\partial t = g(n,T)$$
$$c\rho\partial T/\partial t = \sigma(T)\ nI_o(1-R) \tag{5}$$

The thermal generation rate for pure material is $g(n,T) = \gamma(T)n_i^2(T)$ n, where, for silicon

$$n_i^2(T) = 1.5\times10^{45}T^3\ \exp[-(1.276\times10^4/T)]\ (m^{-6})$$

From equations (5) it is easy to obtain the following equations for carrier density and for temperature evolution in time

$$n = n_o + [c\rho/I_o(1-R)]\int_{T_o}^{T} (\gamma\ n_i^2/\sigma)dt \tag{6}$$

$$t = [c\rho/I_o(1-R)]\int_{T_o}^{T} (1/\sigma\ n)dT$$

Equations (6) were numerically solved using the following values c = 836 J/kgK, ρ = 2330 kg/m^3, 1-R = 0.55, T_o = 300 K, $n_o = n_i(T_o)$ and

$$\sigma(T) = 3.7\times10^{-23}\ T\ (m^2)$$

The temperature dependence of the free-carriers density, as it results from equations (6), is given in Fig. 1, for different power densities. To explain this dependence one must use the Fig. 2 which gives the temperature variation in time. As can be seen, the temperature increases slowly in the range of (300-450)K, but then the increase becomes very fast. If the power density is higher, the temperature rises faster and so the free-carriers density has a lower value at the same temperature.

The Auger recombination is not easily included in the solution (6) because in this case the equations (4) are nonlinear. They from a pair of ordinary differential equations which may be solved numerically. The temperature dependence of the free-carriers density for I_o = 3×10^8 W/m^2 is given in Fig. (1). Contrary to the expression (6), the numerical solution for n does not grow unlimitedly, but shows a rather distinct limitation effect, which is caused

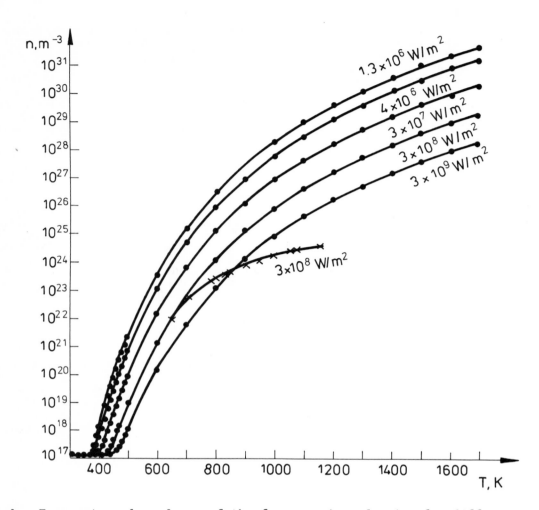

Figure 1. Temperature dependence of the free-carriers density for different power densities

by the Auger recombination. At high excitation levels the recombination balances the thermal generation of the carriers. The increase of the temperature is essentially the same as one can see in Fig. 3.

Discussion

The measured temperatures and amorphous silicon regrowth rates were found to be consistent with the reported solid-state regrowth rate for silicon in furnace annealing. The silicon growth rate in the <100> direction, from Csepregi et al.[7] is

$$v(A/sec) = 1,55 \times 10^{14} exp(-2.3 \ ev/kT) \tag{7}$$

From furnace annealing studies, it was found that the amorphous structure is stable until annealed at a critical temperature when it recrystallizes epitaxially on the existing crystal structure which is a single crystal. The critical temperature is a characteristic of the semiconductor and it is 650°C for silicon. As results from equation (7) above this temperature, the recrystallization rate will be higher as the temperature is higher, near the melting point. At a temperature of 1300 K, an amorphous layer of 2000 Å will recrystallize in about 0.01 s and at 1000 K in 5 s. In the case of cw CO_2 laser irradiation, we can assume the temperature of the sample to be approximately uniform throughout its thickness. Hence, one can see in Fig. 2 that to reach the critical temperature in the silicon sample (650°C) a time of the order of 10 s is necessary for a power density of 3×10^7 W/m² and of 10^2 s for 3×10^9 W/m². To reduce the time of CO_2 laser annealing the sample must be preheated to a temperature in the 200-400°C range. To make this estimation we assumed that the initial carriers density is the intrinsic density at the initial temperature. But it is obvious that a fraction of the implanted impurities will be incorporated in substitutional

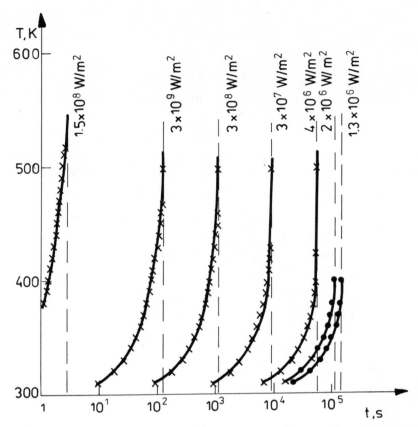

Figure 2. Surface temperature variation in time for different power densities

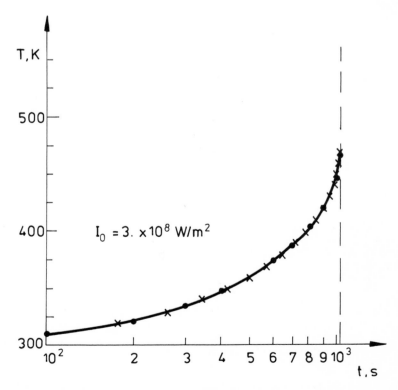

Figure 3. The increase of the surface temperature with (•) and without (x) recombination in equations (4)

sites in the lattice during implantation. Therefore, the time for laser annealing will depend on the dopant concentrations. If we assume the initial carriers density to be $n_o = 1.4 \times 10^{20}$ m^{-3}, for a incident power density of 1.5×10^8 W/m^2, the annealing time from equations (6) is about 3 s, which is in quite good agreement with the experimental results of Ref. 3.

For a better estimate of the time required for cw CO_2 laser annealing at different power densities, equations (1-3) have to be solved. These calculations are useful in understanding the influence of the free carriers density on the annealing process under cw CO_2 laser radiation.

References

1. Van Vechten, J. A., Tsu, R., Saris, F. W. and Hoonhout, D., "Reasons to believe pulsed laser annealing of Si does not involve simple thermal melting", Phys. Lett., vol. 74A, pp. 417-421, 1979

2. Dumke, W. P., "On laser annealing and lattice melting", Phys. Lett., vol. 78A, pp. 477-480, 1980

3. Miyao, M., Ohyu, K., and Tokuyama, T., "Annealing of phosphorus-ion-implanted silicon using a CO_2 laser", Appl. Phys. Lett., vol. 35, pp. 227-229, 1979

4. Tsou, S. C., Tsien, H. P., Takai, M., Röschenthaler, D., Ramin, M., Ryssel, H., and Ruge, I., "Front and back surface cw CO_2-laser annealing of arsenic ion-implanted silicon", Appl. Phys., vol. 23, pp. 163-168, 1980

5. Staflin, T., and Huldt, L., "Infrared absorption spectrum of photogenerated free carriers in silicon", Ark Fys., vol. 20, pp. 527-530, 1962

6. Nilsson, N.G., and Svantesson, K. G., "The role of free carrier absorption in laser annealing of silicon at 1.06 μm", J. Phys. D: Appl. Phys., vol. 13, pp. 39-44, 1980

7. Csepregi, L., Mayer, J. W., and Sigmon, T. W., "Regrowth behavior of ion-implanted amorphous layers on <111> silicon", Appl. Phys. Lett., vol. 29, pp. 92-93, 1976

INDUSTRIAL APPLICATIONS OF LASER TECHNOLOGY

Volume 398

Addendum

The following papers, which were scheduled to be presented at this conference and published in these proceedings, were cancelled.

[398-28] **Moiré interferometric analysis system to perform high sensitivity distortion measurement in composites**
A. R. Hunter, Jr., D. Lapicz, T. Milly, Lockheed Research Lab., USA

[398-35] **Interferometric testing of ballbearing beds**
L. Erdélyi, AOL-Dr. Schuster GmbH, Austria

[398-49] **Detection and measurment of cracks by diffraction and scattered light measurements from hardened steel surfaces**
L. Erdélyi, AOL-Dr. Schuster GmbH, Austria

[398-52] **New automated optical surface contouring system**
N. Balasubramanian, Digital Optics Corp., USA

[398-63] **Overview of laser technology in Bulgarian industry, agriculture, and medicine**
S. K. Dinoev, Z. D. Popova, State Committee of Science, Bulgaria

[398-67] **Influence of laser parameters on cutting of steel sheets**
R. Teti, V. Sergi, V. Tagliaferri, Univ. di Napoli, Italy

[398-68] **Effect of laser treatment on plastic deformation of thin steel sheets**
V. Sergi, V. Tagliaferri, Univ. di Napoli, Italy

The following papers were presented at this conference, but the manuscripts supporting the oral presentations are not available to the Society. SPIE regrets the consequent exclusion of the full text of these papers from these proceedings.

[398-76] **Industrial application of holographic interferometry**
H. Steinbichler, Labor Dr. Steinbichler, West Germany

[398-33] **Automatic evaluation of specklegrams**
H. Kreitlow, P. Steinlein, W. Jüptner, Bremen Institute for Applied Radiation Techniques, West Germany

[398-57] **Laser writing on semiconductor surfaces**
R. P. Salathé, Y. Rytz-Froidevaux, H. P. Weber, Univ. of Berne, Switzerland

AUTHOR INDEX